Lecture Notes in Artificial Intelligence 6441

Edited by R. Goebel, J. Siekmann. and W. Wahlster

Subseries of Lecture Notes 

Longbing Cao
Yong Feng
Jiang Zhong (Eds.)

Advanced Data Mining and Applications

6th International Conference, ADMA 2010
Chongqing, China, November 19-21, 2010
Proceedings, Part II

Volume Editors

Longbing Cao
University of Technology Sydney
Faculty of Engineering and Information Technology
Sydney, NSW 2007, Australia
E-mail: longbing.cao-1@uts.edu.au

Yong Feng
Chongqing University
College of Computer Science
Chongqing, 400030, China
E-mail: fengphd@msn.com

Jiang Zhong
Chongqing University
College of Computer Science
Chongqing, 400030, China
E-mail: zhongjiang@cqu.edu.cn

Library of Congress Control Number: 2010939048

CR Subject Classification (1998): I.2, H.3, H.4, H.2.8, J.1, H.5

LNCS Sublibrary: SL 7 – Artificial Intelligence

ISSN 0302-9743
ISBN-10 3-642-17312-8 Springer Berlin Heidelberg New York
ISBN-13 978-3-642-17312-7 Springer Berlin Heidelberg New York

springer.com

Printed in Germany

Typesetting: Camera-ready by author, data conversion by Scientific Publishing Services, Chennai, India
Printed on acid-free paper 06/3180

Preface

With the ever-growing power of generating, transmitting, and collecting huge amounts of data, information overload is now an imminent problem to mankind. The overwhelming demand for information processing is not just about a better understanding of data, but also a better usage of data in a timely fashion. Data mining, or knowledge discovery from databases, is proposed to gain insight into aspects of data and to help people make informed, sensible, and better decisions. At present, growing attention has been paid to the study, development, and application of data mining. As a result there is an urgent need for sophisticated techniques and tools that can handle new fields of data mining, e.g., spatial data mining, biomedical data mining, and mining on high-speed and time-variant data streams. The knowledge of data mining should also be expanded to new applications.

The 6th International Conference on Advanced Data Mining and Applications (ADMA 2010) aimed to bring together the experts on data mining throughout the world. It provided a leading international forum for the dissemination of original research results in advanced data mining techniques, applications, algorithms, software and systems, and different applied disciplines. The conference attracted 361 online submissions from 34 different countries and areas. All full papers were peer reviewed by at least three members of the Program Committee composed of international experts in data mining fields. A total number of 118 papers were accepted for the conference. Amongst them, 63 papers were selected as regular papers and 55 papers were selected as short papers. The Program Committee worked very hard to select these papers through a rigorous review process and extensive discussion, and finally composed a diverse and exciting program for ADMA 2010. The ADMA 2010 program was highlighted by three keynote speeches from outstanding researchers in advanced data mining and application areas: Kotagiri Ramamohanarao, Chengqi Zhang, and Vladimir Brusic.

September 2010

Longbing Cao
Yong Feng
Jiang Zhong

Organization

ADMA 2010 was organized by Chongqing University, China, and the School of Information Technology and Electrical Engineering, the University of Queensland, Australia, sponsored by the National Natural Science Foundation of China, Chongqing Science and Technology Commission,Chongqing Academy of Science and Technology, and technically co-sponsored by IEEE Queensland Section.

Organizing Committee

Steering Committee Chair

Xue Li — University of Queensland, Australia

Keynote Speakers

Kotagiri Ramamohanarao — University of Melbourne, Australia
Chengqi Zhang — University of Technology Sydney, Australia
Vladimir Brusic — Dana Farber Cancer Institute, USA

General Co-chairs

Charles Ling — The University of Western Ontario, Canada
Shangbo Zhou — Chongqing University, China
Jie Xu — Leeds University, UK

Program Co-chairs

Jinyan Li — Nanyang Technological University, Singapore
Longbing Cao — University of Technology Sydney, Australia
Zhongyang Xiong — Chongqing University, China

Publicity Chair

Xueming Li — Chongqing University, China

Regional Organization Co-chairs

Jiang Zhong — Chongqing University, China
Yong Feng — Chongqing University, China

Finance Chair

Yong Feng — Chongqing University, China

China Registration Chair

Li Wan — Chongqing University, China

China Web Master

Quanji Qiu — Chongqing University, China
Xiaoran Lin — Chongqing University, China

China Secretariat

Min Zheng — Chongqing University, China
Yanyan Zou — Chongqing University, China
Haoyang Ren — Chongqing University, China
Junhui Wang — Chongqing University, China

Program Committee

Hua Li, Canada
Arlindo Oliveira, Portugal
Dragan Gamberger, Croatia
Andre Ponce Leao, Brazil
Andrew Kusiak, America
Wang Shuliang, China
Christophe Giraud-Carrier, USA
Daniel Neagu, UK
Liu Zhen, Japan
Daniel Sanchez, Spain
Dianhui Wang, Australia
Fernando Berzal, Spain
Gang Li, Australia
Jan Rauch, Czech Republic
Jean-Gabriel Ganascia, France
Joseph Roure, UK
Juho Rousu, USA
Junbin Gao, Australia
Paul Vitanyi, USA
Petr Berka, Czech Republic
Rui Camacho, Portugal
Wanquan Liu, Australia
Christophe Rigotti, India
Xiaochun Cheng, UK
Yonghong Peng, UK
Zhaoli Zhu, China
Peng Han, China
Cai Yueping, Japan
Yu Qiao, China
Guang Chen, China
Xinyang Ying, China
Guobin Zhou, China
Yun Li, China
Jun Zhao, China
Hong Tang, China
Hao Wang, China
Hong Yu, China
Li Li, China
Ling Ou, China
Zili Zhang, China
Xingang Zhang, China
Xiaofeng Liao, China
Kaigui Wu, China
Yufang Zhang, China
Hua Li, China
Xiaofan Yang, China
Jiang Zhong, China
Yong Feng, China
Ji Li, China
Li Wan, China
Chengliang Wang, China
Chunxiao Ye, China
Huiping Cao, USA
Yan Qi, USA
Yanchang Zhao, Australia
Sumon Shahriar, Australia

Senthil Kumar, India
Alfredo Cuzzocrea, Italy
Rong Xie, China
Ji Liu, China
Changze Wu, China
Yixiong Chen, China
Qi Xie, China
Pan He, China
BaoHua Qiang, China
Zhixing Li, China
Fenghua Tu, China
Xiaohong Zhang, China
Chuan Li, China

Table of Contents – Part II

III Data Mining Methodologies and Processes

IV Data Mining Applications and Systems

Table of Contents – Part I

I Data Mining Foundations

II Data Mining in Specific Areas

Incremental Learning by Heterogeneous Bagging Ensemble

Qiang Li Zhao, Yan Huang Jiang, and Ming Xu

School of Computer Science, National University of Defense Technology,
Changsha 410073, China
yhjiang@nudt.edu.cn

Abstract. Classifier ensemble is a main direction of incremental learning researches, and many ensemble-based incremental learning methods have been presented. Among them, Learn++, which is derived from the famous ensemble algorithm, AdaBoost, is special. Learn++ can work with any type of classifiers, either they are specially designed for incremental learning or not, this makes Learn++ potentially supports heterogeneous base classifiers. Based on massive experiments we analyze the advantages and disadvantages of Learn++. Then a new ensemble incremental learning method, Bagging++, is presented, which is based on another famous ensemble method: Bagging. The experimental results show that Bagging ensemble is a promising method for incremental learning and heterogeneous Bagging++ has the better generalization and learning speed than other compared methods such as Learn++ and NCL.

Keywords: Incremental Learning, Ensemble learning, Learn++, Bagging.

1 Introduction

An incremental learning algorithm gives a system the ability to learn from new dataset as it becomes available based on the previously acquired knowledge [1]. In incremental learning, the knowledge obtained before will not be discarded. An incremental learning algorithm has to meet the following criteria [2]:

(1)It should be able to learn additional information from new data.

(2)It should not require access to the original data, used to train the existing classifier.

(3)It should preserve previously acquired knowledge (that is, it should not suffer from catastrophic forgetting).

(4)It should be able to accommodate new classes that may be introduced with new data.

According to the number of base classifiers in the target result, the incremental learning algorithms can be divided into two categories: incremental learning of single classifier, and ensemble based incremental learning.

Single-classifier based Incremental learning generates only one base classifier totally, and the structure of the classifier updates each time when new data comes. So this kind of algorithm requires its base training algorithm having incremental learning capabilities. Many incremental learning algorithms belong to this category, such as incremental decision tree [3], incremental SVM [4], ARTMAP (Adaptive Resonance

L. Cao, J. Zhong, and Y. Feng (Eds.): ADMA 2010, Part II, LNCS 6441, pp. 1–12, 2010.

Theory modules Map [5]), EFuNNs (Evolving Fuzzy Neural Networks [6]) etc. In these algorithms, some parameters are needed to control when and how to modify the inner architecture of the classifier. And inappropriate parameters are possible to over-fit the training data, resulting in poor generalization performance.

Ensemble [7] based incremental learning generates one or more classifiers, and the prediction is the combination of the predicted results of all the classifiers generated. Among the algorithms of this category, some of them must work with the specific base training algorithms, such as ENNs (Evolved Neural Networks [8]), SONG (Self-Organizing Neural Grove [9]) and NCL (Negative Correlation Learning for Neural Networks [10]). Some of them has little requirement for base training algorithms and achieves better generality. The typical approach is Learn++ [2].

This paper will discuss the following questions: whether Bagging [11] is a good choice for ensemble based incremental learning? Whether heterogeneous base classifiers can improve the generalization for incremental learning? To answer the problems, an incremental learning algorithms, Bagging++, is proposed in this paper. Bagging++ uses Bagging to construct base classifiers for new dataset. The experimental results show that: 1) Bagging++ achieves better predictive accuracy than learn++ and other compared algorithms, for both homogeneous and heterogeneous versions; 2) the learning process of Bagging++ is much faster than learn++; 3) Heterogeneous base classifiers improve the diversity of the ensembles, which improve the generalization of incremental learning finally.

The structure of this paper is as follows: Section 2 gives an introduction to the related researches. Section 3 proposes two algorithms: Bagging++; Section 4 presents the setup of the comparative experiments. Section 5 discusses the empirical results. Section 6 ends this paper with some conclusions and future works.

2 Related Works

This paper aims at ensemble based incremental learning. Typical approaches in this category include ENN, SONG, NCL and its optimizations, Learn++. The following give an introduction to these approaches.

Seipone and Bullinaria developed an incremental learning approach ENN, which uses an evolutionary algorithm to develop parameters for training neural networks such as the learning rates, initial weight distributions, and error tolerance. After each generation, bad neural networks are discarded, and new ones with different architectures are added into the next generation. The evolutionary process aims to develop the parameters to produce networks with better incremental abilities. One advantage of ENN is that it is possible to tune any of the parameters of the neural network using the evolutionary algorithm, thereby solving the difficult problem of manually determining parameters. ENN can adapt to dynamic environments by appropriately changing architecture and learning rules of neural networks. The weakness of ENN is that the evolutionary algorithm cannot satisfy all the criteria necessary for incremental learning under certain conditions.

SONG is an ensemble-based incremental learning approach that uses ensembles of self-generating neural networks (ESGNN) algorithm to generate base classifiers of self-generating neural trees (SGNTs). SGNT algorithm is defined as the problem of how to construct a tree structure from the given data, which consist of multiple attributes under the condition that the final leaves correspond to the given data. Each SGNT can be learned incrementally. An ESGNN is constructed by presenting different orders of the training set to each of the SGNTs. After the construction of the ensemble, applying a pruning method to compose a SONG is possible. SONG presents the advantages of using an ensemble with the benefit of using a base classifier that is suitable for incremental learning. Except for different input sequences of datasets, SONG does not encourage diversity in the ensemble.

GNCL (Growing NCL) use NCL (Negative Correlation Learning) to generate base classifiers, where NCL is a method to produce diverse neural networks in an ensemble, by inserting a penalty term into the error function of each individual neural network in the ensemble. GNCL performs incremental learning using NCL is to create an ensemble that has initially only one neural network. The neural network is trained with the first available dataset. To each new incoming dataset, a new neural network is inserted in the ensemble. Only the new neural network is trained with the new data set. The other neural networks that were previously inserted in the ensemble do not receive any new training on the new dataset, but their outputs to the new data are calculated in order to interact with the new MLP (Multi-Layer Perceptron), which is trained using NCL. GNCL trains only one neural network with each new dataset, its learning is fast. However, only one neural network to each of the datasets could be a short number to attain a good accuracy.

Another notable approach to incremental learning is Learn++ which is inspired on AdaBoost [12]. In AdaBoost, the distribution of probability is built in a way to give higher priority to instances misclassified only by the last previously created classifier. While in Learn++, the distribution is created considering the misclassification by the composite hypothesis, formed by all the classifiers created so far to that incoming dataset. The advantage of Learn++ is that it creates multiple classifiers when a new dataset is available, which may be a good choice for generalization performance. Although Learn++ creates each of the classifiers to the new dataset using a probability of distribution that is related to all the previous classifiers created to the new dataset, the construction of classifiers to the new dataset does not have interaction with the classifiers created to previous datasets.

Except Learn++, all other approaches mentioned above adopt a special training algorithm for the base classifier individually. The classifiers trained by different training algorithms are all suitable for Learn++, which makes it more flexible than the other ensemble based approaches.

3 Bagging-Based Incremental Learning

In this section we first analyze the advantages and disadvantages of Learn++, then based on the analysis a Bagging-based incremental learning method, Bagging++, will be presented.

3.1 Advantages and Disadvantages of Learn++

Though the original Learn++ proposed by Polikar et al. uses MLP as the base classifiers, it does not limit the types of the base classifiers, any models such as neural network, decision tree, support vector machine, can work with Learn++. This characteristic makes Learn++ having following additional advantages:

(1)Base classifiers of an ensemble can be heterogeneous. Researches showed that an ensemble composed by heterogeneous base classifiers gets higher diversity than a homogeneous ensemble of the same size, which is an efficient way to improve the predictive performance and stability of ensembles.

(2)There are no requirements of incremental abilities for base classifiers. This characteristic makes Learn++ having good flexibility and scalability.

By experiments, several disadvantages of Learn++ are also found:

(1)Learning on new datasets is time-consuming: Learn++ uses weighted sampling to create the training set of the next base classifier. As the courses continue, the training set become more and more difficult, and more and more time is required in training. For some applications, Learn++ may never meet the stop criteria, and can not terminate in a normal way.

(2)Low predictive accuracy: AdaBoost is much suitable for weak training algorithms. For strong training algorithm, it is possible to over-fit training data. The reason is that AdaBoost pay much attention to the difficult instances, and gives high weight for the classifiers which predict these instances correctly. Yet these classifiers may have low predictive abilities for other instances, and give a negative effect for the whole target ensemble.

(3)More and more time is required for prediction: Each time a new data set comes, one or more base classifiers will be added into the target ensemble. The growing ensemble requires more and more time to predict new instances.

From the analyses above, we propose an approach for incremental learning: Bagging++. Bagging++ uses Bagging, a well-known ensemble method, to generate an ensemble for each incremental step.

3.2 Bagging++

Bagging is presented a well-known ensemble method, in which many new training sets of base classifiers are created from the original training set by bootstrap re-sampling. In Bagging, there is no accuracy requirement for each base classifier, so the training time of each base classifier is relatively small. And all base classifiers are equally weighted, so the negative effect of any bad classifier is also limited and the generalization ability of Bagging ensemble is strong and stable. For Bagging, majority voting is used to combine the predicted results of all base classifiers.

In Bagging++, Bagging is adopted to generate an incremental ensemble for each new dataset. Algorithm 1 shows the description of Bagging++.

In Algorithm 1, $\Im$ is a training algorithm. Any training algorithms (such as BPNN, decision tree, SVM, naïve Bayesian) are accepted in step (2.b). If step (2.b) uses the

Algorithm 1. Bagging++
Inputs:
D_{inc} : new dataset
S_e : size of incremental ensemble
E_t : target ensemble
Outputs:
E_t : new target ensemble

(1) Initialize incremental ensemble: E=NULL;
(2) Use Bagging to generate an incremental ensemble of a given size
  (2.a) Get a training set: D_t= Bootstrap(D_{inc});
  (2.b) Learn a base classifier: h=training ($\Im$, D_t);
  (2.c) $E = E \cup \{h\}$;
  (2.d) if $|E|<S_e$, goto (2.a);
(3) Obtain new target ensemble: $E_t = E_t \cup E$;
(4) Return E_t.

same training algorithm to train the classifier at each time, we call the algorithm homogeneous Bagging++. If the base classifiers in the resulted incremental ensemble are learned by different training algorithms, the algorithm is heterogeneous Bagging++.

4 Experimental Setup

We compare six incremental methods: GNCL, Learn++, Bagging++, heterogeneous Learn++, heterogeneous Bagging++ and single BPNN, which is taken as the basis of the experiment. In this part we will answer two questions presented in section 3:

a) Is Bagging ensemble feasible to incremental learning?
b) Can heterogeneous ensemble improve the performance of incremental learning?

There are many methods to create heterogeneous ensembles, but we only want to find if they are better than homogeneous ones, so in the experiment we adopt a quite simple method described in Algorithm 2.

Algorithm 2. Build a base classifier for a heterogeneous ensemble
Input: Training set *T*.
Output: A base classifier.

With For the given *T*, create a BPNN, a C4.5, a SVM and a Naïve Bayes individually with their default parameters;
(1) Get the predictive accuracy of each classifier for *T*. Each classifier predicts on *T*;
(2) Return the classifier with the highest accuracy, abandon others.

4.1 Datasets

Thirty classification datasets from UCI [13] machine learning repository are used to perform the experiments. Table 1 presents a summary of all datasets. It shows the number of instances, the number of input attributes, the number of class labels, the number and size of the training sets used in incremental learning.

Table 1. Data sets used for classification

Dataset	number of Instances	Attributes/ Class labels	# of Inc.	Size of Training Set	Dataset	number of Instances	Attributes/ Class labels	# of Inc.	Size of Training Set
adult	30162	14/2	9	3016	kr-vs-kp	3196	36/2	9	319
mushroom	5644	22/2	9	564	letter-recognition	20000	16/26	9	2000
austra	690	14/2	3	207	optdigits	3823	64/10	6	200
balance-scale	625	4/3	3	187	page	5473	10/5	9	547
breastcancer-wisconsin	683	9/2	3	204	pima	768	8/2	3	230
bupa	345	6/2	3	103	poker-hand	25010	10/10	9	2501
cancer	699	9/2	3	209	segmentation	2100	19/7	9	210
car	1728	6/4	7	222	sick-euthyroid	2000	18/2	9	200
cleveland	303	13/5	3	90	spambase	4601	57/2	9	460
cmc	1473	9/3	6	220	splice	3190	60/3	9	319
dermatology	358	34/6	3	107	tic-tac-toe	958	9/2	4	215
german-numeric	1000	24/2	4	225	transfusion	748	4/2	3	224
glass	214	9/6	3	64	vehicle (statlog)	846	18/4	3	210
imports-85 (autos)	159	25/7	3	47	waveform	5000	21/3	9	500
ionosphere	351	34/2	3	105	yeast	1484	8/10	6	222

4.2 Experimental Methods

In most experiments of incremental learning, researchers seldom consider the time and memory metrics such as training time, size of ensembles. But these metrics are often has important effect for real world applications, for example, in the context real-time applications, the generic algorithm based methods such as ENN and SNCL [14] may not been chosen by the users.

All our tests are performed on 30 datasets, for each dataset a random part is taken out as the test set at first, then the rest instances are equally divided into several incremental training sets. During incremental learning, the training sets are inputted in turn into all the compared methods simultaneously until all the training sets are processed. For each dataset, these courses are repeated for 30 times to eliminate the effect of randomness, all the results listed in the next section are the average results of all 30 times.

We compare all algorithms from the following facets:

Accuracy: After presented an incremental training set is presented, each compared method will create a new classifier (or new ensemble), which will be tested on the test set.

Generalization improvement: It is the improvement of accuracy of from the first to the last incremental steps. A positive improvement implies good incremental ability of the corresponding methods.

Training time: Average of learning time used in each incremental step.

Predicting time: Average of predicting time on the test set after each incremental step. This metric is important for ensemble incremental learning, because when ensemble becomes very large the responding time may not been accepted by the users.

For each incremental training set, Learn++ and Bagging++ create 20 base classifiers. For Learn++ a set of weights are assigned to all base classifiers of the ensemble, while for Bagging++ all base classifier are equally weighted.

In our experiments, the heterogeneous ensembles consist of four types of base classifiers, which are BPNN, C4.5, SVM and Naïve Bayes. All the BPNN (including NCL) classifiers have one hidden layer with number of nodes equaling to the number of the input attributes. And we adopt the fast training algorithm RPROP [15] during training of all BPNNs, for which the stopping criterions are: 1) MSE is less than 0.015; 2) The number of epochs reaches 3000. All the decision trees are built by C4.5 system with the pruning confidence level of 0.25. All the support vector machines are constructed by Libsvm [16], we use radial kernel with γ being 0.001, and complexity parameter ε setting to 0.001.

Our test platform is configured with AMD 4000+, 2G RAM. All above experiments are implemented in C++ programming language and tested on Linux operating system.

4.3 Experimental Results and Analysis

This section presents and analyzes the comparison results of six incremental methods: BPNN, GNCL, homogeneous Learn++, homogeneous Bagging++, heterogeneous and heterogeneous Learn++ (HLearn++), and heterogeneous Bagging (HBagging++).

4.4 Accuracy

Table 2 shows the average accuracy of each approach after the first and last incremental steps in the alphabetic order of datasets. Learn++ (or HLearn++) may not work on some datasets, so in the corresponding cell the results are marked as (-). For fairness purpose, on each dataset all compared methods are ranked according to their prediction accuracy after the last incremental step. And at the last row, the averages of predicted accuracies and the corresponding ranks are also shown. We can find that the single BPNN almost always has the poorest performance, because it suffers much catastrophic forgetting each time when new incremental data arrives. For all of the datasets, homogeneous Bagging++ obtains first rank on 8 datasets, GNCL achieves 3 best results and homogeneous Learn++ only wins on 2 datasets, while heterogeneous Bagging++ and heterogeneous Learn++ receive 12 and 5 best results respectively. According to the average ranks, we can order all the compared methods from best to worst as follows: heterogeneous Bagging++, homogeneous Bagging++, heterogeneous Learn++, GNCL, homogeneous Learn++ and single BPNN. Then it is possible to make the following conclusion about the accuracy of all compared methods:

Table 2. Accuracy results of single BPNN, GNCL, Learn++, Bagging and the heterogeneous version of the last two algorithms

Dataset	Single BPNN	GNCL	Learn++ (Homo.)	Bagging++ (Homo.)	Learn++ (Hetero.)	Bagging++ (Hetero.)
adult	79.88-83.62	80.05-77.57	77.11-83.44	82.55-84.09	77.38-83.46	**82.57-84.15**
mushroom	98.35-99.43	98.87-92.48	**93.90-99.94**	99.21-99.60	96.48-99.94	99.12-99.58
austra	78.31-81.06	81.26-72.95	74.40-81.98	83.19-83.67	74.01-82.22	**85.12-85.89**
balance-scale	82.81-85.52	90.68-66.77	75.05-82.34	88.80-90.26	73.23-82.86	**88.80-90.47**
breast-cancer-wisconsin	95.59-95.82	96.62-97.56	87.84-93.33	97.23-97.51	84.55-96.01	**97.51-97.84**
bupa	62.50-64.81	62.59-68.24	56.30-65.74	66.02-72.04	61.30-66.39	**67.69-72.22**
cancer	96.25-96.62	95.83-96.57	82.55-92.78	96.85-97.18	86.71-95.46	**97.13-97.55**
car	82.53-85.10	85.96-69.41	79.20-85.57	84.44-86.71	**81.51-86.92**	85.02-86.86
cleveland	54.34-58.08	55.76-57.27	48.28-56.97	**57.17-59.39**	49.90-55.86	58.79-58.69
cmc	41.00-47.39	44.95-47.15	42.57-48.65	**47.54-51.42**	41.98-48.67	46.58-51.13
dermatology	86.58-90.99	93.42-86.58	68.83-88.20	93.24-95.50	83.06-91.80	**96.22-96.58**
german-numeric	66.40-71.23	71.10-75.13	65.20-72.63	**74.23-75.73**	65.53-72.53	74.20-75.60
glass	50.61-58.48	61.52-61.52	53.33-59.39	**64.09-63.94**	55.15-62.88	62.27-63.79
imports-85(autos)	42.59-45.74	53.15-56.48	46.48-61.11	51.30-59.81	48.89-55.74	**58.89-62.59**
ionosphere	80.00-83.70	88.89-90.83	70.93-70.74	86.94-89.35	73.52-79.91	**89.63-91.57**
kr-vs-kp	90.36-93.82	**94.39-95.12**	84.59-83.41	93.04-94.39	83.32-86.12	93.10-94.37
letter-recognition	37.12-45.32	40.19-44.77	(-)	45.49-48.92	**63.80-70.85**	64.64-68.93
optdigits	75.86-85.08	83.67-88.58	**75.46-91.00**	84.05-89.00	71.19-88.35	84.60-89.67
page	90.05-89.82	89.77-89.67	90.51-90.96	89.77-89.63	**90.53-93.40**	90.93-91.08
pima	71.32-73.46	72.01-74.02	67.22-71.11	73.33-75.73	65.38-72.69	**73.12-76.20**
poker-hand	45.61-48.18	47.33-49.44	47.02-48.62	48.03-49.40	46.91-48.70	**48.20-49.59**
segmentation	76.76-85.78	85.21-87.48	73.35-88.29	84.30-86.97	**73.21-89.95**	84.84-88.03
sick-euthyroid	90.17-92.92	**92.70-93.68**	74.53-86.88	92.23-93.52	78.08-90.12	92.47-93.40
spambase	90.12-92.76	91.60-93.22	86.62-92.72	**92.15-93.30**	86.14-92.68	92.34-93.16
splice	87.29-94.26	93.06-60.47	75.26-89.05	**92.92-95.28**	81.92-90.47	93.02-95.26
Tic-tac-toe	69.01-72.58	83.10-73.20	71.67-80.65	76.02-76.26	**71.50-82.28**	75.92-77.62
transfusion	73.95-77.68	77.46-78.64	67.15-74.65	**78.77-80.22**	68.90-72.19	78.77-79.65
vehicle(statlog)	66.39-71.65	**73.49-76.27**	62.02-69.78	72.24-74.75	63.98-69.85	72.42-75.37
waveform	83.33-85.75	84.70-86.99	81.81-86.45	85.52-87.02	81.39-86.55	**85.67-87.06**
yeast	48.42-55.75	53.03-58.71	47.50-56.23	**55.96-59.21**	48.55-56.49	55.48-59.19
Accuracy average	77.08	75.56	75.09	79.99	78.38	**81.10**
Rank average	4.67	3.87	4.50	2.33	3.80	**1.83**

1. The generalization of Bagging++ ensemble is better than that of Learn++ in incremental learning. Learn++ gives more weights to the hard instances in each round, then it focus only on these instances while ignoring others (these instances are most of the training set), at last the base classifiers that can correctly predict these instances are given too much weights, so the prediction performance of other instances are affected. This problem is actually a kind of over-fitting. Learn++ is originally designed for working with weak learners, so with strong learner it may suffers more over-fitting.
2. Bagging is quite easy, so it may have a bright future in real applications of incremental learning.
3. Heterogeneous Learn++ and Bagging++ are better than their homogeneous version. Heterogeneous ensemble has advantages in the diversity of the base classifiers, and this can explain why it is better in accuracy.
4. GNCL is better than homogeneous Learn++, and a little worse than heterogeneous Learn++. GNCL is much faster than both of them in during training process, so it is more practical in real applications. But GNCL can only work with the neural networks, so it is less flexible than Bagging++ and Learn++.
5. Homogeneous Learn++ can't finish the work on dataset *letter-recognition*. Considering that we have modified the original Learn++ algorithm to alleviate this problem, this is again shows the restriction of Learn++.

4.5 Generalization Improvement

The generalization improvement is calculated as the average accuracy of the test dataset in the last incremental step minus the average accuracy of the test set in the first incremental step. In this way, a high value indicates a high improvement. A high improvement means that a good incremental ability is attained, while a decrease in the generalization is a sign of poor incremental learning ability. Table 3 presents the generalization improvement of each method for all datasets, and in the last row the average improvements are also shown.

1. Homogeneous Learn++ and heterogeneous Learn++ obtain the best improvement on most of the datasets, which shows that Learn++ has a good incremental ability.
2. The average improvement of GNCL is negative, indicating a poor incremental ability. The problem of GNCL is that it only inserts one neural network into the ensemble for each incremental step, and the small ensemble may suffer too much performance degradation in the next incremental steps.
3. The improvement of homogeneous Bagging++ (and heterogeneous Bagging++) is not as high as Learn++. Considering table 2, the accuracy of Bagging++ in the first incremental step is almost always better than Learn++, so maybe this is the actual reason why its improvement is not so high.

4.6 Training Time

The training time is the total time that a method spends on training classifiers (or ensemble) for all incremental steps. From the results, we can get following conclusions:

Table 3. Generalization Improvement

Dataset	BPNN	GNCL	Learn++	Bagging	HLearn++	HBagging
adult	3.74	-2.48	**6.33**	1.53	6.09	1.58
(mushroom	1.09	-6.39	**6.04**	0.38	3.46	0.46
austra	2.75	-8.31	7.58	0.48	**8.21**	0.77
balance-scale	2.71	-23.91	7.29	1.46	**9.64**	1.67
breastcancer-wisconsin	0.23	0.94	5.49	0.28	**11.46**	0.33
bupa	2.31	5.65	**9.44**	6.02	5.09	4.54
cancer	0.37	0.74	**10.23**	0.32	8.75	0.42
car	2.57	-16.55	**6.38**	2.26	5.40	1.84
cleveland	3.74	1.52	**8.69**	2.22	5.96	-0.10
cmc	6.38	2.20	6.08	3.88	**6.69**	4.55
dermatology	4.41	-6.85	**19.37**	2.25	8.74	0.36
german-numeric	4.83	4.03	**7.43**	1.50	7.00	1.40
glass	**7.88**	0.00	6.06	-0.15	7.73	1.52
imports-85(autos)	3.15	3.33	14.63	**8.52**	6.85	3.70
ionosphere	3.70	1.94	-0.19	2.41	**6.39**	1.94
kr-vs-kp	**3.46**	0.73	-1.19	1.35	2.80	1.27
letter-recognition	**8.20**	4.58	(-)	3.44	7.05	4.28
optdigits	9.22	4.91	**15.55**	4.95	17.16	5.08
page	-0.24	-0.10	0.45	-0.14	**2.87**	0.15
pima	2.14	2.01	**3.89**	2.39	7.31	3.08
poker-hand	**2.57**	2.11	1.59	1.38	1.78	1.40
segmentation	9.02	2.27	**14.94**	2.67	16.75	3.19
sick-euthyroid	2.75	0.98	**12.35**	1.28	12.03	0.93
spambase	2.65	1.63	6.10	1.15	**6.54**	0.82
splice	6.97	-32.59	**13.79**	2.36	8.55	2.24
tic-tac-toe	3.57	-9.90	8.98	0.24	**10.78**	1.70
transfusion	3.73	1.18	**7.50**	1.45	3.29	0.88
Vehicle(statlog)	5.26	2.78	**7.76**	2.52	5.86	2.95
waveform	2.42	2.29	4.64	1.50	**5.16**	1.39
yeast	7.32	5.68	**8.73**	3.25	7.94	3.71
average	3.96	-1.85	**7.53**	2.10	7.44	1.93
negative mprovement	1	**9**	2	2	0	1

Table 4. Training Time

Dataset	BPNN	GNCL	Learn++	Bagging	HLearn++	HBagging
adult	40.67	323.00	4902.42	783.28	5915.11	1381.31
mushroom	0.68	74.98	17.35	9.17	36.74	32.79
austra	0.07	1.11	2.18	0.98	3.75	1.83
balance-scale	0.01	1.05	0.51	0.25	0.85	0.65
breast-cancer-wisconsin	0.01	0.01	3.35	0.16	1.75	0.29
bupa	0.19	0.51	3.88	3.90	3.96	4.05
cancer	0.01	0.01	3.39	0.16	1.53	0.29
car	0.05	5.88	1.38	0.88	2.45	2.09
cleveland	0.02	0.07	0.50	0.37	0.66	0.50
cmc	0.51	8.19	29.25	12.82	30.15	13.77
dermatology	0.09	0.17	3.94	1.50	7.88	3.92
german-numeric	0.28	0.68	3.95	4.20	4.19	4.84
glass	0.03	0.10	1.01	0.75	0.91	0.73
imports-85(autos)	0.04	0.09	1.65	0.69	2.60	1.82
ionosphere	0.20	0.28	5.21	2.77	4.75	2.80
kr-vs-kp	0.84	2.21	26.40	16.07	30.17	20.97
letter-recognition	4.94	7.88	0.00	92.29	454.76	246.30
optdigits	1.03	1.44	8.14	7.87	10.64	11.46
page	0.12	0.32	103.03	2.42	104.32	4.54
pima	0.75	2.21	8.87	14.84	9.23	14.97
poker-hand	153.03	243.58	3630.60	2954.97	3981.91	3317.24
segmentation	0.25	0.44	6.22	4.45	7.96	6.10
sick-euthyroid	0.19	0.34	21.83	3.30	21.62	3.95
spambase	1.03	2.42	174.39	19.48	137.99	28.08
splice	5.01	10.38	928.76	84.33	749.36	189.88
tic-tac-toe	0.05	3.52	1.07	1.07	1.63	1.82
transfusion	0.68	1.40	24.86	13.25	25.16	13.40
vehicle(statlog)	0.28	0.86	4.47	5.30	5.05	5.81
waveform	2.60	7.70	92.62	53.46	99.39	59.05
yeast	1.60	5.96	93.85	31.37	97.66	31.82
average	7.18	23.56	336.84	137.55	391.80	180.24

1. Training time of heterogeneous Learn++ and Bagging++ are always longer than their homogeneous versions for all datasets. This is because our algorithm needs to create four different types of classifiers to add a new base classifier into the ensemble.
2. The average training time of Learn++ is beyond two times that of Bagging++. Learn++ is derived from the idea of AdaBoost, it also have has generalization criteria for each base classifiers of the ensemble, if a base classifier can not meet the criteria, it will be is abandoned and a new base classifier has to be created again. But for Bagging, there is no requirement for the base classifier, so the training time of Bagging is usually proportional to the size of the target ensemble.
3. Training time of heterogeneous Bagging++ is still much less than that of homogeneous Learn++, so the former one is more practical.
4. GNCL is fast in training, because it only creates one NCL for each incremental step. But to create a NCL, it needs the predicted results of other base classifiers, so the average training time of a single NCL is beyond three times that of a normal BPNN.

4.7 Ingredients Components of Heterogeneous Ensemble

For heterogeneous Learn++ and Bagging++, the distributions of each type of base classifiers are shown in table 5. For each dataset, the size of the ensemble in the last incremental step is presented, and the percents of each type of classifiers are shown inside the parentheses (in the order of BPNN, C4.5, SVM and Naïve Bayes). And the average distributions are also shown in the last row of the table.

Table 5. Ingredients of Heterogeneous Ensemble

Dataset	HLearn++	HBagging	Dataset	HLearn++	HBagging
adult	180(99.9/0.1/0/0)	180(100/0/0/0)	kr-vs-kp	180(91.1/0.4/6.3/2.2)	180(99.4/0/0.2/0.4)
mushroom	180(88.8/1.2/2.3/7.7)	180(88.7/0/1.4/9.8)	letter-recognition	180(0/0/0/100)	180(0/0/0/100)
austra	60(87.6/0.3/4.4/7.7)	60(42.3/0/44.7/13)	optdigits	120(18.6/0/0/81.4)	120(72.9/0/0/27.1)
balance-scale	60(95.3/0.2/1.9/2.7)	60(98.6/0/1.4/0)	page	180(85/1.1/0/13.9)	180(30.6/0/0.2/69.3)
breast-cancer-wisconsin	60(54.8/1.6/6.8/36.7)	60(45/0/12.3/42.7)	pima	60(92.8/0.1/0/7.1)	60(88.2/0/0.1/11.8)
bupa	60(81.7/1.2/0.5/17.1)	60(92.7/0/0.1/7.2)	poker-hand	180(100/0/0/0)	180(100/0/0/0)
cancer	60(55.3/2.1/3.9/38.8)	60(42.6/0/12.0/45.4)	segmentation	180(62.1/0.1/0/37.8)	180(79.2/0/0/20.8)
car	140(85.1/0.1/8.7/6.1)	140(99.9/0/0.1/0)	sick-euthyroid	180(80.9/3.3/0.6/15.3)	180(94.5/1.2/0.6/3.8)
cleveland	60(65.7/0/0.1/34.2)	60(55.1/0/0.1/44.8)	spambase	180(97.2/0.5/0/2.4)	180(100/0/0/0)
cmc	120(99.5/0/0.1/0.4)	120(99.9/0/0/0.1)	splice	180(80.7/0.6/3.7/15)	180(60.3/0/0/39.7)
dermatology	60(23.9/0.9/1/74.2)	60(17.1/0/0.1/82.8)	tic-tac-toe	80(94.3/0.3/4.9/0.5)	80(98.2/0/1.8/0)
german-numeric	80(94.8/0/0.7/4.6)	80(89.9/0/0.4/9.7)	transfusion	60(86.6/2.6/0.3/10.6)	60(85.7/6/8.1/0.3)
glass	60(71.3/0/0/28.7)	60(83.8/0/0.1/16.1)	vehicle (statlog)	60(91.2/0/0/8.8)	60(100/0/0/0)
imports-85 (autos)	60(40.6/0/0.1/59.3)	60(42.5/0/0/57.5)	waveform	180(88.6/0/0.1/11.4)	180(99.6/0/0.3/0.1)
ionosphere	60(53.0/3.2/1.4/42.5)	60(36.1/0/0.1/63.9)	yeast	120(96.2/0/0/3.8)	120(96.9/0/0/3.1)
average	114(77.2/0.7/1.5/20.7)	114(77.4/0.2/1.6/20.9)			

According to the average distribution, in both heterogeneous Learn++ and heterogeneous Bagging++ the distributions of each type of classifiers are very similar. This may suggest that the distribution is only related to the characteristics of the training algorithms of base classifiers and is independent of to the ensemble methods.

5 Conclusions and Future Works

Based on massive experiments, in this paper we investigate the use of Bagging ensemble and heterogeneous ensemble in incremental learning. The experimental results show that:

(1)Bagging ensemble is fast and with strong generalization ability, which is a promising method for incremental learning.

(2)Heterogeneous ensemble is an effective way to improve the generalization of original ensemble methods.

There are still many works need to do in the future, which are:

(1)It has been proved that ensemble pruning can improve the generalization and the efficiency of the ensembles at the same time in non-incremental learning. Next we will test whether it has the similar effects in ensemble incremental learning.

(2)In our experiments, we assume the distributions of training data in each incremental step are the same. Next we will redo our test under different data distribution assumptions.

(3)The algorithm used to created heterogeneous ensemble in our experiments is simple and rough. We will improve it in both efficiency and generalization in the future.

Acknowledgments. This work was supported by the National Science Foundation of China under the grant No. 60905032.

References

1. Giraud-Carrier, C.: A Note on the Utility of Incremental Learning. AI Communications 13(4), 215–223 (2000)
2. Polikar, R., Udpa, L., Udpa, S.S., Honavar, V.: Learn++: An incremental learning algorithm for supervised neural networks. IEEE Transactions on Systems, Man, and Cybernetics - Part C: Applications and Reviews 31(4), 497–508 (2001)
3. Utgoff, P.E.: Incremental induction of decision trees. Machine Learning 4, 161–186 (1989)
4. Cauwenberghs, G., Poggio, T.: Incremental and Decremental Support Vector Machine Learning. In: Advances in Neural Information Processing Systems, vol. 12, pp. 409–415. MIT Press, Cambridge (2000)
5. Carpenter, G.A., Grossberg, S., Reynolds, J.H.: ARTMAP: Supervied real-time learning and classification of nonstationary data by a self organizing neural network. Neural Networks 4(5), 565–588 (1991)
6. Kasabov, N.: Evolving fuzzy neural networks for supervised/unsupervised online knowledge-based learning. IEEE Transactions on Systems, Man and Cybernetics - Part B: Cybernetics 31(6), 902–918 (2001)
7. Sewell, M.: Ensemble learning (2008), http://machine-learning.martinsewell.com/ensembles/ensemble-learning.pdf, (unpublished)
8. Seipone, T., Bullinaria, J.: Evolving improved incremental learning schemes for neural network systems. In: Proceedings of the 2005 IEEE Congress on Evolutionary Computing (CEC 2005), Piscataway, NJ, pp. 273–280 (2005)

9. Inoue, H., Narihisa, H.: Self-organizing neural grove and its applications. In: Proceedings of the 2005 International Joint Conference on Neural Networks (IJCNN 2005), Montreal, Canada, pp. 1205–1210 (2005)
10. Minku, F.L., Inoue, H., Yao, X.: Negative correlation in incremental learning. Natural Computing 8, 280–320 (2009)
11. Breiman, L.: Bagging Predictors. Machine Learning 24(2), 123–140 (1996)
12. Schwenk, H., Bengio, Y.: Boosting neural networks. Neural Computation 12, 1869–1887 (2000)
13. Asuncion, D.N.A.: UCI machine learning repository (2007), http://www.ics.uci.edu/mlearn/MLRepository.html (unpublished)
14. Tang, K., Lin, M., Minku, F.L., Yao, X.: Selective Negative Correlation Learning Approach to Incremental Learning. Neurocomputing 72(13-15), 2796–2805 (2009)
15. Riedmiller, M., Braun, H.: RPROP- A fast adaptive learning algorithm. In: Proc. of ISCIS VII (1992)
16. Lin, C.J.: LIBSVM: A Library for Support Vector Machines (2009), http://www.csie.ntu.edu.tw/~cjlin/libsvm/ (unpublished)
17. Demsar, J.: Statistical comparisons of classifiers over multiple data sets. Journal of Machine Learning Research 7, 1–30 (2006)
18. Maloof, M.A., Michalski, R.S.: Incremental learning with partial instance memory. Artificial Intelligence 154(1-2), 95–126 (2004)

CPLDP: An Efficient Large Dataset Processing System Built on Cloud Platform

Zhiyong Zhong, Mark Li, Jin Chang, Le Zhou, Joshua Zhexue Huang, and Shengzhong Feng

Shenzhen Institutes of Advanced Technology
Shenzhen, China
{zhongzy,jj.li,jin.chang,le.zhou}@sub.siat.ac.cn,
zhexuehuang@gmail.com, sz.feng@siat.ac.cn

Abstract. Data intensive applications are widely existed, such as massive data mining, search engine and high-throughput computing in bioinformatics, etc. Data processing becomes a bottleneck as the scale keeps bombing. However, the cost of processing the large scale dataset increases dramatically in traditional relational database, because traditional technology inclines to adopt high performance computer. The boost of cloud computing brings a new solution for data processing due to the characteristics of easy scalability, robustness, large scale storage and high performance. It provides a cost effective platform to implement distributed parallel data processing algorithms. In this paper, we proposed CPLDP (Cloud based Parallel Large Data Processing System), which is an innovative MapReduce based parallel data processing system developed to satisfy the urgent requirements of large data processing. In CPLDP system, we proposed a new method called operation dependency analysis to model data processing workflow and furthermore, reorder and combine some operations when it is possible. Such optimization reduces intermediate file read and write. The performance test proves that the optimization of processing workflow can reduce the time and intermediate results.

Keywords: data processing, cloud computing, MapReduce, performance, workflow.

1 Introduction

Cloud computing offers large scale data storage and computation services delivered through huge data centers. It offers a scalable distributed file system and a programming model for data intensive distributed application.

Adopting cloud computing into data processing, we present CPLDP (Cloud based Parallel Large Dataset Processing System) in this paper. CPLDP is built on Apache Hadoop [9]. Apache Hadoop is an open source project that implements Google File System and MapReduce framework. This is a chance for anyone who wants to build their system on cloud platform. Our system is built on Hadoop distributed file system and MapReduce framework. We firstly implement some basic data processing operations on MapReduce framework. These operations are frequently used in data processing. Every

L. Cao, J. Zhong, and Y. Feng (Eds.): ADMA 2010, Part II, LNCS 6441, pp. 13–20, 2010.

operation indicates a MapReduce job. Upon them, a workflow mechanism is devised to orchestrate them. The reason why we design a workflow system is that we discover that single MapReduce cannot be too complex because of the restriction of its program model. Almost every meaningful result generated from the raw dataset needs a series of operations. We provide a user-friendly GUI to help create workflow of operations through the GUI, user can create data processing workflow that satisfies their complicate requirements. The more MapReduce jobs in a workflow, the more read and write of intermediate results, hence the more execution time it needs. To reduce the intermediate data read and write, an optimization module is designed to optimize workflow before executing them. The optimizing methods include reordering operations and combining some of them. Factors that affect optimization and algorithms that carry out the optimization will be introduced in section 4.

This paper makes the following contributions:

- We propose a new schema of processing large dataset in cloud. In this schema, data processing job is constructed by orchestrating basic operations. This schema is extendable and satisfies most data processing needs.
- We propose operation dependency analysis and operation combination to optimize MapReduce job workflow. With the optimization, the execution time of workflow is reduced significantly.

2 Related Work

Sawzall [11] is a scripting language building atop of GFS and MapReduce framework by Google. It allows user to program the filter and select an aggregator to form their data processing job. The first difference between Sawzall and our system is that our system provides a user friendly interface to let user design their data processing workflow visually, and Sawzall let user "program" their data processing program. Secondly, Sawzall provides a process called "chain" to orchestrate Sawzall jobs. But it doesn't optimize the workflow before running it. Pig Latin [12] is designed by Yahoo! Research atop of HDFS and MapReduce. The data processing workflow is called Pig Latin logical plan in it. It compiles the logical plan into a series of MapReduce primitives, as we compile our workflow into a series of MapReduce jobs. In Pig Latin, it doesn't optimize its MapReduce primitives flow as we do. Dryad [14] is a distributed platform that is developed at Microsoft to provide large-scale, parallel, fault-tolerance execution of processing jobs. And DryadLINQ [15] is the language used to design data processing jobs in it. Unlike MapReduce restricting process into *map/reduce* schema, DryadLINQ allows arbitrary data processing job that can be expressed as directed□acyclic□graphs□(DAG). There is static optimization of DAG before running it and dynamic optimization when running the DAG expressed data processing workflow. It moves the operators which reduce the datasets to front when it is possible, but it doesn't combine operators to reduce the read and write of intermediate files. Instead, it uses Dryad's TCP pipe and in-memory FIFO channels instead of persisting temporary data in files. This is effective when the intermediate file is not that large. When intermediate file is becoming large, it cannot be stored in main memory.

3 CPLDP

Figure 1 shows the architecture of CPLDP. We build our system based on HDFS (Hadoop Distributed File System) and MapReduce framework. Upon them, we implement basic data processing operations.

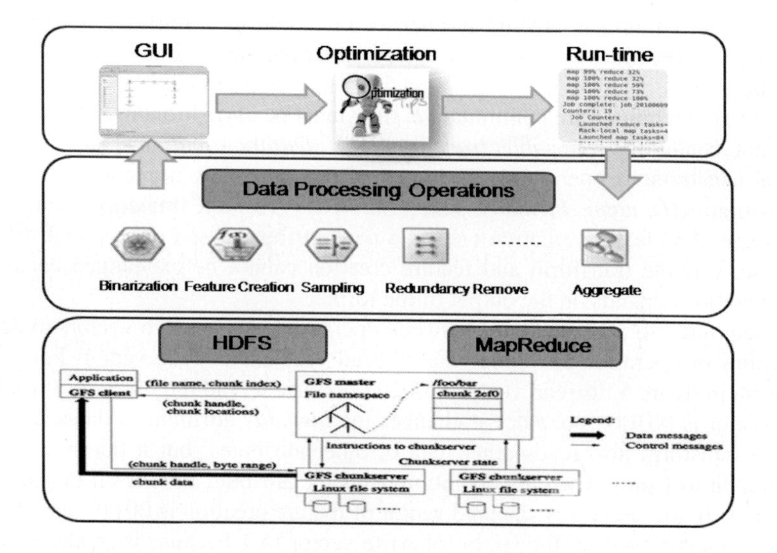

Fig. 1. The architecture of CPLDP

These operations include:

Binarization: transforming continuous values to binary values (1 or 0)
Discretization: transforming continuous values to categorical values
Redundancy Remove: removing the duplicated entries and keeps one
Column Select: selecting columns which are specified
Row Select: selecting rows which are satisfied the condition specified
Feature Creation: creating a new attribute based on the expression specified
Variable Transformation: transforming an attribute according to the expression specified.
Sampling: generating a subset of the dataset with specified size by random sampling
Aggregate: performing an operation on a group of values to get a single result.

The operation includes COUNT, SUM, MAX, MIN, and AVERAGE.

4 Operation Dependency Analysis

We classify operations into two kinds: synchronous operations and asynchronous operations. An operation is synchronous if it needs to check all the input records to complete its operation. Conversely, the asynchronous operations can complete its operation only depend on its own record. Redundancy remove is a synchronous operation because it needs to check all the records to confirm whether one record has replicates. It means that during the execution of redundancy remove, between reading and writing of dataset, there is a barrier that forbids writing after reading a subset of dataset. Synchronous operations cannot exchange with other operations, or there will be writing of dataset beyond the barrier. For example, there is a workflow: *<binarization, redundancy remove>*, if we exchange *binarization* and *redundancy remove* the output of binarization is beyond the barrier of redundancy remove. Aggregate is a synchronous operation, too. On the contrary, other operations in our system are asynchronous.

So if two operations are commutative, they must be asynchronous. But it is not the sufficient condition. *<variable transformation, feature creation>* is a workflow without synchronous operations. The input of this workflow is records of employee information: *<ID, name, birthday, salary, address>*.Variable transform change *birthday* to *age*. And feature creation creates a new attribute based on *age* and *salary*. In this case, variable transform and feature creation cannot be exchanged because the later operation depends on the output of the former.

To determine such dependency between operations, we use two vectors to describe the actions of operation changing dataset: read vector and write vector. For variable transform in figure 4, its read vector is 00100 because it read the *birthday* attribute. Its write vector is 00100, too, since it changes the *birthday* attribute of dataset. We find variable transform also reads other data of other attributes, but it doesn't use them. We set a bit to 1 only when the operation reads the attribute and uses it to execute the operations. In the same way the read vector of feature creation is 00110, and the write vector is 000001. We set the six bit of write vector to 1 because it creates a new attribute and it is added in the rear.

Given two asynchronous operations (A and B) and their vectors (A.r, A.w, B.r and B.w which represents A's read vector, A's write vector and B's read vector and B's write vector respectively), we determine whether they are commutative by calculating the following equations:

$$\mathrm{F1} = \mathrm{A.r} \wedge \mathrm{B.w}\ . \tag{1}$$

$$\mathrm{F2} = \mathrm{A.w} \wedge \mathrm{B.r}\ . \tag{2}$$

B and A are commutative if f1 and F2 are equal to zero. F2 equals to zero means B doesn't read the data that A writes, hence there is no dependency between them. F1 equals to zero means if A and B have been exchanged there is no dependency between them.

Algorithm 1. Reorder data processing operations

```
Procedure LIST REORDERWORKFLOW(LIST workflow)
   for(each asynchronous operation o in workflow) do
      o1 = o's former asynchronous operation;
       if(o1==null)then
       //this means o is at front of workflow.
           continue;
      end if
      if(o1.r^o.w=0 and o1.w^o.r=0and  o.prio>o1.prio)then
```

```
            //o1 and o are commutative and o's priotiy is
            //higher than o1'.
            exchange o and o1;
            goto 3;
        end if
    end for
    return workflow;
end procedure
```

The two factors discussed above only provide the condition that two operations can be exchanged or not. They don't give the information that reflects two operations are necessary to exchange. In fact, in our system, we only let the operation that scales down the dataset execute as early as possible, because it scales down the input dataset of the operation that exchanges with it. We use an attribute called *priority* to describe how much an operation scales down dataset. The higher *priority* is, the more it scales down the dataset. *Sampling* has the highest *priority* because it scales down the dataset most. *Column select* and row select has the second highest and others have normal *priority*. If *variable transform* and *feature creation* in figure 2 are commutative, we exchange them only when the *priority* of *variable transform* is lower than the *priority* of feature *creation*. And in fact, they have equal priority and we don't exchange them even though they are commutative.

With three factors discussed above, the workflow of data processing can be optimized as follows: for every operation in the workflow, if its priority is higher than the priority of the former one, and they are commutative, we exchange them. Then do the same procedure on the new former operation until it could changes with its former one. This process goes as algorithm 1. After reordering, the execution time of workflow is reduced about 3/4. We will discuss it in section 6 in detail.

5 Operation Combination

The reordering of workflow searches the opportunity to reduce the input dataset of some operations. After reordering the operations, we find some linked operations can be combined together to totally avoid the input of the later one. For example, in workflow <*binarization, feature creation*>, the input data is the same as data in section 4. The *binarization* transforms *salary* to 1 if it is higher than a *threshold*, for example 4000, or else transforms it to 0. Feature creation generates a new attribute based on salary and address. In cloud platform, *binarization* reads the input and then executes and writes its output into distributed file system. Then feature creation reads the output of *binarization* and executes its code and then writes its result to distributed file system. The writing output to file system of *binarization* is not necessary because the code of feature creation can execute immediately after the executing of code of *binarization*. The combination of *binarization* and *feature creation* elides the writing and reading of intermediate file. This will reduce a lot of execution time of the whole workflow.

Algorithm 2 describes the procedure of combining operations in already reordered data processing workflow.

Algorithm 2. Combine operations in reordered workflow

```
procedure LIST COMBINE(LIST workflow)
  int i = workflow.length-1;
  while(i>=0) do
     o = workflow.get(i);
     o1=workflow.get(i-1);
     if(o1!=null && o1 and o are asynchronous) then
        o2 = combine(o, o1);
        workflow.set(i-1, o2);
      end if
     i=i-1;
   end while
 end procedure
```

The combination of operations improves the performance significantly. We will discuss it in next section.

6 Experiment and Performance

We evaluate our data processing system by executing a workflow which is described as <*feature creation, variable transform, discretization, column select, row select, sampling, redundancy remove, discretization, aggregate*>. We execute this case with different input file size, from 1000’000*25 records (about 4.5G) to 1000’000*125 records (about 22.9 G) by increasing 1000’000*25 records each time. The original workflow, reordered workflow and combined workflow are executed with different data input scale respectively. Through this process we discuss the factors that influent workflow execution time.

Our test bed is a 12-node Hadoop cluster. One node serves as master (namenode and jobtracker). And the other 11 nodes are slaves (datanode and tasktracker). Each node is equipped with 16 processor (Dual core AMD Opteron(tm) Processor 870, 2 GHz), 32G main memory and 80G hard disk. The hard disk volume is relatively low, that is why the input file size scales from 4.5G to 22.9G in our experiment.

As depicted in figure 2, the optimization process improves the performance significantly. After reordered, the execution time of workflow is reduced about 5/6 except the data is 4.5G. Through our experiment, we found that the sampling size of *sampling*, constraint of *row select* and *column select* influence the execution time of reordered workflow a lot. The original workflow is reordered to <*sampling, column select, row select, feature creation, variable transformation, discretization, redundancy remove, discretization, aggregate*>. Operations that scale down dataset execute as early as possible. This means that the more these operations scale down the dataset, the less time subsequent operations will take. But on the other hand, if these operations don’t scale down dataset with a considerate rate, the execution time after reordering will not decrease significantly. In our experiment, the sampling size is 0.2, which means after *sampling*, only 20% of the dataset remains.

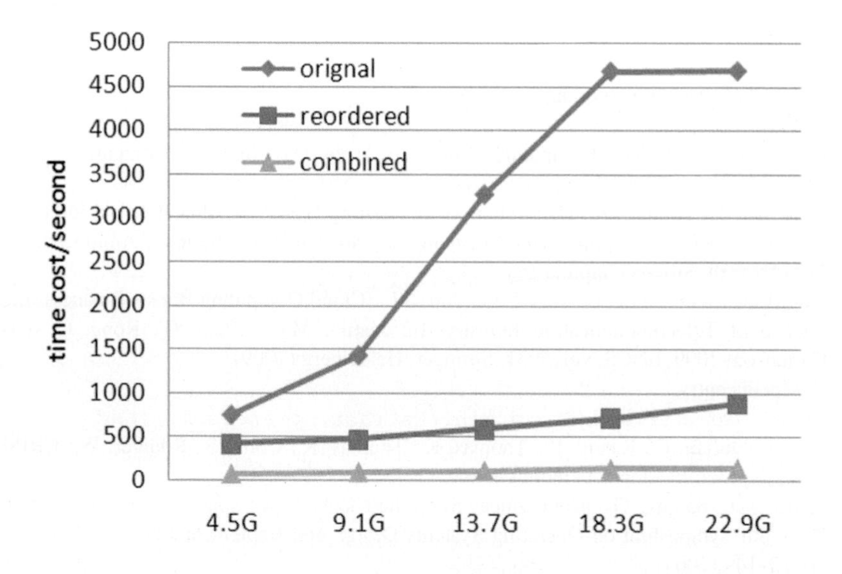

Fig. 2. Experiment result

After combination, the execution time is reduced by about 4/5 compared with the corresponding execution time of reordered workflow. The reordered workflow is combined into two operations: <*sampling, column select, row select, feature creation, variable transformation, discretization, redundancy remove*> and <*discretization, aggregate*>. It seems contradictory with the combination algorithm described in algorithm 3 because here we "combine" the synchronous operations (*redundancy remove* and *aggregate*) with other asynchronous operations.

Our experiment proves that the optimization of MapReduce workflow is necessary and efficient, especially when there are operations scaling down dataset such as *sampling*, *row select* and *column select* in workflow.

7 Conclusion and Future Work

In this paper, we propose a large data processing system CPLDP. Before running data processing workflow, we adopt operation dependency analysis to reorder operations and combine some operations. Such optimization process is effective.

Through our experiment, we find that Hadoop cluster environment influences performance in certain extends. Our future work will focus on the implementation of MapReduce to find other possibilities of optimizing data processing workflow. On the other hand, the MapReduce job configuration is a factor that affects the single job performance. Hence automatically and intelligently configuring MapReduce jobs is another research focus.

References

1. Wikipedia entry: computer data processing, http://en.wikipedia.org/wiki/Data_processing
2. Phillip, B.G.: Data-Rich Computing: Where It's At. In: Data-Intensive Computing Symposium (March 26, 2008), http://research.yahoo.com/news/2104
3. Christian, T., Thomas, R.: Data Intensive Computing How SGI Altix ICE and Intel Xeon Processor 5500 Series[Code-name Nehalem] help Sustain HPC Efficiency Amid Explosive Data Growth. Silicon Graphics Inc. http://www.sgi.com/pdfs/4154.pdf
4. Xu, M., Gao, D., Deng, C., Luo, Z.G., Sun, S.L.: Cloud Computing Boosts Business Intelligence of Telecommunication Industry. In: Jaatun, M.G., Zhao, G., Rong, C. (eds.) CloudCom 2009. LNCS, vol. 5931, Springer, Heidelberg (2009)
5. Wikipedia entry, http://en.wikipedia.org/wiki/Extract,_transform,_load
6. Pete, C., Julian, C., Randy, K., Thomas, K., Thomas, R., Colin, S., Rüdiger, W.: CRISP-DM 1.0 - Step-by-step data mining guide.
7. Jeffrey, D., Sanjay, G.: MapReduce: Simplified Data Processing on Large Cluster. In: Proc. 6th Symposium on Operating Systems Design and Implementation, San Francisco, pp. 13–149 (2004)
8. Sanjay, G., Howard, G., Shunk-Tak, L.: The Google File System. In: Proc.19th Symposium on Operating System Principles, pp. 29–43. Lake George, New York (2003)
9. Hadoop, A.: http://www.hadoop.com
10. Miner, A.: http://www.eti.hku.hk/alphaminer/
11. Pike, R., Dorward, S., Griesemer, R., Quinlan, S.: Interpreting the data: Parallel analysis with Sawzall. Scientific Programming Journal 13(4) (2005)
12. Christopher, O., Benjamin, R., Utkarsh, S., Ravi, K., Andrew, T.: Yahoo! Research Pig Latin: A Not-So-Foreign Language for Data Processing. In: Proceedings of the 2008 ACM SIGMOD International Conference on Management of data (2008)
13. DryadLINQ, http://research.microsoft.com/research/sv/DryadLINQ/
14. Isard, M.: Dryad: Distributed data-parallel programs from sequential building blocks. In: European Conference on Computer Systems (EuroSys), Lisbon, Portugal, pp. 59–72 (2007)
15. Yuan, Y., Michael, I., Dennis, F., Mihai, B., Úlfar, E., Pradeep, K.G., Jon, C.: DryadLINQ: A System for General-Purpose Distributed Data-Parallel Computing Using a High-Level Language. In: Symposium on Operating System Design and Implementation (OSDI), San Diego, CA (2008)

A General Multi-relational Classification Approach Using Feature Generation and Selection

Miao Zou, Tengjiao Wang, Hongyan Li, and Dongqing Yang

School of Electronics Engineering and Computer Science, Peking University, Beijing 100871 China
{zoumiao,tjwang,dqyang}@pku.edu.cn, lihy@cis.pku.edu.cn

Abstract. Multi-relational classification is an important data mining task, since much real world data is organized in multiple relations. The major challenges come from, firstly, the large high dimensional search spaces due to many attributes in multiple relations and, secondly, the high computational cost in feature selection and classifier construction due to the high complexity in the structure of multiple relations. The existing approaches mainly use the inductive logic programming (ILP) techniques to derive hypotheses or extract features for classification. However, those methods often are slow and sometimes cannot provide enough information to build effective classifiers. In this paper, we develop a general approach for accurate and fast multi-relational classification using feature generation and selection. Moreover, we propose a novel similarity-based feature selection method for multi-relational classification. An extensive performance study on several benchmark data sets indicates that our approach is accurate, fast and highly scalable.

Keywords: Multi-relational classification, Feature generation, Feature selection.

1 Introduction

Much real-world data is organized in multiple relations in databases. Mining multi-relational data repositories is an essential task in many applications such as business intelligence. Multi-relational classification is arguably one of the fundamental problems in multi-relational data mining.

Multi-relational classification is challenging. First, there may be a large number of attributes in a multi-relational database where classification is conducted. Since relations are often connected in one way or another, virtually, multi-relational classification has to deal with a very high dimensional search space. Moreover, the structure of a multi-relational database, i.e., the connections among multiple relations, can be very complicated. High complexity in structure of multi-relational data often leads to high cost in feature selection and classifier construction.

To tackle the challenges, the existing approaches, which will be briefly reviewed in Section 2.2, mainly use the inductive logic programming (ILP) techniques to derive hypotheses or extract features for classification. However, those methods are often slow due to the high cost in searching the clause spaces or joining multiple tables. Moreover, those methods often cannot fully utilize information stored in all types of attributes and provide enough information to build effective classifiers using methods such as SVM [1], which constructs models from feature space instead of rules.

L. Cao, J. Zhong, and Y. Feng (Eds.): ADMA 2010, Part II, LNCS 6441, pp. 21–33, 2010.

There are many mature and effective classification methods for data in a single relation, such as SVM and decision trees [2]. Those methods cannot be applied directly for multi-relational classification since they cannot handle multiple relations and their connections. Thus, a natural question is whether we can derive a general feature generation and selection method to build the feature space from multiple relations, so that those existing classification methods for single relations can be easily extended to multi-relational classification.

In this paper, we tackle the problem of multi-relational classification by developing a general approach using feature generation and selection. We make the following contributions. First, we develop a general framework for multi-relational classification such that many existing classification methods on single relation, such as SVM and decision trees, can be applied on multi-relational data. The central idea of the framework is to generate and select features from multi-relational data. Second, we propose a novel similarity-based feature selection method for multi-relational classification by leveraging available features in a large search space effectively. Last, to preserve as much information as possible, we devise a similarity-based feature computation method for data transforming. An extensive performance study on several data sets indicates that our approach is accurate, fast, and highly scalable.

The rest of the paper is organized as follows. In Section 2, we describe the problem of multi-relational classification and review the related work. We present our general framework in Section 3. We develop our feature generation and selection methods in Section 4 and data transforming in Section 5. We report an extensive performance study in Section 6 and conclude the paper in Section 7.

2 Problem Description and Related Work

In this section, we describe the multi-relational classification problem and briefly review the existing approaches.

2.1 Problem Description

A multi-relational data set is a set of relations $R=\{R_1,\ldots,R_n\}$ where every attribute is either numerical or nominal. Among all relations in question, there is a target relation which contains a class-label attribute. The tuples in the target relation are called the target tuples. The other relations are called the background relations.

Example 1 (Problem Description). We take the finance database whose schema is shown in Fig. 1. The data set contains 8 relations. *LOAN* is the target relation, and the attribute *Status* as the class-label attribute. We are interested in building a classifier using the data in the finance database to predict the status of new loans in the future.

Although a loan is recorded as a record in relation *LOAN*, it is highly related to some information stored in the other relations in the database, such as the related transactions through bank accounts. Instead of using only the information in relation *LOAN*, a loan status classifier can be expected more accurate if it can also make good use of the information in the other relations.

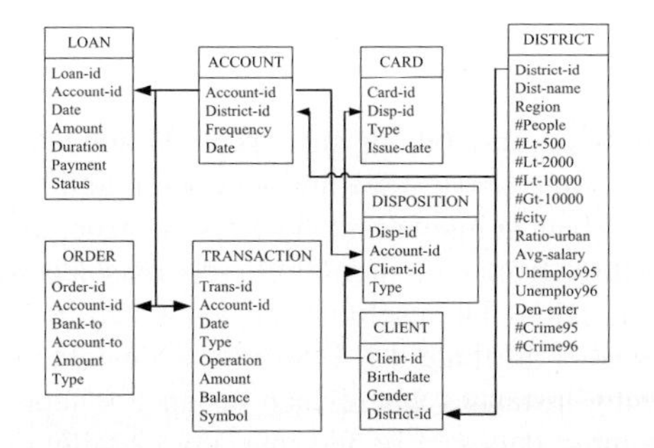

Fig. 1. An example of Financial Database from PKDD CUP 99

2.2 Related Work

Inductive logic programming (ILP for short) [3, 4, 5] becomes a natural approach in multi-relational classification. In general, inductive logic programming takes logic programming [6] as a uniform representation of positive and negative examples, background knowledge and hypotheses, and computes hypotheses that follow all positive and none of the negative examples.

For example, FOIL [7] constructs a set of conjunctive rules which distinguish positive examples against negative examples. The core of rule construction in FOIL is to search for the best predicates iteratively and append those candidates to a rule. PROGOL [8] constructs a most general clause for each example using an A*-like search. Redundant clauses are removed. Mimicking the decision tree methods [2], TILDE [9] adopts the heuristic search of C4.5 [10] and uses a conjunction of literals in tree nodes to express background knowledge. The above ILP systems, however, are well recognized costly on large data sets due to very large search spaces of possible rules or clauses. To handle the efficiency issue of ILP systems, CrossMine [11] develops a tuple ID propagation method to avoid physically joining relations.

Alternative to the inductive logic programming approaches, a possible idea is to transform a multi-relational data set into a "flattened" universal relation. The propositionalization approaches use inductive logic programming to flatten multi-relational data and generate features [12]. Particularly, the features in the universal relation are generated by first-order clauses using inductive logic programming. LINUS [13] is a pioneer propositionalization method, which transforms clauses into propositional expressions as long as all body literals in the clauses are determined. Srinivasan and King [14] addressed the problem of non-determined literals in clauses by constructing boolean features from PROGOL clauses. Compared to the other ILP methods, the propositionalization methods incur high computational cost and heavy information loss in constructing binary features. RelAggs [15] tries to tackle the issues by considering aggregates in feature generation. However, efficiency remains an issue due to the huge universal relation with exponential set of features constructed.

Although good progress has been made on multi-relational classification, a critical problem remains open. Many mature classification methods on single relations such as SVM, which constructs models from feature space instead of rules, cannot be extended to multi-relational data effectively. Although the propositionalization approaches transform multiple relations into a universal relation, the transformation cost on large data set is very high which makes classifier construction very costly.

3 Framework

To enable general classification methods such as SVM applicable on multi-relational data in an effective way, we propose a feature-generation-and-selection framework.

It contains two important components. (1) Feature generation and selection. We generate and select features from data in multiple relations according to the usefulness of the feature in classification. The usefulness is evaluated by the similarity between such feature and class label attribute. (2) Data transformation. We transform training data into a set of training instances where each instance is represented using the features selected. The training data can be fed into any classification method on single relation to build a classifier. When predicting, the test data is transformed similarly.

Let us start with a simple approach using aggregate features. Given a set of relations $R=\{R_1,\ldots,R_n\}$, we can join all relations and generate the universal relation R_U.

Definition 1 (Aggregate feature). For a target tuple *t*, an attribute *A* and an aggregate function *aggr*, the aggregate feature of *t* is defined as $t.A_{aggr} = aggr_{s \in R_U(t)}(s.A)$.

Example 2 (Aggregate features). Consider the finance database in Fig. 1 and target relation *LOAN*. A *t.Account-id* may appear in multiple orders and transactions. For example, for tuples t_1 and t_2 in relation *LOAN*, suppose each has three instances in the universal table R_U. Consider attribute *Type* from *TRANSACTION*. Fig. 2 shows that we can use aggregate functions such as *COUNT* to extract aggregate features.

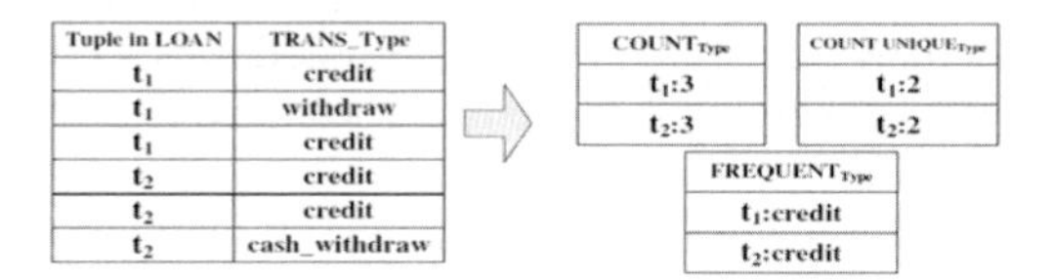

Fig. 2. Extracting aggregate features

While aggregate features are simple, there are some problems in using them in multi-relational classification. First, aggregates such as *COUNT* or *FREQUENT* may not be able to manifest the similarity among training examples in the same class as well as the differences among training examples in different classes, especially on nominal attributes. For example, two instance sets $R_U(t_1)$ and $R_U(t_2)$ with very different distributions on a nominal attribute may have same aggregate feature values using simple aggregates, as demonstrated in Fig. 2. To tackle this issue, authors in [16] introduce target-dependent aggregators to model the distribution of nominal attributes, and use vector distance to construct new features. It brings informative features but also redundant and useless ones without proper feature selection methods.

Second, a universal relation often has a large number of attributes, and thus a large number of aggregate features. Those features form the search space of classification problem. However, some attributes in a multi-relational data set may not be pertinent to the classification task. For example, attribute *Issue-date* in the finance database in Fig. 1 provides little help when determining whether a loan is paid on time. We need to develop an effective approach to select a small set of features effective for multi-relational classification. Ideally, the features should not be redundant to each other.

We will address the above two issues in the next section.

4 Feature Generation and Selection

In this section, we first introduce distribution features in order to extract more detailed information for nominal attributes. Then, we develop an efficient method for selecting pertinent features from both aggregate and distribution features.

4.1 Distribution Features

We observe in Section 3 that simple aggregate attributes may not be able to provide enough information on nominal attributes. Thus, we propose to extend aggregate features to distribution features.

Definition 2 (Distribution feature). For a target tuple t and a nominal attribute A not in the target relation, the distribution feature of t on A, denoted by $t.D_A$, is the distribution of $R_U(t)$ on A. It can be written as a feature vector $t.D_A=\{p_1,\ldots,p_m\}$, where $m=|A|$ and p_i is the occurrence frequency in $R_U(t)$ of the i-th value in A.

Example 3 (Distribution features). Fig. 3 illustrates that distribution features of three target tuples in relation *LOAN* are extracted on attribute *Type*. Clearly, the distribution features provide more information than the aggregate features shown in Fig. 2.

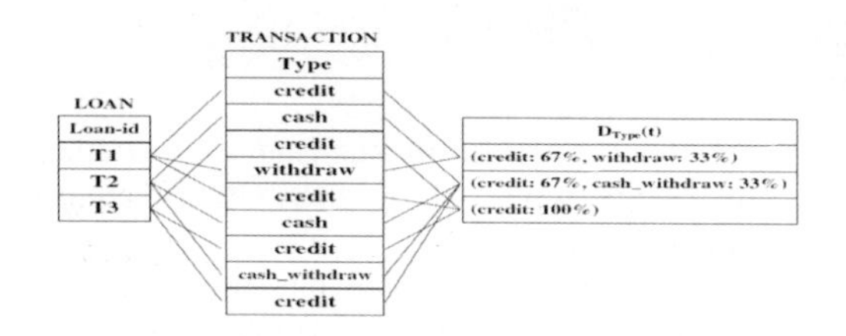

Fig. 3. Extracting distribution features

4.2 Similarity between Features

Many measures are studied in feature selection methods [17]. However, most of them are designed for data mining on one relation. Those methods cannot evaluate features when $R_U(t)$ contains multiple instances with respect to a target tuple t.

To tackle the problem, we introduce a similarity-based feature selection method for multi-relational classification. As the starting point, we adapt the method in [18] to measure the similarity between two features. Later, we will develop a new feature selection method for multi-relational classification based on similarity evaluation.

Definition 3 (Similarity between target tuples). For target tuples t_1 and t_2, an attribute A, and an aggregate feature A_{aggr}, the similarity between t_1 and t_2 on A_{aggr} is defined as

$$sim_{A_{aggr}}(t_1,t_2)=\begin{cases} 0, & if\ |t_1.A_{aggr}-t_2.A_{aggr}|\geq\sigma_{A_{aggr}} \\ 1-\dfrac{|t_1.A_{aggr}-t_2.A_{aggr}|}{\sigma_{A_{aggr}}}, & otherwise \end{cases} \quad (1)$$

where$\acute{o}_{aggr}$ is the standard deviation of $\{ A_{aggr}(t_i)\}$ for all target tuples t_i.

The similarity between two target tuples t_1 and t_2 on a distribution feature D_A is defined as

$$sim_{D_A}(t_1,t_2)=\sum_{i=1}^{|A|} t_1.D_A[i]\times t_2.D_A[i] \quad (2)$$

where $t_1.D_A[i]$ is the i-th element of the feature vector $t_1.D_A$.

Example 4 (Similarity between tuples). Consider the feature vectors on attribute *Type* in Fig. 2. The similarity between t_1 and t_2 on aggregate features $Type_{COUNT}$, $Type_{COUNT\ UNIQUE}$, and $Type_{FREQUENT}$ are all 1. On the distribution feature D_{Type} in Fig. 3, the similarity between t_1 and t_2 is $sim_{D_{Type}} = 0.67 * 0.67 + 0 + 0 = 0.45$.

Definition 3 is important since it gives us a ground to compare how features, either aggregate or distribution ones, manifest the similarity between target tuples. For a feature A_f, we can calculate the similarity $sim_{A_f}(t_1,t_2)$ for each pair of target tuples (t_1, t_2). We define the similarity matrix of the feature using such similarity values.

Definition 4 (Similarity matrix). Let $t_1,\ldots, t_N$ be the set of all target tuples and N be the number of target tuples. For a feature A_f, the similarity matrix of A_f is defined as

$$V^{A_f}=[sim_{A_f}(t_i,t_j)]_{1\leq i,j\leq N}. \quad (3)$$

Example 5 (Similarity matrix). Suppose in our running example, the target relation *LOAN* contains only three target tuples whose distribution feature vectors on *Type* are shown in Fig. 4. The similarity matrix of D_{Type} is $\begin{bmatrix} 0.56 & 0.45 & 0.67 \\ 0.45 & 0.56 & 0.67 \\ 0.67 & 0.67 & 1 \end{bmatrix}$.

Using the similarity matrices, we can measure the feature similarity. Particularly, we are interested in similarity between a feature and the class distribution feature, which is a kind of distribution feature where the vector contains one 1and 0 for others.

Definition 5 (Feature similarity). The similarity between a feature A_f, no matter an aggregate one or a distribution one, and the class distribution feature D_c is defined as

$$sim(D_c, A_f)=\frac{V^{D_c}\cdot V^{A_f}}{|V^{D_c}|\cdot|V^{A_f}|}. \quad (4)$$

It indicates that, in order to achieve a high similarity to a class distribution feature, a feature on an attribute A needs to manifest all tuples in the same class in a small number of values in the domain of A. In other words, those specific values in the domain of A should have a high utility in discriminate a class from the other. Note that the denominator $|V^A_f|$ penalizes features on which tuples in all classes have similar distributions. Such features are thus not informative in classification.

Compared to information gain and some other measures of feature utility in classification, the similarity-based measure considers the distribution of tuples in different classes on the feature, which can effectively support feature selection specific for accurate classification.

4.3 Feature Selection

Based on the feature similarity, we propose a feature selection method (Algorithm 1) for multi-relational classification. Our method iteratively selects features with high utility in discriminating a class from the other. A feature A_f is selected if (1) the similarity between A_f and the class distribution feature is over a threshold *min_sim*, and (2) the similarity between A_f and every pertinent feature selected already is smaller than a threshold *max_select*, in order to avoid redundant features. To evaluate all candidate features efficiently, we adopt a heuristic search and update the candidate feature set and pertinent feature set dynamically. We also adopt the tuple ID propagation technique [11] to avoid joining relations physically. When considering a background relation R_i, for each tuple t in R_i, the IDs of target tuples joinable with t are propagated along the join path and recorded in R_i for further transferring. With these IDs, it is unnecessary to physically join relations and form a universal relation when calculating similarity between features.

Algorithm 1. Feature Selection

Input: target relation R_t and a set of background relations $\{R_1, R_2, \ldots, R_n\}$,
class label attribute A_c
Output: a set of pertinent features

1. pertinent feature set P empty, candidate feature set C empty
2. activate R_t
3. for each relation R_i that joins with R_t directly
4. activate R_i
5. repeat
6. select feature A_f in C of the maximum similarity
7. if $sim(D_c, A_f) <= min\_sim$, then break
8. remove A_f from C
9. if exist A_f' in P that $sim(A_f, A_f') >= max\_select$, then continue
10. add A_f into P
11. for each target tuple t_i
12. if $cover(t_i) <= min\_cov$, then increase the weight of t_i in the weight matrix W
13. if W is updated, then update the similarity of features in C
14. for each inactive R_j that can be appended to a path from R_t to R_i which contains A
15. activate R_j
16. end

return P

Activating a Relation. In Lines 2, 4 and 15 in Algorithm 1, the operation of activating a relation R_i includes the following steps. 1) propagate target tuple IDs to R_i, 2) generate features including both aggregate and distribution ones from each attribute A in R_i, 3) for each extracted feature A_f, calculate $sim(D_c, A_f)$, and 4) add each A_f with its $sim(D_c, A_f)$ into the candidate set C. Once a feature A_f of the maximum similarity is selected, suppose R_i is the relation contains A, and there is a join path from R_t to R_i through l relations $\{R_{i_1},...,R_{i_l}\}$. Then, for each inactive relation R_j which can be appended directly to any relation in $\{R_t, R_{i_1},...,R_{i_l}, R_j\}$, R_j is activate (Line 14 to 15).

Boosting Coverage of Examples. When a feature A_f is selected such that A is not in the target relation, the feature may cover only a portion of target tuples. Although A_f has high utility in classification, it cannot classify target tuples that are not covered by A_f. Therefore, we should pay more attention on the tuples that are not covered well by the features selected already in the progressive feature selection procedure. A weight matrix is to trace how well each target tuple is covered by features selected so far.

Let $t_1,..., t_N$ be the set of all target tuples. The weight matrix W is defined as

$$W = [weight(t_i, t_j)]_{1\le i,j\le N}. \tag{5}$$

The element $W_{i,j}$ is set to 1 as default.

Using the weight matrix, we integrate the coverage information into the similarity calculation between a feature A_f and a class distribution feature D_c as follows.

$$sim(D_c, A_f) = \frac{(W \cdot V^{D_c}) \cdot V^{A_f}}{|W \cdot V^{D_c}| \cdot |V^{A_f}|}. \tag{6}$$

In each iteration, after a feature is selected, suppose there are x features selected so far. For each target tuple t, suppose t is covered by x_i features selected. Then, we calculate $cover(t_i) = x_i/x$. If $cover(t_i)$ is lower than a threshold min_cov $(min\_cov<1)$, we adjust the weight of t_i in the weight matrix W by increasing row i and column i b $(b>1)$ times. Once W is updated, $sim(D_c, A_f)$ for each A_f in the candidate feature set C should be updated. The weight adjustment process is in Lines 11- 13 in Algorithm 1.

5 Data Transformation

A propositional classification method, which builds a classifier on a single relation, takes a set of attribute-value pairs as input. For each aggregate feature A_{aggr} selected, $t.A_{aggr}$ for each target tuple t is the corresponding value. For each distribution feature D_A, $t.D_A$ is a vector $(p_1,..., p_m)$. Thus, the critical issue in data transformation is how to transform $t.D_A$ into an appropriate value for each target tuple t.

Transforming a vector into a value inevitably leads to information loss. We want to reduce the cost in classification accuracy as much as possible. For example, we can use distance-based method [16] to transform the distribution vector into a value. We can compute the centroid O of all target tuples. Then, $t.D_A$ for a target tuple t can be transformed to the Euclidian distance between $t.D_A$ to the centroid O.

Example 6 (Distance-based transformation). Suppose there are in total 8 target tuples in a training set and $t.D_{Type}$ for each tuple is shown in Fig. 4(a). In distance-based transformation, the centroid of distribution feature D_{Type} is (0.75, 0.25). Thus, $t_1.D_{Type}$ is transformed to $\sqrt{(0.75-1)^2+(0.25-0)^2}=0.35$.

In distance-based transformation, all distribution feature vectors of the same distance to the centroid are transformed to the same value. For example, in Fig. 4(a) t_1 and t_3 have the same value on D_A after transformation, though their distribution feature vectors are very different.

To reduce information loss, we propose a new transformation strategy based on similarity analysis among tuples.

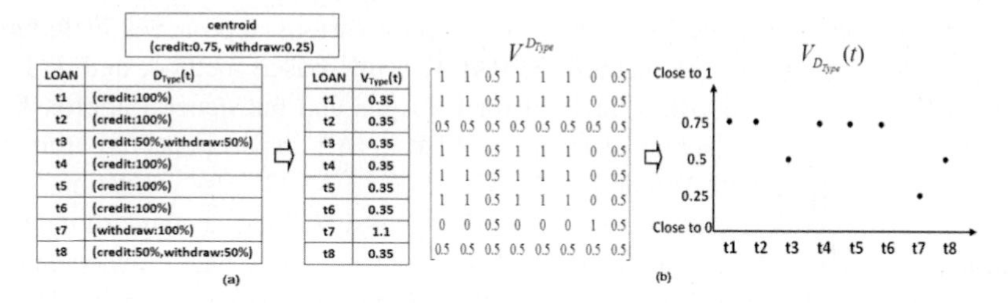

Fig. 4. Transformation with (a) distance-based method and (b) similarity-based method

Definition 6 (Similarity-based transformation). Let $t_1,\ldots, t_N$ be the set of target tuples. We transform a distribution feature vector $t.D_A$ to $v_{D_A}(t)$ such that

$$v_{D_A}(t)=\frac{1}{N}\sum_{i=1}^{N} sim_{D_A}(t,t_i). \tag{7}$$

Example 7 (Similarity-based transformation). Consider the data in Fig. 4(b). The similarity vector $v_{D_{Type}}$ is also shown in the Figure. The similarity transformation transforms $t_1.D_{Type}$ to (1+1+0.5+1+1+1+0+0.5)/8=0.75. The other distribution feature vectors can be transformed similarly and the results are shown in Fig. 4(b).

In similarity-based transformation, $t.D_A$ is transformed to the average of the similarities between $t.D_A$ and the other distribution feature vectors on D_A. If $v_{D_A}(t)$ is close to 1, t has a value distribution on A similar to those of the other target tuples. In other words, t is an ordinary tuple according to A. If $v_{D_A}(t)$ is close to 0, t is different from other target tuples in distribution on A.

The similarity-based transformation strategy works well under the hypothesis that the target tuples in the same class have high similarity to each other and low similarity to tuples in other classes on D_A. This hypothesis is implemented by the evaluation measure used in feature selection procedure where the selected features have high similarity with class distribution feature D_C.

6 Experimental Results

To evaluate the performance of our general approach for multi-relational classification, we first implement it to build a SVM classifier, denoted by MulSVM. We evaluate the effectiveness of the feature generation and selection approach and the feature computation method based on similarity analysis. Then we implement our general approach to build classifiers using other propositional classification methods, and make comparisons with ILP approaches on both accuracy and efficiency.

We use the propositional classification algorithms in WEKA[1]. In the feature selection, *min_sim* is experimentally set to 50% of $sim(D_c, A^0_f)$ with the first selected A^0_f. *max_select* is set to 0.8 and *min_cov* is set to 0.5 experimentally, which means if $cover(t_i)$ for a target tuple t_i is less than 0.5, the elements in row i and column i of W should be increased by b times. The parameter b is set to $\sqrt{2}$ in our experiments.

We compare with a classical propositionalization approach RelAggs [15] and an efficient ILP approach CrossMine [11]. We employ various approaches including 1) SVM, 2) J48, a decision tree method, 3) PART, a rule-based method, and 4) LogitBoost (LB for short), a boosting method for RelAggs and our general approach. All parameters involved are set as default in WEKA. All experiments are run on a 1.5GHz Pentium 4 PC with Windows XP. We adopt a 10-fold cross validation.

We use five real data sets[2], including 1) Mutagenesis (Muta), a standard dataset in relational learning, 2) Financial Database (F-DB), a benchmark back finance database whose schema is shown in Fig. 1, 3) East-West (E-W), a classical relational learning problem in machine learning, 4) Alzheimer toxic (A-t), a relational dataset of disease, and 5) Drug pyrimidines (Drug), a relational dataset of drugs.

6.1 Evaluating MulSVM

In this experiment, we evaluate the effectiveness of the feature generation and selection approach and the feature computation strategy in MulSVM. The following approaches are implemented to make comparison. 1) The naïve solution (Naïve), using all and only aggregate features, without feature selection, 2) The only-aggregate solution (OnlyAggr), using the same feature selection strategy as MulSVM, without distribution feature generated, 3) The distanced-based solution (Distance), using the same feature selection strategy as MulSVM, but using distance-based transformation method on distribution features.

The accuracy and running time of these approaches are given in Table 1. On most data sets, MulSVM has the highest accuracy. It is much faster than the Naive method. It performs better than the Naive solution in most cases, because it introduces distribution features for nominal features. Compared to OnlyAggr and Distance, MulSVM with the similarity-based feature computation method performs better than aggregation only and distance-based methods, expect on date sets Alzheimer and Drug. On those two data sets, most of the methods have the same accuracy because no distribution features are selected.

This experiment indicates that the feature generation method and the similarity-based feature computation strategy in MulSVM improve the accuracy and efficiency.

---

[1] http://www.cs.waikato.ac.nz/ml/weka/

[2] http://www.cs.waikato.ac.nz/ml/proper/datasets.html

Table 1. Accuracy/running time analysis (in percentage/second, best results in bold)

	Muta		F-DB		E-W		A-t		Drug	
Naive	86.2	1.4	87	69	**80**	0.1	89.6	**11.2**	98.1	7.8
OnlyAggr	**87.8**	**0.9**	86.8	**4.1**	75	**<0.1**	**91.2**	55.7	**98.4**	5.4
Distance	87.2	1	86.8	**4.1**	**80**	**<0.1**	**91.2**	61	**98.4**	**4.6**
MulSVM	**87.8**	1	**87.3**	4.4	**80**	**<0.1**	**91.2**	61.4	**98.4**	5.5

6.2 Evaluating the General Approach

We implement the general approach proposed in this paper with some other propositional classification algorithms, denoted by MulJ48, MulPART and MulLB. We compare the performance of these approaches with the corresponding RelAggs-methods and CrossMine. The accuracy and running time are shown in Table 2.

We observe the following results. First, the accuracy of Mul-methods is better than RelAggs-methods on most data sets, especially on Alzheimer and Drug datasets. It is because Mul-methods use distribution features and the similarity-based feature computation strategy, which provide more useful information than RelAggs-methods. On data sets Alzheimer and Drug pyrimidines, Mul-methods have a much higher accuracy than all the other algorithms, with improvement up to 19% and 15%.

Second, only CrossMine is competitive with Mul-methods in efficiency. The Mul-methods prove to be fast even compared to CrossMine on Mutagenesis, Financial and East-West data sets. On Alzheimer data set, Mul-methods are slower than CrossMine because of frequent similarity update due to low tuple coverage. The running time of Mul-methods is only 1%-10% of that of the Relaggs-methods. Both Mul-methods and CrossMine adopt the tuple ID propagation to avoid joining relations physically. This experiment proves the effectiveness of our approach for multi-relational classification.

Table 2. Accuracy/running time analysis (in percentage/second, best results in bold and second with underline)

	Muta		F-DB		E-W		A-t		Drug	
MulSVM	87.8	1	87.3	4.4	80	**<0.1**	91.2	61.4	**98.4**	5.5
RelAggs_SVM	79.8	7.3	82.6	125	80	9	85.7	463	93.4	322
MulJ48	88.8	**0.8**	87	**4.3**	75	**<0.1**	97.6	58.4	98.5	5.3
RelAggs_J48	85.6	7.4	88.2	124	80	9	92.9	462	93.4	324
MulPART	86.2	0.9	84.8	4.5	80	**<0.1**	**98.8**	58.5	**98.4**	5.3
RelAggs_PART	87.2	7.1	89	127	80	9	93.9	465	94.5	334
LB	**89.9**	**0.8**	**89.8**	4.5	**85**	**<0.1**	96.3	59.3	94.2	5.6
RelAggs_LB	85.1	7.3	80	125	**85**	9.1	80.6	465	82.2	324
CrossMine	81.9	1.3	87.3	9.7	80	0.1	94	**0.4**	88.2	**3.1**

In order to evaluate the efficiency and scalability further, we also construct a synthetic database. We generate a relational schema with r relations, including a target one. For each relation, there are a attributes. For the target relation, a nominal attribute is used as the class label attribute. The primary-keys are randomly generated, with the restriction that there is at most one for each relation. For each primary-key there are f corresponding foreign-keys randomly located in the other relations. n tuples are generated for each relation.

First we design a series of databases with the same schema except for the number of tuples. We generate 5 relations for each database, 5 attributes for each relation and 2 foreign-keys for each primary-key (Syn_DB_R5A5F2). We compare the running time of MulSVM (MulSVM represents all Mul-methods because SVM performs slowly on large data set), CrossMine, and RelAggs-methods (RelAggs_J48 represents all RelAggs-methods except RelAggs_SVM). The results are shown in Fig. 5(a).

We design another series of databases with the same schema except for number of relations. We generate 10, 20, 50, 100 and 200 relations, respectively, 5 attributes for each relation and 2 foreign-keys for each primary-key. We fix the number of tuples in each relation to 500 (Syn_DB_A5F2T500). Results on running are shown in Fig. 5(b).

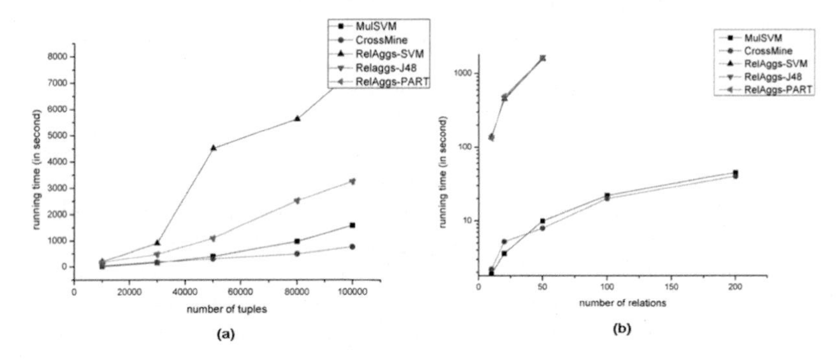

Fig. 5. Running time comparison (a) as number of tuples grows on Syn_DB_R5A5F2, and (b) as number of relations grows on Syn_DB_A5F2T500

When the number of tuples increases, MulSVM is comparative to CrossMine in efficiency, which is much faster than the RelAggs-methods. The running time of RelAggs_SVM grows dramatically compared to the other RelAggs-methods in Fig. 5 because SVM is inefficient on large data sets with excessive features that RelAggs-methods generated. However, MulSVM performs efficiently even on large data sets. On the other hand, as the number of relations increases in Fig. 5(b), MulSVM is also as efficient as CrossMine, which is one of the most efficient ILP algorithms in multi-relational classification. In conclusion, we demonstrate in this experiment that Mul-methods are preferable to classical ILP approaches on both efficiency and accuracy.

7 Conclusions

In this paper, we develop a general approach for multi-relational classification using feature generation and selection. We propose a similarity-based feature selection method for multi-relational classification, and a similarity-based data transformation method to construct feature values. An extensive empirical study verifies that our approach is accurate, fast and scalable. As future work, it is interesting to explore extending our approach for other multi-relational data mining tasks.

References

1. Burges, C.J.C.: A Tutorial on Support Vector Machines for Pattern Recognition. Data Mining Knowledge Discovery 2(2), 121–167 (1998)
2. Quinlan, J.R.: Induction of Decision Trees. Machine Learning 1, 81–106 (1986)
3. Lavrac, N., Dzeroski, S.: Inductive Logic Programming: Techniques and Applications. Ellis Horwood, New York (1994)
4. Muggleton, S.: Inductive Logic Programming. New Generation Computing 8(4), 295–318 (1991)
5. Muggleton, S., Raedt, L.: Inductive Logic Programming: Theory and methods. Journal of Logic Programming 19, 629–679 (1994)
6. Lloyd, J.W.: Foundations of Logic Programming. Springer, New York (1987)
7. Quinlan, J.R., Cameron-Jones, R.M.: Foil: A Midterm Report. In: Brazdil, P.B. (ed.) ECML 1993. LNCS, vol. 667, pp. 3–20. Springer, Heidelberg (1993)
8. Muggleton, S.: Inverse Entailment and Progol. New Generation Computing 13, 245–286 (1995)
9. Blockeel, H., Raedt, L.D.: Top-down Induction of First-order Logical Decision Trees. Artificial Intelligence 101(1-2), 285–297 (1998)
10. Quinlan, J.R.: C4.5 Programs for Machine Learning. Morgan Kaufmann, California (1993)
11. Yin, X.X., Han, J.W., Yang, J., Yu, P.S.: Crossmine: Efficient Classification across Multiple Database Relations. In: ICDE, pp. 399–411 (2004)
12. Kramer, S.: Relational Learning VS Propositionalization: Investigations in Inductive Logic Programming and Propositional machine learning. Technical report, Vienna University of Technology (1999)
13. Lavrac, N.: Principles of Knowledge Acquisition in Expert Systems. PhD thesis, Faculty of Technical Sciences, University of Maribor (1990)
14. Srinivasan, A., King, R.D.: Feature Construction with Inductive Logic Programming: A Study of Quantitative Predictions of Biological Activity Aided by Structural Attributes. Data Mining Knowledge Discovery 3(1), 37–57 (1999)
15. Krogel, M.A.: On Propositionalization for Knowledge Discovery in Relational Databases. PhD thesis, Fakultat fur Informatik, Germany (2005)
16. Perlich, C., Provost, F.: Aggregation-based Feature Invention and Relational Concept Classes. In: Proceedings of the 9th International Conference on Knowledge Discovery and Data Mining, New York, pp. 167–176 (2003)
17. Molina, L.C., Belanche, L., Nebot, A.: Feature Selection Algorithms: A Survey and Experimental Evaluation. In: ICDM 2002 (2002)
18. Yin, X.X., Han, J.W., Yu, P.S.: Cross-Relational Clustering with User's Guidance. In: KDD 2005 (2005)

A Unified Approach to the Extraction of Rules from Artificial Neural Networks and Support Vector Machines

João Guerreiro[1] and Duarte Trigueiros[2]

[1] Department of Information Science and Technology, ISCTE-IUL, Portugal
[2] Faculty of Economics, University of Algarve, Portugal

Abstract. Support Vector Machines (SVM) are believed to be as powerful as Artificial Neural Networks (ANN) in modeling complex problems while avoiding some of the drawbacks of the latter such as local minimæ or reliance on architecture. However, a question that remains to be answered is whether SVM users may expect improvements in the interpretability of their models, namely by using rule extraction methods already available to ANN users. This study successfully applies the Orthogonal Search-based Rule Extraction algorithm (OSRE) to Support Vector Machines. The study evidences the portability of rules extracted using OSRE, showing that, in the case of SVM, extracted rules are as accurate and consistent as those from equivalent ANN models. Importantly, the study also shows that the OSRE method benefits from SVM specific characteristics, being able to extract less rules from SVM than from equivalent ANN models.

Keywords: Data Mining, Support Vector Machines, Artificial Neural Networks, Orthogonal Search-based Algorithm, OSRE, Pedagogical, Decompositional, Rule Extraction.

1 Introduction

Support Vector Machines (SVM) proved to be accurate analytical tools, quite able to predict complex relations in various application fields. Similarly to Artificial Neural Networks (ANN), such accuracy stems from their ability to represent any given function [1] [2] [3] or to define complex decision boundaries. Despite their predictive ability, ANN and SVM have well-known drawbacks such as their black-box approach to modeling and the ensuing lack of transparency. What can be learnt from their systemic underlying knowledge representation is little more than a set of weights, activation functions and optimal parameters, discovered during the Neural Network training, or the kernel function and the optimized parameters of the Support Vector Machine. Hidden inside such complexity is an eventually meaningful relationship between inputs and predicted values.

Given the obvious need to understand the underlying learning mechanisms of ANN, in recent years authors have proposed varied techniques to overcome this missing transparency. However, there is still a demand for a unified rule extraction method that encompasses ANN and SVM, ensuring a compromise between accuracy and fidelity to the original models, together with consistency and comprehensibility, while being highly portable between different algorithms.

L. Cao, J. Zhong, and Y. Feng (Eds.): ADMA 2010, Part II, LNCS 6441, pp. 34–42, 2010.

The paper demonstrates a quite successful application of the Orthogonal Search-based Rule Extraction algorithm (OSRE) [4] to Support Vector Machines modeling. Results obtained with Artificial Neural Networks and the same rule extraction methods are used as benchmarks, showing that the use of OSRE with SVM is capable of maintaining the accuracy of the original classifiers while extracting, in both cases, a consistent set of rules.

2 Rule Extraction

Figure 1 displays chronologically the most relevant algorithms hitherto proposed to extract rules from ANN or other modeling tools. In Figure 1 algorithms are further organized according to the translucency of the rule extraction algorithm [5].

The translucency criterion considers the techniques perception of the learning method, thus creating three broad families of rule extraction approaches: the decompositional approach, the pedagogical approach and the eclectic approach.

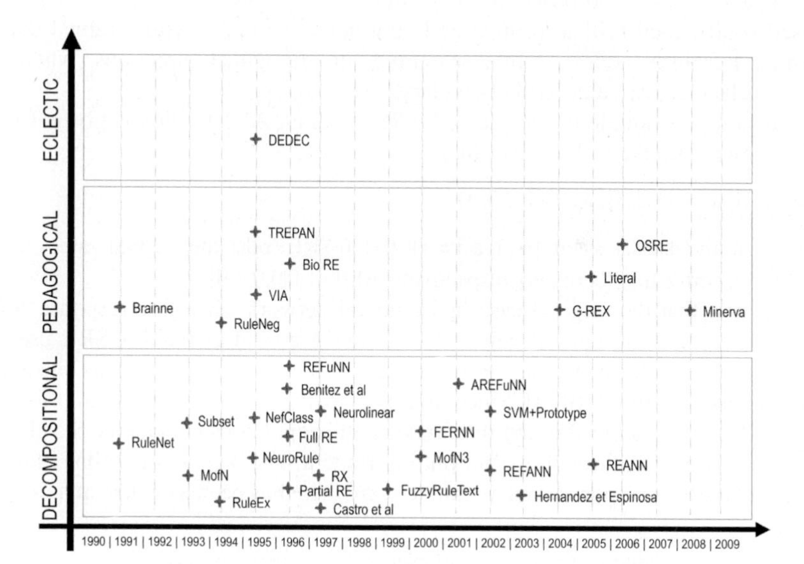

Fig. 1. Rule Extraction Methods organized by broad families and by year

The decompositional approach extracts rules at the level of the individual units by analyzing the activation values, weights and biases of the Neural Networks and the kernel function, vectors and optimization parameters of the Support Vector Machines [5]. Its disadvantage lies in its dependency on the learning mechanism coupled with the inability to accurately derive the logic of the underlying decision surface [4] [5].

The pedagogical approach used considers the trained ANN or SVM as a black box and using the classifier algorithm as an oracle through which it tests its predicted responses [5]. While changing the input values, rules are extracted which express the relationship between inputs and outputs of the Neural Network or Support Vector Machine. The main issue with most pedagogical approaches is that they are exponential in their complexity [4]. The number of rules grows at a rate of k^n, with n being the number of input variables with k possible values. Nevertheless, they are highly portable due to their ability to operate with all types of classifiers.

Finally, the eclectic approach incorporates elements of both decompositional and pedagogical rule extraction techniques. The algorithms of the eclectic type use the internal architecture of the trained ANN or SVM to complement a symbolic learning algorithm [4] [5].

3 The Use of OSRE with ANN

The Orthogonal Search-based Rule Extraction algorithm (OSRE) from Etchells and Lisboa [4] is a successful pedagogical methodology often applied in biomedicine (see [6] [7] [8] [9] and others). OSRE possesses the attractive characteristic of reducing the problem from exponential to linear in terms of the number of inputs [4] and is based on a formalism proposed by Tsukimoto [11]. OSRE extends the algorithm proposed by Ruleneg [10] to ordinal and continuous variables using trained data to perform a 1-from-N coding, while searching, in orthogonal directions, where the decision surface crosses a decision boundary.

An illustrative example of the use of OSRE (with an ANN) follows: given 3 input variables, each with the following values:

$a_1 = [1,2]$, $a_2 = [1,2,3]$, $a_3 = [1,2,3,4]$.

In one case in the dataset where $[a_1,a_2,a_3] = [1,2,2]$, OSRE codes the original values into 1-from-N form, converting them into, respectively, [0,1|0,1,0|0,0,1,0].

Considering that the original case had a neural network activation response > 0.5, then while stepping through all values of a_1 leaving a_2 and a_3 fixed, OSRE uses the classifier as an oracle inspecting its response as shown in Table 1. This method is called stepwise negation [10] [4]. Since, in this case, there is no change in a_1, this input variable will not be included in the rule. Indeed, this input does not change the classifiers response. Stepping through a_2 and putting aside a_1 and a_3, the algorithm again inspects the classifiers response, and as can be seen, there has been a change at a_2 =2.

Table 1. Artificial Neural Network activation response

Stepping through a_1	Stepping through a_2	Stepping through a_3
Net (01|010|0010) > 0.5	Net (01|001|0010) > 0.5	Net (01|010|0001) < 0.5
Net (10|010|0010) > 0.5	Net (01|010|0010) > 0.5	Net (01|010|0010) > 0.5
	Net (01|100|0010) < 0.5	Net (01|010|0100) > 0.5
		Net (01|010|1000) > 0.5

Therefore, a first rule is extracted: $a_2 \leq 2$. Finally, stepping through a_3, the algorithm finds a change at $a_3 = 2$. A second rule will be $a_3 \geq 2$. In brief, the network states that when $a_2 \leq 2$ and $a_3 \geq 2$ the response of the network is > 0.5 i.e. in class. So the rule for this data point is $(a_2 \leq 2) \wedge (a_3 \geq 2)$.

The resulting set of rules can then be refined by deleting repeated rules and those which fall below a predetermined specificity. The last refinement considers reducing the conjunctions and determining whether there has been a drop in specificity. Where specificity remains the same, conjunctions are eliminated thus defining the final set.

4 The Use of OSRE with SVM

OSRE is now applied to the extraction of rules from Support Vector Machines, and we will highlight here the successful results obtained. Following the steps used above to demonstrate the use of OSRE with Artificial Neural Networks, the first step consists of searching for a change in the classification result of the SVM, while stepwise negating the Boolean space of each input in the training dataset as we did in Table 1.

In order to discuss the ability of OSRE to extract rules from SVM, the original benchmark datasets from the Etchells and Lisboa study [4] are used here, namely the three Monks [12] datasets, the Wisconsin Breast Cancer [13] dataset and the Iris [14] dataset. Given that the relationships underlying these datasets have been widely studied and debated, namely using ANN, it is now possible to compare those published results with the SVM case highlighted in this paper, thereby evaluating the set of rules extracted by such method against those obtained from ANN. The training of SVM was carried out using SMO (Sequential Minimal Optimization) [15] and the kernel parameters were chosen to attain or surpass the accuracy of the original Artificial Neural Networks used in the OSRE literature. Results by dataset are as follows:

Monks. The Monks data consists of three artificial generated datasets each having 2 classes and 6 input variables:

$a_1 = [1,2,3]$, $a_2 = [1,2,3]$, $a_3 = [1,2]$, $a_4 = [1,2,3]$, $a_5 = [1,2,3,4]$, $a_6 = [1,2]$.

Monks-1. The known rule for these dataset is:

$$(a_5{=}1)\ (a_1{=}a_2)\ . \qquad (1)$$

Using a SVM with a polynomial kernel we were able to get a trained model with 99.8% accuracy and the set of rules represented in Table 2.

The results show that the rules extracted from the SVM are consistent with the known rule and identical to the ones obtained with the ANN [4].

Table 2. OSRE SVM Extracted Rules - Monks-1

Specificity	Sensitivity	Rules
1	0.4928	$[a_5{=}1]$
1	0.2338	$[a_1{=}3, a_2{=}3]$
1	0.2266	$[a_1{=}2, a_2{=}2]$
1	0.2050	$[a_1{=}1, a_2{=}1]$

Monks-2. For the second dataset, the known rule is:
Exactly two of:

$$\{a_1 = 1, a_2 = 1, a_3 = 1, a_4 = 1, a_5 = 1, a_6 = 1\}. \tag{2}$$

Table 3 shows that it is possible to extract from a SVM, trained with a polynomial kernel, identical rules to the ANN OSRE results, which represent all the permutations of the known rule.

Table 3. OSRE SVM Extracted Rules - Monks-2

Specificity	Sensitivity	Rules
1	0.1601	[$2 \le a_1 \le 3$, $2 \le a_2 \le 3$, $a_3 = 1$, $2 \le a_4 \le 3$, $2 \le a_5 \le 4$, $a_6 = 1$]
1	0.1019	[$a_1 = 1$, $2 \le a_2 \le 3$, $a_3 = 1$, $2 \le a_4 \le 3$, $2 \le a_5 \le 4$, $a_6 = 2$]
1	0.0970	[$2 \le a_1 \le 3$, $a_2 = 1$, $a_3 = 1$, $2 \le a_4 \le 3$, $2 \le a_5 \le 4$, $a_6 = 2$]
1	0.0825	[$2 \le a_1 \le 3$, $2 \le a_2 \le 3$, $a_3 = 1$, $a_4 = 1$, $2 \le a_5 \le 4$, $a_6 = 2$]
1	0.0825	[$a_1 = 1$, $2 \le a_2 \le 3$, $a_3 = 2$, $2 \le a_4 \le 3$, $2 \le a_5 \le 4$, $a_6 = 1$]
1	0.0776	[$2 \le a_1 \le 3$, $2 \le a_2 \le 3$, $a_3 = 2$, $a_4 = 1$, $2 \le a_5 \le 4$, $a_6 = 1$]
1	0.0728	[$2 \le a_1 \le 3$, $2 \le a_2 \le 3$, $a_3 = 2$, $2 \le a_4 \le 3$, $a_5 = 1$, $a_6 = 1$]
1	0.0728	[$2 \le a_1 \le 3$, $a_2 = 1$, $a_3 = 2$, $2 \le a_4 \le 3$, $2 \le a_5 \le 4$, $a_6 = 1$]
1	0.0485	[$2 \le a_1 \le 3$, $2 \le a_2 \le 3$, $a_3 = 1$, $2 \le a_4 \le 3$, $a_5 = 1$, $a_6 = 2$]
1	0.0485	[$2 \le a_1 \le 3$, $a_2 = 1$, $a_3 = 2$, $a_4 = 1$, $2 \le a_5 \le 4$, $a_6 = 2$]
1	0.0388	[$a_1 = 1$, $a_2 = 1$, $a_3 = 2$, $2 \le a_4 \le 3$, $2 \le a_5 \le 4$, $a_6 = 2$]
1	0.0339	[$a_1 = 1$, $2 \le a_2 \le 3$, $a_3 = 2$, $a_4 = 1$, $2 \le a_5 \le 4$, $a_6 = 2$]
1	0.0291	[$a_1 = 1$, $2 \le a_2 \le 3$, $a_3 = 2$, $2 \le a_4 \le 3$, $a_5 = 1$, $a_6 = 2$]
1	0.0291	[$2 \le a_1 \le 3$, $a_2 = 1$, $a_3 = 2$, $2 \le a_4 \le 3$, $a_5 = 1$, $a_6 = 2$]
1	0.0242	[$2 \le a_1 \le 3$, $2 \le a_2 \le 3$, $a_3 = 2$, $a_4 = 1$, $a_5 = 1$, $a_6 = 2$]

Results thus confirm that the use of a different smooth classifier such as the SVM, does not change the rules that define the problem boundaries, as extracted by OSRE.

Monks–3. In this case, both ANN and SVM classified cases with a 98.91% accuracy and again, rules extracted using OSRE converge towards the known rule.
The known rule is:

$$(a_2 \neq 3)\ (a_5 \neq 4) \vee (a_4 = 1)\ (a_5 = 3)\ . \tag{3}$$

The rules from the SVM are shown in Table 4. They mimic the known rule.

Table 4. OSRE-SVM Extracted Rules – Monks-3

Specificity	Sensitivity	Rules
0.9812	0.9479	[$1 \le a_2 \le 2$, $1 \le a_5 \le 3$]
0.9962	0.1493	[a_4=1, a_5=3]

Wisconsin Breast Cancer. This data set contains nine variables with discrete values between 1 and 10 and two outcome classes showing whether the patient's cancer is benign or malign. An ANN correctly classifies 96% of cases in the training phase and 96.9% in the validation phase. SVM obtained identical results. Rules extracted from ANN and SVM using OSRE were discarded where specificity fell below 90%. Rules extracted from ANN are represented in Table 5.

Table 5. OSRE – ANN Extracted Rules – Wisconsin Breast Cancer

Specificity	Sensitivity	Rules
0.9913	0.7261	$[2 \leq a_2 \leq 10, 2 \leq a_3 \leq 10, 3 \leq a_5 \leq 6 \vee 8 \leq a_5 \leq 10, a_8 = 1 \vee 3 \leq a_8 \leq 6 \vee 8 \leq a_8 \leq 10]$
0.9137	1	$[2 \leq a_2 \leq 10, 2 \leq a_3 \leq 10]$
0.9568	0.8690	$[2 \leq a_1 \leq 10, 2 \leq a_2 \leq 10, 2 \leq a_3 \leq 10, 2 \leq a_6 \leq 10]$

Figure 2 shows the boundaries of the extracted rule with higher sensitivity:

$$(a_2 \geq 2)(a_3 \geq 2)\,. \tag{4}$$

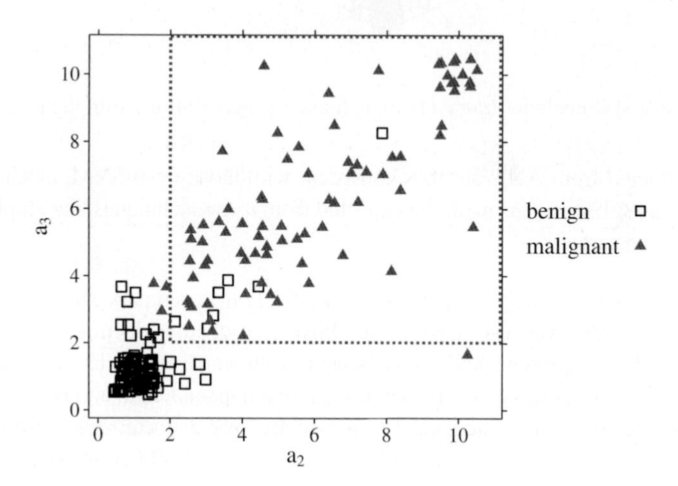

Fig. 2. Decision Boundaries (*dotted border*) found by the ANN rule with higher sensitivity

Using a polynomial kernel, OSRE extracted from SVM rules as in Table 6.

Table 6. OSRE – SVM Extracted Rules – Wisconsin Breast Cancer

Specificity	Sensitivity	Rules
0.9568	0.8690	$[2 \leq a_1 \leq 10, 2 \leq a_2 \leq 10, 2 \leq a_3 \leq 10, 2 \leq a_6 \leq 10]$
0.9568	0.8928	$[2 \leq a_2 \leq 10, 2 \leq a_3 \leq 10, 2 \leq a_6 \leq 10]$

OSRE extracted from SVM less rules than with ANN and again, the rule with higher sensitivity is similar to the rule extracted from ANN but with one more variable (a_6) to explain the problem.

$$(a_2 \geq 2)(a_3 \geq 2)(a_6 \geq 2)\,. \quad (5)$$

Replacing the input variables with their description, the extracted rule can be represented by decision boundaries shown in Figure 3.

$$(\text{Uniformity of Cell Size} \geq 2)(\text{Uniformity of Cell Shape} \geq 2)(\text{Bare Nuclei} \geq 2)\,. \quad (6)$$

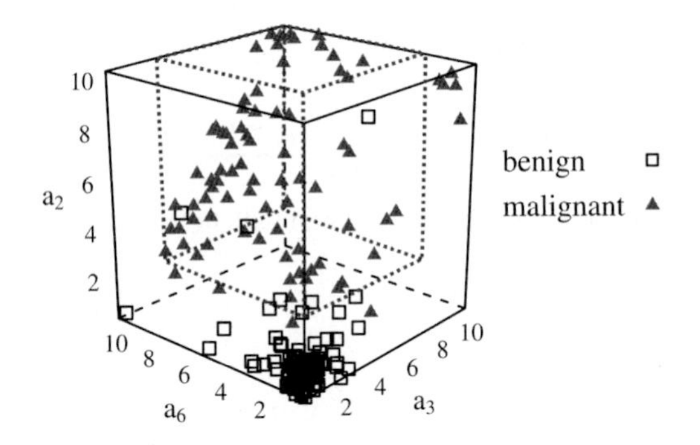

Fig. 3. Decision Boundaries (*dotted border*) found by the SVM rule with higher sensitivity

Rules extracted from ANN are thus consistent with those from SVM, albeit minor differences are also observed, as might be expected from using techniques that employ distinct types of decision surfaces.

Iris. The Iris dataset, widely viewed as a classification benchmark since Ronald Fisher introduced it in 1936, was also used in the OSRE introductory document [16] to test the ability of OSRE to represent ANN models using orthogonal rules. The Iris data has 150 items with 4 attributes, namely sepal length, sepal width, petal length and petal width, and 3 classes: Iris Setosa, Iris Virginica and Iris Versicolor. We extracted rules from ANN and SVM trained to represent the Iris Versicolor. Both ANN and SVM correctly classify 97.3% of cases in the training and in the validation phase. Rules extracted from ANN are shown in Table 7.

Table 7. OSRE – ANN Extracted Rules – Iris Dataset

Specificity	Sensitivity	Rules
0.9803	0.96	$[2 \leq a_3 \leq 8,\ 2 \leq a_4 \leq 8]$
0.9803	0.96	$[3 \leq a_4 \leq 5,\ 7 \leq a_4 \leq 8]$
0.9803	0.96	$[4 \leq a_4 \leq 5]$
0.9411	0.89	$[3 \leq a_3 \leq 8,\ 1 \leq a_4 \leq 5 \lor 7 \leq a_4 \leq 8]$

Using a polynomial kernel, OSRE extracted from SVM the rules shown in Table 8.

Table 8. OSRE – SVM Extracted Rules – Iris Dataset

Specificity	Sensitivity	Rules
0.9803	0.96	$[3 \le a_4 \le 5, 7 \le a_4 \le 8]$

Once again, experimental analysis shows that OSRE extracted less rules from SVM than from ANN, thus improving the interpretability of the final solution.

5 Conclusions and Future Research

The study shows that rules can indeed be extracted from SVM models. Moreover, such rules are identical or similar to those extracted from ANN models. It is thus concluded that users of SVM, when trying to improve the interpretability of their models, can expect the support already available to users of ANN. Orthogonal Search-based Rule Extraction (OSRE) can be used to extract consistent and accurate rules from both SVM and ANN, adding to Data Mining tasks the interpretability and comprehensibility these tools lack.

Moreover, since SVM are capable of unique solutions [17] and unique solutions may mean simpler decision surfaces, it follows that OSRE may be able to extract less rules from SVM than from equivalent ANN models, as is indeed the case of two of the instances in the paper. Less rules, in turn, mean improved pattern interpretability. Indeed, the paper opens up an interesting line of research, namely that aimed at ascertaining whether there exists some type of intrinsic adequacy between model interpretation algorithms such as OSRE and model building algorithms such as SVM.

It would also be interesting to extend the application of OSRE to larger and more challenging datasets trained using other smooth classifiers, in order to test its unifying abilities beyond ANN and SVM. Extracted rules that stand out among different classifiers may represent a more generic description of the problem's solution.

References

1. Fisher, D.H., McKusick, K.B.: An Empirical Comparison of ID3 and Back-Propagation. In: 11th International Joint Conference on Artificial Intelligence, vol. 1, pp. 788–793. Morgan Kaufmann, Michigan (1989)
2. Shavlik, J.W., Mooney, R.J., Towell, G.G.: Symbolic and Neural Learning Algorithms: An Experimental Comparison. Machine Learning 6, 111–143 (1991)
3. Weiss, S.M., Kapouleas, I.: An Empirical Comparison of Pattern Recognition, Neural Nets, and Machine Learning Classification Methods. In: 11th International Joint Conference on Artificial Intelligence, vol. 1, pp. 781–787. Morgan Kaufmann, Michigan (1989)
4. Etchells, T.A., Lisboa, P.J.G.: Orthogonal Search-based Rule Extraction (OSRE) for Trained Neural Networks: A Practical and Efficient Approach. IEEE Transactions on Neural Networks 17, 374–384 (2006)

5. Andrews, R., Diederich, J., Tickle, A.B.: Survey and Critique of Techniques for Extracting Rules from Trained Artificial Neural Networks. Knowledge-Based Systems 8, 373–389 (1995)
6. Aung, M.S., Lisboa, P.J., Etchells, T.A., Testa, A.C., Calster, B., Huffel, S., Valentin, L., Timmerman, D.: Comparing Analytical Decision Support Models Through Boolean Rule Extraction: A Case Study of Ovarian Tumour Malignancy. In: Liu, D., Fei, S., Hou, Z., Zhang, H., Sun, C. (eds.) ISNN 2007. LNCS, vol. 4492, pp. 1177–1186. Springer, Heidelberg (2007)
7. Lisboa, P.J.G., Jarman, I.H., Etchells, T.A., Ramsey, P.: A Prototype Integrated Decision Support System for Breast Cancer Oncology. In: 9th International Work Conference on Artificial Neural Networks, pp. 996–1003. Springer, San Sebastián (2007)
8. Jarman, I.H., Etchells, T.A., Martín, J.D., Lisboa, P.J.G.: An Integrated Framework for Risk Profiling of Breast Cancer Patients Following Surgery. Artificial Intelligence in Medicine 42, 165–188 (2008)
9. Lisboa, P.J.G., Etchells, T.A., Jarman, I.H., Aung, M.S.H., Chabaud, S., Bachelot, T., Perol, D., Gargi, T., Bourdès, V., Bonnevay, S., Négrier, S.: Time-to-event Analysis with Artificial Neural Networks: An Integrated Analytical and Rule-Based Study for Breast Cancer. Neural Networks 21, 414–426 (2008)
10. Pop, E., Hayward, R., Diederich, J.: RULENEG: Extracting Rules from a Trained ANN by Stepwise Negation. Queensland University of Technology, Australia (1994)
11. Tsukimoto, H.: Extracting Rules from Trained Neural Networks. IEEE Transactions on Neural Networks 11, 377–389 (2000)
12. Thrun, S.B., Bala, J., Bloedorn, E., Bratko, I., Cestnik, B., Cheng, J., De Jong, K., Deroski, S., Fahlman, S.E., Fisher, D., Hamann, R., Kaufman, K., Keller, S., Kononenko, I., Kreuziger, J., Michalski, R.S., Mitchell, T., Pachowicz, P., Reich, Y., Vafaie, H., Van de Welde, W., Wenzel, W., Wnek, J., Zhang, J.: The MONK's Problems: A Performance Comparison of Different Learning Algorithms. Technical Report CS-91-197, Computer Science Department, Carnegie Mellon University, Pittsburgh (1991)
13. Wisconsin Breast Cancer, `http://archive.ics.uci.edu/ml/datasets.html`
14. Iris Dataset, `http://archive.ics.uci.edu/ml/datasets.html`
15. Platt, J.C.: Fast Training of Support Vector Machines using Sequential Minimal Optimization. In: Platt, J.C. (ed.) Advances in Kernel Methods: Support Vector Learning, pp. 185–208. MIT Press, Cambridge (1999)
16. Etchells, T.A.: Rule Extraction from Neural Networks: A Practical and Efficient Approach. Ph.D Dissertation. John Moores University, Liverpool (2003)
17. Burges, C., Crisp, D.: Uniqueness of the SVM Solution. In: Advances in Neural Information Processing Systems, vol. 12, pp. 223–229. MIT Press, Cambridge (2000)

A Clustering-Based Data Reduction for Very Large Spatio-Temporal Datasets

Nhien-An Le-Khac[1], Martin Bue[2], Michael Whelan[1], and M-Tahar Kechadi[1]

[1] School of Computer Science and Informatics, University College Dublin, Belfield, Dublin 4, Ireland

[2] Ecole Polytechnique Universitaire de Lille, Villeneuve d'Ascq cedex, France

{an.lekhac,michael.whelan,tahar.kechadi}@ucd.ie, Martin.Bue@polytech-lille.net

Abstract. Today, huge amounts of data are being collected with spatial and temporal components from sources such as meteorological, satellite imagery etc. Efficient visualisation as well as discovery of useful knowledge from these datasets is therefore very challenging and becoming a massive economic need. Data Mining has emerged as the technology to discover hidden knowledge in very large amounts of data. Furthermore, data mining techniques could be applied to decrease the large size of raw data by retrieving its useful knowledge as representatives. As a consequence, instead of dealing with a large size of raw data, we can use these representatives to visualise or to analyse without losing important information. This paper presents a new approach based on different clustering techniques for data reduction to help analyse very large spatio-temporal data. We also present and discuss preliminary results of this approach.

Keywords: spatio-temporal datasets, data reduction, centre-based clustering, density-based clustering, shared nearest neighbours.

1 Introduction

Many natural phenomena present intrinsic spatial and temporal characteristics. Besides traditional applications, recent concerns about climate change, the threat of pandemic diseases, and the monitoring of terrorist movements are some of the newest reasons why the analysis of spatio-temporal data has attracted increasing interest. With the recent advances in hardware, high-resolution spatio-temporal datasets are collected and stored to study important changes over time, and patterns of specific events. However, these datasets are often very large and grow at a rapid rate. So, it becomes important to be able to analyse, discover new patterns and trends, and display the results in an efficient and effective way.

Spatio-temporal datasets are often very large and difficult to analyse [1][2][3]. Fundamentally, visualisation techniques are widely recognised to be powerful in analysing these datasets [4], since they take advantage of human abilities to perceive visual patterns and to interpret them [5]. However spatial visualisation techniques currently provided in the existing geographical applications are not adequate for decision-support systems when used alone. For instance, the problems of how to visualise

L. Cao, J. Zhong, and Y. Feng (Eds.): ADMA 2010, Part II, LNCS 6441, pp. 43–54, 2010.

the spatio-temporal multi-dimensional datasets and how to define effective visual interfaces for viewing and manipulating the geometrical components of the spatial data [6] are the challenges. Hence, alternative solutions have to be defined. Indeed, new solutions should not only include a static graphical view of the results produced during the data mining (DM) process, but also the possibility to dynamically and interactively obtain different spatial and temporal views as well as interact in different ways with them. DM techniques have been proven to be of significant value for analysing spatio-temporal datasets [7][8]. It is a user-centric, interactive process, where DM experts and domain experts work closely together to gain insight on a given problem. In particular, spatio-temporal data mining is an emerging research area, encompassing a set of exploratory, computational and interactive approaches for analysing very large spatial and spatio-temporal datasets. However, several open issues have been identified ranging from the definition of techniques capable of dealing with the huge amounts of spatio-temporal datasets to the development of effective methods for interpreting and presenting the final results.

Analysing a database of even a few gigabytes is an arduous task for machine learning techniques and requires advanced parallel hardware and algorithms. Huge datasets create combinatorially explosive search spaces for DM algorithms which may make the process of extracting useful knowledge infeasible owing to space and time constraints. An approach for dealing with the intractable problem of learning from huge databases is to select a small subset of data for mining [2]. It would be convenient if large databases could be replaced by a small subset of representative patterns so that the accuracy of estimates (e.g., of probability density, dependencies, class boundaries) obtained from such a reduced set should be comparable to that obtained using the entire dataset.

Traditionally, the concept of data reduction has received several names, e.g. editing, condensing, filtering, thinning, etc, depending on the objective of the reduction task. Data reduction techniques can be applied to obtain a reduced representation of the dataset that is much smaller in volume, yet closely maintains the integrity of the original data. That is, mining on the reduced dataset should be more efficient yet produce the same analytical results. There has been a lot of research into different techniques for the data reduction task which has lead to two different approaches depending on the overall objectives. The first one is to reduce the quantity of instances, while the second is to select a subset of features from the available ones.

In this paper, we will focus on the first approach to data reduction which deals with the reduction of the number of instances in the dataset. This approach can be viewed as similar to sampling, a technique that is commonly used for selecting a subset of data objects to be analysed. There are different techniques for this approach such as the scaling by factor [9][10], data compression [11], clustering [12], etc.

Often called numerosity reduction or prototype selection, instance reduction algorithms are based on a distance calculation between instances in the dataset. In such cases selected instances, which are situated close to the centre of clusters of similar instances, serve as the reference instances. In this paper we focus on spatio-temporal clustering technique. Clustering is one of the fundamental techniques in DM [2]. It groups data objects based on information found in the data that describes the objects and their relationships. The goal is to optimise similarity within a group of objects and the dissimilarity between the groups in order to identify interesting structures in the underlying data. Clustering is used on spatio-temporal data to take advantage of the fact that, objects that are close together in space and/or in time can usually be grouped together. As a consequence, instead of dealing with a large size of raw data, we can use these cluster representatives to visualise or to analyse without losing important information.

The rest of the paper is organised as follows. In Section II we discuss background and related work. Section III describes in detail our data reduction technique based on clustering. Section IV we evaluate the results of our data reduction technique as a pre-processing step on a very large spatio-temporal dataset. In Section V we discuss future work and conclude.

2 Background

2.1 Data Reduction

Sampling. The simplest approach for data reduction, the idea is to draw the desired number of random samples from the entire dataset. Various random, deterministic and density biased sampling strategies exist in literature [7][13]. However, naive sampling methods are not suitable for real world problems with noisy data, since the performance of the algorithms may change unpredictably and significantly. The random sampling approach effectively ignores all the information present in the samples not chosen for membership in the reduced subset. An advanced data reduction algorithm should include information from all samples in the reduction process [14][15].

Discretisation. Data discretisation techniques can be used to reduce the number of values for a given continuous attribute by dividing the range of the attribute into intervals. Interval labels can then be used to replace actual data values. Replacing numerous values of a continuous attribute by a small number of interval labels thereby reduces and simplifies the original data. In [9] the data reduction consists of discretising numeric data into ordinal categories. The process starts by first being given a number of points (called split points or cut points) to split the entire attribute range, and then repeats this recursively on the resulting intervals. The problem with this approach taken in [9] is that the interval selection process drops a large percentage of the data while trying to reduce the range of values a dimension can have.

2.2 Spatio-Temporal Data Mining

Spatio-temporal DM represents the junction of several research areas including machine learning, information theory, statistics, databases, and geographic visualisation. It includes a set of exploratory, computational and interactive approaches for analysing very large spatial and spatio-temporal datasets. Recently various projects have been initiated in this area ranging from formal models [4][16] to the study of the spatio-temporal data mining applications [5][16]. In spatio-temporal data mining the two dimensions "spatial" and "temporal" have added substantial complexity to the traditional DM process. It is worth noting, while the modelling of spatio-temporal data at different levels of details presents many advantages for both the application and the system. However it is still a challenging problem. Some research has been conducted to integrate the automatic zooming of spatial information and the development of multi-representation spatio-temporal systems [17][18][19]. However, the huge size of datasets is an issue with these approaches. In [8][9], the authors proposed a strategy that is to be incorporated in a system of exploratory spatio-temporal data mining, to improve its performance on very large spatio-temporal datasets. This system provides a DM engine that can integrate different DM algorithms and two complementary 3-D visualisation tools. This approach reduces their datasets by scaling them by a factor F; it simply runs through the whole dataset taking one average value for the F^3 points inside each cube of edge F. This reducing technique has been found to be inefficient as a data reduction method which may lose a lot of important information contained in the raw data.

2.3 Related Work

To the best of our knowledge, there is only our recent work [12] which proposed a knowledge-based data reduction method. This method is based on clustering [1] to cope with the huge size of spatio-temporal datasets in order to facilitate the mining of these datasets. The main idea is to reduce the size of that data by producing a smaller representation of the dataset, as opposed to compressing the data and then uncompressing it later for reuse. The reason is that we want to reduce and transform the data so that it can be managed and mined interactively. Clustering technique used in this approach is K-Medoids [12]. The advantage of this technique is simple; its representatives (medoids points) cannot however reflect adequately all important features of the datasets. The reason is that this technique is not sensitive to the shape of the datasets (convex).

3 Knowledge-Based Data Reduction

In this section, we present a new approach to improve our previous data reducing method [12]. We summarise firstly a framework of spatio-temporal data mining where our reducing method will be applied. Next, we describe our knowledge-based approach in an analytical way with an algorithm and the discussion on its issues.

3.1 Spatio-Temporal Data Mining Framework

As described in [8][9], our spatio-temporal data mining framework consists of two layers: mining and visualisation. The mining layer implements a mining process along with the data preparation and interpretation steps. For instance, the data may need some cleaning and transformation according to possible constraints imposed by specific tools, algorithms, or users. The interpretation step consists of using the selected models returned during the mining to effectively study the application's behaviour. The visualisation layer contains different visualisation tools that provide complementary functionality to visualise and interpret mined results. More details on the visualisation tools can be found in [8][9].

In the first layer, we applied the two-pass strategy. The reason is that the raw spatial-temporal dataset is too large for any algorithm to process; the goal of this strategy is to reduce the size of that data by producing a smaller representation of the dataset, as opposed to compressing the data and then uncompressing it later for reuse. Furthermore, we aim to reduce and transform the data so that it can be managed and mined interactively. In the first pass, the data objects are grouped according to their close similarity and then these groups are analysed by using different DM techniques based on specific objectives in the second pass. The objective of the first pass is to reduce the size of the initial data without losing any relevant information. On the other hand, the purpose of the second pass is to apply mining technique such as clustering, association rules on the tightly grouped data objects to produce new knowledge and ready for evaluation and interpretation. In the first implementation of this strategy, a scaling-based approach was used for the first past. Although it is simple and easy to implement, it loses a lot of important information contained in the raw data.

3.2 Clustering for Data Reduction

In this paper, we propose a new data reduction method based on clustering to help with the mining of the very large spatio-temporal dataset. Clustering is one of the fundamental techniques in DM. It groups data objects based on the characteristics of the objects and their relationships. It aims at maximising the similarity within a group of objects and the dissimilarity between the groups in order to identify interesting structures in the underlying data. Some of the benefits of using clustering techniques to analyse spatio-temporal datasets include a) the visualisation of clusters can help with understanding the structure of spatio-temporal datasets, b) the use of simplistic similarity measures to overcome the complexity of the datasets including the number of attributes, and c) the use of cluster representatives to help filter (reduce) datasets without losing important/interesting information.

We have implemented a combination of density-based and graph-based clustering in our approach. We have chosen a density-based method rather than other clustering method such as centre-based because it is efficient with spatial datasets as it takes into account the shape (convex) of the data objects [1]. However, it would be a performance issues when a simple density-based algorithm applied on huge amount of spatial datasets including differences in density. Indeed, the running times as well as the choice of suitable parameters are performance impacts of complex density-based algorithms [2]. In our algorithm, a modification of DBSCAN is used because DBSCAN algorithm [20] is simple; it is also one of the most efficient density-based algorithms, applied not only in research but also in real applications.

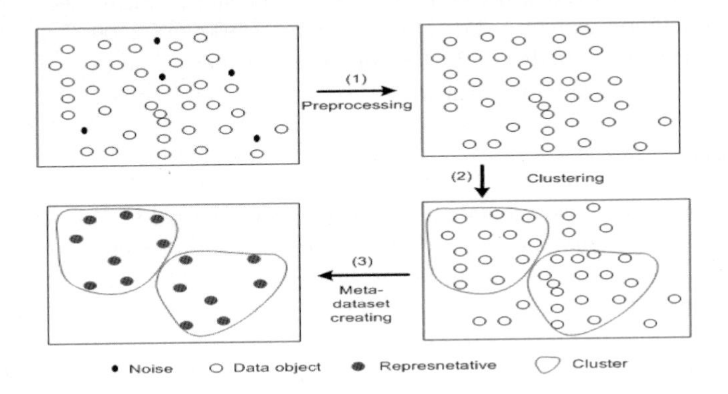

Fig. 1. Step by step view of the first pass of the strategy

In order to cope with the problem of differences in density, we combine DBSCAN with a graph-based clustering algorithm. Concretely, the Shared Nearest Neighbor Similarity (SNN) algorithm [21] is used to firstly build a similarity graph. Next, DBSCAN will be carried out based on the similarity degree. The advantage of SNN is that it address the problems of low similarity and differences in density. Another approach with a combination of SNN and DBSCAN was proposed in [2]. However, it is not in the context of data reduction and it did not take into account the problem of the huge size of the datasets in the context of memory constraint.

Fig.1 shows an overview of our reduction method including four steps: (1) a simple pre-processing is applied on raw datasets to filter NULL values. (2) SNN similarity graph is built for all datasets. Similarity degree of each data object is also computed in this step. The two parameters *Eps* and *Minpts* are selected based on these similarity degrees. Then, DBSCAN-based algorithm is carried out on the datasets to determine *core objects, specific core objects, density-reachable objects, density-connected objects*. These objects are defined in [20][22].

Clusters are also built based on *core objects* and *density-reachable* features in this step. Data objects which do not belong to any cluster will be considered as noise objects. *Core objects* or *specific core objects* are selected as cluster representatives that form a new (meta-) dataset (3). This dataset can then be analysed and produce useful information (i.e. models, patterns, rules, etc.) by applying other DM techniques (second pass of our framework). It is important to note that data objects that have a very high similarity between each other can be grouped together in the same clusters. As a result of this pass, the new dataset is much smaller than the original data without losing any important information from the data that could have an adverse effect on the result obtained from mining the data at a later stage.

4 Evaluation and Analysis

In this section, we study the feasibility of data reducing for spatio-temporal datasets by using DM techniques described in Section III. We compare also this approach with two other ones: scaling [8] (cf. 2.2) and K-Medoids clustering [2]. The dataset is the Isabel hurricane data [23] produced by the US National Centre for Atmospheric Research (NCAR). It covers a period of 48 hours (time-steps). Each time-step contains several atmospheric variables. The grid resolution is 500×500×100. The total size of all files is more than 60GB (~1.25 GB for each time-step). The experimentation details and a discussion are given below.

4.1 Experiments

The platform of our experimentation is a PC of 3.4 GHz Dual Core CPU, 3GB RAM using Java 1.6 on Linux kernel 2.6. Datasets of each time-step include 13 non-spatio attributes, so-called dimensions. In this evaluation, QCLOUD is chosen for analysis; it is the weight of the cloud water measured at each point of the grid. The range of QCLOUD value is [0…0.00332]. Different time-steps are chosen to evaluate: 2, 18, and 42. Totally, the testing dataset contains around 25 million data points of 4 dimensions X, Y, Z, QCLOUD for each time step.

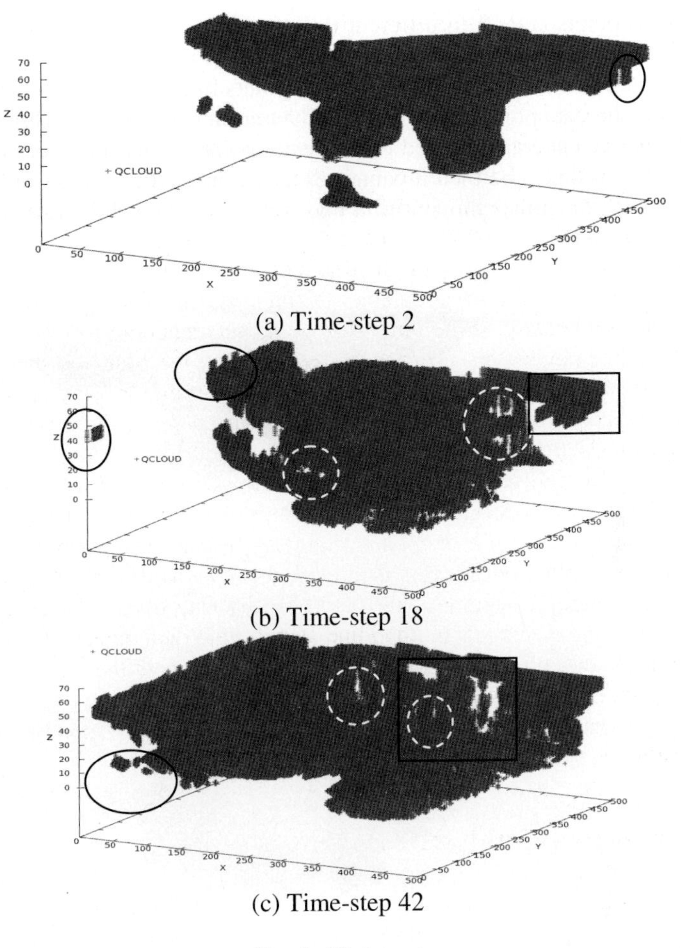

(a) Time-step 2

(b) Time-step 18

(c) Time-step 42

Fig. 2. All dataset

Fig.2 shows all data points before processing by any data reduction technique for testing dataset. Fig.3 shows the scaling results of our testing dataset in the grid coordinate at the selected time-steps. The scaling factor chosen is 50 ×50 × 50 for X, Y, Z i.e. we obtain 125000 data points after scaling as representative points. Fig.4 shows the results of our density-based clustering approach presented in Section 3 on the testing dataset. We also show representatives (*specific core point*) of each cluster. The number of representatives is approximate 120000. Fig.5 shows our testing dataset after the reducing process by a K-Medoids clustering. The number of clusters is 2000. We only show 50 data points including the medoid point of each cluster as representatives. We have totally 100000 data points for this case.

4.2 Analysis

As shown in Fig. 3, 4 and 5, representative points (or representative, in brief) could reflect the general shape of hurricane based on (X,Y,Z,QCLOUD) comparing to their whole original points (Fig.2) in different time-steps. By observing these figures, we recognise that the scaling approach cannot reflect the border of data points in details. For instance, it cannot show the shape of data points in the upper left corner (Fig.2a, Fig.3a and Fig.4a, data points in the circle). The reason is that in the scaling approach, all data points in a cube are reduced to the centre point. If this centre point is outside the dense areas of this cube then it cannot exactly represent the dense feature of this cube. Moreover, sometimes this position may get Null value while other positions in this cube are not null.

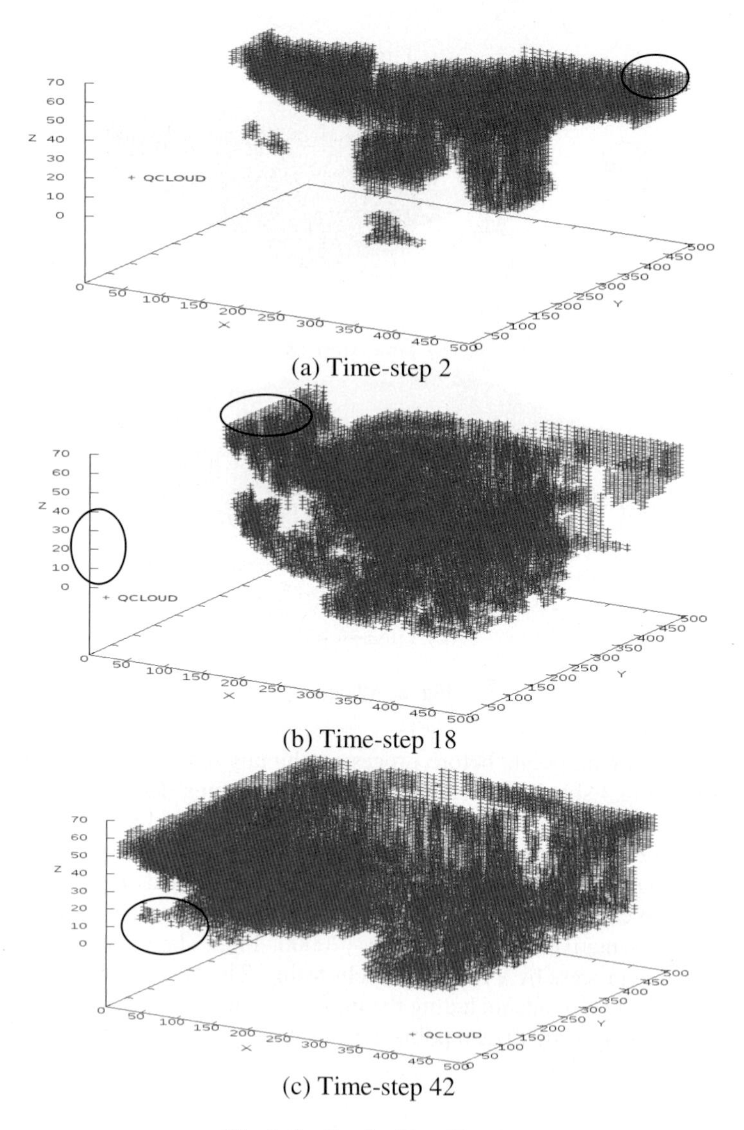

(a) Time-step 2

(b) Time-step 18

(c) Time-step 42

Fig. 3. Scaling by 50 × 50 × 50

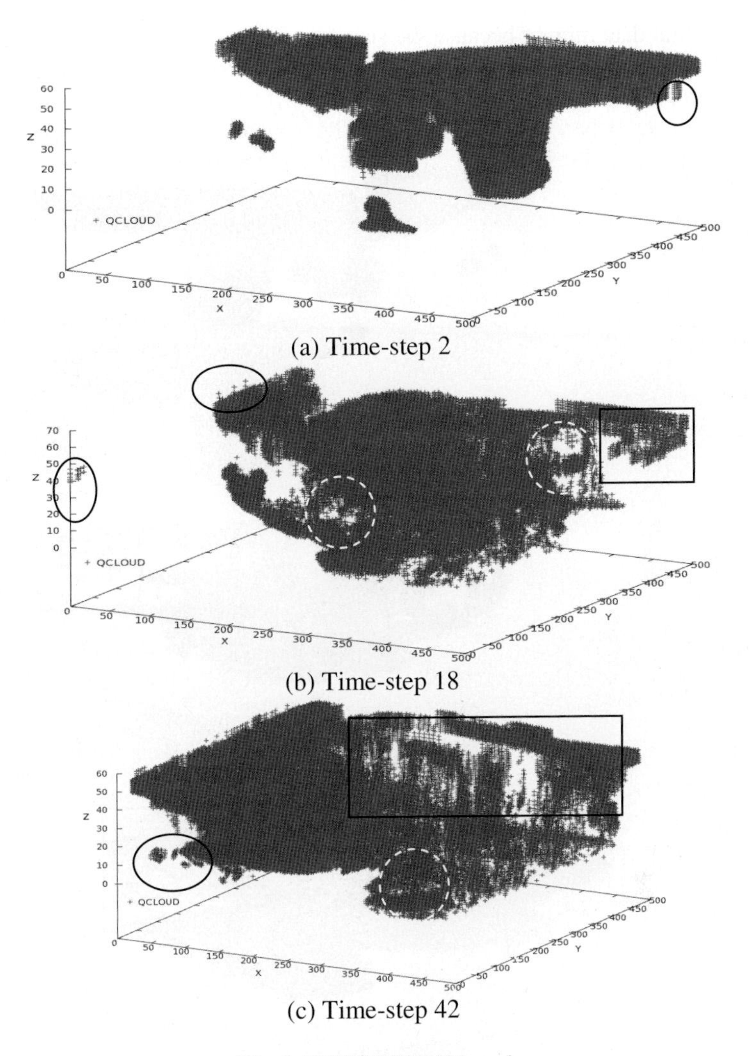

(a) Time-step 2

(b) Time-step 18

(c) Time-step 42

Fig. 4. SNN-DBSCAN-based

Fig.2b, Fig.3b and Fig.4b for the time-step 18 also confirm our observation. The scaling approach (Fig.3b) cannot show the data points on the extreme left of figure (near the Z-axis). Other circles in Fig2, Fig.3 and Fig.4 remark positions where this approach cannot represent efficiently. Indeed, the density-based approach (Fig.4) can show different holes (remark by dash circle in Fig.4 and Fig.2) clearer than the original ones (Fig.2) because in our approach, redundant data points are eliminated. Note that these holes are normally very important to geography experts to study the important features of a hurricane. They are sometimes not clear in early hours of the hurricane (Fig.2a, Fig.4a) Furthermore, representatives in our approach form different clusters and further analysis can base on this feature to extract more hidden knowledge. It is clearly that we can significantly reduce running time in following stages of spatio-temporal data mining because the size of representatives is less than 10% of all dataset. Besides, these representatives also reflect the movement of hurricane as their whole origin points do.

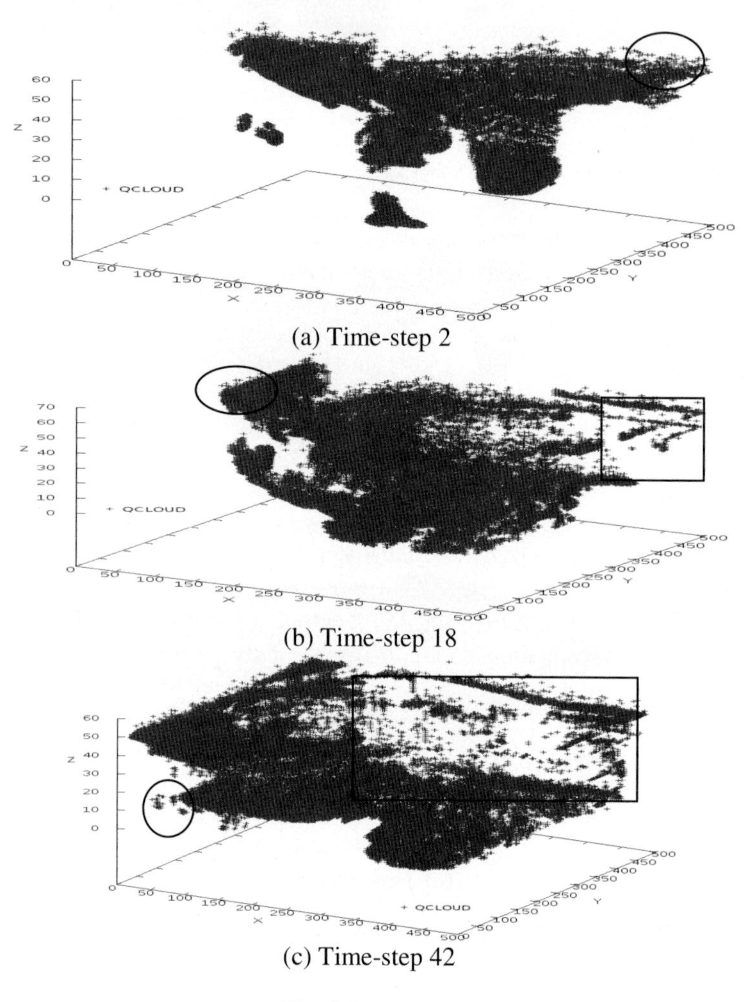

(a) Time-step 2

(b) Time-step 18

(c) Time-step 42

Fig. 5. K-Medoids

In our approach, specific core points are used instead of core points. The reason is that the number of core points is approximately 90% of all dataset in our experiments. Thus, the reduction does not gain in term of size of dataset. On the other hand, if there is a limitation of main memory, then the multi-partition approach will be applied. For instance, we carried out our approach with 10 parts in a limited memory and it gives an approximate result in representatives comparing to the case where whole dataset loaded in memory. The reason is that we combine SNN algorithm with our approach and it can deal with the different densities in the datasets. Consequently, it can reduce the effect of the partition of the dataset. However, running time as well as choosing efficient parameters (e.g. number of partitions) for this approach is also a performance issue. Besides, K-medoid approach gives a better performance comparing to the scaling one (Fig.3,5 e.g. data points in the circle). However, it cannot efficiently reflect the shape of data comparing to the SNN-DBSCAN based (Fig.4,5 e.g. data points in the rectangular). As a brief conclusion, these experiments above show that with our new approach, the use of simple DM techniques can be applied to reduce the large size of spatio-temporal datasets and preserve their important knowledge used by experts.

5 Conclusion and Future Work

In this paper, we study the feasibility of using DM techniques for reducing the large size of spatio-temporal datasets. As there are many reducing techniques presented in literature such as sampling, data compression, scaling, etc., most of them are concerned with reducing the dataset size without paying attention to their geographic properties. Hence, we propose to apply a clustering technique to reduce the large size without losing important information. We apply a density-based clustering that is a combination of SNN and DBSCAN-based algorithm on different time-steps. The experimental results show that knowledge extracted from mining process can be used as efficient representatives of huge datasets. Furthermore, we do not lose any important information from the data that could have an adverse effect on the result obtained from mining the data at a later stage. Besides, a solution for the limitation of memory is also proposed. We have reported some of these preliminary visual results for QCLOUD in its space X,Y,Z for three different time-steps (2, 18 and 42).

A more extensive evaluation is on-going. In the future we intend to analyse different combinations of dimensions over more time steps to try and find hidden information on their relationships with each other. Indeed, we are currently testing with a hybrid approach where density-based and centre-based clusterings are used to increase the performance in terms of running time and representative positions.

References

1. Dunham, M.H.: Data Mining: Introductory and Advanced Topics. Prentice-Hall, Englewood Cliffs (2003)
2. Tan, P.N., Steinbach, M., Kumar, V.: Introduction to Data Mining. Addison-Wesley, Reading (2006)
3. Ye, N. (ed.): The Handbook of Data Mining. Lawrence Erlbaum. Associates Publishers, Mahwah (2003)
4. Johnston, W.L.: Model visualisation. In: Information Visualisation in Data Mining and Knowledge Discovery, pp. 223–227. Morgan Kaufmann, Los Altos (2001)
5. Andrienko, N., Andrienko, G., Gatalsky, P.: Exploratory Spatio-Temporal Visualisation: an Analytical Review. Journal of Visual Languages and Computing, special issue on Visual Data Mining 14(6), 503–541 (2003)
6. Liu, H., Motoda, H.: On Issues of Instance Selection. Data Mining Knowledge Discovery 6, 2, 115–130 (2002)

7. Roddick, J.F., Hornsby, K., Spiliopoulou, M.: An Updated Bibliography of Temporal, Spatial, and Spatio-temporal Data Mining Research. In: Proceedings of the First International Workshop on Temporal, Spatial, and Spatio-Temporal Data Mining-Revised Papers, pp. 147–164 (2000)
8. Roddick, J.F., Lees, B.G.: Paradigms for Spatial and Spatio-Temporal Data Mining. Geographic Data Mining and Knowledge Discovery (2001)
9. Compieta, P., DiMartino, S., Bertolotto, M., Ferrucci, F., Kechadi, T.: Exploratory Spatio-Temporal Data Mining and Visualization. Journal of Visual Languages and Computing 18, 3, 255–279 (2007)
10. Bertolotto, M., DiMartino, S., Ferrucci, F., Kechadi, T.: Towards a Framework for Mining and Analysing Spatio-Temporal Datasets. International Journal of Geographical Information Science 21, 8, 895–906 (2007)
11. Sayood, K.: Introduction to Data Compression, 2nd edn. Morgan Kaufmann, San Francisco (2000)
12. Whelan, M., Le-Khac, N.A., Kecahdi, M.T.: Data Reduction in Very Large Spatio-Temporal Data Sets. In: IEEE International Workshop On Cooperative Knowledge Discovery and Data Mining 2010 (WETICE 2010), Larissa, Greece (2010)
13. Kivinen, J., Mannila, H.: The Power of Sampling in Knowledge Discovery. In: Proceedings of the ACM SIGACT-SIGMOD-SIGART, Minneapolis, Minnesota, US, pp. 77–85 (1994)
14. Lewis, D.D., Catlett, J.: Heterogeneous Uncertainty Sampling for Supervised Learning. In: Proceedings of the 11th International Conference on Machine Learning, New Brunswick, pp. 148–156. US: Morgan Kaufmann Publishers, San Francisco (1994)
15. Cohn, D., Atlas, L., Ladner, R.: Improving Generalization with Active Learning. Machine Learning 15, 2, 201–221 (1994)
16. Costabile, M.F., Malerba, D.: Special Issue on Visual Data Mining. Journal of Visual Languages and Computing 14, 499–510 (2003)
17. Bettini, C., Dyreson, C.E., Evans, W.S., Snodgrass, R.T.: A glossary of time granularity concepts. In: Etzion, O., Jajodia, S., Sripada, S. (eds.) Dagstuhl Seminar 1997. LNCS, vol. 1399, pp. 406–413. Springer, Heidelberg (1998)
18. Bettini, C., Jajodia, S., Wang, X.: Time Granularities in Databases, Data Mining, and Temporal Reasoning. Springer, Heidelberg (2000)
19. Cattel, R. (ed.): The Object Database Standard: ODMG 3.0. Morgan Kaufmann, San Francisco (1999)
20. Ester, M., Kriegel, H.P., Sander, J., Xu, X.: A Density-Based Algorithm for Discovering clusters in Large Spatial Databases with Noise. In: Proc. 2nd Int. Conf. on Knowledge Discovery and Data Mining (KDD 1996), Portland, USA, pp. 226–231 (1996)
21. Jarvis, R.A., Patrick, E.A.: Clustering using a similarity Measure Based on shared Nearest Neighbours. IEEE Transactions on Computers C-22(11), 1025–1034 (1973)
22. Januzaj, E., Kriegel, H.P., Pfeifle, M.: DBDC: Density-Based Distributed Clustering. In: Bertino, E., Christodoulakis, S., Plexousakis, D., Christophides, V., Koubarakis, M., Böhm, K., Ferrari, E. (eds.) EDBT 2004. LNCS, vol. 2992, pp. 88–105. Springer, Heidelberg (2004)
23. National Hurricane Center, `http://www.tpc.ncep.noaa.gov/2003isabel.shtml`

Change a Sequence into a Fuzzy Number

Diana Domańska[1] and Marek Wojtylak[2]

[1] Institute of Computer Science, University of Silesia,
Będzińska 39, 41200 Sosnowiec, Poland
ddomanska@poczta.onet.pl
[2] Institute of Meteorology and Water Management (IMGW),
Bratków 10, 40045 Katowice, Poland
monitoring.katowice@imgw.pl

Abstract. In general in the literature practitioners transform of real numbers into fuzzy numbers to the median or average, so they follow the probabilistic path. However, theoreticians do not investigate transformations of real numbers into fuzzy numbers when they analyse fuzzy numbers. They usually operate only on the fuzzy data. In the paper we describe an algorithm for transforming a sequence of real numbers into a fuzzy number. The algorithms presented are used to transform multidimensional matrices constructed from times series into fuzzy matrices. They were created for a special fuzzy number and using it as an example we show how to proceed. The algorithms were used in one of the stages of a model used to forecast pollution concentrations with the help of fuzzy numbers. The data used in the computations came from the Institute of Meteorology and Water Management (IMGW).

Keywords: Algorithm, Fuzzy number, Fuzzy model.

1 Introduction

In recent years many prediction approaches, such as statistical [1], fuzzy [2], neural networks [3] [4] and neuro-fuzzy predictors [5] have emerged. During the tests of the models which are based on fuzzy numbers real data is used. In this work we want to present a simplified description of a sequence of data, numbers which are concentrated, i.e. are close to each other. Fuzzy sets theory is helpful [6] [7] [8]. However, we could ask: How can we obtain a fuzzy information from real information? One of the possible paths is a probabilistic distribution, however we know from experience we know that we can fit many distributions for a set of numbers. Another possibility is to replace the raw data with a fuzzy number. The fuzzy number obtained in this way gives the possibility of standarisation, which further allows us to obtain one number. In the proposed model of air pollution forecasting (APFM) we use real data for the testing purposes. The individual stages of APFM [9] [10] arise based on the fuzzy numbers. Also in APFM the selection of this method is obvious, if we take into account the data on which the computations take place. Weather forecasts, meteorological situations and pollution concentrations have a chaotic character. Therefore in order to describe

L. Cao, J. Zhong, and Y. Feng (Eds.): ADMA 2010, Part II, LNCS 6441, pp. 55–62, 2010.

natural phenomena we use fuzzy numbers, which made it necessary to develop algorithms which allow a real number to be transformed into a fuzzy number.

The paper is organized in the following way. In Section 2 we describe the basic terms. In Section 3 we present the algorithms. In Section 4 we discuss the gradient method and we give some examples. In Section 5 we describe experiments for time series and in Section 6 we present the conclusions.

2 Basic Definitions

In the literature there are many interpretations of a fuzzy number [11] [12]. In our work we use the interpretation of the form (1).

Definition 1. *The fuzzy set* $A \in FS(\mathbb{R})$*, whose membership function:*

$$\mu_A : \mathbb{R} \to [0,1] , \tag{1}$$

satisfies the following conditions:

1. $\exists_{x \in \mathbb{R}} \mu_A(x) = 1$,
2. *for any elements* $x_1, x_2 \in \mathbb{R}$ *and* $\lambda \in [0,1]$ *we have* $\mu_A[\lambda x_1 + (1-\lambda)x_2] \geqslant \min\{\mu_A(x_1), \mu_A(x_2)\}$,
3. μ_A *is an interval continuous function,*

we will be called a fuzzy number.

In APFM the algorithms arise based on a specific fuzzy number (2). This fuzzy number was chosen based on our own calculations and based on paper [13].

$$\mu_G(x) = \begin{cases} \exp(\frac{-(x-m_1)^2}{2 \cdot \sigma_1^2}) & \text{if } x \le m_1 , \\ 1 & \text{if } x \in (m_1, m_2) , \\ \exp(\frac{-(x-m_2)^2}{2 \cdot \sigma_2^2}) & \text{if } x \ge m_2 , \end{cases} \tag{2}$$

where $m_1 \le m_2 , \sigma_1 > 0, \sigma_2 > 0$ for $m_1, m_2, \sigma_1, \sigma_2 \in \mathbb{R}$.

Let us denote the family of fuzzy sets on $\mathbb{R}$ as $FS(\mathbb{R})$ and let us denote the fuzzy numbers set as FN.

In the algorithms we use the discrete form of the fuzzy number which is defined in the following way.

Definition 2. *Let us assume that* $A \in FS(\mathbb{R})$ *and* $\mu_A(x_i) > 0$ *for* $i = 1, \ldots, n$. *We will call a set of pairs* $B = \{(x_1, \mu_A(x_1)), \ldots, (x_n, \mu_A(x_n))\}, n \in \mathbb{N}$ *with membership functions* μ_B *a discrete fuzzy number. Let us denote as d-*FN *the set of all discrete fuzzy numbers.*

3 Algorithms

We use algorithms to change the sequence into a fuzzy number. We divide this change into two algorithms. At first in Section 3.1 we change the number sequence to a discrete fuzzy number B, then in Section 3.2 we change B into a fuzzy number of the form defined by (2).

3.1 Algorithm 1

Input: number sequence $X = \{\xi_1, \ldots, \xi_m\}, m \in \mathbb{N}$.
Output: $B \in d\text{–}FN$.

Having the sequence X, we define:

$$\alpha = \min X \,, \beta = \max X \,, h = \frac{\beta - \alpha}{n - 1}, \tag{3}$$

where $n \in \mathbb{N}, n > 1$ is a parameter.
Let us define:

$$\begin{aligned} \gamma_i &= \alpha + h \cdot (i - 0.5) \text{ for } i = 0, \ldots, n \,, \\ a_i &= \{\xi \in X : \gamma_{i-1} \leq \xi < \gamma_i\} \text{ for } i = 1, \ldots, n \,, \\ x_i &= \alpha + (i - 1) \cdot h \text{ for } i = 1, \ldots, n \,. \end{aligned} \tag{4}$$

Let us assign $(x_i, \mu_B(x_i))$, where $\mu_B(x_i) = |a_i| / \max\{|a_1|, \ldots, |a_p|\}$ is the membership grade of x_i for $i = 1, \ldots, n$. We eliminate all x_i for which $\mu_B(x_i) = 0$ and we obtain the discrete fuzzy number B. The histograms of the data are usually multimodal and flattened.

3.2 Algorithm 2

Input: $x_1, \ldots, x_n$ and values $y_1, \ldots, y_n$ and weights $\omega_1, \ldots, \omega_n, \omega_i > 0$. In particular $\omega_i = 1$ or $\omega_i = \frac{1}{y_i}$.
Output: number of the form (2). Let us assume that the set of pairs $B = \{(x_i, y_i), i = 1, \ldots, n\}$, where $x_i \in \mathbb{R}, y_i = \mu_B(x_i)$ is a discrete fuzzy number. We will determine the fuzzy number (2) in the mean-square sense from the discrete fuzzy number B. Function (2) contains four parameters. For simplicity, let us denote them as $p_1 = m_1$, $p_1 = \sigma_1$, $p_3 = m_2$, $p_4 = \sigma_2$. To compute parameters p_1, p_2, p_3, p_4 we use the mean-square approximation. Practically, we can write that function (2) is dependent on 5 parameters: x and p_1, p_2, p_3, p_4.

$$\mu_G(x) = \mu_G(x; p_1, p_2, p_3, p_4) = \mu_G(x; p). \tag{5}$$

Afterwards we compute the minimum value of this function. For this purpose we use the gradient method.

Therefore, we define function (6).

$$\chi^2(p) = \frac{1}{2} \sum_{i=1}^{n} \omega_i \cdot (y_i - \mu_G(x_i; p))^2, \tag{6}$$

Function χ^2 is a $C^1_{\mathbb{R}}$ class because μ_G is a $C^1_{\mathbb{R}}$ class. To determine zero approximation in an iterated process, we implement the following algorithm. At first we have to select parameters p_1, p_3. Let us define a set:

$$M = \{x_i : \mu_B(x_i) = 1; \; i = 1, \ldots, n\}\,. \tag{7}$$

From definition 2, we know that set B is normal, so M is not empty. Let us define:

$$p_1 = \min_{x \in M}(x) \text{ and } p_3 = \max_{x \in M}(x), \tag{8}$$

$$p_2 = \frac{p_1 - x_1}{2} \text{ and } p_4 = \frac{x_n - p_3}{2}. \tag{9}$$

To find the minimum of the function χ^2 we compute partial derivatives and then we equal them to 0 (a necessary condition for the existence of the minimum). Additional difficulties of finding the minimum of χ^2 are the dependencies between the parameters which come from the form of the function μ: $p_1 \leqslant p_3, p_2 \geqslant 0, p_4 \geqslant 0$.

The following equalities arise.

$$\Phi_i(p) = \frac{\partial \chi^2}{\partial p_i} = \sum_{j=1}^{n} (\mu_G(x_j;p) - y_j) \frac{\delta \mu_G}{\delta p_i}|_{x=x_i}, i = 1, \ldots, 4 \tag{10}$$

Let us calculate:

$$\frac{\partial \chi^2}{\partial p_1} = \left(\sum_{j=1}^{n} (\mu_G(x_j;p) - y_j) \right) \cdot \frac{\partial \mu_G}{\partial p_1},$$

where

$$\frac{\partial \mu_G}{\partial p_1} = \begin{cases} \exp(\frac{-(x-p_1)^2}{2 \cdot p_2^2}) \cdot \frac{(x-p_1)}{p_2^2} & \text{if} \quad x \leqslant p_1, \\ 0 & \text{if} \quad x \in (p_1, \infty). \end{cases} \tag{11}$$

$$\frac{\partial \chi^2}{\partial p_2} = \left(\sum_{j=1}^{n} (\mu_G(x_j;p) - y_j) \right) \cdot \frac{\partial \mu_G}{\partial p_2},$$

where

$$\frac{\partial \mu_G}{\partial p_2} = \begin{cases} \exp(\frac{-(x-m_1)^2}{2 \cdot p_2^2}) \cdot \frac{(x-p_2)^2}{p_2^3} & \text{if} \quad x \leqslant p_1, \\ 0 & \text{if} \quad x \in (p_1, \infty). \end{cases} \tag{12}$$

The derivatives $\partial\chi^2/\partial p_1$, $\partial\chi^2/\partial p_2$ at point p_1 are continuous. Like p_1 we calculate $\partial\chi^2/\partial p_3$ and like p_2 we calculate $\partial\chi^2/\partial p_4$. The derivatives are used to compute a minimum of the function (6) in an iterated process. Thereby, we get a function of the form (2) with the proper parameters p_1, p_2, p_3, p_4.

4 Gradient Method

Using algorithms 1 and 2 we obtain consecutive approximations of the fuzzy numbers. Below is an example of changing a given time series into a fuzzy number using algorithms 1 and 2. In APFM we operate on matrices [14], but for better understanding we will give examples on time series for each of the attributes separately. At first we take the humidity data from 28 January 2006,

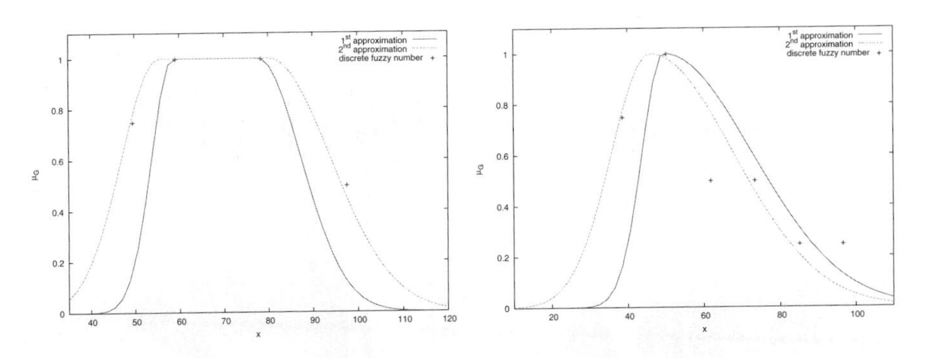

Fig. 1. The results of algorithms 1 and 2: for the humidity attribute at 28 January 2006, hour 11 (left) and at 28 January 2006, hour 13 (right)

hour 11.00. For $n = 7$ and $X = \{50, 49, 54, 51, 79, 76, 93, 60, 57, 59, 77, 76, 97\}$. Using algorithm 1 we obtain a discrete fuzzy number (13):

$$B = \{(49.0, 0.75), (58.6, 1.00), (77.8, 1.00), (97.0, 0.50)\} . \tag{13}$$

Using the consecutive approximations in algorithm 2 we obtain values shown in Fig. 1. We advance for all the input data in this way. The second approximation is better, gives a more accurate approximation of the fuzzy number to the number sequence than the first approximation and the probabilistic methods. Because of the large amount of data in APFM, the computations are made on multidimensional matrices with the dimensions 36×9. Having properly grouped the matrices for each attribute and each hour separately, we create sequences of numbers. Next, we use algorithms 1 and 2 to transform selected sequences of numbers into fuzzy numbers. Individual time series which consist of fuzzy numbers for the selected attributes are presented in Fig. 2.

5 Experiments

Using the gradient method for each attribute i and hour t we obtain a fuzzy number with the parameters p_1, p_2, p_3, p_4.

Next, for each t we standardise the obtained fuzzy number and obtain one value d. For this purpose we use two methods called: avg (14) and avgM (15). Let us define $T = \{t = i \cdot \Delta t : i = 1, \ldots, n_T\}$, $\Delta t > 0$, where Δt means a time step (usually $\Delta t = 1$ hour).

$$\forall_{t \in T} d_{avg}(t) = \frac{\sum_{i=0}^{m} \xi_i}{m} , \tag{14}$$

$$\forall_{t \in T} d_{avgM}(t) = \frac{1}{2}(p_1 + p_3) . \tag{15}$$

In Fig. 3 we have the obtained time series runs of the data computed with the avg, avgM method, real meteorological data and the forecast data for the

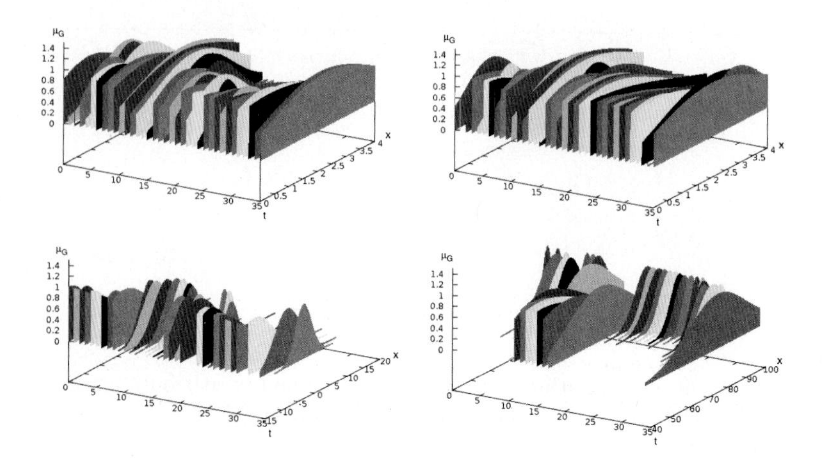

Fig. 2. Time series runs of the fuzzy numbers (2): for the wind speed attribute at 27 January 2006 (top-left), for the wind speed attribute at 28 January 2006 (top-right), for the temperature attribute at 28 January 2006 (bottom-left) and for the humidity attribute at 28 January 2006 (bottom-right)

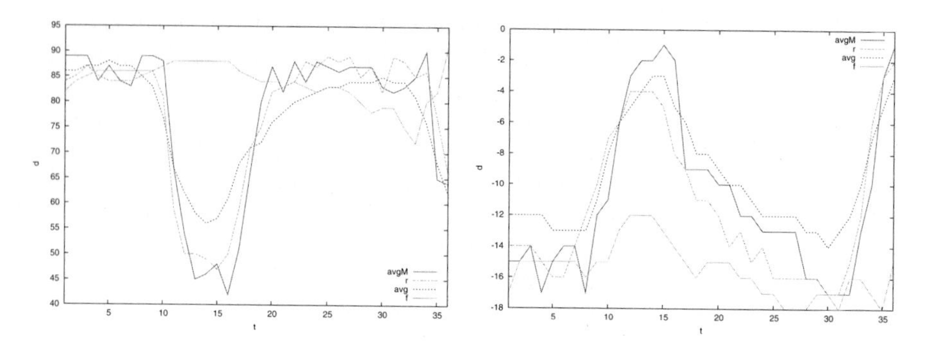

Fig. 3. Comparison of data obtained with the avg, avgM method and of the real data and forecast data: for the humidity attribute at 28 January 2006 (left) and for the temperature attribute 26 January 2006 (right)

humidity attribute at 28 January 2006 and for the temperature attribute at 26 January 2006.

We use the Mean Absolute Error [15] to estimate the verifiability of our algorithms for forecasting pollution concentrations. For each attribute i and hour t the Mean Absolute Error between the standardised computed data d and the real meteorological situations r is given by:

$$M_{d_i,r_i} = \frac{1}{n_T} \sum_{t=1}^{n_T} |d_{i,t} - r_{i,t}| \tag{16}$$

where $|x|$ is the absolute value of x.

We denote the Mean Absolute Error for the computed data from (14) and r as $M_{d_{avg_{i,t}},r_{i,t}}$. For the differences between the computed data from (15) and r we use the notation $M_{d_{avgM_{i,t}},r_{i,t}}$ for the Mean Absolute Error. Finally, we denote the Mean Absolute Error for r and the predicted weather forecast from IMGW f as $M_{f_{i,t},r_{i,t}}$.

Table 1. Results of the tests for the wind speed attribute

Date	$M_{d_{avgM_{i,t}},r_{i,t}}$	$M_{d_{avg_{i,t}},r_{i,t}}$	$M_{f_{i,t},r_{i,t}}$
26 January 2006	0.64	1.17	1.58
27 January 2006	0.68	1.01	1.75
28 January 2006	0.65	0.76	1.25

Table 2. Results of the tests for the temperature attribute

Date	$M_{d_{avgM_{i,t}},r_{i,t}}$	$M_{d_{avg_{i,t}},r_{i,t}}$	$M_{f_{i,t},r_{i,t}}$
26 January 2006	1.81	2.28	4.19
27 January 2006	1.99	2.73	3.74
28 January 2006	1.69	2.82	3.46

Table 3. Results of the tests for the humidity attribute

Date	$M_{d_{avgM_{i,t}},r_{i,t}}$	$M_{d_{avg_{i,t}},r_{i,t}}$	$M_{f_{i,t},r_{i,t}}$
26 January 2006	5.77	5.65	12.73
27 January 2006	4.67	6.55	13.79
28 January 2006	3.72	4.87	11.51

In Tabs. 1, 2, 3 we compile the values of the Mean Absolute Error between the computed data from the algorithms, real meteorological situations and the forecast. In Tab. 1 we see that there are very small differences between the real meteorological data and the computed data with the avg and avgM method for algorithms 1 and 2. avgM method gave the best results. Using existing algorithms from the COSMO LM model, we obtain results more distant from the real. We see the results In Tabs. 2 and 3.

6 Conclusions

In the paper we have proposed and discussed in detail the algorithms which transform a sequence of numbers into a fuzzy number. The experiments were performed on data from IMGW. Probabilistic methods which have not given satisfactory results were used. However, too many functions could be fitted to the number sequence. To make the fitting, better algorithms 1 and 2 were created.

Comparing the data computed with the help of our algorithms with the real data and the forecast data, we see very small differences between the computed data and the measured data. In particular in the standardisation method the best results were obtained for the avgM method. There are bigger differences between the forecast data and the measured data. Using algorithms to forecast the weather and pollution concentrations will improve the obtained results.

References

1. Rigatosa, G., Zhangb, Q.: Fuzzy Model Validation Using the Local Statistical Approach. Fuzzy Sets and Systems 160(7), 882–904 (2009)
2. Lee, C., Liu, A., Chen, W.: Pattern Discovery of Fuzzy Time Series for Financial Prediction. IEEE Trans. Knowl. Data Eng. 18(5), 613–625 (2006)
3. Jiang, W., Wang, P.: Research on Interval Prediction of Nonlinear Chaotic Time Series Based on New Neural Networks. In: Proc. 6th World Congress Intell. Control and Automation, pp. 2835–2839 (2006)
4. Gautam, A.K., Chelani, A.B., Jain, V.K., Devotta, S.: A New Scheme to Predict Chaotic Time Series of Air Pollutant Concentrations using Artificial Neural Network and Nearest Neighbor Searching. Atmospheric Environ. 42(18), 4409–4417 (2008)
5. Aznarte, M.J.L., Benítez Sánchez, J.M., Nieto Lugilde, D., de Linares Fernández, C., Díaz de la Guardia, C., Alba Sánchez, F.: Forecasting Airborne Pollen Concentration Time Series with Neural and Neuro-Fuzzy Models. Expert Systems with Applications 32(4), 1218–1225 (2007)
6. Zadeh, L.A.: Fuzzy Sets. Inf. and Control 8, 338–353 (1965)
7. Klir, G., Folger, T.: Fuzzy Sets, Uncertainty and Information. Prentice Hall PTR, Englewood Cliffs (1988)
8. Zimmerman, H.J.: Fuzzy Set Theory and Its Applications, 2nd edn. Kluwer, Dordrecht (1991)
9. Domańska, D., Wojtylak, M.: Development Data Serving to Forecasting Pollution Concentrations. In: Decision Support System, Katowice, pp. 351–359 (2009) (in Polish)
10. Domańska, D., Wojtylak, M.: Selection Criteria of Forecast Pollution Concentrations Using Collateral Informations. In: Computer Methods and Systems, Kraków, pp. 213–218 (2009)
11. Dubois, D., Prade, H.: Ranking Fuzzy Numbers in the Setting of Possibility Theory. Information Sciences 30, 183–224 (1983)
12. Sanchez, E.: Solutions of Fuzzy Equations with Extended Operations. Fuzzy Sets and Systems 12, 237–248 (1984)
13. Ośródka, L., Wojtylak, M., Krajny, E., Dunal, R., Klejnowski, K.: Application Data Mining for Forecasting of High-level Air Pollution in Urban-industrial Area in Southern Poland. In: Proc. of the 10th Int. Conf. on Harmonisation within Atmospheric Dispers. Modelling for Regulatory Purposes, pp. 664–668 (2005)
14. Domańska, D., Wojtylak, M.: Air Pollution Forecasting Model Control. Accepted for the Journal of Medical Informatics & Technologies
15. Hyndman, R.J., Koehler, A.B.: Another Look at Measures of Forecast Accuracy. International Journal of Forecasting 22(4), 679–688 (2006)

Multiple Kernel Learning Improved by MMD

Jiangtao Ren[1,⋆], Zhou Liang[2], and Shaofeng Hu[3]

[1] School of Software, Sun Yat-sen University, China
issrjt@mail.sysu.edu.cn
[2] School of Software, Sun Yat-sen University, China
liangzhou.sysu@gmail.com
[3] Department of Computer Science, Sun Yat-sen University, China
hugoshatzsu@gmail.com

Abstract. When training and testing data are drawn from different distributions, the performance of the classification model will be low. Such a problem usually comes from sample selection bias or transfer learning scenarios. In this paper, we propose a novel multiple kernel learning framework improved by Maximum Mean Discrepancy (MMD) to solve the problem. This new model not only utilizes the capacity of kernel learning to construct a nonlinear hyperplane which maximizes the separation margin, but also reduces the distribution discrepancy between training and testing data simultaneously, which is measured by MMD. This approach is formulated as a bi-objective optimization problem. Then an efficient optimization algorithm based on gradient descent and quadratic programming [13] is adopted to solve it. Extensive experiments on UCI and text datasets show that the proposed model outperforms traditional multiple kernel learning model in sample selection bias and transfer learning scenarios.

Keywords: Kernel Learning, Maximum Mean Discrepancy.

1 Introduction

In traditional supervised learning, training and testing data are assumed to be drawn from identical and independent distribution (i.i.d.). However, this assumption doesn't hold in many real world application, such as bioinformatics, text mining, sensor network data mining and low quality data mining [15]. Such situations are often be related to sample selection bias or transfer learning problems in machine learning [4,3,8,14]. For example, in text classification, the training text data may come from newspapers while testing data may come from blogs, the distribution of training and testing data are different in some degree and not i.i.d. Usually in such cases the performance of the classifier will decrease according to the degree of the distribution discrepancy. How to handle this problem is a key challenge in sample selection bias and transfer learning tasks.

⋆ The work is supported by the National Natural Science Foundation of China under Grant No. 60703110.

L. Cao, J. Zhong, and Y. Feng (Eds.): ADMA 2010, Part II, LNCS 6441, pp. 63–74, 2010.

Recently multiple kernel learning (MKL) has received more and more attention as a new kernel method[1,12]. [6] proposes an inductive learning algorithm which can be formulated into an SDP or QCQP optimization problem. Some faster algorithms are also explored, including SILP [11] and gradient based algorithm [10]. The objective of kernel learning is to find a best kernel space where the separation margin for labeled training data will be maximized. In spite of its nonlinear capacity that fits well for sophisticated data distributions, MKL doesn't perform well in all cases. For example, if the distribution of unlabeled testing data is different from the labeled training data, the learned kernel space will not be suitable for the classification of testing data, which will result in poor classification performance. Such problems often happen in scenarios such as sample bias and transfer learning.

In this paper, we propose a new model to solve this problem based on the MKL framework. The main idea of this model is to learn a kernel transformation which simultaneously maximizes the separation margin and minimizes the distribution discrepancy between training and testing data. As a popular metric for measuring the distance between distributions in kernel space, Maximum Mean Discrepancy (MMD) is used in our model. By adding MMD into the MKL framework, the original single-objective optimization problem of MKL becomes a bi-objective one, which will trade off between bringing two distribution together and obtaining a hyperplane that maximizes the separation margin. So our new model is a new multiple kernel learning method improved by MMD, called MKL-MMD. We derive optimization formulas similar to MKL where MMD is integrated naturally as a regularizer of original MKL. Then the efficient optimization algorithm based on gradient descent plus QP [13] are used to solve the new optimization problem proposed in this paper with slight changes. The empirical experimental results on UCI and text datasets show that our new multiple kernel learning model improved by MMD outperforms traditional MKL model significantly.

The rest of the paper is organized as follows. Section 2 briefly reviews the related works. Details of our model are discussed in Section 3, including the problem definition, framework and optimization algorithm. In Section 4, experimental results are presented. Finally, we give a conclusion in Section 5.

2 Related Works

2.1 Sample Selection Bias and Transfer Learning

In the standard formulation of supervised machine learning problems, we assume that training and testing data are drawn from the same distribution. Then classifier learned from the training data can be used to predict the testing data with high accuracy. However, this assumption often does not hold in practice. The training data often include some information irrelevant for classification or even diverse from the testing set. These may be due to some practical reasons such as the cost of data labeling or acquisition. The problem occurs in many areas such as astronomy, econometrics, and species habitat modeling. In [3,4], some formal explanations are presented, and this is called the sample selection bias problem. The main correction technique used in sample selection bias is to re-weight the cost of training point errors to more closely reflect that of the test distribution.

In the transfer learning scenario [8], we focus on the learning method using the training data coming from related domains, whose distribution is different from that of testing data. One of the major challenges in this field is how to handle the distribution difference between the target and source domains(or the in and out domains). [9] models distribution distance using MMD metric which we also adopt in our paper. The main difference between our approach and [9] is that [9] learns a nonlinear classifier in kernel space with predefined kernel matrix while ours is based on kernel learning. [7] also utilizes the MMD metric to obtain a kernel space that brings the two distribution together. However, it employs dimension reduction and is based on kernel principle component analysis (KPCA). Therefore, our MKL-MMD approach is different from the approaches of [7] and [9].

2.2 Maximum Mean Discrepancy

In the sample selection bias problem or transfer learning problem, we need to evaluate the difference in distribution between two domains given finite observations of $\{x_i\}$ and $\{x'_i\}$ (or training set and testing set). By reducing the difference between out domain and in domain, we can got well learning performance in transfer learning. For this purpose, a non-parametric distance estimate between distributions called Maximum Mean Discrepancy(MMD) is introduced for minimizing the difference between out domain and in domain. MMD is based on the Reproducing Kernel Hilbert Space (RKHS) distance, given by the following formula:

$$MMD^2 = \left\| \frac{1}{N_1} \sum_{i=1}^{N_1} \phi(x_i) - \frac{1}{N_2} \sum_{i=1}^{N_2} \phi(x'_i) \right\|^2 . \tag{1}$$

For the sample selection bias problem, MMD is used for reweighing the training examples to correct bias between the training set and testing set in [5]. For transfer learning, the MMD measurement is used for Dimensionality Reduction [7], and it can also be introduced into SVM as a regularizer to solve the transfer learning problem [9]. Considering the kernel method, we change the MMD formula to its kernel form:

$$MMD^2 = \frac{1}{N_1^2} \sum_{i,j=1}^{N_1} K(x_i, x_j) + \frac{1}{N_2^2} \sum_{i,j=1}^{N_2} K(x'_i, x'_j) - \frac{2}{N_1 \cdot N_2} \sum_{i,j=1}^{N_1, N_2} K(x_i, x'_j). \tag{2}$$

then the MMD formula can be changed to:

$$MMD^2 = \sum_{i,j=1} D_{ij} K_{ij}, \tag{3}$$

where $K_{ij} = K(x_i, x_j)$ and D_{ij}:

$$D_{ij} = \begin{cases} \frac{1}{N_1^2} & \text{when } x_i, x_j \in D_{out} \\ \frac{1}{N_2^2} & \text{when } x_i, x_j \in D_{in} \\ \frac{-2}{N_1 \cdot N_2} & \text{otherwise} \end{cases}$$

2.3 Kernel Learning

Multiple Kernel Learning (MKL) tries to learn a suitable kernel for a given data classification problem by learning a combination weight vector for a series of basis kernels. Formally, we are given positive definite basis kernels K_m, where $m = 1, 2, \cdots, D$. Each K_m is associated with a RKHS $\mathcal{H}_m$, and $K_m \in \mathcal{H}_m$. The goal is to find a best linear combination weight vector d for these basis kernels, which will result in a target kernel given in the following formula:

$$K_d = \sum_{m=1}^{D} d_m K_m, \qquad \textit{with} \quad d \geq 0.$$

In this formula, $d = (d_1, d_2, \cdots, d_D)$ is the the kernel combination weight vector. An initial kernel learning framework was proposed by [6]. In [6], SVM dual variable and kernel combination weight parameters are learned simultaneously with a group of predefined basic kernels. This SVM based kernel learning framework can be formulated as:

$$\begin{aligned} \min_{\mathbf{d}} \ & \omega(K_d) && (4) \\ & = \max_{\alpha} \quad 2\alpha^T \mathbf{e} - \alpha^T G(K_d)\alpha \\ \textit{subject to} \ & 0 \leq \alpha \leq C, \alpha^T \mathbf{y} = 0, \\ & K_d \succeq 0, \textit{trace}(K_d) = c, && (5) \\ & K_d = \sum_{m=1}^{D} d_m K_m, \mathbf{d} \succeq 0 && (6) \end{aligned}$$

where $G(K_d) = Y K_d Y$ and Y is a diagonal matrix of labels. We denote α as a Lagrange coefficients vector and $\mathbf{e}$ as all ones vector. Although the above formula is straightforward and proved to be efficient, it turns out to be an SDP optimization problem, which is time consuming. Recently, faster algorithms have been proposed to solve this problem. [11] uses the SILP algorithm to solve an objective function similar to Eq.(4). [10] proposed a multiple kernel learning framework from a different prime problem formulation of SVM. Its dual form shares much similarity with formula (4) and worths to be shown here:

$$\begin{aligned} \min_{\mathbf{d}} \max_{\alpha} \ & 2\alpha^T \mathbf{e} - \alpha^T G(K_d)\alpha && (7) \\ \textit{subject to} \ & 0 \leq \alpha \leq C, \alpha^T \mathbf{y} = 0, \\ & \sum_{m=1}^{D} d_m = 1, \mathbf{d} \succeq 0 \end{aligned}$$

Denoted that Eq.(7) reduce constraints of Eq.(5) and Eq.(6) by imposing semi-positive constraint on basic kernel. Moreover, after simplification, a two step optimization, including gradient descent as outer iteration and SVM based quadratic programming as inner iteration, can be applied to solve Eq.(7), which is a lot faster than SDP. Recently, another optimization of kernel parameters with gradient descent and SVM solvers is introduced in [13] by using formula:

$$\begin{aligned} \min_{\mathbf{d}} \max_{\alpha} \ & 2\alpha^T \mathbf{e} - \alpha^G (K_d) \alpha + r(\mathbf{d}) \\ \textit{subject to} \ & 0 \leq \alpha \leq C, \alpha^T \mathbf{y} = 0, \\ & \mathbf{d} \succeq 0 \end{aligned} \tag{8}$$

where

$$r(\mathbf{d}) = \|\mathbf{d}\|_p, p = 1, 2$$

$r(d)$ represents the regularizer of the kernel parameter $\mathbf{d}$, while the first two terms are no different from the dual form of standard SVM. In practice, L_1 and L_2 regularizers of $\mathbf{d}$ are often used because of their popularity and simplicity. As a result, we restrict our discussion on the L_1 or L_2 norm of $\mathbf{d}$ as the regularizer function $r(d)$.

3 Multiple Kernel Learning Improved by MMD

In the previous section, we have introduced some kernel learning models and their optimization methods. These inductive learners learn the target kernel matrix from labeled training data, which will transform the original data into a new kernel space where the separation margin will be maximized. However, if the distribution of unlabeled testing data is different from the labeled training data, the learned kernel space will not be suitable for the classification of testing data, resulting in poor classification performance. Such situations arises frequently in sample selection bias and transfer learning scenarios. To solve this problem, it is natural to consider how to reduce the distribution discrepancy when we learn the kernel transformation. As discussed before, MMD is a good measure of distribution discrepancy in kernel space. Therefore, we can extend the optimization objective function of kernel learning by simultaneously optimizing separation margin and MMD, which will transform the data into a new kernel space where training data and testing data are closer and separated better, and will improve the classification performance in sample selection bias and transfer learning situations.

3.1 Problem Definition

For machine learning scenario, $\mathcal{X}$ denote instance space, while $\mathcal{Y}$ denote output space. In the case of classification, $\mathcal{Y} \in \{-1, +1\}$ and $\mathcal{X} \in R^d$. We define D as a data set from domain space $\mathcal{X} \times \mathcal{Y}$. Given $D_{trn}\{(x_i, y_i)\}_{i=1}^N$ as a training set and $D_{tst}\{(x'_i, y'_i)\}_{i=1}^M$ as a testing set, our task is to learn a classifier from D_{trn} and use it to predict the label y_i' in D_{tst}. The gram matrix K is defined as the inner product of corresponding projection $\phi(x)$, i.e., $K = \langle \phi(x), \phi(x) \rangle$. If $f : \mathcal{X} \rightarrow \mathcal{Y}$ is the function for learning task, our objective is to learn a kernel K_d, where $f(x) = \sum_i \alpha_i y_i K_d(x_i, x) + b$ and $K_d(x_i, x_j) = \sum_m d_m K_m(x_i, x_j)$. Note that $K_m \in \mathcal{H}_m, m \in \{1, 2, \cdots, D\}$. This is a group of basic kernels given in advance. For both sample selection bias and transfer learning scenario, D_{trn} and D_{tst} are not drawn according to the same distribution. In such cases, a bias between D_{trn} and D_{tst} exists and the estimation from conventional MKL is not reliable any more. To measure the distribution difference, We define a discrepancy measure of the D_{trn} and D_{tst} in kernel space, and give it in the form of MMD metric. Our basic idea is to reformulate the MKL algorithms to handle sample selection bias and transfer learning application with empirical MMD estimation.

3.2 Framework

In section 2.3, we introduced some multiple kernel learning models. The latter two are much faster and simpler than the first one. Since our problem is to embedded the MMD metric into MKL framework, we choose the second one for its simplicity. The formula of Eq.(7) can be extended as follows by considering the reduction of distribution discrepancy in the kernel space :

$$\min_{\mathbf{d}} \ \mathcal{J}(\mathbf{d}) + \lambda \cdot dis(D_{trn}, D_{tst})^2 \tag{9}$$

$$subject \ \ to \ \sum_{m=1}^{D} d_m = 1, \mathbf{d} \succeq 0$$

where

$$\mathcal{J}(\mathbf{d}) = \alpha^{*T}\mathbf{e} - \frac{1}{2}\alpha^{*T} G(K_d)\, \alpha^*. \tag{10}$$

Noted that α^* is the optimal SVM solution of a fixed K_d and K_d is the combination of basic kernels given weight vector $\mathbf{d}$. The first part of formula (9) is the traditional optimization objective function of kernel learning, the second part is the measure of distribution discrepancy between training and testing data. It is clear that formula (9) represents a bi-objective optimization problem, and $\lambda > 0$ is the parameter for trading off between the two objectives. As we can see from formula (10), $\mathcal{J}(\mathbf{d})$ is a function of decision vector $\mathbf{d}$. We are going to transform the whole formula (9) into a function of $\mathbf{d}$, so our next problem is to build the connection between $\mathbf{d}$ and $dis(D_{trn}, D_{tst})^2$.

From Eq.(3) and $K_d = \sum_{m=1}^{D} d_m K_m$ is the combination term of basis kernels; $dis(D_{trn}, D_{tst})^2$ can be reformulated as:

$$\begin{aligned} dis(D_{trn}, D_{tst})^2 &= \sum_{i,j=1} D'_{ij} K_{ij} \\ &= \sum_{i,j=1} D'_{ij} \sum_{m=1}^{D} d_m K_{mij} \end{aligned}$$

$$= \sum_{m=1}^{D} d_m \cdot (\sum_{i,j=1} D'_{ij} K_{mij})$$
$$= \mathcal{K}_{dis}^T d, \tag{11}$$

where

$$D'_{ij} = \begin{cases} \frac{1}{N^2} & \text{when } x_i, x_j \in D_{trn} \\ \frac{1}{M^2} & \text{when } x_i, x_j \in D_{tst} \\ \frac{-2}{N \cdot M} & \text{otherwise} \end{cases}$$

and $\mathcal{K}_{dis}$ is a vector defined as:

$$\mathcal{K}_{dis} = (\sum D'_{ij} K_{1ij}, \sum D'_{ij} K_{2ij}, \cdots, \sum D'_{ij} K_{Dij})^T$$

Substituting Eq.(11) into Eq.(9), we obtain a new version of our objective function, which is purely a function of **d**.

$$\min_{d} \quad W(d) \tag{12}$$
$$subject \quad to \sum_{m=1}^{D} d_m = 1, \mathbf{d} \succeq 0$$

where

$$W(d) = \alpha^{*T} \mathbf{e} - \frac{1}{2} \alpha^{*T} G(K_d) \alpha^* + \lambda \cdot \mathcal{K}_{dis}^T d. \tag{13}$$

Now we focus our discussion on $\mathcal{K}_{dis}$. As discussed before and based on Eq.(11), it is clear that the inner product of $\mathcal{K}_{dis}$ and **d** reflects the distribution discrepancy between training and testing data in the learned kernel space. Unlike traditional techniques in sample selection bias, we reweight the basis kernels related to the data to reduce the bias in kernel space. As we will see later, a two step optimization method is proposed in the next section, which is similar to the SVM based gradient descent algorithm in [13].

3.3 Optimization Algorithm

In this section, we derive an optimization algorithm based on the gradient descent method and Quadratic Programming (QP) to solve the optimization problem defined in Eq.(12) effectively.

We first examine Eq.(7) and Eq.(8) in Section 2.3. The main difference between them is that the constraint $\sum_{m=1}^{D} d_m = 1$ in Eq.(7) is replaced by the regularizer $r(\mathbf{d})$ in Eq.(8). It is observed that formula (13) is similar with formula (8). The only difference is that the regularizar $r(d)$ is replaced by the MMD metric $\mathcal{K}_{dis}^T d$ derived from Eq.(11). The MMD metric $\mathcal{K}_{dis}^T d$ acts like an L_1 regularizer, so we don't need an extra regularizer with objective function, to keep **d** simple, we borrow the same idea from Eq.(8) to remove the equality constraint $\sum_{m=1}^{D} d_m = 1$ from the optimization problem. After this relaxation, we get a simpler form of the optimization problem than Eq.(12) as follow:

$$\min_{\mathbf{d}} \quad W(\mathbf{d}) \qquad (14)$$
$$subject \quad to \quad \mathbf{d} \succeq 0$$

$W(d)$ is the same as that given in Eq.(13). Eq.(14) shares a similar form discussed in [13]. Therefore we adopt the same optimization strategy in [13] to handle this problem.

In order to simplify the optimization procedure, we firstly relax the only constraint in Eq.(14). Now Eq.(14) turns into an unconstraint nonlinear optimization problem. In this case, the gradient descent method is an efficient optimization method to handle it. A regular gradient descent algorithm updates the variable iteratively by the following equation:

$$\mathbf{d}^{n+1} = \mathbf{d}^n - s \cdot \nabla_d W, \qquad (15)$$

s is a step size of regular gradient algorithm and $\nabla_d W$ is the gradient of $W(d)$. For $W(d)$ is differentiable and we need to calculate $\nabla_d W$, we use a straight forward extension of Lemma 2 in [2], then we conclude that $W(d)$ have derivatives given by:

$$\frac{\partial W}{\partial d_m} = \lambda \cdot \sum D'_{ij} K_{mij} - \frac{1}{2} \alpha^{*T} \frac{\partial G(K_d)}{\partial d_m} \alpha^*. \qquad (16)$$

The matrix D' and K_m are constructed as described in the previous sections. For classification, $G(K_d) = Y K_d Y$ and its derivative is straightforward as K_d is an affine function of $\mathbf{d}$.

Now this algorithm can be summarized as a two step optimization procedure: firstly, given a fixed $\mathbf{d}$, we can obtain an optimal α^* by SVM optimization procedure; secondly, α^* is used to calculate $\nabla_d W$ based on (16). This procedure is repeated until convergence. For the constraints $\mathbf{d} \succeq 0$ we have relaxed before, we can simply take the strategy of $d_m \leftarrow max(0, d_m)$ to ensure it is satisfied. We refer to this algorithm as Multiple Kernel Learning Improved by MMD (MKL-MMD), which is presented in Algorithm 1 in details.

4 Experiments

To demonstrate the effectiveness of the framework proposed in this paper, we carry out experiments on real-world data collections. The performance is compared with traditional MKL classifiers introduced in previous sections.

4.1 Data Sets Description

To study the performance of our algorithm in sample selection bias and transfer learning scenario, we use UCI data sets and text data sets respectively in our experiments.

For the scenario of sample selection bias, we use data sets selected from the UCI machine learning repository, which are all collected from real world scenario.

Algorithm 1. Multiple Kernel Learning Improved by MMD

Input: sample data set D_{trn} and D_{tst}
Output: kernel parameter d
$n \leftarrow 0$
Initialize d^0 randomly;
repeat
  $K \leftarrow k(d^n)$;
  calculate D' using (3) with D_{trn} and D_{tst}
  Use an SVM solver to solve the single kernel problem with kernel K and obtain α^*.
  Calculate $\mathbf{d}^{n+1}$ using (15) and (16)
  $d_m^{n+1} = max(0, d_m^{n+1})$;
  $n \leftarrow n + 1$;
until converged

Table 1. UCI data set and Text data set

name	D_{trn}	D_{tst}	dim	class	name	D_{trn}	D_{tst}	dim	class
haberman	78	228	3	2	chessboard	14	86	2	2
SPECTF	28	239	44	2	comp vs sci	201	264	6773	2
magic	65	189	10	2	auto vs aviation	134	220	5991	2
madelon	40	360	500	2	real vs simulated	202	201	6161	2
wine	36	142	13	2	orgs vs people	155	242	4771	2
sonar	43	165	60	2	orgs vs places	102	105	4415	2

For example, the haberman dataset contains cases on the survival of patients who had undergone surgery for breast cancer in a study conducted between 1958 and 1970 at the University of Chicago's Billings Hospital.

We also should consider some special cases of data sets in experiments for our kernel learning problem. For example, the wine data is a dataset with 3 class labels. Since traditionally kernel learning solve binary classification problem, we select one class of data as the positive class, and the other two classs as negative class to transform it into a two-class data set. Both the magic and madelon datasets are relatively large and it is difficult for kernel learning because of the high computation complexity. So we reduce their size to a suitable level. In the following experiments we select 254 instances from magic and 400 instances from madelon. To simulate the sample selection bias scenario, we choose a small portion of the data examples (about 10% to 25%) as training set D_{trn} and the rest as testing set D_{tst}. Detailed information of these selected UCI datasets are summarized in Table 1.

For the transfer learning scenario, we select text data from 20-NewsGroups, SRAA and Reuters-21578. The 20-NewsGroups corpus contains approximately 20,000 Usenet articles from 7 top categories. Two top categories, "comp", "sci" which have a two-level hierarchy are selected for study. Each transfer learning task involves a top category classification problem but the training and testing data are drawn from different sub-categories. For example, in comp and sci, the training set D_{trn} is drawn from "comp.graphics" and "sci.crypt", while the testing set D_{tst} is drawn from "comp.windows.x" and "sci.space". So the distributions of the training and test data are different but related. The SRAA corpus contains 73218 UseNet articles from four discussion groups about simulated autos, simulated aviation, real autos and real aviation. The Reuters-21758 corpus contains 21758 Reuters news articles in 1987 organized into 5 top categories. We select three top categories, "orgs", "people" and "places" in this experiment. Both the SRAA and Reuters-21758 have a two-level hierarchy similar to 20-NewsGroups.

Text datasets are commonly modeled as sets of instances with high dimensionality as shown in Table 1. In our multiple kernel learning setup, a basis kernel is constructed from each feature, so a high dimensionality will result in a high dimensional combination weight vector d and make it hard to learn in the multiple kernel learning framework. A simple but effective approach for this problem is to perform dimension reduction on text data. Principal component analysis (PCA) is a standard tool for dimension reduction. Then we use PCA to perform dimension reduction on text data to make sure the MKL algorithm has practical performance. By PCA, the dimensions of all the text datasets are reduced to 50. Detailed information of the selected text datasets are listed in Table 1.

4.2 Setup of Experiments

We compare the performance of our MKL-MMD algorithm with GMKL proposed in [13]. GMKL is a recently developed supervised kernel learning method, which generalizes the combination way of basis kernels by combining them as sums or other suitable ways. We compare our model with GMKL with conventional MKL to emphasize the advantages of our model in the scenarios of sample selection bias and transfer learning. On the other hand, because MMD acts as a regularizer function in the kernel learning framework, we compare our model with GMKL using L_1 and L_2 regularizer function simultaneously. There are several important experiment at settings: firstly, as the combination term of basis kernel is not important for comparison, we just set the composite kernel as $K_d(\mathbf{x}_i, \mathbf{x}_j) = \sum_{m=1}^{D} d_m exp(-\frac{(x_{im}-x_{jm})^2}{\sigma})$ on both MKL-MMD and GMKL. Here D is the number of dimension and x_{im}, x_{jm} are the values in the m dimension of the examples x_i, x_j. σ is a kernel parameter and is set to 1. Secondly, as there is a positive constraint $\mathbf{d} \succeq 0$ for variable $\mathbf{d}$, we simplify the L_1 regularization equivalently as $r(d) = \mathbf{e}^T d$ and L_2 regularization $r(d) = d^T d$ of GMKL, where $\mathbf{e}$ is an all one vector. Thirdly, for the cost parameter C in the SVM part of both the MKL-MMD and GMKL models, we fixed it to 10 for both of the algorithms. Finally, to compare the performance of the different models, the evaluation criterion is the accuracy.

4.3 Experiments Results

We present our experiment results in this section. Table 2 summarize the performance comparison between MKL-MMD and GMKL on UCI and text datasets

Table 2. Experiment results on UCI data task

task name	λ	MKL-MMD	GMKL l_1	GMKL l_2
Haberman	10	72.37%	71.49%	70.61%
SPECTF	10^{-6}	79.08%	76.15%	78.66%
magic	$1.7 \cdot 10^4$	70.37%	66.67%	67.20%
madelon	10^{-1}	58.06%	55.56%	51.67%
wine	10^{-3}	95.77%	94.37%	92.25%
sonar	10^{-2}	81.21%	80.00%	80.61%
chessboard	10^5	62.79%	60.47%	54.65%
comp vs sci	10	74.24%	68.56%	67.05%
auto vs aviation	10^{-3}	75.45%	70.91%	64.09%
real vs simulated	$9 \cdot 10^{-4}$	65.67%	61.19%	57.21%
orgs vs people	10^{-4}	61.16%	59.92%	58.68%
orgs vs places	10^{-2}	63.81%	60.00%	60.95%

respectively. Accuracy of MKL-MMD presented in these tables are the values according to the best value of parameter λ, which is also shown in the tables.

In the experiments, UCI datasets are in a setup with small training set size and regarded as with sample selection bias. From Table 2, it is obvious that MKL-MMD preforms better than GMKL even in the situation that the size of the training dataset is small. For example, for the 'magic' dataset, MKL-MMD achieves an accuracy of around 70.37% while the best accuracy of GMKL is 67.20%. It also can be seen that for 'wine', MKL-MMD yields an accuracy of around 95.77% while the best accuracy of GMKL is 94.37%. In summary, though GMKL can perform well in sample selection bias scenarios, MKL-MMD can perform better.

For the transfer learning tasks based on text datasets, MKL-MMD also performs better than GMKL. The accuracy is significantly improved in most of the tasks of text classification learned from MMD-MKL comparing with GMKL. For example, in **auto vs aviation** task, the best accuracy performance that MMD-MKL can achieve is 75.45%, while GMKL L_1 and GMKL L_2 only achieve 70.91% and 64.09% respectively.

5 Conclusions

A common difficulty in both sample selection bias and transfer learning is the distribution discrepancy between training and test data. In this paper, we try to solve this problem in multiple kernel learning framework by introducing a distribution distance metric in kernel space called MMD. The main idea is to learn a suitable kernel transformation in which the separation margin and the closeness between the distributions can be simultaneously optimized. Experiments on UCI and text datasets show that our method outperforms traditional MKL method with L_1 or L_2 regulation. We can conclude that our new method can improve the robustness of the MKL technique to handle transfer learning or sample selection bias problem.

References

1. Bach, F.R., Lanckriet, G.R.G., Jordan, M.I.: Multiple kernel learning, conic duality, and the smo algorithm. In: ICML (2004)
2. Chapelle, O., Vapnik, V., Bousquet, O., Mukherjee, S.: Choosing multiple parameters for support vector machines. Machine Learning 46(1-3), 131–159 (2002)
3. Cortes, C., Mohri, M., Riley, M., Rostamizadeh, A.: Sample selection bias correction theory. In: Freund, Y., Györfi, L., Turán, G., Zeugmann, T. (eds.) ALT 2008. LNCS (LNAI), vol. 5254, pp. 38–53. Springer, Heidelberg (2008)
4. Fan, W., Davidson, I.: On sample selection bias and its efficient correction via model averaging and unlabeled examples. In: Jonker, W., Petković, M. (eds.) SDM 2007. LNCS, vol. 4721, Springer, Heidelberg (2007)
5. Huang, J., Smola, A.J., Gretton, A., Borgwardt, K.M., Schölkopf, B.: Correcting sample selection bias by unlabeled data. In: NIPS, pp. 601–608 (2006)
6. Lanckriet, G.R.G., Cristianini, N., Bartlett, P.L., Ghaoui, L.E., Jordan, M.I.: Learning the kernel matrix with semi-definite programming. In: ICML, pp. 323–330 (2002)
7. Pan, S.J., Kwok, J.T., Yang, Q.: Transfer learning via dimensionality reduction. In: AAAI, pp. 677–682 (2008)
8. Pan, S.J., Yang, Q.: A survey on transfer learning. Tech. Rep. HKUST-CS08-08, Department of Computer Science and Engineering, Hong Kong University of Science and Technology, Hong Kong, China (November 2008), `http://www.cse.ust.hk/~sinnopan/publications/TLsurvey_0822.pdf`
9. Quanz, B., Huan, J.: Large margin transductive transfer learning. In: CIKM. pp. 1327–1336 (2009)
10. Rakotomamonjy, A., Bach, F., Canu, S., Grandvalet, Y.: More efficiency in multiple kernel learning. In: ICML, pp. 775–782 (2007)
11. Sonnenburg, S., Rätsch, G., Schäfer, C., Schölkopf, B.: Large scale multiple kernel learning. Journal of Machine Learning Research 7, 1531–1565 (2006)
12. Szafranski, M., Grandvalet, Y., Rakotomamonjy, A.: Composite kernel learning. In: ICML, pp. 1040–1047 (2008)
13. Varma, M., Babu, B.R.: More generality in efficient multiple kernel learning. In: ICML, p. 134 (2009)
14. Zhong, E., Fan, W., Peng, J., Zhang, K., Ren, J., Turaga, D.S., Verscheure, O.: Cross domain distribution adaptation via kernel mapping. In: KDD, pp. 1027–1036 (2009)
15. Zhu, X., Khoshgoftaar, T.M., Davidson, I., Zhang, S.: Editorial: Special issue on mining low-quality data. Knowl. Inf. Syst. 11(2), 131–136 (2007)

A Refinement Approach to Handling Model Misfit in Semi-supervised Learning

Hanjing Su[1], Ling Chen[2], Yunming Ye[1,*], Zhaocai Sun[1], and Qingyao Wu[1]

[1] Department of Computer Science, Shenzhen Graduate School,
Harbin Institute of Technology, Shenzhen 518055, China
{suhangjing2008,wuqingyao.china}@gmail.com,
yeyunming@hit.edu.cn, zhcsun@hotmail.com

[2] QCIS, Faculty of Engineering and Information Technology,
University of Technology, Sydney 2007, Australia
Ling.chen@uts.edu.au

Abstract. Semi-supervised learning has been the focus of machine learning and data mining research in the past few years. Various algorithms and techniques have been proposed, from generative models to graph-based algorithms. In this work, we focus on the *Cluster-and-Label* approaches for semi-supervised classification. Existing cluster-and-label algorithms are based on some underlying models and/or assumptions. When the data fits the model well, the classification accuracy will be high. Otherwise, the accuracy will be low. In this paper, we propose a refinement approach to address the model misfit problem in semi-supervised classification. We show that we do not need to change the cluster-and-label technique itself to make it more flexible. Instead, we propose to use successive refinement clustering of the dataset to correct the model misfit. A series of experiments on UCI benchmarking data sets have shown that the proposed approach outperforms existing cluster-and-label algorithms, as well as traditional semi-supervised classification techniques including Self-training and Tri-training.

Keywords: Semi-supervised learning, model misfit, classification.

1 Introduction

In many practical classification applications, the acquisition of labeled training examples is both costly and time consuming, where human effort is usually required. Therefore, *semi-supervised learning* that exploits unlabeled examples in addition to labeled ones has become a hot topic in the past years. Various algorithms and techniques have been proposed, including generative models [1, 2], graph-based methods [3, 4], as well as transductive SVMs [5]. A literature survey of semi-supervised learning can be found in [6].

Most recent semi-supervised learning algorithms work by formulating the assumption that "nearby" points, and points in the same structure (*e.g.*, cluster), should

---

[*] This research is supported in part by NSFC under Grant no. 61073195.

L. Cao, J. Zhong, and Y. Feng (Eds.): ADMA 2010, Part II, LNCS 6441, pp. 75–86, 2010.

have similar labels [7]. Models based on such an assumption may achieve high classification accuracy when the data distribution complies with the assumption. However, when the models do not fit the data well, the performance of the resulting semi-supervised classifiers can be quite poor. In this paper, we aim to address the model misfit problem to improve the performance of semi-supervised classification. In particular, we focus on the *Cluster-and-Label* techniques [8, 9].

In the process of Cluster-and-Label, data are clustered using an unsupervised learning technique, and the class of unknown data is predicted based generated clusters. As noted in [6], cluster-and-label methods can perform well if the particular clustering algorithms match the true data distribution. In other words, the performance of cluster-and-label algorithms suffers from the model misfit problem. The straightforward solution might be replacing the clustering model with one which matches the true data distribution appropriately. However, in the scenario of semi-supervised learning where only a small portion of examples are labeled, the true distribution of data is not easy to be found.

In this paper, we propose a novel yet simple method to deal with the model misfit problem of cluster-and-label techniques. Our method does not need to change the underlying clustering model itself. Instead, we successively refine the generated clusters based on the cluster purity, using the same clustering model. The resulting classifier, called *purity refinement cluster tree* (PRC-Tree), is a hierarchical clustering model which splits the labeled and unlabeled training data successively into nested clusters with higher purity. PRC-Tree enables the discovery of clusters in different subspaces; the combination of nested clusters of PRC-Tree is expected to match the true data distribution better. Consequently, PRC-Tree can be more effective and robust. The predication of test data is made by using an efficient path-based strategy which searches the PRC-Tree for the best cluster close to the unknown data.

To evaluate the performance of the proposed approach, we conduct a series of experiments on the benchmarking UCI data sets. The experimental results show that the generalization ability of PRC-Tree outperforms both cluster-and-label techniques, as well as traditional semi-supervised learning algorithms including Self-training [10] and Tri-training [12].

The rest of the paper is organized as follows. Section 2 briefly reviews existing research related to our work. Section 3 presents the hierarchical semi-supervised classification model PRC-Tree. Performance evaluation of the proposed method is presented in Section 4. Finally, Section 5 concludes this paper and discusses future work.

2 Related Work

In the past few years, many semi-supervised learning algorithms have been proposed, including self-training [10], co-training [11, 12, 13], generative methods [1, 2], semi-supervised support vector machines [5], and graph-based methods [3, 4]. Most recent existing semi-supervised learning algorithms are developed based on the assumption that "nearby" points, and points in the same structure (*e.g.*, cluster), should have similar labels. For example, the graph-based methods construct a graph structure where the nodes are labeled and unlabeled examples in the dataset, and edges reflect the "closeness" of examples. Graph-based methods then propagate the labels from known examples to unknown examples based on the above assumption, which is similar to the "homophily" principle observed in social networks. Therefore, the "homophily" assumption made by most recent semi-supervised learning algorithms inherently decides the resulting performance. When the data distribution complies with the assumption, satisfactory classification accuracy can be achieved. Otherwise, the performance may suffer. We aim to address the model misfit problem in semi-supervised learning. In particular, we focus on the *cluster-and-label* methods, which are similarly developed based the "homophily" assumption. We review some representative cluster-and-label methods as follows.

Demiriz et al. [8] introduce an early work of cluster-and-label. In their approach, data are segmented using an unsupervised learning technique that is biased toward producing clusters as pure as possible in terms of class distribution. To do this, they modify the objective function of an unsupervised techniques (*e.g.*, K-means clustering), to minimize both the within cluster variance and a measure of cluster impurity based on class labels. A genetic algorithm optimizes the objective function to produce clusters, which are then used to predict the class of future points. Our approach is different from their work because we consider the measure of within cluster variance and the measure of cluster impurity separately. We perform a sequence of nested clustering, where each clustering minimizes the within cluster variance only. The cluster impurity measure is used to decide whether a further partition of a cluster is needed. Instead of generating a plain set of clusters, our method delivers a hierarchical clustering model, which enables the discovery of clusters from subspaces with higher purity. Furthermore, our method is more efficient since no genetic algorithms are needed to optimize the objective function.

Dara et al. [9] map labeled and unlabeled data to nodes in self-organizing feature map (SOM). Based on the similar assumption that data of the same class should be mapped to the same SOM node, the method trains the SOM first with labeled data, and then supplies the model with unlabeled data. Unlabeled data are predicted according to the mapping SOM nodes. Next, it iteratively re-trains the SOM with labeled data and predicted data until no new labels can be assigned to unlabeled data. They observe the model misfit problem as "*non-labeling*" SOM nodes, to which data with different labels are assigned. It is impossible to induce a label for unlabeled data clustered to non-labeling SOM nodes. Our algorithm which further splits a cluster if it is not pure enough avoids the generation of non-labeling clusters with properly defined purity threshold. Their method also generates "*undefined*" SOM nodes, to which no labeled data are clustered. Undefined nodes cannot be used to predict labels for unlabeled data also. Since our approach organizes clusters hierarchically, the problem can be addressed gracefully by using the label of a parent (or an ancestor) cluster.

In summary, our algorithm is different from existing cluster-and-label algorithms in the way that existing methods cluster the training data for once, while we perform nested clustering on the training data. In contrast to a plain set of clusters generated by existing methods, we generate a hierarchical clustering model which discovers clusters from subspaces with higher purity, and improves the model flexibility to match the true data distribution.

3 The PRC-Tree

In this section, we firstly discuss some preliminaries. Then, the proposed purity refinement cluster tree is described. The reason why PRC-Tree can address the model misfit problem is analyzed finally.

3.1 Preliminaries

In semi-supervised classification problem, one is given a training set $\boldsymbol{D}$ consisting of a labeled subset $\boldsymbol{D}_L$ and an unlabeled subset $\boldsymbol{D}_U$, $\boldsymbol{D} = \boldsymbol{D}_L \cup \boldsymbol{D}_U$. $\boldsymbol{D}_L = \{(x_1, y_1), (x_2, y_2), \ldots (x_L, y_L)\}$ where $x_i \in \Re^{\mathrm{d}}$ is a feature vector of d dimensions, and $y_i \in \boldsymbol{C}$ is the corresponding class label. $\boldsymbol{D}_U = \{x_1', x_2', \ldots, x_U'\}$ where $x_i' \in \Re^{\mathrm{d}}$. Given another set of test data $\boldsymbol{T} = \{v_1, v_2, \ldots, v_T\}$ where $v_j \in \Re^{\mathrm{d}}$, the objective of semi-supervised classification is to learn a function $f{:}\boldsymbol{T} \rightarrow \boldsymbol{C}$, which sufficiently approximates the mapping from the vectors in $\boldsymbol{T}$ to classes in $\boldsymbol{C}$.

As mentioned above, our approach aims to address the model misfit problem by performing nested clustering to create a hierarchical clustering model. We define some related notions as below.

Definition 1. (nested clustering) A clustering $\boldsymbol{S}^j$ with k clusters is said to be nested in the clustering $\boldsymbol{S}^i$, which contains r ($< k$) clusters, if for any cluster C_q in $\boldsymbol{S}^j$, there is a cluster C_p in $\boldsymbol{S}^i$ such that $C_q \subseteq C_p$. And there exists at least one cluster in $\boldsymbol{S}^j$, which holds $C_q \subset C_p$, $C_q \neq C_p$.

Definition 2. (cluster tree) A cluster tree is a sequence of nested clustering, so that $\forall\, i < j$, $\forall\, C_q \in \boldsymbol{S}^j$, there is a $C_p \in \boldsymbol{S}^i$, such that $C_q \subseteq C_p$. Then, we call C_p an ancestor cluster of C_q. Partically, if $i = j - 1$, then we call C_p a parent cluster of C_q, and C a child of C_p.

For example, Fig. 1 shows an example cluster tree. The root node represents the complete set of training data, which is partitioned into 3 clusters: C_1^1, C_2^1, and C_3^1, where the superscript is used to denote the level of the tree and the subscript denotes the cluster number. The cluster C_2^1 is further partitioned into 3 clusters: C_1^2, C_2^2 and C_3^2.

3.2 PRC-Tree Classification

Basically, our PRC-Tree classification method can be decomposed into two steps: 1) creating a cluster tree from the training dataset; and 2) classifying a new test example using the cluster tree. We describe the two steps sequentially as follows.

Creating a cluster tree. The input of this step is the set of training data, consisting of both labeled and unlabeled examples. The pseudo-code of this step is described in Algorithm 1. We initially create the root of the cluster tree with all training examples. A base clustering model is used to cluster the root. For each generated cluster, we measure its *purity* in terms of class distribution. If the purity of the cluster is greater than some pre-defined threshold θ, the cluster is left as a leaf node of the cluster tree. Otherwise, it is further partitioned using the base clustering model.

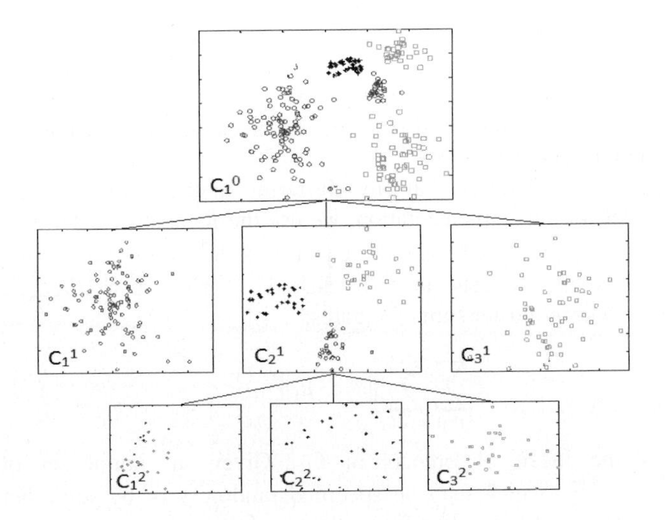

Fig. 1. Example of a cluster tree

The design of a purity measure is an important issue of our method. Although different purity metrics has been proposed, such as the Gini index, we find that measuring the purity of a cluster in terms of the percentage of examples belonging to the majority class suffices in our method.

```
Input:        D = D_L ∪ D_U, the cluster purity threshold θ.
Output:   A cluster tree
Description: BuildTree(D, θ)
1.  Let current node C = D.
2.  IF purity(C) ≥ θ
3.          Assign C as a leaf node
4.  ELSE
5.       {C_1, C_2, ..., C_k} = Cluster(C).
6.          FOR each C_i
7.                Assign C_i as a child node of C
8.                BuildTree(C_i, θ)
9.          ENDFOR
10. ENDIF
```

Algorithm 1. Creating a cluster tree

Let N be the number of examples in a cluster C. Let N_i be the number of examples with a label of class i. Then, the purity of C, denoted as $P(C)$, is computed as,

$$P(C) = \begin{cases} \frac{\max\{N_1, N_2, \ldots, N_k\}}{N} & if\ C\ has\ labeled\ examples \\ 1 & otherwise \end{cases} \quad (1)$$

The higher the value, the more certain the class distribution of a cluster. In order to generate a cluster tree expected to achieve higher classification accuracy, a higher value of the purity threshold θ is preferred. Particularly, in order to avoid the generation of "non-labeling" clusters [9] as discussed in Section 2, the purity threshold should be set to 1.

Our approach is general so that any clustering method can be used as the base clustering model. In our implementation, we use the *K*-means clustering. Given a set of objects $(x_1, x_2, \ldots, x_n)$, where each object $x_i \in \Re^d$ is a d-dimensional vector, *k*-means partitions the objects into *k* clusters $(k < n)$, $C = \{C_1, C_2, \ldots, C_k\}$, by minimizing the within-cluster sum of squares:

$$\sum_{i=1}^{k} \sum_{x_j \in C_i} \|x_j - \mu_i\|^2 , \tag{2}$$

where μ_i is the mean of objects in C_i. Given an initial set of k means $\mu_1{}^{(1)}, \mu_2{}^{(1)}, \ldots, \mu_k{}^{(1)}$, which may be specified randomly or by some heuristics, the algorithm proceeds by iteratively refining them as follows:

1: Assign each object to the cluster with the closest mean.

$$C_i{}^{(t)} = \{x_j : \|x_j - \mu_i{}^{(t)}\| \le \|x_j - \mu_l{}^{(t)}\|,\ l = 1, \ldots, k\}. \tag{3}$$

2: Calculate the new means to be the centroid of the observations in the cluster.

$$\mu_i{}^{(t+1)} = \frac{1}{|C_i{}^{(t)}|} \sum_{x_j \in C_i{}^{(t)}} x_j . \tag{4}$$

The algorithm is deemed to have converged when the assignments no longer change. Moreover, two issues are involved when using *k*-means clustering as the base clustering model: (1) How to decide the parameter *k* at each node of the cluster tree? (2) How to select the initial centroids of *k*-means at each node of the cluster tree? Thanks to the labeled examples in the training data, we solve the problems as follows. For the first issue, we calculate the percentage of labeled examples of each class, and return *k* as the number of classes with a percentage higher than certain threshold. For example, if the threshold is set to 0, we set *k* as the number of classes existing in the labeled training data. For issue 2, rather than using randomly assigned initial centroids, we aim to improve the performance of *k*-means by calculating the means of labeled examples belonging to each class.

Classifying a test data. The second step is to predict the label for a test example, using the cluster tree created in the first step. The straightforward solution might be comparing the test example with all leaf clusters of the cluster tree, and using the closest cluster to predict the label. However, it may not be fair to compare the distance between the test example with clusters on different levels of the tree. For example, consider the set of training examples in Fig. 1, which is redrawn in Fig. 2. Let the filled circle be the test example. If directly considering the leaf clusters, as shown in Fig. 2 (a), the example is closer to the centroid of cluster $C_2{}^2$, although it's probably better to be clustered to $C_1{}^1$. Therefore, in our approach, we use a path-based strategy to search the best cluster from the cluster tree.

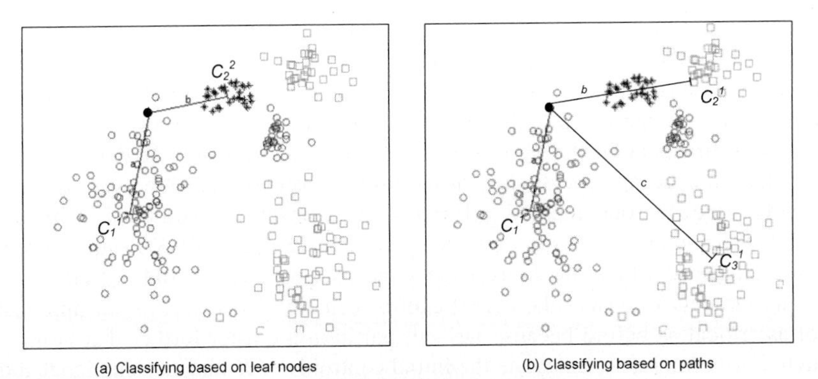

Fig. 2. Leaf based search *v.s.* path based search

For example, Fig. 2 (b) shows the distance between the test example and the clusters on the first level. Since it has the shortest distance with cluster C_1^1, and C_1^1 is a leaf cluster of the tree, the test example will be predicted using C_1^1.

After finding the closest leaf cluster C, we assign the label of the test example as the label of the cluster, *label*(C), which is defined as:

$$\text{Label}(C) = \begin{cases} \text{argmax}_i N_i & \text{if } C \text{ has labeled examples} \\ \text{Label}(C.\text{parent}) & \text{otherwise} \end{cases}, \quad (5)$$

Where N_i is the number of labeled examples belonging to class i. If a cluster does not have any labeled example, we define the label of a cluster as the label of its parent (ancestor). Therefore, there exists no "undefined" clusters [9] such that no label can be induced if a test example is clustered to the cluster. The pseudo-code of this step is shown in Algorithm 2.

Input: The cluster tree $\boldsymbol{T}$, the test example v_j
Output: The label of v_j
Description: ***Classify***$(\boldsymbol{T}.root,\ v_j)$
1. Let current node $C = \boldsymbol{T}.root$.
2. **WHILE** C is a not leaf node
3. Compute the distance between v_j and child clusters of C,
4. Let C_i be the most close cluster
5. **Classify**$(C_i,\ v_j)$
6. **ENDWHILE**
7. **Assign** v_j.label = *Label*(C)

Algorithm 2. Classifying a test example

3.3 Discussion

When the model does not fit the true data distribution, the data points of the same class hardly form natural clusters, *e.g.*, the red circles in Fig. 1. Most of existing clustering model may not be able to identify a class of data which are similar in some subspaces. For example, the K-means clustering method cannot discover such clusters because it assumes hyper-spherical clusters. Since our method recursively clusters the training data, it enables the discovery of clusters in subspaces. Therefore, the successive refinement clustering model is more flexible to match the data distribution which does not comply with the "homophily" assumption. Moreover, the nested clustering used by our method alleviates the impact of parameter setting. For example, the number of clusters k is an important parameter of K-means clustering. Existing cluster-and-label methods using k-means clustering should decide the value of k appropriately to ensure the classification accuracy. However, in our approach, it is not as critical as before because we will partition a cluster further if it is not pure enough. Furthermore, we calculate the initial centroids of K-means using labeled data, which can be viewed as supervised clustering.

4 Experiments

In this section, we evaluate the performance of the proposed method by comparing it with the existing cluster-and-label method, as well as traditional semi-supervised classifiers. We firstly describe the experiment setup. Next, we present and analyze the experimental results. All experiments were conducted on an Intel(R) Xeon(R), 1.60 GHZ computer with 4GB memory.

4.1 Experiment Setup

Datasets. 15 datasets from the UCI machine learning repository[1] are used in our experiments. The characteristics of the data sets are summarized in Table 1.

Table 1. Data Characteristics

Data set	#Instances	#Attributes	#Classes
balance	625	4	3
breast-w	699	9	2
clean1	476	166	2
ecoli	336	7	8
glass	214	9	7
haberman	306	3	2
heart	270	13	2
ionosphere	351	34	2
iris	150	4	3
machine	209	7	8
madelon	2000	500	2
page-blocks	5473	10	5
segment	2310	19	7
waveform	5000	40	3
wine	178	13	3

[1] http://archive.ics.uci.edu/ml/

For each data set, we divide it into three portions with equal size. One of the portions is used as the test data, and the other two serve as the training data. We further split the training data as labeled examples and unlabeled examples. Let ð be the percentage of labeled training examples. We vary the parameter by setting ð=10%, 30%, and 50% respectively. All data splits are performed in a stratified way.

Competing methods. We compare our PRC-Tree against the existing cluster-and-label technique, denoted as Standard Cluster-and-Label (SCL). SCL clusters the training data into using *K*-means. Each cluster is labeled by the majority of labeled examples in the cluster. If a cluster does not have any labeled examples, it will not be used to classify a test data. In such a case, test data in this cluster will be classified using nearest neighbor (NN) based on the labeled examples only.

We also compare our method with traditional semi-supervised classifiers, including Self-training [10], Tri-training [12] and Co-forest [13]. For Self-training, NN is used as the base classifier. Self-training adds, in each round, the most confidently predicted examples into the labeled set. We use the unlabeled example which has the shortest distance with labeled examples as the most confidently predicted example. For Tri-training, the decision tree J48 is used as the base classifier. For Co-forest, 10 random decision tree classifiers are used.

4.2 Experimental Results

The experimental results of all competing methods on datasets with different rates of labeled examples are shown in Tables 2, 3, 4 respectively. Each result is averaged over 20 runs of the corresponding method, with dataset randomly shuffled. For the proposed method, we set the purity threshold θ to be 0.8, 0.9 and 1.0 respectively. For each dataset, we highlight the best classification accuracy achieved by the methods. The last row of each table shows the average accuracy of each method on all datasets.

Table 2. Classification accuracy achieved on datasets with ð=10%

Data sets	SCL	PRC-Tree			Self-training	Tri-training	Co-forest
		θ = 0.8	θ = 0.9	θ = 1.0			
balance	67.54	72.74	72.28	71.84	68.15	72.40	**73.73**
breast-w	**96.05**	**96.05**	94.88	94.01	94.40	91.63	94.52
clean1	54.39	67.17	68.42	**68.65**	58.91	62.88	64.42
ecoli	78.92	**80.23**	79.43	78.64	49.90	67.42	74.81
glass	47.50	52.05	51.98	51.98	45.51	50.66	**52.13**
haberman	68.60	69.40	69.90	69.89	64.65	**70.60**	66.55
heart	**80.96**	78.01	73.80	73.80	73.12	70.56	71.02
ionosphere	66.21	82.34	82.21	82.34	69.43	77.86	**84.69**
iris	80.62	81.14	82.91	82.91	**85.93**	79.68	82.81
machine	74.38	**76.53**	73.53	73.46	66.38	68.76	71.76
madelon	**57.71**	56.27	56.27	56.27	50.43	51.09	49.88
pageblock	91.35	92.23	92.67	92.81	92.26	94.81	**95.16**
segment	68.68	83.98	85.00	85.13	85.18	88.90	**91.21**
waveform	73.58	76.61	74.92	73.65	61.55	72.79	**79.68**
wine	**96.57**	96.05	95.96	95.96	88.24	67.45	84.29
Avg.	73.54	**77.40**	76.94	76.76	70.26	72.50	75.78

Table 3. Classification accuracy achieved on datasets with ð=30%

Data sets	SCL	PRC-Tree			Self-training	Tri-training	Co-forest
		θ = 0.8	θ = 0.9	θ = 1.0			
balance	72.06	**78.22**	77.93	76.67	72.03	76.77	77.30
breast-w	**96.37**	**96.37**	96.11	95.02	94.95	92.95	95.36
clean1	55.09	74.61	75.99	**76.12**	71.85	69.61	73.58
ecoli	81.35	**82.24**	81.58	80.46	78.13	80.04	80.51
glass	51.69	58.01	57.57	57.57	56.76	56.25	**63.08**
haberman	62.10	59.34	56.70	56.50	68.19	**69.60**	66.05
heart	**82.72**	82.50	78.46	76.42	76.64	74.82	77.95
ionosphere	69.73	85.39	85.08	84.56	80.26	87.56	**90.82**
iris	82.91	87.81	91.14	91.97	**92.60**	92.39	91.45
machine	75.76	78.84	78.84	77.61	75.76	79.23	**81.92**
madelon	58.82	**59.15**	**59.15**	**59.15**	50.59	54.59	52.96
pageblock	91.35	93.03	93.85	94.11	94.37	96.13	**96.42**
segment	68.35	87.39	89.12	89.83	92.20	92.69	**95.41**
waveform	77.18	76.81	75.47	74.12	63.95	74.81	**81.32**
wine	**95.87**	95.70	95.70	94.99	93.50	85.35	92.01
Avg.	72.06	79.69	79.51	79.00	77.45	78.85	**81.08**

Table 4. Classification accuracy achieved on datasets with ð=50%

Data sets	SCL	PRC-Tree			Self-training	Tri-training	Co-forest
		θ = 0.8	θ = 0.9	θ = 1.0			
balance	70.65	**79.22**	78.64	77.91	75.55	77.76	78.37
breast-w	**96.37**	**96.37**	**96.37**	94.64	95.24	94.02	95.60
clean1	55.25	80.00	81.69	81.25	77.88	76.18	**83.42**
ecoli	**83.55**	80.84	78.41	76.86	79.25	79.67	82.00
glass	52.20	63.67	62.13	61.98	60.29	63.82	**69.92**
haberman	60.85	62.39	61.59	61.29	65.84	**69.90**	65.85
heart	**82.55**	**82.55**	79.14	75.90	76.47	73.86	80.90
ionosphere	70.21	85.13	86.39	86.00	82.95	88.95	**92.17**
iris	82.91	88.85	90.83	91.14	93.12	92.08	**93.54**
machine	75.53	81.46	81.30	80.07	80.07	84.15	**85.92**
madelon	58.74	62.40	**62.56**	**62.56**	51.03	56.45	55.61
pageblock	91.26	93.26	94.23	94.47	95.16	96.57	**96.90**
segment	68.22	88.74	90.70	91.56	93.69	94.29	**96.87**
waveform	77.14	77.56	75.85	74.69	67.85	75.48	**81.58**
wine	**96.05**	**96.05**	95.87	96.31	94.21	88.59	95.70
Avg.	74.76	81.23	81.05	80.44	79.24	80.78	**83.62**

Comparing PRC-Tree and SCL, our method improves over SCL on almost all of the experimental settings, which indicates that the successive purity refinement based clustering strategy address the model misfit problem very well. Comparing all competing methods, we summarize the results from the following two aspects:

1) Consider the number of times a classifier achieves best performance on datasets, which is shown in parenthesis, the methods are ranked as follows:

ð=10%	Co-Forest (6), PRC-Tree (4), SCL (4), Self-Training (1), Tri-Training (1)
ð=30%	Co-Forest (6), PRC-Tree (5), SCL (3), Self-Training (1), Tri-Training (1)
ð=50%	Co-Forest (8), PRC-Tree (5), SCL (4), Self-Training (0), Tri-Training (1)

2) Consider the average accuracy achieved by a classifier on all datasets (in parenthesis), the methods are ordered as follows:

ð=10%	PRC-Tree (77.40), Co-Forest (75.78), SCL (73.54), Tri-Training (72.50), Self-Training (70.27)
ð=30%	Co-Forest (81.08), PRC-Tree (79.69), Tri-Training (78.85), Self-Training (77.45), SCL (72.06)
ð=50%	Co-Forest (83.62), PRC-Tree (81.23), Tri-Training (80.78), Self-Training (79.24), SCL (74.76)

Again, from both aspects, our method clearly outperforms the standard cluster-and-label method. Our PRC-Tree also performs better than traditional semi-supervised classifiers of Self-Training and Tri-training. Our method achieves competitive performance when compared with the ensemble classifier Co-Forest. When the rate of labeled examples is low (e.g., π=10%), the average accuracy achieved by our method is even higher than Co-Forest. However, our method is more efficient than the ensemble classifier.

When varying the purity threshold èfrom 0.8 to 1.0, we notice that our method achieves the best performance when the threshold is set to 0.8. This is because, when the threshold is set to 1.0, the refinement strategy used by our method leads to the overfitting problem by making sure each cluster consists of labeled examples from one class only.

5 Conclusion

In this paper, we address the problem of model misfit evident in cluster-and-label based semi-supervised learning algorithms. We propose a method, called PRC-Tree, to successively refine the clustering of training data, based on the cluster purity in terms of class distribution. The resulting hierarchical cluster tree enables the discovery of clusters from subspaces, and the combination of nested clusters is expected to match the heterogeneous data distribution better. After creating the cluster tree, our method employs a path-based searching strategy to find the best cluster efficiently for a test example to predict its label. We conduct a series of experiments on the UCI benchmarking data sets to evaluate the performance of our approach. The results demonstrate that our method deals with the model misfit problem very well by improving over the existing cluster-and-label method on most of experimental scenarios. The generalization ability of our method is also better than tradition semi-supervised algorithms such as Self-training and Tri-training. The classification accuracy of our method is comparable with that of the ensemble method Co-Forest.

We notice from the experimental results, that the purity threshold plays an important role in the learning performance. Selecting an appropriate threshold value is an important issue of our method. As an ongoing work, we are interested in deciding the threshold value automatically based on some statistical test methods (*e.g.*, Chi-square test).

References

1. Nigam, K., McCallum, A.K., Thrun, S., Mitchell, T.: Text classification from labeled and unlabeled documents using EM. Machine Learning 39, 103–134 (2000)
2. Fujino, A., Ueda, N., Saito, K.: A hybrid generative/discriminative approach to semi-supervised classifier design. In: The Twentieth National Conference on Artificial Intelligence, pp. 764–769. AAAI Press/ MIT Press, Pennsylvania (2005)
3. Blum, A., Chawla, S.: Learning from labeled and unlabeled data using graph mincuts. In: The 18th International Conference on Machine Learning, pp. 16–26. Morgan Kaufmann, MA (2001)
4. Zhu, X., Ghahramani, Z., Lafferty, J.: Semi-supervised learning using Gaussian fields and harmonic functions. In: The 20th International Conference on Machine Learning, pp. 912–919. AAAI Press, Washington (2003)
5. Joachims, T.: Transductive inference for text classification using support vector machines. In: The 16th International Conference on Machine Learning, pp. 200–209. Morgan Kaufmann, Slovenia (1999)
6. Zhu, X.: Semi-Supervised Learning Literature Survey. Tech. Rep. 1530, University of Wisconsin – Madison, Madison, WI (2006)
7. Krishnapuram, B., Williams, D., Xue, Y., Hartemink, A., Carin, L.: On Semi-supervised Classification. In: Saul, L.K., Weiss, Y., Bottou, L. (eds.) Advances in Neural Information Processing Systems, vol. 17, MIT Press, Cambridge (2005)
8. Demiriz, A., Bennett, K., Embrechts, M.: Semi-supervised clustering using genetic algorithms. The Artificial Neural Networks in Engineering (1999)
9. Dara, R., Kremer, S., Stacey, D.: Clustering unlabeled data with SOMs improves classification of labeled real-world data. In: Proceedings of the World Congress on Computational Intelligence, Honolulu (2002)
10. Yarowsky, D.: Unsupervised word sense disambiguation rivaling supervised methods. In: The 33rd Annual Meeting of the Association for Computational Linguistics, pp. 189–196. Morgan Kaufmann Publishers, Massachusetts (1995)
11. Avrim, B., Mitchell, T.M.: Combining Labeled and Unlabeled Data with Co-Training. In: The 11th annual conference on Computational learning Theory, pp. 92–100. Morgan Kaufmann Publishers, San Francisco (1998)
12. Zhou, Z.H., Li, M.: Tri-training: Exploiting unlabeled data using three classifiers. IEEE Transaction on Knowledge and Data Engineering 17, 1529–1541 (2005)
13. Li, M., Zhou, Z.H.: Improve computer-aided diagnosis with machine learning Techniques using undiagnosed samples. IEEE Transactions on Systems, Man and Cybernetics – Part A: Systems and Information Systems 37, 1088–1098 (2007)

Soft Set Approach for Selecting Decision Attribute in Data Clustering

Mohd Isa Awang[1], Ahmad Nazari Mohd Rose[1], Tutut Herawan[2,3], and Mustafa Mat Deris[3]

[1] Faculty of Informatics
Universiti Sultan Zainal Abidin, Terengganu, Malaysia
[2] Department of Mathematics Education
Universitas Ahmad Dahlan, Yogyakarta, Indonesia
[3] Faculty of Information Technology and Multimedia
Universiti Tun Hussein Onn Malaysia, Johor, Malaysia
isa@udm.edu.my, anm@udm.edu.my, tutut81@uad.ac.id,
mmustafa@uthm.edu.my

Abstract. This paper presents the applicability of soft set theory for discovering a decision attribute in information systems. It is based on the notion of a mapping inclusion in soft set theory. The proposed technique is implemented with example test case and one UCI benchmark data; US Census 1990 dataset. The results from test case show that the selected decision attribute is equivalent to that under rough set theory.

Keywords: Information system, Soft set theory, Decision attribute.

1 Introduction

Partitioning a set of heterogeneous data objects into smaller homogeneous classes of data sets is a fundamental operation in data mining. This can be done by introducing a decision attribute in an information system. To this, one practical problem is faced: for many candidates in a database, we need to select only one that is the best attribute to partition the objects. The theory of soft set proposed by Molodtsov 1999 is a new method for handling uncertain data. Soft sets are called (binary, basic, elementary) neighborhood systems. Molodtsov pointed out that one of the main advantages of soft set theory is that it is free from the inadequacy of the parameterization tools, like in the theories of fuzzy set, probability and interval mathematics. As for standard soft set, it may be redefined as the classification of objects in two distinct classes, thus confirming that soft set can deal with a Boolean-valued information system. The theory of soft set has been applied to data reduction and decision support systems [[2], [3], [4]]. The existing researches rely heavily on Boolean-valued information system. However, to date, only scarce researchers deal with multi-valued information system. The concept of decision attribute selection in a multi-valued information system is another area which purportedly supports data classification.

By applying the concept of a mapping inclusion in soft set theory, we propose an alternative technique for decision attribute selection. The main purpose of the proposed

L. Cao, J. Zhong, and Y. Feng (Eds.): ADMA 2010, Part II, LNCS 6441, pp. 87–98, 2010.

technique is to ensure that the process of decision attribute selection is used in transforming an information system into a decision system.

One example test case was considered to evaluate the accuracy of a decision attribute as compared to a technique called Maximum Dependency Attributes (MDA) [6] using rough set theory. The results show that the selected decision attribute is equivalent to that under rough set theory. The experiment was also conducted with a set of data from public dataset; UCI benchmark dataset – US Census 1990 in order to show that the proposed technique can also be implemented with large datasets.

The rest of this paper is organized as follows. Section 2 describes the fundamental concept of information systems. Section 3 describes the concept of soft set theory. Section 4 describes soft set-based approach for decision attribute selection in multi-valued information systems. Finally, the conclusion of our work is described in section 5.

2 Information System

An information system is a 4-tuple (quadruple), $S=(U,A,V,f)$, where U is a non-empty finite set of objects, A is a non-empty finite set of attributes, $V=\bigcup_{a\in A}V_a$, V_a is the domain (value set) of attribute a, $f:U\times A\to V$ is a total function such that $f(u,a)\in V_a$, for every $(u,a)\in U\times A$, called information (knowledge) function. An information system is also called a knowledge representation systems or an attribute-valued system that can be intuitively expressed in terms of an information table (as shown in Table 1).

Table 1. An information system

U	a_1	a_2	$\cdots$	a_k	$\cdots$	$a_{\|A\|}$
u_1	$f(u_1,a_1)$	$f(u_1,a_2)$	$\cdots$	$f(u_1,a_k)$	$\cdots$	$f(u_1,a_{\|A\|})$
u_2	$f(u_2,a_1)$	$f(u_2,a_2)$	$\cdots$	$f(u_2,a_k)$	$\cdots$	$f(u_2,a_{\|A\|})$
u_3	$f(u_3,a_1)$	$f(u_3,a_2)$	$\cdots$	$f(u_3,a_k)$	$\cdots$	$f(u_3,a_{\|A\|})$
$\vdots$	$\vdots$	$\vdots$	$\ddots$	$\vdots$	$\ddots$	$\vdots$
$u_{\|U\|}$	$f(u_{\|U\|},a_1)$	$f(u_{\|U\|},a_2)$	$\cdots$	$f(u_{\|U\|},a_k)$	$\cdots$	$f(u_{\|U\|},a_{\|A\|})$

In many applications, there is an outcome of classification that is known. This knowledge, which is known as *a posteriori* knowledge is expressed by one (or more) distinguished attribute called decision attribute. This process is known as supervised learning. An information system of this kind is called a decision system. A *decision system* is an information system of the form $D=(U,A\cup\{d\},V,f)$, where $d\notin A$ is the decision attribute. The elements of A are called *condition attributes*.

In the following sub-section, we propose an idea of decomposition of a multi-valued information system $S=(U,A,V,f)$ into $|A|$ numbers of Boolean-valued information system $S^i=(U,a_i,V_{\{0,1\}},f)$, where $|A|$ is the cardinality of A.

2.1 Decomposition of a Multi-valued Information System

The decomposition of $S=(U,A,V,f)$ is based on decomposition of $A=\{a_1,a_2,\cdots,a_{|A|}\}$ into the disjoint-singleton attribute $\{a_1\},\{a_2\},\cdots,\{a_{|A|}\}$. Here, we only consider for complete information systems. Let $S=(U,A,V,f)$ be an information system such that for every $a\in A$, $V_a=f(U,A)$ is a finite non-empty set and for every $u\in U$, $|f(u,a)|=1$. For every a_i under i th-attribute consideration, $a_i\in A$ and $v\in V_a$, we define the map $a_v^i:U\rightarrow\{0,1\}$ such that $a_v^i(u)=1$ if $f(u,a)=v$, otherwise $a_v^i(u)=0$. The next result, we define a binary-valued information system as a quadruple $S^i=(U,a_i,V_{\{0,1\}},f)$. The information systems $S^i=(U,a_i,V_{\{0,1\}},f)$, $1\le i\le|A|$ is referred to as a decomposition of a multi-valued information system $S=(U,A,V,f)$ into $|A|$ binary-valued information systems, as depicted in Figure 1. Every information system $S^i=(U,a_i,V_{ai},f)$, $1\le i\le|A|$ is a deterministic information system since for every $a\in A$ and for every $u\in U$, $|f(u,a)|=1$ such that the structure of a multi-valued information system and $|A|$ number of binary-valued information systems give the same value of attribute related to objects.

3 Soft Set Theory

Throughout this section U refers to an initial universe, E is a set of parameters, $P(U)$ is the power set of U and $A\subseteq E$.

Definition 1. (See [1].) *A pair (F,A) is called a soft set over U, where F is a mapping given by*

$$F:A\rightarrow P(U).$$

In other words, a soft set over U is a parameterized family of subsets of the universe U. For $\varepsilon\in A$, $F(\varepsilon)$ may be considered as a set of ε-elements of the soft set (F,A) or as the set of ε-approximate elements of the soft set. Clearly, a soft set is not a (crisp) set. For illustration, Molodtsov considered several examples in [1]. In addition, in this section we present the relation between rough and soft sets, i.e., every rough set can be considered as a soft set as stated the following proposition.

Based on the definition of an information system and a soft set, in this section we show that a soft set is a special type of information systems, i.e., a binary-valued information system.

Proposition 2. *If* (F, A) *is a soft set over the universe U, then* (F, A) *is a binary-valued information system* $S = (U, A, V_{\{0,1\}}, f)$.

Proof. Let (F, A) be a soft set over the universe U, we define a mapping

$$F = \{f_1, f_2, \cdots, f_n\},$$

where

$$f_i : U \to V_i \text{ and } f_i(x) = \begin{cases} 1, & x \in F(a_i) \\ 0, & x \notin F(a_i) \end{cases}, \text{ for } 1 \le i \le |A|.$$

Hence, if $V = \bigcup_{a_i \in A} V_{a_i}$, where $V_{e_i} = \{0,1\}$, then a soft set (F, A) can be considered as a binary-valued information system $S = (U, A, V_{\{0,1\}}, f)$.

From Proposition 2, it is easily understand that a binary-valued information system can be represented as a soft set. Thus, we can make a one-to-one correspondence between (F, E) over U and $S = (U, A, V_{\{0,1\}}, f)$.

Definition 3. (See [2].) *The class of all value sets of a soft set* (F, E) *is called value-class of the soft set and is denoted by* $C_{(F,E)}$.

3.1 Multi-soft Sets in Information Systems

The idea of multi-soft sets is based on a decomposition of a multi-valued information system $S = (U, A, V, f)$, into $|A|$ number of binary-valued information systems $S = (U, A, V_{\{0,1\}}, f)$, where $|A|$ denotes the cardinality of A. Consequently, the $|A|$ binary-valued information systems define *multi-soft sets* $(F, A) = \{(F, a_i) : 1 \le i \le |A|\}$.

Based on the notion of a decomposition of a multi-valued information system in the previous section, in this sub-section we present the notion of multi-soft set representing multi-valued information systems. Let $S = (U, A, V, f)$ be a multi-valued information system and $S^i = (U, a_i, V_{ai}, f)$, $1 \le i \le |A|$ be the $|A|$ binary-valued information systems. From Proposition 2, we have

$$S = (U, A, V, f) = \begin{cases} S^1 = (U, a_1, V_{\{0,1\}}, f) & \Leftrightarrow (F, a_1) \\ S^2 = (U, a_2, V_{\{0,1\}}, f) & \Leftrightarrow (F, a_2) \\ \vdots & \vdots \quad \vdots \\ S^{|A|} = (U, a_{|A|}, V_{\{0,1\}}, f) & \Leftrightarrow (F, a_{|A|}) \end{cases}$$

$$= ((F, a_1), (F, a_2), \cdots, (F, a_{|A|}))$$

We define $(F, A) = ((F, a_1), (F, a_2), \cdots, (F, a_{|A|}))$ as a *multi-soft sets* over universe U representing a multi-valued information system $S = (U, A, V, f)$.

4 Soft Set Approach for Decision Attribute Selection

In this section, we describe the application of soft set theory for selecting a decision attribute. The proposed technique is based on the notion of a mapping inclusion in soft set theory.

Definition 4. *Let (F, E) be a multi-soft sets over U representing an information system $S = (U, A, V, f)$. Let $(F, e_i), (F, e_j) \subseteq (F, E)$ be two distinct soft sets. The cohesiveness of soft set (F, e_i) with respect to soft set (F, e_j) is defined by*

$$\operatorname{coh}_{e_j}(e_i) = \frac{\sum_{i=1}^{|C_{(F,e_i)}|} \left(\bigcup |X| : X \subseteq Y_i / \bigcup |X| : X \cap Y_i \neq \phi \right)}{|C_{(F,e_i)}|}, \tag{1}$$

where $X \subseteq C_{(F,e_j)}$ and $Y_i \subseteq C_{(F,e_i)}$.

The decision to select the best partitioning attribute that is represented by the soft set in the multi-soft set can be made based on total cohesiveness of the soft set.

Definition 5. *Let (F, E) be a multi-soft sets over U representing an information system, $S = (U, A, V, f)$. Total cohesiveness of soft set (F, e_i) with respect to all soft set (F, e_j), where $j \neq i$ is obtained by calculating the mean cohesiveness of the soft set and is given by the following equation*

$$\operatorname{tcoh}(e_i) = \frac{\sum_{i=1}^{|A|-1} \operatorname{coh}(e_i)}{|A|-1}, \quad 1 \leq i, j \leq |A| \text{ and } i \neq j. \tag{2}$$

The highest total cohesiveness of the soft set will be the most suitable candidate to be selected as a decision attribute for data clustering. An algorithm and pseudo codes for the whole process of obtaining the total cohesiveness for decision attribute selection will be described in the following Figure.

```
1. Obtain multi-soft set (F,E) over U
2. For each soft set in the multi-soft sets, obtain
   the cohesiveness of the soft set (F,e_i) with
   respect to soft set (F,e_j), for 1 ≤ i, j ≤ |A| and
   i ≠ j .
3. For each soft set (F,e_i), obtain total
   cohesiveness for 1 ≤ i ≤ |A|
```

Fig. 1. The algorithm to obtain total cohesiveness of the soft sets

There are three main steps that need to be accomplished in order to select the most appropriate attribute to partition objects in an information system. The pseudo-codes as in Fig. 2 are used to accomplish the task.

The process starts with a determination of value-class of the multi-soft set (F,E) and is denoted by $C_{(F,E)}$ as stated in Definition 3. Let n be a number of objects and a be a number of attributes in an information system, the value-class of multi-soft set is obtained using pseudo codes as shown below.

Applying soft set in selecting decision attribute, further may be used to partition objects in data clustering technique. However, there are not much literatures found on this topic.

In the following examples, we present the computational activities through one small datasets and present the result obtain by both MDA [6] and the proposed technique.

Test Case Example. A small-sized dataset, The Credit Card Promotion in [5]. The dataset contains 10 objects with five categorical-valued attributes, i.e. Magazine Promotion (MP), Watch Promotion (WP), Life Insurance Promotion (LIP), Credit Card Insurance (CCI) and Sex (S). We use the formula of total cohesiveness in equation (2). The higher the total cohesiveness is the higher the accuracy of the selecting decision attribute.

MDA technique proposed by Herawan et. al. [6] uses the dependency of attributes in rough set theory to determine dependency degree of each attribute. The attribute with a maximum degree of dependency will be selected as a clustering attribute. The value of MDA for all attributes is summarized in Table 3 below.

```
// Pseudo code 1 - to obtain a value-class of multi-soft set over U
  Let i = 1;
  While i<=noOfAttributes=a {
    find  all value-classes for each attribute value;
    i=i+1;
  } // end while
```

```
// Pseudo code 2 - to obtain cohesiveness of the soft set
  Let i=1;
  For each soft set i up to soft set a {
    Let  j=1;
    For each soft set j up to soft set a {
      if soft set i is equal to soft set j then {
        j=j+1; loop;
      } // end if
     Calculate lower inclusion of all predicate in soft
      set i with respect to soft set j;
     Calculate upper inclusion of all predicate in soft
      set I with respect to soft set j;
      j=j+1;
    } // end for each j
    Cohesiveness =(ΣLower inclusion / Upper inclusion)
                  / no. of predicate;
    i=i+1;
  } // end for each i
```

```
// Pseudo code 3 - to obtain total cohesiveness of the soft set
Let i=1;
For each soft set i up to soft set a { // to determine
// total cohesiveness (TC) of soft set i
   TC = ΣCohesiveness/(no. soft set=a - 1);
   i=i+1;
} // end for each i
```

Fig. 2. Pseudo codes

Table 2. A Credit Card Promotion dataset from [5]

Person	MP	WP	LIP	CCI	S
1	yes	no	no	no	male
2	yes	yes	yes	no	female
3	no	no	no	no	male
4	yes	yes	yes	yes	male
5	yes	no	yes	no	female
6	no	no	no	no	female
7	yes	no	yes	yes	male
8	no	yes	no	no	male
9	yes	no	No	no	male
10	yes	yes	yes	no	female

Table 3. The degree of dependency of all attributes in Table 2 using MDA technique (from [6])

Attribute (Depends on)	Degree of dependency				MDA
MP	WP	LIP	CCI	S	0.5
	0	0.5	0.2	0	0.2
WP	MP	LIP	CCI	S	0
	0	0	0	0	
LIP	MP	WP	CCI	S	0.3
	0.3	0	0.2	0	
CCI	MP	WP	LIP	S	0.5
	0.3	0	0.5	0.4	0.4
S	MP	WP	LIP	CCI	0.2
	0	0	0	0.2	

From Table 3 above, the attributes MP and CCI has the same maximum degree of dependency, therefore, based on the algorithm, the next degree of dependency of the attributes will be considered, until the tie is broken. For this case, the second degree corresponding to attribute CCI, i.e. 0.4 is higher than that of MP, i.e. 0.2. Therefore, attribute CCI is selected as a clustering attribute.

Our proposed technique called Total Cohesiveness, start by transforming an information system (Table 2) into a multi soft sets as shown in Fig. 3 below.

$$(F,E)=\left\{\begin{array}{l}(F,\mathrm{MP})=\{\mathrm{yes}=1,2,4,5,7,9,10\},\{\mathrm{no}=3,6,8\}\\(F,\mathrm{WP})=\{\mathrm{yes}=2,4,8,10\},\{\mathrm{no}=1,3,5,6,7,9\}\\(F,\mathrm{LIP})=\{\mathrm{yes}=2,4,5,7,10\},\{\mathrm{no}=1,3,6,8,9\}\\(F,\mathrm{CCI})=\{\mathrm{yes}=4,7\},\{\mathrm{no}=1,2,3,5,6,8,9,10\}\\(F,\mathrm{S})=\{\mathrm{male}=1,3,4,7,8,9\},\{\mathrm{female}=2,5,6,10\}\end{array}\right\}$$

Fig. 3. The multi soft sets from Table 2

Based on multi-soft sets in Fig. 3, we use formula from equations (1) and (2) to obtain the total cohesiveness for each soft set with respect to all other soft sets in the multi soft set as shown in Table 4.

Before cohesiveness of subset of U is computed, the value-class for each soft set is obtained and is shown as in Fig. 4 below.

The cohesiveness of subsets of U based on attribute MP with respect to attribute WP is computed using equation (1) as below.

a. MP with respect to WP

$$\mathrm{coh}_{\mathrm{WP}}(\mathrm{MP})=\frac{\left(|\phi|/|\{1,2,3,4,5,6,7,8,9,10\}|\right)+\left(|\phi|/|\{1,2,3,4,5,6,7,8,9,10\}|\right)}{|\{\{1,2,4,5,7,9,10\},\{3,6,8\}\}|},$$

$$=\frac{0+0}{2}=0,$$

$$C_{(F,\mathrm{MP})} = \{\{1,2,4,5,7,9,10\},\{3,6,8\}\}.$$
$$C_{(F,\mathrm{WP})} = \{\{2,4,8,10\},\{1,3,5,6,7,9\}\}.$$
$$C_{(F,\mathrm{LIP})} = \{\{2,4,5,7,10\},\{1,3,6,8,9\}\}.$$
$$C_{(F,\mathrm{CCI})} = \{\{4,7\},\{1,2,3,5,6,8,9,10\}\}.$$
$$C_{(F,\mathrm{S})} = \{\{1,3,4,7,8,9\},\{2,5,6,10\}\}.$$

Fig. 4. The class values of soft sets from Fig. 3

By applying similar procedure, the cohesiveness of subsets of U having different values of attribute MP with respect to attributes LIP, CCI and S are computed and are given below.

b. MP with respect to LIP

$$\mathrm{coh}_{\mathrm{LIP}}(\mathrm{MP}) = \frac{0.5+0}{2} = 0.25,$$

c. MP with respect to CCI

$$\mathrm{coh}_{\mathrm{CCI}}(\mathrm{MP}) = \frac{0.2+0}{2} = 0.1,$$

d. MP with respect to S

$$\mathrm{coh}_{\mathrm{CCI}}(\mathrm{MP}) = \frac{0+0}{2} = 0.$$

Finally, we obtain the total cohesiveness of soft set MP with respect to the set of all soft set $(F, E \setminus \{\mathrm{MP}\})$ using the formula (2) as follows;

$$\mathrm{tcoh}(\mathrm{MP}) = \frac{0+0.25+0.1+0}{4} = 0.0875.$$

The total cohesiveness (TC) of all soft set from Fig. 4 can be summarized as in Table 4 below.

As shown in Table 4, the highest value of total cohesiveness (TC) is obtain from the soft set (F, CCI), and therefore, the attribute CCI (i.e. Credit Card Insurance) is selected as a decision attribute. The decision attribute, CCI, selected by TC technique is equivalent to the one selected by MDA technique [6] using rough set theory. However, by using the same set of data, TC is able to select decision attribute without the need to look at the next degree of dependency of attributes as needed by the MDA technique.

Table 4. The value of TC of all soft sets

Soft set (with respect to)	Cohesiveness				TC
(F,MP)	(F,WP)	(F,LIP)	(F,CCI)	(F,S)	0.0875
	0	0.25	0.1	0	
(F,WP)	(F,MP)	(F,LIP)	(F,CCI)	(F,S)	0
	0	0	0	0	
(F,LIP)	(F,MP)	(F,WP)	(F,CCI)	(F,S)	0.0625
	0.15	0	0.1	0	
(F,CCI)	(F,MP)	(F,WP)	(F,LIP)	(F,S)	0.15
	0.15	0	0.25	0.2	
(F,S)	(F,MP)	(F,WP)	(F,LIP)	(F,CCI)	0.025
	0	0	0	0.1	

5 Experiment Tests with Public Dataset

This section discusses the experiment that was done using the proposed technique, Total Cohesiveness technique. One UCI benchmark datasets will be tested using this technique. The technique is implemented in Java language using Windows XP operating system with processor Intel® core 2 duo 2.0 MHz and memory 4GB.

US Census 1990 Data dataset

The dataset was obtained from (U.S. Department of Commerce) Census Bureau and was collected as part of the 1990 census. There are originally 2458285 Instances and 68 categorical attributes in this dataset, however, for the purpose of experimentation test, only 31 objects and 15 attributes was extracted from the original dataset. The list of attributes that was selected include; Age (Ag), Ancstry1 (An1), Ancstry2 (An2), Avail (Av), Citizen (Ci), Class (Cl), Depart (De), Disabl1 (Di1), Disabl2 (Di2), English (Eg), Feb55 (Fe), Fertil (Fr), Hispanic (Hi), Hour89 (H8) and Hours (Hs).

The list of soft sets from multi-soft set generated by the proposed technique is shown in Table 5 below.

The above result shows that there are two soft sets that possess the highest TC value that is equal to 1 and it belong to the attributes Avail (Av) and Feb55 (Fe). However, these attributes cannot be considered to be selected as a decision attribute because it (TC=1) is generated from the equivalent soft set that is consist of only one predicate and all subsets of object of the universe, U , is a value-class of that predicate. Therefore, for this experiment, the soft set with second highest TC value will be selected as a decision attribute. TC value for soft set (F,Ci) is equal to 0.3364 is the second highest value listed in Table 5 above, so the attribute Citizen from the US Census dataset is the most appropriate attribute to be selected as a decision attribute to cluster the data.

Table 5. The value of TC of all soft sets from experiment

Soft set (with respect to all other soft set)	Total Cohesiveness
(F, Ag)	0.0340
$(F, An1)$	0.0327
$(F, An2)$	0.0302
(F, Av)	1.0000
(F, Ci)	0.3364
(F, Cl)	0.0411
(F, De)	0.0576
$(F, Di1)$	0.1881
$(F, Di2)$	0.1881
(F, Eg)	0.1665
(F, Fe)	1.0000
(F, Fr)	0.0290
(F, Hi)	0.2149
$(F, H8)$	0.0584
(F, Hs)	0.0444

6 Conclusion

In this paper we have presented the applicability of soft set theory for selection of decision attributes in data clustering analysis. Based on the fact that every rough set can be considered as a soft set, we have successfully used the notion of inclusion of a mapping in soft sets for discovering decision attributes in information systems. The results obtained are equivalent to the rough set-based techniques. The experiment conducted with the US Census 1990 dataset has shown that the proposed technique is suitable to be used with large datasets.

References

1. Molodtsov, D.: Soft set theory-first results. Computers and Mathematics with Applications 37, 19–31 (1999)
2. Maji, P.K., Roy, A.R., Biswas, R.: An application of soft sets in a decision making problem. Computers and Mathematics with Applications 44, 1077–1083 (2002)

3. Chen, D., Tsang, E.C.C., Yeung, D.S., Wang, X.: The Parameterization Reduction of Soft Sets and its Applications. Computers and Mathematics with Applications 49, 757–763 (2005)
4. Kong, Z., Gao, L., Wang, L., Li, S.: The normal parameter reduction of soft sets and its algorithm. Computers and Mathematics with Applications 56, 3029–3037 (2008)
5. Roiger, R.J., Geatz, M.W.: Data Mining: A Tutorial-Based Primer. Addison Wesley, Reading (2003)
6. Herawan, T., Deris, M.M., Abawajy, J.H.: A Rough Set Approach for Selecting Clustering Attribute. Knowledge-Based Systems 23(3), 220–231 (2010)

Comparison of BEKK GARCH and DCC GARCH Models: An Empirical Study

Yiyu Huang[1,2], Wenjing Su[2], and Xiang Li[1,*]

[1] Key Laboratory of Geographical Information Science, Ministry of Education, East China Normal University, 200062 Shanghai, China

[2] Department of Statistics, School of Economics and Management, Lund University, Box 743 SE-22007 Lund, Sweden

helen_huang@ecnu.cn, wenjing7su@gmail.com, xli@geo.ecnu.edu.cn

Abstract. Modeling volatility and co-volatility of a few zero-coupon bonds is a fundamental element in the field of fix-income risk evaluation. Multivariate GARCH model (MGARCH), an extension of the well-known univariate GARCH, is one of the most useful tools in modeling the co-movement of multivariate time series with time-varying covariance matrix. Grounded on the review of various formulations of multivariate GARCH model, this paper estimates two MGARCH models, BEKK and DCC form, respectively, based on the data of three AAA-rated Euro zero-coupon bonds with different maturities (6 months/1 year/2 years). Post-model diagnostics indicates satisfying fitting performance of these estimated MGARCH models. Moreover, this paper provides comparison on the goodness of fit and forecasting performances of these forms by adopting the mean absolute error (MAE) criterion. Throughout this application, the conclusion can be drawn that significant fitting and forecasting performances originate from the trade-off between parsimony and flexibility of the MGARCH models.

Keywords: Volatility, Multivariate GARCH Models, BEKK/DCC Form, Quasi – Maximum Likelihood Method, Zero-Coupon Bonds.

1 Introduction

With the increase in the complexity of the instruments in the risk management field, huge demands for the various models which can simulate and reflect the characteristics of the financial time series have expanded. One of the significant features of financial data that has won much attention is the volatility; because it is a numerical measure of the risk faced by individual investors and financial institutions. It is well known that the volatility of financial data often varies over time and tends to cluster in periods, i.e., high volatility is usually followed by high volatility, and low volatility by low volatility. This phenomenon corresponds to the fluctuating volatility. The Generalized Autoregressive Conditional Heteroskedasticity (GARCH) model and its extensions have been proved to be able to capture the volatility clustering and predict volatilities in the future.

* Corresponding author.

L. Cao, J. Zhong, and Y. Feng (Eds.): ADMA 2010, Part II, LNCS 6441, pp. 99–110, 2010.

Specifically, when analyzing the co-movements of financial returns, it is always essential to construct, estimate, evaluate, and forecast the co-volatility dynamics of asset returns in a portfolio. This task can be fulfilled by multivariate GARCH (MGARCH) models. The development of MGARCH models could be thought as a great breakthrough against the curse of dimensionality in the financial modeling. Many different formulations have been constructed parsimoniously and still remain necessary flexibility. The application fields that MGARCH models can extend to include asset pricing, portfolio theory, VaR estimation and risk management or diversification, which require the volatilities and co-volatilities of several markets [1].

In this paper, MGARCH models are estimated and evaluated for volatility and co-volatility of three zero coupon bond prices with different maturities. The data is provided by the website of the European Central Bank (ECB) which is the institution of the European Union tasked with administrating the monetary policy of the EU member states taking part in the Euro zone.

A zero coupon bond is a non-coupon-bearing bond that pays face value at the time of maturity even though it is bought at a price lower than its face value. It has no reinvestment risk and is more sensitive to interest rate change than coupon-bearing bonds. Due to these features, zero-coupon bonds can be easily used to create any type of cash flow stream and thus match asset cash flows with liability cash flows (e.g. to provide for college expenses, house-purchase down payment, or other liability funding.), and used by pension funds and insurance companies to offset, or immunize the interest rate risk of these firms' long-term liabilities.

Moreover, the return of zero coupon bond, referred to as zero rate, is a fundamental element in the field of fix-income pricing and risk evaluation. By using cash-flow-mapping method [2], any fixed cash flow can be mapped to a portfolio consisting of a few representative zero coupon bonds, which match the cash flow's return and volatility. This viewpoint exemplifies how to generalize the specific zero coupon bond volatilities into a general case. It also motivates our study to model volatility and co-volatility of three zero-coupon bonds with three conventional maturities of 6 months, 1 year and 2 years.

The reminder of this paper is organized as follows. Section 2 reviews MGARCH models, including its different forms, diagnostics techniques and the forecasting strategy. In section 3 we present the BEKK and DCC MGARCH models of volatility and co-volatility of ECB zero coupon bond data sets. Conclusions are detailed in section 4.

2 Model Specification and Estimation Methodology

At the beginning of reviewing different formulations of MGARCH models, one should consider what specification of an MGARCH model should be imposed in contrast to the univariate case. On the one hand, it should be flexible enough to state the dynamics of the conditional variances and covariances. On the other hand, as the number of parameters in an MGARCH model increases rapidly along with the dimension of the model, the specification should be parsimonious to simplify the model estimation and also reach the purpose of easy interpretation of the model parameters. However, parsimony may reduce the number of parameters, in which situation the relevant dynamics in the covariance matrix cannot be captured. Another feature that multivariate GARCH models must satisfy is that the covariance matrix should be positive definite. Bearing these specifications in mind, one can get a review of the following formulations of multivariate GARCH models and comprehend each of their relative competence and drawbacks.

2.1 Formulations of Multivariate GARCH Models

VEC-GARCH Models

The first MGARCH model was introduced by Bollerslev, Engle and Wooldridge in 1988, which is called VEC model. In the VEC model, every conditional variance and covariance is a function of all lagged conditional variances and covariances, as well as lagged squared returns and cross-products of returns. The model can be expressed below:

$$vech(H_t) = c + \sum_{j=1}^{q} A_j vech(\varepsilon_{t-j}\varepsilon'_{t-j}) + \sum_{j=1}^{p} B_j vech(H_{t-j}) \,, \tag{1}$$

where $vech(\cdot)$ is an operator that stacks the columns of the lower triangular part of its argument square matrix, H_t is the covariance matrix of the residuals, N presents the number of variables, t is the index of the lth observation, c is an $N(N+1)/2\times1$ vector, A_j and B_j are $N(N+1)/2 \times N(N+1)/2$ parameter matrices and ε is an $N\times1$ vector.

The condition for H_t to be positive definite for all t is not restrictive. In addition, the number of parameters equals $(p+q)\times(N(N+1)/2)2+N(N+1)/2$, which is large. Furthermore, it demands a large quantity of computation.

BEKK-GARCH Models

To ensure positive definiteness, a new parameterization of the conditional variance matrix H_t was defined by Baba, Engle, Kraft and Kroner (1990) and became known as the BEKK model, which is viewed as a restricted version of the VEC model. It achieves the positive definiteness of the conditional covariance by formulating the model in a way that this property is implied by the model structure.

The form of the BEKK model is as follows

$$H_t = CC' + \sum_{j=1}^{q}\sum_{k=1}^{K} A'_{kj}\,\varepsilon_{t-j}\varepsilon'_{t-j}\,A_{kj} + \sum_{j=1}^{p}\sum_{k=1}^{K} B'_{kj}H_{t-j}B_{kj} \,, \tag{2}$$

where A_{kj}, B_{kj}, and C are $N\times N$ parameter matrices, and C is a lower triangular matrix.

The purpose of decomposing the constant term into a product of two triangular matrices is to guarantee the positive semi-definiteness of H_t. Whenever $K > 1$ an identification problem would be generated for the reason that there are not only a single parameterization that can obtain the same representation of the model.

The first-order BEKK model is

$$H_t = CC' + A'\varepsilon_{t-1}\varepsilon'_{t-1}A + B'H_{t-1}B \,. \tag{3}$$

Estimation of a BEKK model still bears large computations due to several matrix transpositions. The number of parameters of the complete BEKK model is $(p+q)KN^2+N(N+1)/2$. Even in the diagonal one, the number of parameters soon reduces to $(p+q)KN+N(N+1)/2$, but it is still large. The BEKK form is not linear in parameters, which makes the convergence of the model difficult. However, the strong point lies in that the model structure automatically guarantees the positive definiteness of H_t. Under the overall consideration, it is typically assumed that $p = q = K = 1$ in BEKK form's application.

Constant Conditional Correlation Models

The Constant Conditional Correlation model was introduced by Bollerslev in 1990 to primarily model the conditional covariance matrix indirectly by estimating the conditional correlation matrix. The conditional correlation is assumed to be constant while the conditional variances are varying. Obviously, this assumption is impractical for real financial time series. Then certain modifications were made grounded on this form [3].

Dynamic Conditional Correlation Models

The Dynamic Conditional Correlation model was proposed by Engle in 2002. It is a nonlinear combination of univariate GARCH models and it is also a generalized version of the CCC model. The form of Engle's DCC model is as follows:

$$H_t = D_t R_t D_t \, , \tag{4}$$

where

$$D_t = diag(h_{11t}^{1/2}, \cdots, h_{NNt}^{1/2}) \, , \tag{5}$$

and each h_{iit} is described by a univariate GARCH model. Further,

$$R_t = diag(q_{11t}^{1/2}, \cdots, q_{NNt}^{1/2}) Q_t diag(q_{11t}^{1/2}, \cdots, q_{NNt}^{1/2}) \, , \tag{6}$$

where $Q_t = (q_{ijt})$ is the $N\times N$ symmetric positive definite matrix which has the form:

$$Q_t = (1-\alpha-\beta)\overline{Q} + \alpha u_{t-1} u_{t-1}' + \beta Q_{t-1} \, . \tag{7}$$

Here, $u_{it} = \varepsilon_{it} / \sqrt{h_{iit}}$, and are non-negative scalars that + <1, $\overline{Q}$ is the $N\times N$ unconditional variance matrix of u_t.

The number of parameters to be estimated is $(N+1)\times(N+4)/2$, which is relatively smaller than the complete BEKK form with the same dimension when N is small. When N is large, the estimation of the DCC model can be performed by a two-step procedure which decreases the complexity of the estimation process. In brief, in the first place, the conditional variance is estimated via univariate GARCH model for each variable. The next step is to estimate the parameters for the conditional correlation. The DCC model can make the covariance matrix positive definite at any point in time. The shortcoming of the model is that all conditional correlations follow the same dynamic structure.

2.2 Estimation of Multivariate GARCH Models

Let $H_t(\theta)$ be a positive definite $N\times N$ conditional covariance matrix of some $N\times 1$ residual vector ε_t, parameterized by the vector θ. Denoting the available information at time t by ξ_t, we have

$$E_{t-1}[\varepsilon_t \mid \xi_{t-1}] = 0 , \tag{8}$$

$$E_{t-1}[\varepsilon_t \varepsilon_t' \mid \xi_{t-1}] = H_t(\theta) . \tag{9}$$

Generally the conditional covariance matrix $H_t(\theta)$ is well specified based on a certain MGARCH model. Suppose there is an underlying parameter vector θ_0 which one wants to estimate using a given sample of T observations. The quasi maximum likelihood approach estimates θ_0 by maximizing the Gaussian log likelihood function

$$\log L_T(\theta) = -\frac{N \cdot T}{2}\log(2\pi) - \frac{1}{2}\sum_{t=1}^{T}\log|H_t| - \frac{1}{2}\sum_{t=1}^{T}\varepsilon_t' H_t^{-1}\varepsilon_t . \tag{10}$$

2.3 Diagnostics of Multivariate GARCH Models

The check of the adequacy of MGARCH models is essential in identifying whether a well specified MGARCH model can obtain reliable estimates and inference.

Graphical diagnostics for MGARCH models can be fulfilled by examining plots of the sample autocorrelation (ACF) and the sample cross correlation functions (XCF). To ensure the inference from the estimated parameters in the MGARCH model is enough valid, the residuals should be exhibited as a set of white noise with features like expected zero mean vector, no autocorrelations, constant variance, and normal distribution of the residuals.

The autocorrelation and cross correlation functions for the squared process are shown to be useful in identifying and checking time series behavior in the conditional variance equation of the GARCH form.

In the literature, several tests have been developed to test the autocorrelation no matter in univariate form. Box and Pierce derived a goodness-of-fit test, called the portmaneau test.

But still, the fact is that very few tests are adaptable to multivariate models even though there are many diagnostic tests dealing with univariate models.

To summarize, once the model is assumed to catch the dynamics of the time series, the standardized residual $\hat{z}_t = \hat{H}_t^{-1}\hat{\varepsilon}_t$ should satisfy the following conditions [4]:

1) $E(\hat{z}_t \hat{z}_t') = I_N$, (11)

2) $Cov(\hat{z}_{it}^2, \hat{z}_{jt}^2) = 0$, for all pairs of the variable index $i \neq j$, (12)

3) $Cov(\hat{z}_{it}^2, \hat{z}_{j,t-k}^2) = 0$, for $k > 0$. (13)

Testing 1) would find the misspecification in the conditional mean; testing 2) is to verify whether the conditional distribution is Gaussian; the purpose of testing 3) is to check the adequacy of the dynamic specification of H_t even without knowing the validity of the assumption on the distribution of z_t.

3 Construction of Multivariate GARCH Models

The original data is provided by the European Central Bank (ECB) website. It contains daily zero rates of AAA-rated euro area central government bonds, from 01/01/2007 to 30/04/2010. The data of 2007, 2008 and first half of 2009, totally 635 observations, is used to estimate MGARCH models, and the rest data, from Jul/2009, is used to evaluate model forecasting.

With given ZR_{it}, the zero rate at time t, and maturity T, the zero coupon bond price p_{it} is calculated as

$$p_{it} = S \times e^{-ZR_{it} \cdot T} \ , i = 1, 2, 3. \tag{14}$$

where S is the par value, in our case taking the value 100. The daily log return r_t is calculated as follows:

$$r_{it} = \ln(p_{it} / p_{i,t-1}) \ , i = 1, 2, 3. \tag{15}$$

Three variables (*var*1/*var*2/*var*3) correspond to three daily returns with different maturities (6*m*/1*y*/2*y*). Their descriptive statistics are given in Table 1. By viewing the value of kurtosis, one can conclude that the return series all have fat tails relative to the normal distribution, which indicates a much more possibility of extreme movements. Moreover, the result of ARCH effect [5] test proposed by Engle of each return series is given in Table 2, where "*H*" being 1 indicates rejecting of null hypothesis that there is no ARCH effect. One may see that each variable/return has significant ARCH effect.

Table 1. Descriptive Statistics of Return Series with Different Maturities (6*m*/1*y*/2*y*)

	Mean	Max	Min	Std Dev.	Skewness	Kurtosis	Jarque-Bera	Prob
*var*1	0.0023	0.1374	-0.0581	0.0166	2.5324	19.6906	8049.379	0.0000
*var*2	0.0045	0.2190	-0.1944	0.0403	0.3058	6.7716	386.2683	0.0000
*var*3	0.0082	0.3533	-0.4522	0.0979	-0.1392	5.1290	121.9801	0.0000

Table 2. GARCH Effect Testing of Return Series

	var 1		*var* 2		*var* 3	
Lag	H	pValue	H	pValue	H	pValue
1	1	0.0357	1	0.1887×10^{-5}	1	0.2385×10^{-5}
2	1	0.0121	1	0.4053×10^{-5}	1	0.1165×10^{-5}
3	1	0.0000	1	0.0157×10^{-5}	1	0.0279×10^{-5}
4	1	0.0000	1	0.0002×10^{-5}	1	0.0108×10^{-5}
5	1	0.0000	1	0.0006×10^{-5}	1	0.0086×10^{-5}

3.1 Multivariate-GARCH Modeling

As the BEKK-GARCH and DCC-GARCH models are the two most widely used multivariate GARCH models, we will restrict to model the volatility and co-volatility of the three variables by using BEKK and DCC forms.

Next we present the estimated model, and their diagnostics and forecasting are provided in the following subsections.

The estimation process is performed in our case by the econometrics software package RAT 7.0. The optimization algorithm used for the maximum likelihood estimation is BFGS proposed independently by Broyden, Fletcher, Goldfarb and Shanno in 1970. Convergence is assumed to occur if the change in the coefficients to be estimated is less than the convergence criterion option *cvcrit* specified.

The estimated BEKK-GARCH model can be obtained by substituting the following matrices into Equation 3.

$$A = \begin{pmatrix} 0.3400 & 0.1881 & 0.6015 \\ -0.0415 & -0.1082 & 0.5756 \\ 0.0245 & 0.1713 & 0.5036 \end{pmatrix}, \tag{16}$$

$$B = \begin{pmatrix} 1.2262 & 0.3803 & -0.1912 \\ -0.3316 & 0.6331 & 0.1752 \\ 0.0997 & 0.0802 & 0.9035 \end{pmatrix}, \tag{17}$$

$$C = \begin{pmatrix} 6.534\times10^{-4} & 0 & 0 \\ 3.545\times10^{-3} & -4.80\times10^{-7} & 0 \\ 7.689\times10^{-3} & -1.374\times10^{-6} & 1.21\times10^{-7} \end{pmatrix}. \tag{18}$$

The estimated DCC-GARCH model can be obtained by substituting the following numerical information into Equations 4, 5 and 7.

$$h_{11t} = 1.2142\times10^{-6} + 0.2145\varepsilon_{1,t-1}^2 + 0.8259h_{11,t-1} \cdot \tag{19}$$

$$h_{22t} = 4.6008\times10^{-7} + 0.1692\varepsilon_{2,t-1}^2 + 0.8566h_{22,t-1} \cdot \tag{20}$$

$$h_{33t} = -5.1693\times10^{-6} + 0.1502\varepsilon_{3,t-1}^2 + 0.8713h_{33,t-1} \cdot \tag{21}$$

$$Q_t = (1-0.0934-0.8971)\overline{Q} + 0.0934u_{t-1}u_{t-1}' + 0.8971Q_{t-1} \cdot \tag{22}$$

$$\overline{Q} = \begin{pmatrix} 0.9934 & 0.8301 & 0.6979 \\ 0.8301 & 1.01244 & 0.9683 \\ 0.6979 & 0.9683 & 1.0034 \end{pmatrix}. \tag{23}$$

Except for several constant terms, all the other estimated variables are statistically significant.

3.2 Model Diagnostics

Residual-based diagnostics are conducted to test the residual pattern implied by the deviation of the estimated model from underlying assumptions. As the model estimation method employed here is the Gaussian quasi MLE method. One of its assumptions is that the residuals should follow a Gaussian distribution. Hence, to test whether the estimations of the model parameters are robust, we can check whether the residuals of the estimated process are white noise.

Tables 3 and 5 show the testing results of GARCH effect on the standardized residuals of the BEKK model and the DCC model respectively. $H = 0$ represents the acceptance of the null hypothesis that no GARCH effects exist. In contrast with Table 2, we can conclude that GARCH effect has been eliminated quite a lot. The Ljung-Box test results in Tables 4 and 6 based on the autocorrelation plot test the randomness at each distinct lag. $H = 0$ means that we tend to accept the null hypothesis that the series is random.

Table 3. GARCH Effect Testing of each Standardized Residuals of the BEKK Model

	var 1		*var* 2		*var* 3	
Lag	H	pValue	H	pValue	H	pValue
1	0	0.2117	0	0.1537	0	0.8706
2	0	0.3649	0	0.3154	0	0.8924
3	0	0.4380	0	0.5108	0	0.9291
4	1	0.0309	0	0.3234	0	0.6651
5	1	0.0163	0	0.3108	0	0.6329

Table 4. LBQ Test of each Standardized Residuals of the BEKK Model

	var 1		*var* 2		*var* 3	
Lag	H	pValue	H	pValue	H	pValue
1	0	0.9488	0	0.0935	0	0.0507
2	0	0.8372	0	0.1144	0	0.1412
3	0	0.9459	0	0.2220	0	0.2655
4	0	0.9164	0	0.2025	0	0.2158
5	0	0.9477	0	0.2984	0	0.3276

Table 5. GARCH Effect Testing of each Standardized Residuals of the DCC Model

	var 1		*var* 2		*var* 3	
Lag	H	pValue	H	pValue	H	pValue
1	0	0.0500	0	0.4501	0	0.5339
2	0	0.1458	0	0.5963	0	0.6506
3	0	0.2669	0	0.1113	0	0.7538
4	0	0.3477	0	0.1851	0	0.8506
5	0	0.4898	0	0.2481	0	0.9222

Table 6. LBQ Test of each Standardized Residuals of the DCC Model

	var 1		*var* 2		*var* 3	
Lag	H	pValue	H	pValue	H	pValue
1	0	0.1951	0	0.0292	0	0.0830
2	0	0.4152	0	0.0928	0	0.2226
3	0	0.5695	0	0.1779	0	0.3635
4	0	0.2261	0	0.1891	0	0.2889
5	0	0.2339	0	0.2262	0	0.4050

3.3 Comparison of BEKK and DCC Models

The empirical measure of logarithmic daily return variability is called the realized volatility. It is calculated using the subsequent 10 observations on the log-returns in our case. In contrast realized volatility constructed from high-frequency returns with the restrictive parametric multivariate GARCH models, link between realized volatility and the diagonal elements of the conditional covariance matrix would be established [6].

In our case, the estimated volatility follows the dynamics of the realized volatility. Additionally, the volatility clustering and the relation between maturity and volatility can be clearly indicated in line graphs (e.g. Figure 1). And there is a horizontal lag between these two lines for the reason that we calculate the realized volatility by using the next ten observations.

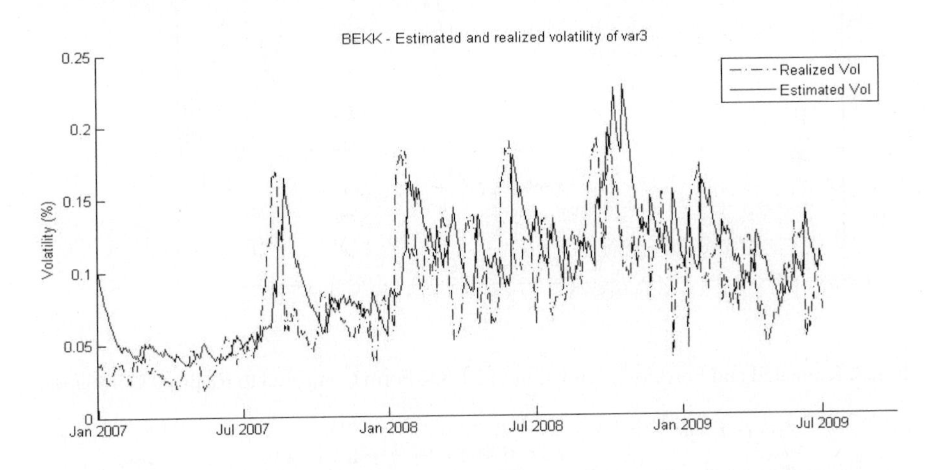

Fig. 1. Estimated Volatility of *var* 3 in the BEKK Model Compared to Realized Volatility

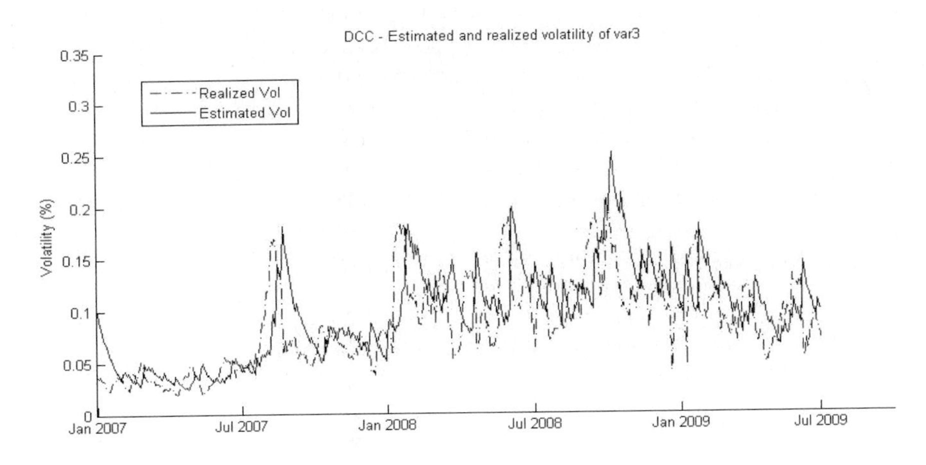

Fig. 2. Estimated Volatility of *var* 3 in the DCC Model Compared to Realized Volatility

Figure 3 presents the poor forecasting performance of the BEKK – GARCH model through the comparison of the realized correlations and the forecast correlations just on the right side of the vertical line. The forecasting performance of the DCC – GARCH model (shown in Figure 4) looks better than that of the BEKK – GARCH model. As the number of parameter estimated by BEKK – GARCH models is much more than that of DCC – GARCH models, the summation of the error accumulated by each parameter of the BEKK form tends to be greater than that of the DCC form.

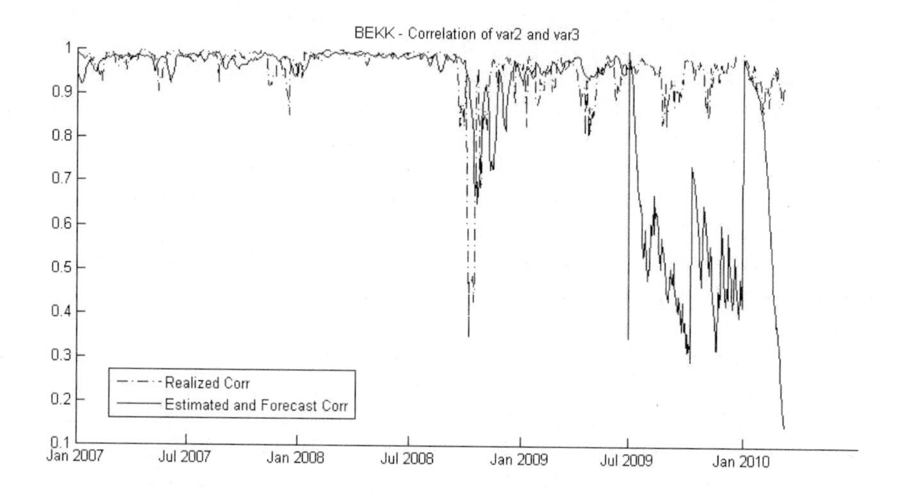

Fig. 3. Estimated and Forecast Correlation of BEKK Form Compared to Realized Correlation

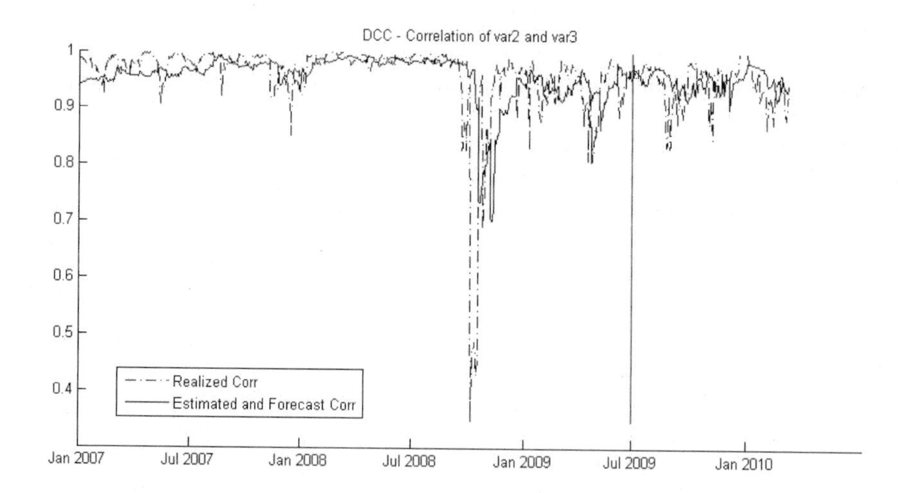

Fig. 4. Estimated and Forecast Correlation of DCC Form Compared to Realized Correlation

The mean absolute error (MAE) [7] can measure how close the estimated variables are to the realized values. It is also called the mean average error. In our case MAE is calculated by

$$MAE_{vi} = \frac{1}{n}.\sum_{k=1}^{n} \left| \sigma_{ik} - \hat{\sigma}_{ik} \right| , \tag{24}$$

for volatility where n is the total number of observations or

$$MAE_{vi} = \frac{1}{n}.\sum_{k=1}^{n} \left| \rho_{ijk} - \hat{\rho}_{ijk} \right| , \tag{25}$$

for correlation where $i, j = 1, 2, 3$.

Table 7. MAE in Volatility and Correlation of the BEKK Model

Average error in volatility		Average error in correlation	
*MAE v*1	0.0056	*MAE*12	0.1317
*MAE v*2	0.0132	*MAE*13	0.1990
*MAE v*3	0.0306	*MAE*23	0.0324

Table 8. MAE in Volatility and Correlation of the DCC Model

Average error in volatility		Average error in correlation	
*MAE v*1	0.0062	*MAE*12	0.1354
*MAE v*2	0.0142	*MAE*13	0.2027
*MAE v*3	0.0309	*MAE*23	0.0352

The values of the mean absolute error between these models suggest that the parameter estimation of the BEKK model is more accurate than that given by the DCC model through the magnitude of the difference between their corresponding MAEs.

4 Conclusion

The research focuses on constructing and diagnosing two formulations of multivariate GARCH models- the BEKK and DCC forms. The estimation process is fulfilled in the software package RATS 7.0 through the maximum likelihood method. As the implementation is conducted under the premise that the residual terms follow the Gaussian distribution, the diagnostics in evaluating the adequacy of modeling is operated by checking whether such assumption is credible enough.

By comparing the goodness of fit through the mean absolute error, we find that the fitting performance of the BEKK – GARCH form is much sounder than DCC – GARCH form in our example. The distinction may due to the relatively more number of parameters in BEKK – GARCH forms compared to DCC – GARCH forms. In this sense, the BEKK – GARCH model can process a well-warranted capability in explaining the information hidden in the history data. On the opposite, the DCC – GARCH model owns an advantage over BEKK – GARCH model in the area of forecasting as the DCC form is more parsimonious than the BEKK form so that the

DCC form won't possibly accumulate errors as much as the BEKK form does. To conclude, it is crucially important to balance parsimony and flexibility when modeling multivariate GARCH models.

Acknowledgments. This research was sponsored by National Natural Science Foundation of China, Ref. Nos. 40730526 and 40701142.

References

1. Bauwens, L., Laurent, S., Rombouts, J.V.K.: Multivariate GARCH Models: A Survey. Journal of Applied Econometrics 21, 79–109 (2006)
2. Hull, J.C.: Options, Futures and other Derivatives. Prentice Hall, New York (2005)
3. Annastiina, S., Timo, T.: Multivariate GARCH Models. SSE/EFI Working Paper Series in Economics and Finance 669 (2008)
4. Bauwens, L., Laurent, S., Rombouts, J.V.K.: Multivariate GARCH Models: A Survey. Journal of Applied Econometrics 21, 79–109 (2006)
5. Walter, E.: Applied Econometric Time Series, 3rd edn. John Wiley & Sons, Inc., Chichester (2009)
6. Andersen, T., Bollerslev, T., Diebold, F.X., Labys, P.: Modeling and forecasting realized volatility. Econometrica 71, 529–626 (2003)
7. Engle, R.F.: Dynamic Conditional Correlation – A Simple Class of Multivariate GARCH Models. Journal of Business and Economic Statistics 20(3), 339–350 (2002)

Adapt the mRMR Criterion for Unsupervised Feature Selection

Junling Xu

School of Computer Science and Engineering,
Southeast University, Nanjing 210096, China
jlxu@seu.edu.cn

Abstract. Feature selection is an important task in data analysis. mRMR is an equivalent form of the maximal statistical dependency criterion based on mutual information for first-order incremental supervised feature selection. This paper presents a novel feature selection criterion which can be considered as the unsupervised version of mRMR. The concepts of relevance and redundancy are both concerned in the feature selection criterion. The effectiveness of the new unsupervised feature selection criterion is confirmed by the theoretical proof. Experimental validation is also conducted on several popular data sets, and the results show that the new criterion can select features highly correlated with the latent class variable.

Keywords: Feature selection, Unsupervised feature selection, Mutual information.

1 Introduction

Data analysis often deals with large data sets containing not only a huge amount of instances but also a significant number of features. Some of the features are redundant, while some are irrelevant and noisy. Therefore, it is an important step to preprocess the data to remove the noisy and redundant features when analyzing high-dimensional data sets. This process is commonly termed as feature selection. Generally, feature selection methods can be categorized as supervised and unsupervised, based on the class information of the data. When class labels of the data are available, supervised feature selection can be used, otherwise, unsupervised feature selection is the right option. Feature selection methods can also be categorized as filter and wrapper, based on their dependence on the learning algorithm that will finally use the selected subset [1]. Filter methods are independent of the learning algorithm, whereas wrapper methods use the learning algorithm as the evaluation function.

The objective of unsupervised feature selection is to select important features which are representative and can characterize the main property of all the original features in the absence of class labels. There have been some works on it [2,3,4,5,6,7,8]. The algorithm described in [2] evaluates the clustering tendency of each feature by an entropy index, which is based on the observation that data

L. Cao, J. Zhong, and Y. Feng (Eds.): ADMA 2010, Part II, LNCS 6441, pp. 111–121, 2010.

with clusters has very different point-to-point distance histogram from data without clusters. In [3], a maximum information compression index is used to measure feature similarity so that feature redundancy is detected. In [4], a forward orthogonal search algorithm by maximizing the overall dependency is proposed to detect significant features and select a subset of all the original features. However, all of these methods are designed to deal with numerical features. In [5], weights are assigned to different feature spaces for k-means clustering based on within-cluster and between-cluster matrices. Feature saliency is integrated in EM algorithm in [6] so that feature selection is performed simultaneously with clustering process. A wrapper criterion for clustering [7], which evaluates the quality of clusters using normalized cluster separability (for k-means) or normalized likelihood (for EM clustering), was proposed by Dy and Brodley. Recently, Li *et al.* adapted their scatter separability criterion to localized feature selection [8]. All of the above-mentioned four methods are all wrapper methods which are computationally very expensive and do not scale well to large datasets.

In order to design a unsupervised-filter feature selection method which can simultaneously deal with numerical and non-numerical features, this paper propose a novel feature selection criterion based on mutual information (UmRMR). The motivation for considering MI is derived from its capability to measure a general dependence between two features. Though mutual information has been used in supervised feature selection [9,10], to the best of our knowledge, it is the first time which has been used as an evaluation measure in unsupervised feature selection. The process of our feature selection is a sequential forward search which ranks features according to UmRMR. Experiments on several popular data sets show the effectiveness of the proposed method.

The rest of the paper is organized as follows. In the next section, we present the details of the proposed feature selection criterion, and reveal the relationship with mRMR. The experimental study, as well as the results, is described in Section 3. Section 4 concludes the paper and outlines directions for future work.

2 Unsupervised Feature Selection Criterion

This section briefly introduces the mRMR feature selection criterion; and then presents the unsupervised feature evaluation criterion UmRMR which can be considered as an unsupervised version of mRMR; finally, the relationships of UmRMR and mRMR are discussed.

2.1 The mRMR Criterion

The process of feature selection used here is a sequential forward search which ranks the features according to an evaluation measure. Assume that at step $m-1$ the set U of unselected features, and a feature subset S_{m-1}, consisting of $m-1$ features has been determined. How should the mth significant feature $\mathbf{y}_m$ be chosen? According to the "minimal-redundancy-maximal-relevance" (mRMR) criterion proposed in [10], the mth feature is selected based on the following formula:

$$\ell_m = arg \max_{1 \leq i \leq n, \mathbf{x}_i \in U} \{I(\mathbf{x}_i; \mathbf{c}) - \frac{1}{m-1} \sum_{\mathbf{x}_j \in S_{m-1}} I(\mathbf{x}_i; \mathbf{x}_j)\}. \quad (1)$$

The first item in (1), the maximal relevance (Max-Relevance) condition, tends to select the feature which has the largest dependency on the target class $\mathbf{c}$. It is likely that features selected according to Max-Relevance could have rich redundancy, i.e., the dependency among these features could be large. When two features highly depend on each other, the respective class-discriminative power would not change much if one of them was removed. Therefore, the minimal redundancy (Min-Redundancy) condition is added to select mutually exclusive features.

2.2 Unsupervised mRMR Criterion (UmRMR)

Since the class information is unavailable in unsupervised feature selection process, we need to give new definitions for relevance and redundancy. We now first define the relevance of a feature.

Definition 1 (Relevance). *The relevance of a feature $\boldsymbol{x}_i$ is its average mutual information to the whole feature set:*

$$Rel(\boldsymbol{x}_i) = \frac{1}{n} \sum_{j=1}^{n} I(\boldsymbol{x}_i; \boldsymbol{x}_j) = \frac{1}{n}(H(\boldsymbol{x}_i) + \sum_{1 \leq j \leq n, j \neq i} I(\boldsymbol{x}_i; \boldsymbol{x}_j)). \quad (2)$$

In the definition of the relevance of a feature, $H(\mathbf{x}_i)$ indicates the information content contained in feature $\mathbf{x}_i$: the larger $H(\mathbf{x}_i)$ is, the more information it can supply the learning algorithm; and $\sum_{1 \leq j \leq n, j \neq i} I(\mathbf{x}_i; \mathbf{x}_j)$ is the amount of information decreased from the information content contained in all the other features due to the knowledge of $\mathbf{x}_i$: the larger $\sum_{1 \leq j \leq n, j \neq i} I(\mathbf{x}_i; \mathbf{x}_j)$ is, the less new information other features can supply the learning algorithm; If we select feature $\mathbf{x}_i$ which has the maximal $Rel(\mathbf{x}_i)$, then it can lead to the loss of information to the least extent.

In order to define the redundancy of a feature, we assume that the $Rel(\mathbf{x}_i)$ of a feature $\mathbf{x}_i$ is proportional to its entropy $H(\mathbf{x}_i)$ (i.e. the relevance value provided by per information unit is a constant $\mathscr{C}_{\mathbf{x}_i}$ for a certain feature $\mathbf{x}_i$, while it may be various for different features). If a feature $\mathbf{x}_i$ in U is considered to be selected, then for any feature $\mathbf{y}_j$ in S_{m-1}, its conditional information content on $\mathbf{x}_i$ is $H(\mathbf{y}_j \mid \mathbf{x}_i)$. Obviously, the information supplied by $\mathbf{y}_j$ decreases due to the existence of $\mathbf{x}_i$, so does the relevance of $\mathbf{y}_j$. Based on our assumption, we have the definition of conditional relevance:

Definition 2 (Conditional Relevance). *The conditional relevance of feature $\boldsymbol{y}_j$ on feature $\boldsymbol{x}_i$ is*

$$Rel(\boldsymbol{y}_j \mid \boldsymbol{x}_i) = \frac{H(\boldsymbol{y}_j \mid \boldsymbol{x}_i)}{H(\boldsymbol{y}_j)} Rel(\boldsymbol{y}_j) = \mathscr{C}_{\boldsymbol{y}_j} H(\boldsymbol{y}_j \mid \boldsymbol{x}_i). \quad (3)$$

As can be seen from (3), the relevance of feature $\mathbf{y}_j$ decreases due to the selection of feature $\mathbf{x}_i$, as $H(\mathbf{y}_j \mid \mathbf{x}_i) \leq H(\mathbf{y}_j)$. So we can define the redundancy of $\mathbf{x}_i$ to $\mathbf{y}_j$ as follows:

Definition 3 (Redundancy). *The redundancy of feature $\boldsymbol{x}_i$ relative to feature $\boldsymbol{y}_j$ is given by*

$$Red(\boldsymbol{x}_i; \boldsymbol{y}_j) = Rel(\boldsymbol{y}_j) - Rel(\boldsymbol{y}_j \mid \boldsymbol{x}_i). \tag{4}$$

To select the mth significant feature $\mathbf{y}_m$, both the relevance of the feature to all the original feature set and the redundancy of the feature to the already-selected features should be considered, we define the feature selection criterion UmRMR as follows:

$$\ell_m = arg \max_{1 \leq i \leq n, \mathbf{x}_i \in U} \{Rel(\mathbf{x}_i) - \max_{\mathbf{y}_j \in S_{m-1}} Red(\mathbf{x}_i; \mathbf{y}_j)\} \tag{5}$$

or

$$\ell_m = arg \max_{1 \leq i \leq n, \mathbf{x}_i \in U} \{Rel(\mathbf{x}_i) - \frac{1}{m-1} \sum_{\mathbf{y}_j \in S_{m-1}} Red(\mathbf{x}_i; \mathbf{y}_j)\}. \tag{6}$$

The mth significant feature can be selected as $\mathbf{y}_m = \mathbf{x}_{\ell_m}$, which decreases the uncertainty about other features with a higher percentage, compared with other single feature in the feature set U, and brings little redundant information.

2.3 Relationships of UmRMR and mRMR

The experiments in [10] showed that the mRMR incremental selection scheme provides a better way to maximize the dependency of the selected features and the target class. But for unsupervised feature selection, the class distribution underlying the data sets is unknown, can we still maximize the dependency of the selected features and the underlying target class without the information about the class? After studying the relationship of UmRMR and mRMR, we found that UmRMR can solve this problem to some extent. First we will show the relationship between the two definitions of relevance.

Proposition 1. *The relevance of $\boldsymbol{x}_i$ in UmRMR is a lower bound of the relevance condition of mRMR under the naive bayes assumption, i.e. $Rel(\boldsymbol{x}_i) \leq I(\boldsymbol{x}_i; \boldsymbol{c})$.*

Proof: Let $\mathbf{c}$ be the target class of instances, then it can be viewed as a function of features $\mathbf{x}_1, \cdots, \mathbf{x}_n$. Because $\mathbf{x}_1, \cdots, \mathbf{x}_n$ can be seen as random variables, so is $\mathbf{c}$. All of these variables here are assumed to be of discrete type, otherwise, they will be discretized first. We have

$$H(\mathbf{x}_i|\mathbf{c}) = \sum_{c} \sum_{x_i} P(x_i, c) log \frac{1}{P(x_i|c)}$$

$$= \sum_{c} \sum_{x_i} \sum_{x_j} P(x_i, x_j) \frac{P(x_i, x_j, c)}{P(x_i, x_j)} log \frac{1}{P(x_i|c)}$$

$$= \sum_{x_i}\sum_{x_j} P(x_i, x_j) \sum_c \frac{P(x_i, x_j, c)}{P(x_i, x_j)} log \frac{1}{P(x_i|c)}$$

$$\leq \sum_{x_i}\sum_{x_j} P(x_i, x_j) log[\sum_c \frac{P(x_i, x_j, c)}{P(x_i, x_j)} \frac{1}{P(x_i|c)}] \tag{7}$$

$$= \sum_{x_i}\sum_{x_j} P(x_i, x_j) log[\frac{1}{P(x_i, x_j)} \sum_c \frac{P(x_i, x_j, c)}{P(x_i|c)}]$$

$$= \sum_{x_i}\sum_{x_j} P(x_i, x_j) log[\frac{1}{P(x_i|x_j)} \frac{1}{P(x_j)} \sum_c \frac{P(x_i, x_j, c)}{P(x_i|c)}]$$

$$= \sum_{x_i}\sum_{x_j} P(x_i, x_j)[log \frac{1}{P(x_i|x_j)} + log \sum_c \frac{P(x_i, x_j, c)}{P(x_i|c)P(x_j)}]$$

$$\leq H(\mathbf{x}_i|\mathbf{x}_j) + log \sum_{x_i}\sum_{x_j}\sum_c \frac{P(x_i, x_j)P(x_i, x_j, c)}{p(x_j)P(x_i|c)}. \tag{8}$$

(7) and (8) are attained due to the using of the Jensen's Inequality. The naive bayes assumption supposes that the features are not independent but conditionally independent given the value of $\mathbf{c}$. Under the naive bayes assumption, $P(x_i, x_j, c) = P(c)P(x_i|c)P(x_j|c)$, substituting for $P(x_i, x_j, c)$ in (8) we get

$$H(\mathbf{x}_i|\mathbf{c})$$

$$\leq H(\mathbf{x}_i|\mathbf{x}_j) + log \sum_{x_i}\sum_{x_j}\sum_c \frac{P(x_i, x_j)P(c)P(x_i|c)P(x_j|c)}{p(x_j)P(x_i|c)}$$

$$= H(\mathbf{x}_i|\mathbf{x}_j) + log \sum_{x_i}\sum_{x_j} P(x_i, x_j) \sum_c \frac{P(x_j, c)}{p(x_j)}$$

$$= H(\mathbf{x}_i|\mathbf{x}_j) + log \sum_{x_i}\sum_{x_j} P(x_i, x_j)$$

$$= H(\mathbf{x}_i|\mathbf{x}_j). \tag{9}$$

According to (2) and (9),

$$Rel(\mathbf{x}_i) = \frac{1}{n}\sum_{j=1}^{n} I(\mathbf{x}_i; \mathbf{x}_j)$$

$$= \frac{1}{n}\sum_{j=1}^{n}(H(\mathbf{x}_i) - H(\mathbf{x}_i|\mathbf{x}_j))$$

$$\leq \frac{1}{n}\sum_{j=1}^{n}(H(\mathbf{x}_i) - H(\mathbf{x}_i|\mathbf{c}))$$

$$= I(\mathbf{x}_i; \mathbf{c}). \tag{10}$$

Proposition 1 means that maximizing $Rel(\mathbf{x}_i)$ can increase the value of $I(\mathbf{x}_i; \mathbf{c})$ to some extent.

The redundancy of mRMR is defined as the dependance of the the feature to be selected and the already selected features. From (4), It seems that our definition of redundance only considered the relevance and conditional relevance of the selected features, which is different from the definition in the mRMR criterion. Now we will show the relationship between the two definitions of redundancy.

Proposition 2. *The redundancy of $\boldsymbol{x}_i$ relative to $\boldsymbol{y}_j$ in UmRMR equals the redundancy in mRMR times a constant.*

Proof

$$\begin{aligned} Red(\mathbf{x}_i;\mathbf{y}_j) &= Rel(\mathbf{y}_j) - Rel(\mathbf{y}_j \mid \mathbf{x}_i) \\ &= Rel(\mathbf{y}_j) - \frac{H(\mathbf{y}_j \mid \mathbf{x}_i)}{H(\mathbf{y}_j)} Rel(\mathbf{y}_j) \\ &= \frac{I(\mathbf{x}_i,\mathbf{y}_j)}{H(\mathbf{y}_j)} Rel(\mathbf{y}_j) \\ &= \mathscr{C}_{\mathbf{y}_j} I(\mathbf{x}_i,\mathbf{y}_j). \end{aligned} \quad (11)$$

3 Experimental Results

In this section, we first test whether UmRMR can select features highly correlated with the latent class. Then we test the effectiveness of UmRMR on improving the performance of the learning algorithm. Finally we compare UmRMR with other two unsupervised feature selection criteria. The process of feature selection used here is a sequential forward search which ranks the features according to UmRMR, so we call this algorithm as SFS-UmRMR. Data sets used here are all taken from the UCI machine learning repository [11], they originally are either with discrete features or continuous features. For continuous features they were discretized by supervised method provided in [12] before the feature selection process.

3.1 Can SFS-UmRMR Select Features Highly Correlated with the Latent Classes?

Here we compare the orderly list of features produced by SFS-UmRMR and IG (information gain), a supervised feature selection method which is widely employed in machine learning. For each of the seven data sets shown in Table 1 (ecoli, iris, lymph, dermatology, breast-w, spambase and haberman), two orderly list of features were generated as shown in Table 2. The **bold** numbers are features with the same order in the two lists or features occurred in both lists within the top given number of features. It can be seen from Table 2 that important features certificated by IG can often be ranked in front by SFS-UmRMR, though the results attained by SFS-UmRMR did not using the class information as IG (except in the process of discretization).

Table 1. List of data sets

Data Set	Number of Features	Number of Instances	Number of Classes
ecoli	7	336	8
iris	4	150	2
lymph	18	148	4
dermatology	34	366	6
breast-w	9	699	2
spambase	57	4601	2
haberman	3	306	2

Table 2. Feature ranking by IG and SFS-UmRMR

Data Set	Ranking by IG	Ranking by SFS-UmRMR
ecoli	{**6,7,1,2,5,3,4**}	{**6,7,1,2,5,3,4**}
iris	{**3,4,1,2**}	{**3,4,1,2**}
lymph	{**13**,*18*,**15,14**,*2*,**10**...}	{**14,13,10**,*12*,**15**,*5*...}
dermatology	{**21,20,22**,*33*,*29*,**27**...}	{**20,27,21**,*16*,**22**,*9*...}
breast-w	{**2,3,6,7,5**...}	{**2,7,3,5,6**...}
spambase	{**52,53,56**,7,**21**...}	{57,**56,53,21,52**...}
haberman	{**3,2,1**}	{**3,2,1**}

3.2 How Effective Is SFS-UmRMR?

To demonstrate the efficacy of our algorithm from another view, we test SFS-UmRMR on more data sets to inspect its effectiveness. Since data sets, of which the features are all non-numerical are relatively few, the data sets used here originally are either with discrete features or continuous features. For continuous features they are discretized by supervised method provided in [12] before the feature selection process. Eight data sets are considered here (the information about the data sets is shown in Table 3). The desired number of features is not given, so the output of SFS-UmRMR is an orderly list of features. A wrapper method with k-nearest-neighbor (k-NN) algorithm is then used to seek the minimal feature subset, which can attain the classification accuracy provided by the complete data, and the optimal feature subset, which can attain the best classification accuracy. The classification accuracy is calculated by performing the 10-fold cross-validation procedure, and then the average classification accuracy of 10 runs of the k-NN algorithm is calculated. Features are selected one by one to form a subset according to their order in the output of SFS-UmRMR. The value of k, in the k-NN rule, is chosen by performing many experiments for different values of k, where $1 \leq k \leq \sqrt{N_{tr}}$ and N_{tr} is the number of the samples in the training set, and k is chosen as the one that gives the best classification performance.

A minimal feature subset and an optimal feature subset for each of the eight data sets including ionosphere, zoo, sponge, sonar, glass, arrhythmia, lung-cancer, and vote are selected. The numbers of features in the minimal subsets and in the optimal subsets for the eight data sets are {10, 12, 17, 22, 5, 2, 3, 2} and {10, 12, 32, 54, 6, 48, 5, 7} respectively. A comparison between the classification accuracy based on the complete data and the reduced data for the eight data sets is reported in Table 4. It can be seen that the classification accuracy based on the selected subsets outperformes those based on the complete data. This means that the selected feature subsets are representative and informative and thus can be used to replace the complete data for pattern classification.

Table 3. List of data sets

Data Set	Number of Features	Number of Instances	Number of Classes
ionosphere	33	351	2
zoo	17	101	7
sponge	45	76	3
sonar	60	208	2
glass	9	214	7
arrhythmia	279	452	16
lung-cancer	56	32	2
vote	16	435	2

Table 4. Classification accuracy over the complete data and reduced data

	No. of Features			Accuracy(%)		
Dataset	C	M	O	C	M	O
ionosphere	33	10	10	89.77±0.71{2}	90.57±1.14{2}	90.57±1.14{2}
zoo	17	12	12	96.14±0.50{1}	98.02±0.00{1}	98.02±0.00{1}
sponge	45	17	32	92.50±0.66{3}	92.63±0.66{2}	93.68±1.32{2}
sonar	60	22	54	86.44±1.44{1}	86.88±1.92{1}	88.08±2.64{2}
glass	9	5	6	70.00±2.10{3}	75.75±2.34{1}	77.57±0.94{1}
arrhythmia	279	2	48	59.27±0.55{6}	59.60±1.33{4}	67.61±0.55{3}
lung-cancer	56	3	5	79.69±1.56{3}	85.00±1.56{1}	88.44±1.56{2}
vote	16	2	7	93.15±0.46{4}	95.17±0.00{1}	95.59±0.92{1}

C/M/O: Complete/Minimal/Optimal Data. {}: the value of k used in k-NN rule

3.3 Comparison with Other Feature Selection Methods

We compare the performance of SFS-UmRMR with two unsupervised feature selection methods ENTROPY [2] and FOS-MOD [4]. Now, we first give a brief introduction of the two feature selection methods.

ENTROPY (entropy-based ranking) is proposed by Dash and Liu in [2]. The entropy is defined as the equation: $E(\mathbf{x}) = -\sum_{i=1}^{N}\sum_{j=1}^{N}(S_{i,j} \times logS_{i,j}) + (1 - S_{i,j}) \times log(1 - S_{i,j})$, where $S_{i,j}$ is the similarity value between the ith and the jth instances. $S_{i,j}$ is defined as the equation: $S_{i,j} = e^{\alpha \times dist_{i,j}}, \alpha = -ln(0.5)/\overline{dist}$, where $dist_{i,j}$ is the distance between the ith and the jth instances after the feature $\mathbf{x}$ is removed, $\overline{dist}$ is the average distance among the instances after the feature $\mathbf{x}$ is removed. Features are ranked in an ascending order according to the value of E.

FOS-MOD is an unsupervised forward orthogonal search algorithm. In this method, features are selected in a stepwise way, one at a time, by estimating the capability of each specified candidate feature subset to represent the overall features in the measurement space. The dependency between features is measured by a squared correlation function. The squared-correlation coefficient between feature $\mathbf{x}$ and $\mathbf{y}$ is $sc(\mathbf{x}, \mathbf{y}) = (\mathbf{x}^T\mathbf{y})^2/[(\mathbf{x}^T\mathbf{x})(\mathbf{y}^T\mathbf{y})]$. The concept of relevance in FOS-MOD is very similar to (2), where mutual information is replaced with squared-correlation coefficient. The concept of redundancy in FOS-MOD is implicity considered by an orthogonalization process, that is, the relevance of the mth significant feature is computed after it is orthogonalized with the $m-1$ previous selected features.

Table 5. List of data sets

Data Set	Number of Features	Number of Instances	Number of Classes
liver-disorders	6	345	2
glass	9	214	7
segment	19	2310	7
sonar	60	208	2
vehicle	18	846	4
ionosphere	33	351	2

The six data sets used here are all with continuous features (the information about the data sets is shown in Table 5). For SFS-UmRMR, data sets are discretized with simple equal-width binning where the number of bins is chosen automatically by maximizing the likelihood via leave-one-out cross-validation before the feature selection process. First, all of the three feature selection methods are applied on each data set without giving the desired number of features, and thus we can attain three orderly lists of features for each data set; then features are selected one by one to form a subset according to their order in the output of each algorithm. k-NN algorithm is used to attain the classification accuracy with different sizes of the reduced feature subset for each data set, and the best classification performance for a certain k is chosen. A comparison in terms of the k-NN classification accuracy for different sizes of the reduced feature subset is shown in Fig. 1.

As can be seen from Fig. 1, SFS-UmRMR outperforms the other two methods almost on all of the six data sets. SFS-UmRMR can attain as good performance as the complete data when the feature number is relatively small, while the other two methods seem to prefer more features. Noticeably, the three classification accuracy curves have distinct tendencies. For SFS-UmRMR, as the feature number increases, the accuracy constantly increases and then converges at some point or declines. On the contrary, the accuracies for ENTROPY and FOS-MOD almost constantly increase as more and more features are used, indicating that redundant features or even irrelevant features are thought to be important by the two methods.

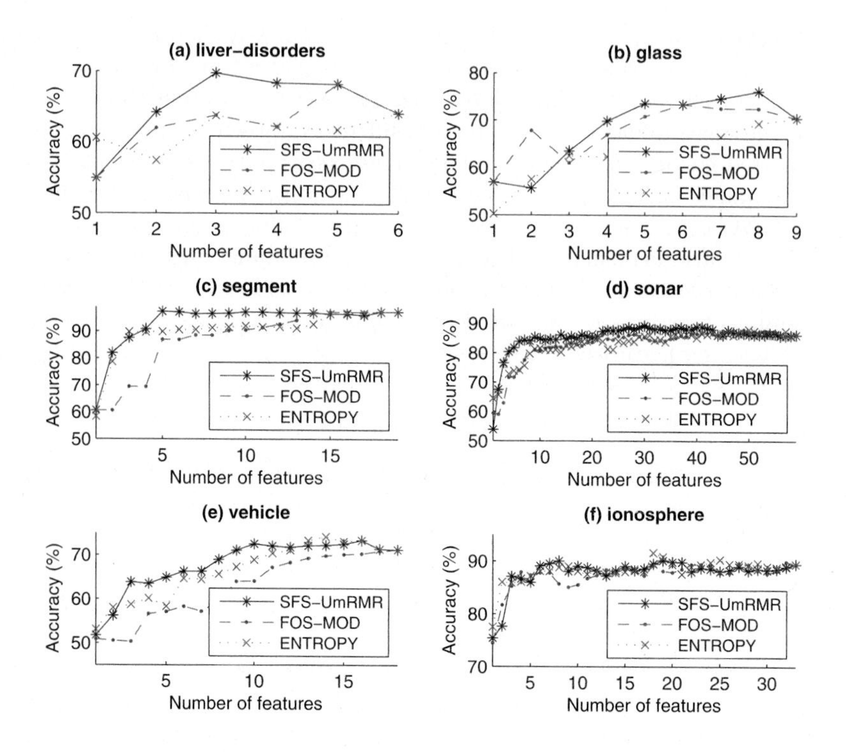

Fig. 1. Performance comparison on several data sets

4 Conclusion

In cases of the class information of the data set is unavailable or lost, unsupervised feature selection is a hard work. In this paper, we have proposed a novel unsupervised feature selection criterion UmRMR based on mutual information where the concepts relevance and redundancy of features are both defined from a standpoint of information theory. We also discussed the relationship between UmRMR with the famous mRMR criterion, which can be seen as a theoretical proof for the effectiveness of UmRMR. Experimental result also conformed that UmRMR can select features highly correlated with the latent class.

Unlike the method proposed in [4] which assumes a linear relationship exists between sample features, UmRMR can deal with general relationship between features. In many cases, where features are not linked by linear relationship, UmRMR can be competent for the feature selection task. In the future, we will compare the performance of UmRMR with more unsupervised feature selection methods on more data sets.

Acknowledgments. This work is supported by the National Natural Science Foundation of China (Grant No. 60503033, 60703086).

References

1. Langley, P.: Selection of Relevant Features in Machine Learning. In: AAAI Fall Symposium on Relevance, pp. 1–5. AAAI Press, New Orleans (1994)
2. Dash, M., Choi, K., Scheuermann, P., Liu, H.: Feature Selection for Clustering - a Filter Solution. In: 2nd IEEE International Conference on Data Mining, pp. 115–122. IEEE Press, Washington (2002)
3. Mitra, P., Murthy, C.A., Pal, S.K.: Unsupervised Feature Selection Using Feature Similarity. IEEE Trans. Pattern Analysis and Machine Intelligence 24, 301–312 (2002)
4. Wei, H.L., Billings, S.A.: Feature Subset Selection and Ranking for Data Dimensionality Reduction. IEEE Trans. Pattern Analysis and Machine Intelligence 29, 162–166 (2007)
5. Modha, D.S., Spangler, W.S.: Feature Weighting in k-Means Clustering. Machine Learning 52, 217–237 (2003)
6. Law, M.H.C., Figueiredo, M.A.T., Jain, A.K.: Simultaneous Feature Selection and Clustering Using Mixture Models. IEEE Trans. Pattern Analysis and Machine Intelligence 26, 1154–1166 (2004)
7. Dy, J.G., Brodley, C.E.: Feature Selection for Unsupervised Learning. Journal of Machine Learning Research 5, 845–889 (2004)
8. Li, Y.H., Dong, M., Hua, J.: Localized Feature Selection for Clustering. Pattern Recognition Letters 29, 10–18 (2008)
9. Battiti, R.: Using Mutual Information for Selecting Features in Supervised Neural Net Learning. IEEE Trans. Neutral Networks 5, 537–550 (1994)
10. Peng, H., Long, F., Ding, C.: Feature Selection Based on Mutual Information: Criteria of Max-Dependency, Max-Relevance, and Min-Redundancy. IEEE Trans. Pattern Analysis and Machine Intelligence 27, 1226–1238 (2005)
11. Blake, C., Merz, C.: UCI Repository of Machine Learning Database, `http://www.ics.uci.edu/~mlearn/MLRepository.html`
12. Fayyad, U.M., Irani, K.B.: Multi-interval Discretization of Continuous-valued Attributes for Classification Learning. In: 13th International Joint Conference on Articial Intelligence, pp. 1022–1027. Morgan Kaufmann, Chambery (1993)

Evaluating the Distance between Two Uncertain Categorical Objects

Hongmei Chen, Lizhen Wang*, Weiyi Liu, and Qing Xiao

Department of Computer Science and Engineering, School of Information Science and Engineering, Yunnan University, Kunming 650091, China
lzhwang@ynu.edu.cn

Abstract. Evaluating distances between uncertain objects is needed for some uncertain data mining techniques based on distance. An uncertain object can be described by uncertain numerical or categorical attributes. However, many uncertain data mining algorithms mainly discuss methods of evaluating distances between uncertain numerical objects. In this paper, an efficient method of evaluating distances between uncertain categorical objects is presented. The method is used in nearest-neighbor classifying. Experiments with datasets based on UCI datasets and the plant dataset of "Three Parallel Rivers of Yunnan Protected Areas" verify the method is efficient.

Keywords: Uncertain data mining, Uncertain categorical object, Expected semantic distance, Nearest-neighbor classifying.

1 Introduction

Data uncertainty is an inherent property in various applications, such as location-based services and sensor monitoring, due to measurement inaccuracy, sampling discrepancy, outdated data sources, or other errors. When data mining techniques are applied to these data, their uncertainty has to be considered to obtain high quality results [3].

Uncertain data mining has recently attracted interests from researchers. Many traditional data mining algorithms are extended to handle uncertain data [1, 3-12].

Evaluating distances between uncertain objects is needed for some uncertain data mining techniques based on distance, such as clustering analysis, outlier detection, K-nearest-neighbors classifying, and top-K queries.

An uncertain object can be described by uncertain numerical or categorical attributes. However, many uncertain data mining algorithms mainly discuss methods of evaluating distances between uncertain numerical objects [3-9].

In this paper, we will discuss how to efficiently evaluate distances between uncertain categorical objects.

Usually, an uncertain categorical object can be characterized by probability distribution vectors over domains of uncertain categorical attributes [1, 9, 11]. Distances between categorical values in the domain of an uncertain categorical attribute can be

* Corresponding author.

L. Cao, J. Zhong, and Y. Feng (Eds.): ADMA 2010, Part II, LNCS 6441, pp. 122–133, 2010.

evaluated by semantic distances [2]. So, distances between uncertain categorical objects can be expressed by expected semantic distances. Further, the computation of expected semantic distances can be optimized.

Generally, the main contributions of this paper can be summarized as follows:

- An efficient method of evaluating distances between uncertain categorical objects is presented.
- The method is used in nearest-neighbor classifying.
- Experiments with datasets based on UCI datasets and the plant dataset of "Three Parallel Rivers of Yunnan Protected Areas" verify the method is efficient.

The rest of the paper is organized as follows: Section 2 introduces related works. Section 3 describes the method of evaluating distances between uncertain categorical objects. Section 4 gives the application of the method in nearest-neighbor classifying. Section 5 shows the results of experiments. Section 6 concludes the paper.

2 Related Works

Methods of evaluating distances between uncertain numerical objects have been developed [3-9]. However, methods about uncertain categorical objects have received little attention.

Some methods to compare uncertain categorical objects are investigated in [1].

The first is computing the probability that uncertain categorical objects are equal by probability distribution vectors over domains of uncertain categorical attributes. This method is naïve. This will be discussed in subsection 3.3.

The second is computing distances between probability distribution vectors. This method gives the similarity of probability distribution vectors, and does not consider the similarity of values between uncertain categorical objects.

These methods are not adequate in some cases. For example, there are three uncertain objects o_1, o_2, o_3 with an uncertain categorical attribute *grade*. The domain of *grade* is {*excellent, good, pass, fail*}. Their probability distribution vectors over the domain of *grade* are $o_1.P$=(0.9, 0.1, 0, 0), $o_2.P$=(0, 0, 0.9, 0.1) and $o_3.P$=(0, 0, 0.1, 0.9). According to the first or second method, we cannot get which is closest to o_1 among o_2 and o_3. In fact, we can know o_1 is closest to o_2 than o_3 from semantic point of view.

This paper will discuss how to evaluate distances between uncertain categorical objects from semantic point of view.

3 Evaluating Distances between Uncertain Categorical Objects

3.1 Expected Semantic Distances between Uncertain Categorical Objects

Definition 1. An attribute A_i is called an uncertain categorical attribute, if its domain is a set of categorical values $D^{A_i} = \{d_1^{A_i}, ..., d_{N^{A_i}}^{A_i}\}$, and the attribute value of an object o_u is characterized by the probability distribution vector $o_u.P^{A_i} = (o_u.p_1^{A_i}, ..., o_u.p_{N^{A_i}}^{A_i})$ over D^{A_i}, where $\sum_{s=1}^{N^{A_i}} o_u.p_s^{A_i} \leq 1$.

Definition 2. An object is called an uncertain categorical object, if it is described by one uncertain categorical attribute at least.

For the sake of simplicity, consider a set of uncertain categorical objects $O=\{o_1,\ldots,o_n\}$, in which o_u is described by a set of uncertain categorical attributes $A=\{A_1,\ldots,A_m\}$, and $\sum_{s=1}^{N^{A_i}} o_u.p_s^{A_i} = 1(1 \le u \le n, 1 \le i \le m)$, although there is no such restriction.

Example 1. Distributions of four plants over eight locations in the Three Parallel Rivers of Yunnan Protected Areas are shown in Table 1.

Table 1. Uncertain categorical objects *plant* with an uncertain categorical attribute *location*

plant	*location*							
	X.G.L.L.	*D.Q.*	*L.J.*	*H.P.*	*L.S.*	*G.S.*	*D.L.*	*Y.L.*
o_1	0.7	0.1	0.1	0	0	0.1	0	0
o_2	0	0	1	0	0	0	0	0
o_3	0.1	0.4	0.1	0	0	0.4	0	0
o_4	0	0	0.2	0	0	0	0.6	0.2

Generally, distances between categorical values in the domain of an uncertain categorical attribute can be evaluated from semantic point of view, though it can not be measured directly. For example, categorical values *excellent, good, pass, fail* can be ordered, i.e., *excellent* is best, *fail* is worst, and *good* is better than *pass*. Then, *good* is closest to *excellent* than *pass* or *fail*, and so on. Distances between categorical values in the domain of *location* can be also evaluated from different aspects. According to regionalism, *X.G.L.L.* is closet to *D.Q.*, and so on. According to other criterions, such as latitude, longitude, altitude, precipitation rain fall, exposed to the sun, or shady, distances between them can be reevaluated.

When getting semantic distances between categorical values, distances between probability distribution vectors can be expressed by expected semantic distances. Further, expected semantic distances between uncertain categorical objects can be defined.

Definition 3. The expected semantic distance between $o_u.P^{A_i}$ and $o_v.P^{A_i}$ is defined as:

$$esd(o_u.P^{A_i}, o_v.P^{A_i}) = \sum_{s,t=1}^{N^{A_i}} o_u.p_s^{A_i} \times o_v.p_t^{A_i} \times sd(d_s^{A_i}, d_t^{A_i}), \tag{1}$$

Where $sd(d_s^{A_i}, d_t^{A_i})$ is the semantic distance between $d_s^{A_i}$ and $d_t^{A_i}$.

Definition 4. The expected semantic distance between o_u and o_v is defined as:

$$ESD(o_u, o_v) = \sqrt[q]{\sum_{i=1}^{m} (esd(o_u.P^{A_i}, o_v.P^{A_i}))^q}. \tag{2}$$

When q=1, $ESD(o_u, o_v)$ is the Manhattan distance. When q=2, $ESD(o_u, o_v)$ is the Euclidean distance. When q>2, $ESD(o_u, o_v)$ is the Minkowski distance.

3.2 Computing Expected Semantic Distances

The first step of computing expected semantic distances between uncertain categorical objects is getting semantic distances between categorical values. Generally, this can be done by domain experts. There are some methods to evaluate semantic distances between categorical values, too [2]. Two simple methods are used in this paper.

Firstly, if categorical values can be ordered from semantic point of view, they will be numbered according to their orders. Semantic distances between categorical values can be evaluated according to their numbers.

Example 2. Categorical values *excellent, good, pass, fail* can be ordered. So they will be numbered by 1, 2, 3, 4 respectively. Semantic distances *sd(excellent, excellent)*=|1-1|=0, *sd(excellent, good)*=|1-2|=1, and so on.

Secondly, if categorical values can not be ordered, the method which is similar to that in [2] will be adopted. Base on the concept hierarchy tree of an attribute, the semantic proximity between two concepts or values is presented in [2]. Similarly, semantic distances between categorical values can be defined based on the concept hierarchy tree.

Definition 5. [2] The depth *d(nd)* of a node *nd* in a concept hierarchy tree *T* is the length of the path from the root to *nd*. The height *h(nd)* is the length of the longest path from *nd* to leaves in the sub-tree whose root is *nd*. The height *h(T)* of *T* is the height of the root. The level *l(nd)* is *l(nd)*= *h(T)*- *d(nd)*.

Definition 6. The nearest common ancestor of two nodes nd_1 and nd_2 in a concept hierarchy tree *T* is denoted by $a(nd_1, nd_2)$. The semantic distance between nd_1 and nd_2 is defined as:

$$\begin{cases} sd(nd_1, nd_2) = 0 & nd_1 = nd_2 \\ sd(nd_1, nd_2) = l(a(nd_1, nd_2)) & nd_1 \neq nd_2 \end{cases}. \quad (3)$$

Example 3. Review Example 1. According to regionalism, the concept hierarchy tree of *location* is shown in Fig.1. Semantic distances between categorical values in the domain of *location* are shown in Table 2.

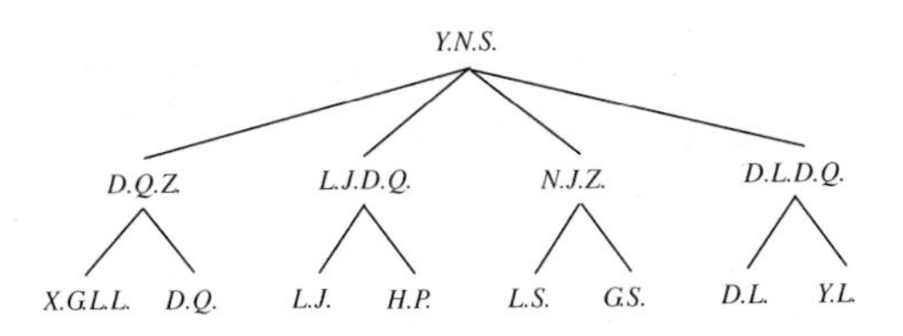

Fig. 1. The concept hierarchy tree of *location*

Table 2. Semantic distances between categorical values in the domain of *location*

sd	*X.G.L.L.*	*D.Q.*	*L.J.*	*H.P.*	*L.S.*	*G.S.*	*D.L.*	*Y.L.*
X.G.L.L.	0	1	2	2	2	2	2	2
D.Q.	\	0	2	2	2	2	2	2
L.J.	\	\	0	1	2	2	2	2
H.P.	\	\	\	0	2	2	2	2
L.S.	\	\	\	\	0	1	2	2
G.S.	\	\	\	\	\	0	2	2
D.L.	\	\	\	\	\	\	0	1
Y.L.	\	\	\	\	\	\	\	0

The second step of computing expected semantic distances between uncertain categorical objects is computing expected semantic distances between probability distribution vectors.

The domain of an uncertain categorical attribute is a finite set. The maximum number of different semantic distances between categorical values is $C_{N^A}^2+1=\frac{N^A(N^A-1)}{2}+1$. In fact, according to methods used in this paper, the maximum number is N^A (the first method) or $h(T)$+1 (the second method). So definition 3 can be rewritten as definition 7.

Definition 7. Let the number of different semantic distances between categorical values be nsd^{A_i}, the order be $0=sd_0^{A_i}<sd_1^{A_i}<...<sd_{nsd^{A_i}-1}^{A_i}$. The expected semantic distance between $o_u.P^{A_i}$ and $o_v.P^{A_i}$ is defined as:

$$esd(o_u.P^{A_i},o_v.P^{A_i})=\sum_{r=0}^{nsd^{A_i}-1}o_{u,v}.p_r^{A_i}\times sd_r^{A_i}=\sum_{r=1}^{nsd^{A_i}-1}o_{u,v}.p_r^{A_i}\times sd_r^{A_i}, \tag{4}$$

where $o_{u,v}.p_r^{A_i}=\sum_{sd(d_s^{A_i},d_t^{A_i})=sd_r^{A_i}}o_u.p_s^{A_i}\times o_v.p_t^{A_i}$.

According to definition 7, the key of computing $esd(o_u.P^{A_i},o_v.P^{A_i})$ is computing $o_{u,v}.p_r^{A_i}$. When computing $o_{u,v}.p_r^{A_i}$, an index table t^{A_i} of A_i can be used to index $o_u.p_s^{A_i}$ and $o_v.p_t^{A_i}$ which are related with $o_{u,v}.p_r^{A_i}$. All pairs (s,t) which satisfy $sd(d_s^{A_i},d_t^{A_i})=sd_r^{A_i}$ are stored in the rth linked list of t^{A_i}.

Example 4. Review Example 1. The index table of *location* is shown in Fig.2. Expected semantic distances between probability distribution vectors, which are also expected semantic distances between uncertain categorical objects, are shown in Table 3.

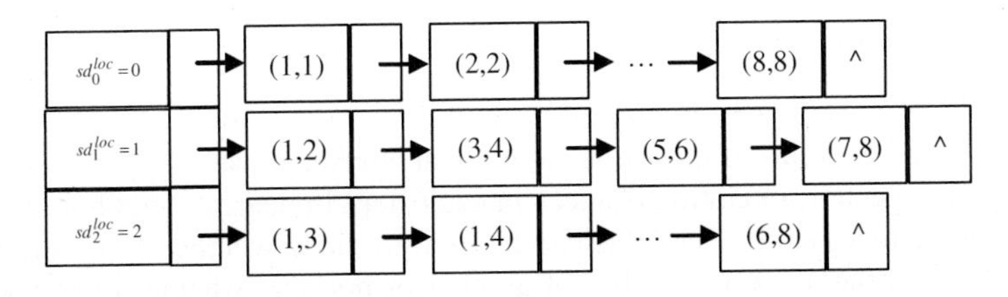

Fig. 2. The index table of *location*(1:*X.G.L.L*,2:*D.Q.*,3:*L.J.*,4:*H.P.*,5:*L.S.*,6:*G.S.*,7:*D.L.*,8:*Y.L.*)

Table 3. Expected semantic distances between probability distribution vectors and expected semantic distances between uncertain categorical objects

esd(*ESD*)	o_1	o_2	o_3	o_4
o_1	0.82	1.8	1.39	1.96
o_2	\	0	1.8	1.6
o_3	\	\	1.24	1.96
o_4	\	\	\	0.88

3.3 Discussion

Above expected semantic distances between uncertain categorical objects have some properties.

Firstly, the expected semantic distance between $o_u.P^{A_i}$ and itself may be not zero. Further, it may be not smaller than the expected semantic distance between $o_u.P^{A_i}$ and $o_v.P^{A_i}$.

Secondly, when values of categorical objects are certain, the expected semantic distance is still applicable because certain values $o_u.d_s^{A_i}$ can be characterized by probability distribution vectors $o_u.P^{A_i} = (o_u.p_s^{A_i} = 1)$.

Thirdly, though two simple methods of evaluating semantic distances between categorical values are used in this paper, it is difficult to get the order of categorical values or the concept hierarchy tree of an uncertain categorical attribute sometime. A naïve method is to suppose all semantic distances between different categorical values are same. Then the expected semantic distance between $o_u.P^{A_i}$ and $o_v.P^{A_i}$ is decided by $\sum_{s=1}^{N^{A_i}} o_u.p_s^{A_i} \times o_v.p_s^{A_i}$. This is the first method mentioned in Section 2.

4 Nearest-Neighbor Classifying on Uncertain Categorical Objects

Classification is one of the most important tasks in data mining. Nearest-neighbor classifier computes distances between an instance and train samples, and selects the label of the nearest sample as the label of the instance. Now, we extend nearest-neighbor classifying on uncertain categorical objects, where the label of an uncertain categorical object is certain.

4.1 Optimization

When computing expected semantic distances between uncertain categorical objects according to index tables of uncertain categorical attributes, we can get a better lower boundary of expected semantic distances in each step. In nearest-neighbor classifying on uncertain categorical objects, this can be used to judge whether an uncertain categorical object is the nearest-neighbor of another or not, and whether remainder steps continue or not.

Theorem 1. After computing $o_{u,v}.p_k^{A_i} = \sum_{sd(d_s^{A_i}, d_t^{A_i}) = sd_k^{A_i}} o_u.p_s^{A_i} \times o_v.p_t^{A_i}$ according to the kth linked list of t^{A_i} in the kth step, the current lower boundary of the expected semantic distance between $o_u.P^{A_i}$ and $o_v.P^{A_i}$ is $\sum_{r=0}^{k} o_{u,v}.p_r^{A_i} \times sd_r^{A_i} + (1 - \sum_{r=0}^{k} o_{u,v}.p_r^{A_i}) \times sd_{k+1}^{A_i}$, i.e., $esd(o_u.P^{A_i}, o_v.P^{A_i}) \geq \sum_{r=0}^{k} o_{u,v}.p_r^{A_i} \times sd_r^{A_i} + (1 - \sum_{r=0}^{k} o_{u,v}.p_r^{A_i}) \times sd_{k+1}^{A_i}$, where $0 \leq k < nsd^{A_i} - 1$.

Proof:

$$esd(o_u.P^{A_i}, o_v.P^{A_i}) - (\sum_{r=0}^{k} o_{u,v}.p_r^{A_i} \times sd_r^{A_i} + (1 - \sum_{r=0}^{k} o_{u,v}.p_r^{A_i}) \times sd_{k+1}^{A_i})$$

$$= \sum_{r=0}^{nsd^{A_i}-1} o_{u,v}.p_r^{A_i} \times sd_r^{A_i} - \sum_{r=0}^{k} o_{u,v}.p_r^{A_i} \times sd_r^{A_i} - (1 - \sum_{r=0}^{k} o_{u,v}.p_r^{A_i}) \times sd_{k+1}^{A_i}$$

$$= \sum_{r=k+1}^{nsd^{A_i}-1} o_{u,v}.p_r^{A_i} \times sd_r^{A_i} - \sum_{r=k+1}^{nsd^{A_i}-1} o_{u,v}.p_r^{A_i} \times sd_{k+1}^{A_i}$$

$$= \sum_{r=k+1}^{nsd^{A_i}-1} o_{u,v}.p_r^{A_i} \times (sd_r^{A_i} - sd_{k+1}^{A_i})$$

$$\geq 0$$

□

According to Theorem 1, we can get the current lower boundary of the expected semantic distance between o_u and o_v after getting all current lower boundaries of the expected semantic distance between $o_u.P^{A_i}$ and $o_v.P^{A_i}$ $(1 \leq i \leq m)$ in the kth step.

Example 5. Review Example 1. Suppose we already got $ESD(o_1, o_3)$=1.39. Now, we judge which is the nearest-neighbor of o_1 among o_2, o_3, and o_4.

k=0: $o_{1,2}.p_0^{loc} = 0.1$, $o_{1,4}.p_0^{loc} = 0.02$

$ESD(o_1, o_2) = esd(o_1.P^{loc}, o_2.P^{loc}) \geq (1 - o_{1,2}.p_0^{loc}) \times sd_1^{loc} = (1 - 0.1) \times 1 = 0.9$

$ESD(o_1, o_4) = esd(o_1.P^{loc}, o_4.P^{loc}) \geq (1 - o_{1,4}.p_0^{loc}) \times sd_1^{loc} = (1 - 0.02) \times 1 = 0.98$

k=1: $o_{1,2}.p_1^{loc} = 0$, $o_{1,4}.p_1^{loc} = 0$

$$ESD(o_1,o_2)=esd(o_1.P^{loc},o_2.P^{loc})$$
$$\geq o_{1,2}.p_1^{loc}\times sd_1^{loc}+(1-o_{1,2}.p_0^{loc}-o_{1,2}.p_1^{loc})\times sd_2^{loc}=0\times 1+(1-0.1-0)\times 2=1.8$$
$$ESD(o_1,o_4)=esd(o_1.P^{loc},o_4.P^{loc})$$
$$\geq o_{1,4}.p_1^{loc}\times sd_1^{loc}+(1-o_{1,4}.p_0^{loc}-o_{1,4}.p_1^{loc})\times sd_2^{loc}=0\times 1+(1-0.02-0)\times 2=1.96$$

At this time, we can judge o_3 is the nearest-neighbor of o_1 among o_2, o_3, and o_4, and it is not needed to continue remainder steps.

4.2 Algorithm

Algorithm: the nearest-neighbor classifying algorithm on uncertain categorical objects using expected semantic distances
Input: an uncertain categorical object o; n uncertain categorical objects $o_1,\ldots,o_n$; m index tables $t^{A_1},\ldots,t^{A_m}$ of m uncertain categorical attributes $A_1,\ldots,A_m$
Output: the label of o
Steps:

1. For i=1 to m do

$$esd(o.P^{A_i},o_1.P^{A_i})=\sum_{r=1}^{nsd^{A_i}-1} o_{,1}.p_r^{A_i}\times sd_r^{A_i} \qquad \text{//according to index table } t^{A_i}$$

2. Computing $ESD(o, o_1)$ //according to definition 4
3. $ESD_{\min} = ESD(o, o_1)$
4. o.label = o_1.label
5. For u = 2 to n do

5.1 For i=1 to m do

$$esd(o.P^{A_i},o_u.P^{A_i})=0$$

$o_{,u}.p^{A_i}=0$ // $o_{,u}.p^{A_i}$ is $\sum_{r=0}^{k} o_{,u}.p_r^{A_i}$. Initially, $o_{,u}.p^{A_i}$ is 0.

5.2 For k=0 to $\operatorname*{Max}_{i=1}^{m}(nsd^{A_i}-1)$ do

5.2.1 For i=1 to m do

If $k\leq(nsd^{A_i}-1)$ then

$$o_{,u}.p_k^{A_i}=\sum_{sd(d_s^{A_i},d_t^{A_i})=sd_k^{A_i}} o.p_s^{A_i}\times o_u.p_t^{A_i} \qquad \text{//according to index table } t^{A_i}$$

$$esd(o.P^{A_i},o_u.P^{A_i})=esd(o.P^{A_i},o_u.P^{A_i})+o_{,u}.p_k^{A_i}\times sd_k^{A_i}$$

$$o_{,u}.p^{A_i}=o_{,u}.p^{A_i}+o_{,u}.p_k^{A_i}$$

5.2.2 Computing $ESD(o, o_u)$ //according to definition 4

5.2.3 If $ESD(o,o_u)\geq ESD_{\min}$ then Break

5.2.4 For i=1 to m do

If $k<(nsd^{A_i}-1)$ then

$$lowerboundary\_esd(o.P^{A_i},o_u.P^{A_i})=esd(o.P^{A_i},o_u.P^{A_i})+(1-o_{,u}.p^{A_i})\times sd_{k+1}^{A_i}$$

Else

$$lowerboundary\_esd(o.P^{A_i}, o_u.P^{A_i}) = esd(o.P^{A_i}, o_u.P^{A_i})$$

5.2.5 Computing $LowerBoundary\_ESD(o,o_u)$ //according to definition 4

5.2.6 If $LowerBoundary\_ESD(o,o_u) \geq ESD_{\min}$ then Break

5.3 If $k > \operatorname{Max}_{i=1}^{m}(nsd^{A_i} - 1)$ and $ESD(o, o_u) < ESD_{\min}$ then

$ESD_{\min} = ESD(o, o_u)$

o.label = o_u.label

The time complexity of Algorithm is $O(mn \operatorname{Max}_{i=1}^{m} nsd^{A_i})$. Optimization of Algorithm lies in steps 5.2.4, 5.2.5 and 5.2.6.

5 Experiments

Goals of experiments are to evaluate the effect and the efficiency of the expected semantic distance instead of the naïve method, which is the first method mentioned in section 2, when evaluating distances between uncertain categorical objects in nearest-neighbor classifying on uncertain categorical objects.

Codes are written in C#. Data are stored in Access. Programs are executed on the machine with a 2 GHz Core 2 Duo CPU, a 1 GB Memory, and running the operating system of windows XP.

Firstly, we generate uncertain datasets based on UCI datasets [13] and the plant dataset of "Three Parallel Rivers of Yunnan Protected Areas". For a precise object o_u^{pre}, we generate an uncertain object o_u, where $o_u.p_t^{A_i} = p$ and $\sum_{s \neq t} o_u.p_s^{A_i} = 1 - p$ if A_i of o_u^{pre} is $d_t^{A_i}$, according to given the parameter $p(0.5<p<1)$ [11].

The description of datasets is showed in Table 4. Uncertain datasets are coded as *no.-p*, where *no.* is the no. of datasets, *p* is the parameter.

Table 4. The description of datasets

No.	Dataset	Number of attribute	Number of instance
1	Balance-scale	4	625
2	Car	6	1728
3	Hayes-roth	4	132
4	Plant	8	10000
5	Tic-tac-toe	9	958

Secondly, we construct manually index tables of uncertain categorical attributes according to methods used in this paper. We suppose categorical values can be ordered except that in No. 5 dataset, and give two strategies about index tables in No.5 dataset: the first is $sd(b,b)=sd(o,o)=sd(x,x)=0$, $sd(o,x)=sd(x,o)=1$, $sd(b,o)=sd(o,b)=sd(b,x)=sd(x,b)=2$ (labeled by 1 in Table 5); the second is $sd(b,b)=sd(o,o)=sd(x,x)=0$, $sd(b,o)=sd(o,b)=sd(b,x)=sd(x,b)=1$, $sd(o,x)=sd(x,o)=2$ (labeled by 2 in Table 5).

Table 5. Results of experiments about the classification accuracy

No. of uncertain dataset	Algorithm (using expected semantic distances)	another algorithm (using the naïve method)
1-0.6*	70.56%	26.08%
1-0.7*	70.88%	26.08%
1-0.8*	77.08%	27.68%
1-0.9*	78.56%	26.08%
2-0.6	67.72%	61.28%
2-0.7	67.02%	58.78%
2-0.8	68.06%	59.13%
2-0.9	66.03%	63.48%
3-0.6	66.92%	65.38%
3-0.7	66.92%	66.15%
3-0.8	69.23%	65.38%
3-0.9	69.23%	66.15%
4-0.6	52.65%	51.55%
4-0.7	52.3%	51.85%
4-0.8	53.05%	51.3%
4-0.9	52.95%	51.8%
5-0.6*	76.75%[1] 100%[2]	67.75%
5-0.7*	66.58%[1] 100%[2]	49.95%
5-0.8*	67.75%[1] 100%[2]	49.95%
5-0.9*	82.62%[1] 100%[2]	50.06%

Thirdly, we compare the classification accuracy of Algorithm (using expected semantic distances) with another algorithm (using the naïve method) by experiments with uncertain datasets. In experiments, we adopt the five-fold cross validation method. Results of experiments are showed in Table 5.

In Table 5, we can see the classification accuracy of Algorithm (using expected semantic distances) is higher than that of another algorithm (using the naïve method). Especially, advantages are obvious in 1-*ps* and 5-*ps* (labeled by *). We can also see the classification accuracy can be improved if semantic distances between categorical values can be well evaluated from 5-*ps* (labeled by 1 and 2).

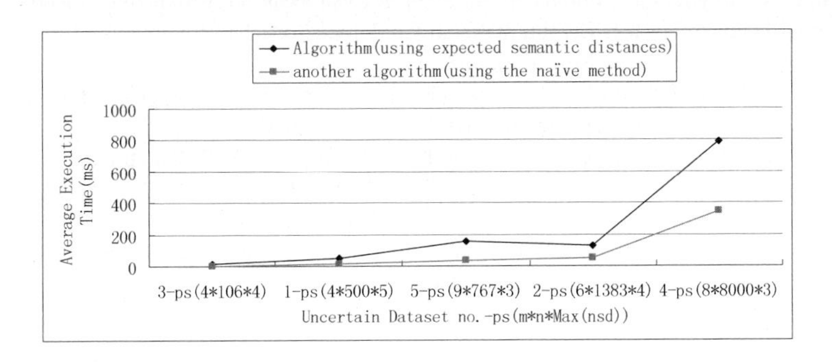

Fig. 3. Results of experiments about the time complexity

Fourthly, we compare the time complexity of Algorithm (using expected semantic distances) with another algorithm (using the naïve method) by experiments with uncertain datasets. Results of experiments are showed in Fig.3. Due to sizes of uncertain datasets *no.-ps* based on a dataset *no.* are same, only average execution times are given in Fig.3. And average execution times are ordered by sizes of uncertain datasets.

Fig. 3 shows the execution time of Algorithm (using expected semantic distances) is more than that of another algorithm (using the naïve method), but it is reasonable.

6 Conclusions

Evaluating distances between uncertain categorical objects is needed for some uncertain data mining techniques based on distance. In this paper, we present a method based on the expected semantic distance. We give methods of evaluating semantic distances between categorical values, index tables of uncertain categorical attributes to compute expected semantic distances between probability distribution vectors, and the method of optimizing computation of expected semantic distances between uncertain categorical objects in nearest-neighbor classifying on uncertain categorical objects. Experiments with datasets based on UCI datasets and the plant dataset of "Three Parallel Rivers of Yunnan Protected Areas" verify the method is effective and efficient.

This paper also leaves some research issues, for example, how to evaluate semantic distances between categorical values, how to apply the method to other data mining technologies. We will attempt to address the above issues in our future works.

Acknowledgments. This research is supported by the National Natural Science Foundation of China(No. 61063008), the Research Foundation of the Educational Department of Yunnan Province(No. 09Y0048) and the Research Foundation of Yunnan University(No. 2009F29Q).

References

1. Singh, S., Mayfield, C., Prabhakar, S., Shah, R., Hambrusch, S.: Indexing Uncertain Categorical Data. In: 23th IEEE International Conference on Data Engineering, pp. 616–625. IEEE Computer Society, New York (2007)
2. Wang, L.Z.: A Method of the Abstract Generalization on the Basis of the Semantic Proximity. Chinese Journal of Computers 10, 1114–1121 (2000)
3. Chau, M., Cheng, R., Kao, B.: Uncertain Data Mining: A New Research Direction. In: The Workshop on the Sciences of the Artificial, Taiwan (2005)
4. Ngai, W.K., Kao, B., Chui, C.K., Cheng, R., Chau, M., Yip, K.Y.: Efficient Clustering of Uncertain Data. In: 6th IEEE International Conference on Data Mining, pp. 436–445. IEEE Computer Society, New York (2006)
5. Lee, S.D., Kao, B., Cheng, R.: Reducing UK-means to K-means. In: 7th IEEE International Conference on Data Mining Workshops, pp. 483–488. IEEE Computer Society, New York (2007)

6. Kao, B., Lee, S.D., Cheung, D.W., Ho, W.S., Chan, K.F.: Clustering Uncertain Data Using Voronoi Diagrams. In: 8th IEEE International Conference on Data Mining, pp. 333–342. IEEE Computer Society, New York (2008)
7. Aggarwal, C.C., Yu, P.S.: Outlier Detection with Uncertain Data. In: 8th SIAM International Conference on Data Mining, pp. 483–493. SIAM, Philadelphia (2008)
8. Beskales, G., Soliman, M.A., Llyas, I.F.: Efficient Search for the Top-k Probable Nearest Neighbors in Uncertain Databases. In: Very Large Data Base, pp. 326–339. ACM, New York (2008)
9. Aggarwal, C.C., Yu, P.S.: A Survey of Uncertain Data Algorithms and Applications. IEEE Transactions On Knowledge And Data Engineering 21, 609–623 (2009)
10. Leung, C.K.-S., Carmichael, C.L., Hao, B.: Efficient Mining of Frequent Patterns from Uncertain Data. In: 7th IEEE International Conference on Data Mining, pp. 489–494. IEEE Computer Society, New York (2007)
11. Qin, B., Xia, Y., Prabhakar, S., Tu, Y.: A Rule-Based Classification Algorithm for Uncertain Data. In: 25th IEEE International Conference on Data Engineering, pp. 1633–1640. IEEE Computer Society, New York (2009)
12. Wang, L.Z., Zhou, L.H., Lu, J., Yip, J.: An Order-clique-based Approach for Mining Maximal Co-locations. Information Sciences 179, 3370–3382 (2009)
13. UCI Machine Learning Repository, `http://archive.ics.uci.edu/ml/`

Construction Cosine Radial Basic Function Neural Networks Based on Artificial Immune Networks

YongJin Zeng and JianDong Zhuang

College of Computer Science and Technology, Jimei University, 361021, XiaMen, China
jameszyj@jmu.edu.cn

Abstract. In this paper, we propose a novel Intrusion Detection algorithm utilizing both Artificial Immune Network and RBF neural network. The proposed anomaly detection method using multiple granularities artificial immune network algorithm to get the candidate hidden neurons firstly, and then, we training a cosine RBF neural network base on gradient descent learning process. The principle interest of this work is to benchmark the performance of the proposed algorithm by using KDD Cup 99 Data Set, the benchmark dataset used by IDS researchers. It is observed that the proposed approach gives better performance over some traditional approaches.

Keywords: Intrusion Detection Algorithm, RBF Neural Network, Multiple Granularities Immune Network.

1 Introduction

At its heart network intrusion detection is a discrimination problem. Radial basis function (RBF) neural networks have received considerable applications in nonlinear approximation and pattern classification. It is generally believed that RBFNNs are inferior to feed forward neural networks (FFNNs) in terms of their accuracy and generalization ability.

RBFNNs are often trained in practice by hybrid learning algorithms. Such learning algorithms employ a supervised scheme for updating the weights that connect the RBFs with the output units and an unsupervised clustering algorithm for determining the centers of the RBFs, which remain fixed during the supervised learning process. Alternative learning algorithms relied on forward subset selection methods, such as the orthogonal least squares (OLS) algorithm [1]. The relationship between the performance of RBFNNs and their size motivated the development of network construction and/or pruning procedures for autonomously selecting the number of RBFs [2-5]. The problems of determining the number, shapes, and locations of the RBFs are essentially related to and interact with each other. Solving these problems simultaneously was attempted by developing a multi-objective evolutionary algorithm [6].

An alternative set of approaches to training RBFNNs relied on gradient descent to update all their free parameters [7]. This approach reduces the development of reformulated RBFNNs to the selection of admissible generator functions that determine the form of the RBFs. Linear generator functions of a special form produced cosine RBFNNs, that is, a special class of reformulated RBFNNs constructed by cosine

L. Cao, J. Zhong, and Y. Feng (Eds.): ADMA 2010, Part II, LNCS 6441, pp. 134–141, 2010.

RBFs. Cosine RBFs have some attractive sensitivity properties, which make them more suitable for gradient descent learning than Gaussian RBFs. An alternative set of approaches to training RBFNNs relied on gradient descent to update all their free parameters. Training RBFNNs by a fully supervised learning algorithm based on gradient descent is sensitively dependent on the properties of the RBFs.

In this paper we proposed a novel algorithm, it firstly use multiple granularities artificial immune network to find the candidate hidden neurons, then it refine the RBF neural network with all candidate hidden neurons and employ preserving criterion to remove some redundant hidden neurons. This new algorithm takes full advantage of the class label information and starting with a small neural network; hence it is likely to be more efficient and is except to generalize well.

This paper is organized as follows. In Section II, multiple granularities immune network algorithm is developed to get candidate hidden neurons. In Section III, a cosine RBF neural network training process is introduced. Experiment studies about the network intrusion detection are presented in Section IV, and concluding remarks are given in Section V.

2 Multiple Granularities Immune Network Algorithm

In order to get the neurons of the hidden layer of RBF network efficiently, it could utilize usually some clustering algorithm such as K-Means, SOM and AIN (artificial immune network). Here we employ a variation of AIN algorithm to construct the original hidden layer of RBF network. The original AIN method is an unsupervised algorithm [9,10], so it is difficult to confirm the optimal number of the neuron based on the class label information. The most problem of original AIN algorithm for hidden layer is that it is a computation under the same granularity but the classification is under different granularities. In this section, we give a multiple granularities AIN algorithm for hidden neurons. Immune clone operation, immune mutation operation and immune suppression operation are defined in [9]. The multiple granularities immune network algorithm (MGIN) is described as following:

Input : data set $\boldsymbol{X}$, and the descend factor a of granularity

Output : the candidate hidden neurons, $\boldsymbol{H}$

Step1: Calculation the radius r of the dataset hyper sphere, and let $\boldsymbol{r}$ be the immune suppression parameter, let $H = \Phi$, let $X' = X$.

Step2: Construct artificial immune network $\boldsymbol{M}$ based on X' ;

Step3: Let $X' = \Phi$, Let $\boldsymbol{M}$ be the cluster centers and partition the samples based on Gaussian radial basis function and width parameter is its suppression parameter $\boldsymbol{r}$. If partition $\boldsymbol{i}$ contains only one class data points and $\boldsymbol{M_i}$ is the center of it, let $R = R \cup M_i$; otherwise add the data points of partition $\boldsymbol{i}$ into X' .

Step4: If $X' \neq \Phi$, let $r = r \times a$ and go to Setp2; otherwise return $\boldsymbol{H}$ as hidden neurons and stop.

Algorithm 1. MGIN for candidate hidden neurons

The entire algorithm1 has time complexity $O(m * N)$ [10], where N is the number of training samples and m is the maximum size of $\boldsymbol{H}$. According to the property of $\boldsymbol{H}$, a neighborhood classifier could be built based the hidden neurons $\boldsymbol{H}$, where the distance function is Gaussian radial basis function.

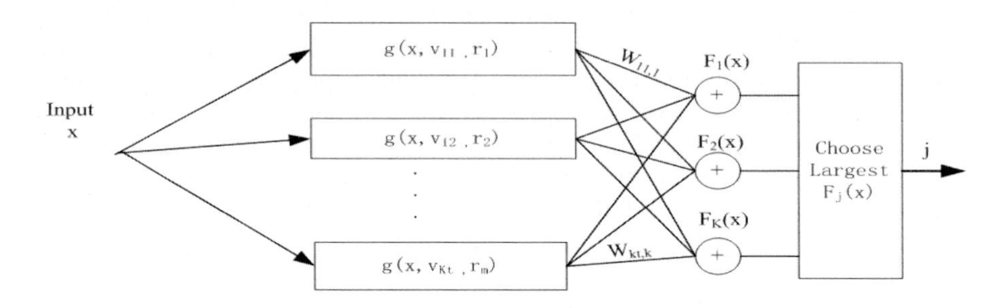

Fig. 1. RBF network architecture

Theorem 1: Let $\boldsymbol{V}$ be the centers of a neighborhood classifier, then a RBF network classifier can be constructed based on $\boldsymbol{V}$.

Proof: suppose the number of classes is $\boldsymbol{K}$, and the number of output neurons of the RBF network classifier is $\boldsymbol{K}$, the data point of $\boldsymbol{V}$ is m.

We construct a RBF network classifier as fig.1. Let v_j be a neuron of the hidden layer and be one center of class $\boldsymbol{i}$, let $W_{j,t}$ be the weight between neuron v_j and output neuron $\boldsymbol{t}$, and let $W_{j,t} = \begin{cases} 1 & \text{, if class}(v_j) = t \\ -1, & \text{otherwise} \end{cases}$. We will prove that this RBF network can classify the data correctly.

Let x_i be an arbitrarily sample of the dataset, which's class label is k, and $d_{ij} = \| x_i - v_j \|$, $ds_i = \arg\min_{class(v_j)=k} (\| x_i - v_j \|)$, $dd_i = \arg\min_{class(v_j)\neq k} (\| x_i - v_j \|)$, $\Delta_i = dd_i - ds_i$, $\Delta = \arg\min_{i=1,\ldots,n}(\Delta_i)$.

Hence Δ may be seen as the minimum separation between different classes in the nearest neighbor classification.

$$F_k(x_i) = \sum_{(\forall v_j)class(v_j)=k} e^{-d_{ij}/\sigma} - \sum_{(\forall v_f)class(v_f)\neq k} e^{-d_{if}/\sigma}$$

$$F_k(x_i) \geq e^{-ds_i/\delta} - \sum_{(\forall v_f)class(v_f)\neq k} e^{-d_{if}/\sigma} \geq e^{-ds_i/\delta} - (m-1)e^{-dd_i/\delta}$$

$$= e^{-ds_i/\delta}(1-(m-1)e^{-(dd_i-ds_i)/\delta}) \geq e^{-ds_i/\delta}(1-(m-1)e^{-\Delta/\delta})$$

If the width parameter δ of the radial basis function is satisfied $\delta \leq (\nabla / \lg(m-1))$, then $F_k(x_i) > 0$ and $F_f(x_i) < 0$ when $f \neq k$. According to the class decision criterion, the output class label must be *k*.

After the RBF network classifier has been constructed, we can employ gradient descent learning process to training a cosine RBFNN and remove some redundant neurons in the next section.

3 Training Cosine RBF Neural Network

Consider an RBFNN with inputs from R^m, c RBFs and $\boldsymbol{K}$ output units. Let $v_j \in R^m$ be the prototype that is center of the *j*th RBF and $w_i = [\mathrm{w}_{i1}, \mathrm{w}_{i2}, ..., \mathrm{w}_{ic}]^{\mathrm{T}}$ be the vector containing the weights that connect the *i*th output unit to the RBFs. Define the sets $V = \{v_i\}$ and $W = \{w_j\}$ and let also $A = \{a_i\}$ be a set of free parameters associated with the RBFs. An RBFNN is defined as the function $N : R^m \rightarrow R^n$ that maps $\mathbf{x} \in R^m$ to $N(V, W, A; \mathbf{x})$, such that

$$\prod_i N(V,W,A;\mathbf{x}) = f(\sum_{j=1}^{c} w_{ij} g_j(\left\|\mathbf{x} - v_j\right\|^2) + w_{i0}) \tag{1}$$

where $f(x) = 1/(1+e^{-x})$ used in this paper, g_j is represents the response of the RBF centered at the prototype v_j. Using this notation, the response of the *i*th output unit to the input x_k is

$$\tilde{y}_{i,k} = \prod_i N(x_k) = f(\sum_{j=1}^{c} w_{ij} g_{j,k} + w_{i0}) \tag{2}$$

where $g_{j,k}$ represents the response of the RBF centered at the prototype v_j to the input vector x_k. Unlike the traditional RBFNN using the exponential functions, in this paper, we using a cosine function [7] for $g_{j,k}$ is

$$g_{i,k} = a_j / (\left\|x_k - v_j\right\|^2 + a_j^{\ 2})^{1/2} \tag{3}$$

Cosine RBFNNs can be trained by the original learning algorithm, which was developed by using “stochastic” gradient descent to minimize [14]

$$E_k = 1/2 \sum_{i=1}^{n} (\tilde{y}_{i,k} - y_{i,k})^2 \tag{4}$$

for $k = 1,2,...,M$. For sufficiently small values of the learning rate, sequential minimization of E_k, leads to a minimum of the total error $E = \sum_{k=1}^{m} E_k$. After an example

(x_k, y_k) is presented to the RBFNN, the new estimate $w_{i,k}$ of each weight vector w_i, is obtained by incrementing its current estimate by the amount $\Delta w_{i,k} = -\beta \nabla_{w_i} E_k$, where β is the learning rate.

$$w_{i,k} = w_{i,k-1} + \Delta w_{i,k} = w_{i,k-1} + \beta\, g_{i,k}\, \tilde{y}_{i,k} (1 - \tilde{y}_{i,k})(y_{i,k} - \tilde{y}_{ik}) \tag{5}$$

The new estimate $a_{j,k}$ of each reference distance a_j, can be obtained by incrementing its current estimate by the amount as $\Delta a_{j,k} = -\beta \partial E_k / \partial a_j$ [7].

$$\begin{aligned} a_{i,k} &= a_{i,k-1} + \Delta a_{i,k} = a_{i,k-1} + \beta g_{j,k} (1 - g_{j,k}^2) \varepsilon_{j,k}^h / a_{i,k-1} \\ \varepsilon_{j,k}^h &= (g_{j,k}^3 / a_j^2) \sum_{i=1}^{c} f'(\tilde{y}_{i,k})(y_{i,k} - \tilde{y}_{i,k}) w_{i,j} \end{aligned} \tag{6}$$

According to (3), the jth cosine RBF can be eliminated during the training process if its reference distance a_j approaches zero.

Hence we can get new algorithm to training RBF classifier. We use the multiple granularities artificial immune network algorithm to get the candidate hidden neurons firstly, and then, we training a cosine RBF neural network base on gradient descent learning process descried in this section.

4 Experiment Results

In this section, we choose the 1999 KDD intrusion detection contest dataset to test the new classifier. The 1999 KDD intrusion detection contest used 1998 DARPA intrusion detection dataset to construct the connection records and extract the object features. 1998 DARPA intrusion detection dataset was acquired from nine weeks of raw TCP dump data for a local-area network (LAN) simulating a typical U.S. Air Force LAN and peppered with four main categories of attacks: DoS, Probe, U2R, R2L. A connection record is a sequence of TCP packets starting and ending at some well defined times, between which data flows to and from a source IP address to a target IP address under some well defined protocol. Each connection is labeled as either normal, or as an attack, with exactly one specific attack type. For each TCP/IP connection, 41 various quantitative and qualitative features were extracted.

4.1 Data Preparing and Normalization

Ten new data sets are set up to perform the algorithm. Each data set contains 1900 normal instances and 100 intrusion instances, all of which are selected at random from the normal data set and abnormal data set respectively.

There is a problem when processing instances whose different features are on different scales. This will cause bias toward some features over other ones. To solve this problem, in this paper these raw data sets are normalized as follows:

$$f_j = \frac{1}{m}\sum_{i=1}^{m} x_{ij} \tag{7}$$

$$\delta_j = \sqrt{\frac{1}{m-1}\sum_{i=1}^{m}(x_{ij} - f_j)} \tag{8}$$

$$x_{ij}' = (x_{ij} - f_j)/\delta_j \tag{9}$$

where f_j denotes the average feature instance of X_j, δ_j denotes the standard deviation feature instance of X_j, and x_{ij}' denotes the feature of the normalized instance. Here, all features are equally weighted in order to enhance the algorithm's generality.

4.2 Discussion and Comparison with Other Methods

Several different kinds of classification methods are compared with our proposed MGIN based RBF network classifier on the KDD intrusion detection data set. In the experiment, we compare this new method with the traditional OLS RBF classifier and BP neural network classifier.

Table 1. Experimental Results for the new approach

a (Descend Factor for AIN)	0.5	0.6	0.75	0.8	0.85	0.95	0.98
Initial RDFs Count c_{ini}=H	89	86	78	64	63	53	53
Final RDFs Count c_{fin}	57	54	53	52	51	50	51
Detection Rate(%)	97.26	97.33	97.48	97.26	97.14	97.28	97.17
False positive (%)	0.31	0.33	0.32	0.22	0.21	0.18	0.19

The BP algorithm were trained with n_h hidden neurons, n_h was varied from 10 to 50, and the maximum training cycles is 5000. The width parameter of radial function is the most important to the OLS RBF classifier; it varied from 1 and 4 in this paper and the maximum RBFs is 100. For the new algorithm, the learning rate used for updating the output weights and prototypes of the cosine RBFNNs was $\beta = 0.01$, the descend factor of the granularity for the AIN was varied from $a = 0.5$ to $a = 0.98$ and maximum adaptation cycles is 100.

The results from the table 1 shows that, in despite of the variant of the descend factor influence the number of the original RBFNNs, this classifier keeps a good performance that the average detection rate is higher than 96.78% and the false positive is lower than 0.4%. Also it is showed that the false positive is slight down with the descend factor become bigger, and it means that the MGIN could find the better prototypes for the dataset with a longer training process.

Table 2. Experimental Results for BP

n_h **(hidden neurons)**	**10**	**20**	**30**	**35**	**40**	**45**	**50**
Detection Rate(%)	98.21	98.23	98.35	98.26	99.10	99.25	99.16
False positive (%)	0.46	0.36	0.31	0.35	0.32	0.26	0.31

Table 3. Experimental Results for OLS RBF

δ **(Width Parameter for RBF)**	**1.0**	**1.5**	**2.0**	**2.5**	**3.0**	**3.5**	**4.0**
Detection Rate(%)	86.26	85.85	86.07	90.7	91.11	92.44	94.04
False positive (%)	0.22	0.34	0.54	0.84	0.52	0.65	0.71

According to testing results, we found that, the BP network has the best result at the most time for the false positive rate and detection rate. However, the network structure of BP neural network is difficult to be determined for the higher dimensional pattern classification problems and cannot be proved to converge well. Also it found that the detection rate of the new classifier is increased obviously than the traditional RBF network classifier. The positive false rate is lower also lower than OLS RBF.

5 Conclusions

This paper proposes MGIN based RBF neural-network classifier, which contains two stages: employing multiple granularities immune network to find the candidate hidden neurons; and then use some removing criterion to delete the redundant neurons and adjusting the weight between hidden neurons and output units. Experimental results indicate that the new classifier has the best detection ability for the network intrusion detection when compared with other conventional classifiers for our tested pattern classification problems.

Acknowledgement. This work is supported by Natural Science Foundation Project of CQ CSTC(2008BB2195).

References

1. Chen, S., Cowan, C.F., Grant, P.M.: Orthogonal least squares learning algorithms for redial basis function networks. IEEE Trans. Neural Networks 2(2), 302–309 (1991)
2. Mao, K.Z., Huang, G.B.: Neuron Selection for RBF Neural Network Classifier Based on Data Structure Preserving Criterion. IEEE Trans. Neural Networks 16(6), 1531–1540 (2005)
3. Huang, G.B., Saratchandran, P.: A Generalized Growing and Pruning RBF (GGAP-RBF) Neural Network for Function Approximation. IEEE Trans. Neural Networks 16(1), 57–67 (2005)

4. Lee, S.J., Hou, C.L.: An ART-Based Construction of RBF Networks. IEEE Trans. Neural Networks 13(6), 1308–1321 (2002)
5. Lee, H.M., Chen, C.M.: A Self-Organizing HCMAC Neural-Network Classifier. IEEE Trans. Neural Networks 14(1), 15–27 (2003)
6. Gonzalez, J., Rojas, I., Ortega, J.: Multiobjective evolutionary optimization of the size, shape, and position parameters of radial basis function networks for function approximation. IEEE Trans. Neural Networks 14(10), 1478–1495 (2003)
7. Karayiannis, N.B.: Reformulated radial basis neural networks trained by gradient descent. IEEE Trans. Neural Networks 10(5), 657–671 (1999)
8. Miller, D., Rao, A.V.: A Global Optimization Technique for Statistical Classifier Design. IEEE Trans. on Signal Processing 44(12), 3108–3122 (1996)
9. : Timmis: Artificial immune system: an novel data analysis technique inspired by immune network theory. Wales:Wales university (2001)
10. Zhong, J., Wu, Z.F.: A Novel Dynamic Clustering Algorithm Based on Immune Network and Tabu Search. Chinese Journal of Electronics 14(2), 285–288 (2005)

Spatial Filter Selection with LASSO for EEG Classification

Wenting Tu and Shiliang Sun

Department of Computer Science and Technology, East China Normal University
500 Dongchuan Road, Shanghai 200241 P.R. China
w.tingtu@gmail.com, slsun@cs.ecnu.edu.cn

Abstract. Spatial filtering is an important step of preprocessing for electroencephalogram (EEG) signals. Extreme energy ratio (EER) is a recently proposed method to learn spatial filters for EEG classification. It selects several eigenvectors from top and end of the eigenvalue spectrum resulting from a spectral decomposition to construct a group of spatial filters as a filter bank. However, that strategy has some limitations and the spatial filters in the group are often selected improperly. Therefore the energy features filtered by the filter bank do not contain enough discriminative information or severely overfit on small training samples. This paper utilize one of the penalized feature selection strategies called LASSO to aid us to construct the spatial filter bank termed LASSO spatial filter bank. It can learn a better selection of the spatial filters. Then two different classification methods are presented to evaluate our LASSO spatial filter bank. Their excellent performances demonstrate the stronger generalization ability of the LASSO spatial filter bank, as shown by the experimental results.

Keywords: Brain-computer interface, Common spatial patterns, Extreme energy ratio, Feature extraction, Feature selection, LASSO.

1 Introduction

Brain-computer interfaces (BCIs) based on electroencephalogram (EEG) signals aim to provide their users communication and control capabilities that do not depend on the brain's normal output channels of peripheral nerves and muscles. They have wide usage such as text input programs, electrical wheelchairs or neuroprostheses. BCI technology relies on the ability of individuals to voluntarily and reliably produce changes in their EEG signal activities. There are four basic components in a general EEG-based BCI: EEG-signal acquisition, feature extraction, pattern classification and device control. The contribution of this paper focus on feature extraction of EEG signals from the viewpoint of machine learning. Raw EEG scalp potentials are proven to have a poor spatial resolution because of volume conduction, so spatial filtering is important for signal processing. Extreme energy ratio (EER) [1] is a recently proposed method to learn a feature extractor. It constructs a group of spatial filters to discover source signals whose average energy features of two conditions are most different. In other words, it learns spatial filters maximizing the variance of band-pass filtered EEG signals under one condition while minimizing it for the other condition. Though bearing the same

L. Cao, J. Zhong, and Y. Feng (Eds.): ADMA 2010, Part II, LNCS 6441, pp. 142–149, 2010.

motivation with common spatial patterns (CSP) [2], it simplifies the CSP algorithm to a Rayleigh quotient formulation. After solving an eigenvalue decomposition problem, it obtains several eigenvectors as candidate spatial filters and selects m eigenvectors from top and end of the eigenvalue spectrum respectively to construct the spatial filter bank, where m is a parameter defined by user or learned by the cross-validation technique.

However, the filter bank constructed by the selection strategy mentioned above has some obvious limitations. First, whatever m is, the filter bank includes the first and last eigenvectors of the eigenvalue spectrum. However, these two spatial filters may overfit on the small training set [3]. Second, the previous method is not flexible since the number of spatial filters in the group is usually even. Third, the parameter m is often defined improperly: If m is too small, the constructed filter bank would fail to fully capture the discrimination between different classes. On the other hand, the classifier could severely overfit if m is too large [4]. Due to these limitations, it is desirable to improve the construction of spatial filter bank to obtain discriminatory, generalizing energy features for classification.

Here one penalized feature selection strategy is used to help us build the spatial filter bank. A general penalized feature selection method often includes a classification objective function and a penalty terms. LASSO [5] is a computationally effective method with the penalty is defined as the L1-norm, and the error is the residual sum of squares. It can lead to the sparsest among the solutions with highest prediction power, and the sparsity solution can offer a deeper insight of the features which are most informative to the classification task. Thus, we can select the spatial filters which are corresponding the features selected by LASSO to build our filter bank which is called LASSO spatial filter bank.

When the LASSO spatial filter bank is constructed, we present two strategies to learn a classifier for performing the classification task to the test samples: One is to train an independent classifier, and the other is to use the output of LASSO as a linear classifier. The former method is more flexible owing to its rich choices about the kind of the classifier. The later one is more computationally effective.

In the next section we describe previous work on EER algorithm. Subsequently, in Section 3, we first analyze the limitations of previous spatial filter bank. Then we present our proposed LASSO spatial filter bank and two classification strategies. Section 4 outlines the experiments we performed. Moreover, the results and performance analysis are also presented. Finally, we show our conclusion and recommendations for future work in Section 5.

2 EER Algorithm: A Brief Review

Denote an observed EEG sample as an $N \times T$ matrix X, where N is the number of recording electrodes and T is the number of total points during the recording period. The spatially filtered signal $S \in \mathbb{R}^{T \times 1}$ with a spatial filter for X denoted as $\phi \in \mathbb{R}^N$ can be defined as

$$S = \phi^T X. \tag{1}$$

EER is a recently proposed method to learn a group of spatial filters maximizing the variance of band-pass filtered EEG signals under one condition while minimizing it for the other condition. Though having the same motivation as CSP, it simplifies the CSP algorithm to a Rayleigh quotient:

$$\max / \min \frac{\phi^T \bar{C}_A \phi}{\phi^T \bar{C}_B \phi}. \tag{2}$$

Here $\bar{C}_A$ or $\bar{C}_B$ are the covariances for specific class which can be computed as the average of all single covariances so as to get a more accurate and stable estimated covariance:

$$\bar{C}_A = \frac{1}{T_A} \sum_{i=1}^{T_A} \frac{X_i X_i^T}{tr(X_i X_i^T)}, \tag{3}$$

$$\bar{C}_B = \frac{1}{T_B} \sum_{j=1}^{T_B} \frac{X_j X_j^T}{tr(X_j X_j^T)}, \tag{4}$$

and T_A, T_B are respectively the number of trails belongs to condition A and B.

For classification, by optimizing (2) we can obtain two optimal spatial filters ϕ^*_{max} and ϕ^*_{min} which maximizes and minimizes the objective function in (2). It turns out that ϕ^*_{max} and ϕ^*_{min} are two eigenvectors respectively corresponding to the maximal and minimal eigenvalues of the matrix $(\bar{C}_B^{-1} \bar{C}_A)$. The energy values of the EEG sample spatially filtered by ϕ^*_{max} and ϕ^*_{min} are two parts of the energy feature of the sample. When we wish to extract m sources, EER will seek $2m$ spatial filters. Half of them maximize the objective function (2) while the other half minimize it. Thus, ϕ^*_{max} consists of m generalized eigenvectors of the matrix pair $(\bar{C}_A, \bar{C}_B)$ which correspond to the m maximal eigenvalues: $\phi^*_{max} \triangleq [\phi_1, \cdots, \phi_m]$. Similar, the m entries of ϕ^*_{min} are m generalized eigenvectors of the matrix pair $(\bar{C}_A, \bar{C}_B)$ whose eigenvalues are minimal. For a new EEG sample, it can be filtered by $2m$ spatial filters coming from two filter banks ϕ^*_{max} and ϕ^*_{min}. Thus, the energy feature vector consists of the $2m$ energy values.

3 LASSO Spatial Filter Bank

3.1 Limitations of the previous EER Method

As method in Section 2, EER always selects m eigenvectors from top and end of the eigenvalue spectrum respectively to construct the spatial filter bank. In other words, the previous spatial filter bank includes $2m$ spatial filters half of which are eigenvectors from top of the eigenvalue spectrum and the other half are ones from end of the eigenvalue spectrum. The spatial filter bank constructed by this strategy has some obvious limitations: First, it always includes the first and the last eigenvectors of the eigenvalue spectrum. However, since the non-stationary nature of the brain signals and the existence of outliers, those two spatial filters may overfit on training set and thus are not suitable to be in the filter bank. Second, the number of spatial filters in the bank is always even, which cause the method not flexible. Third, the number of the spatial filters in the bank is always to be defined unsuitable: Suppose the filter bank includes the spatial filters which are most suitable to be selected and the number of those spatial filters is k, it is always true that $2m$ is much larger than k , thus the builded filter bank may severely overfit on the training set. On the other hand, if we choose m is enough small to avoid overfitting, the filter bank may not include the spatial filters which are most suitable to be included in it, thus the features obtained by it will not have enough discriminative ability.

3.2 LASSO Spatial Filter Bank

Here we make use of a feature selection strategy to help us build the spatial filter bank. Feature selection methods can be classified into three categories, depending on their strategies to combine the feature selection search with the construction of the classification model [7]. Filter approach separates feature selection from classifier construction. Wrapper approach evaluates classification performance of selected features and keeps searching until certain accuracy criterion is satisfied. Embedded approach embeds feature selection within classifier construction. Among them, embedded methods have a significant advantage that they include interaction with the classification model, while at the same time being far less computationally intensive than wrapper methods. Recently, one of the embedded methods called penalized feature selection has aroused intentions in bioinformatics [8]. A penalized feature selection method includes a classification objective function and a penalty term. An algorithm called LASSO has been proposed with the penalty is defined as the L1-norm, and the error is the residual sum of squares. Here we employ it to construct the spatial filter bank, which is called LASSO spatial filter bank (see Fig. 1 for illustration).

Suppose we have n EEG training samples. After processed by all the candidate spatial filters, they can be expressed by their energy features expression corresponding all N candidate spatial filters: $X_1, X_2, \cdots, X_n \subseteq \mathbb{R}^{N\times 1}$. Denote their corresponding conditions is $Y_1, Y_2, \cdots, Y_n \subseteq \{-1, 1\}$. The LASSO solution β is to the optimization problem of minimizing

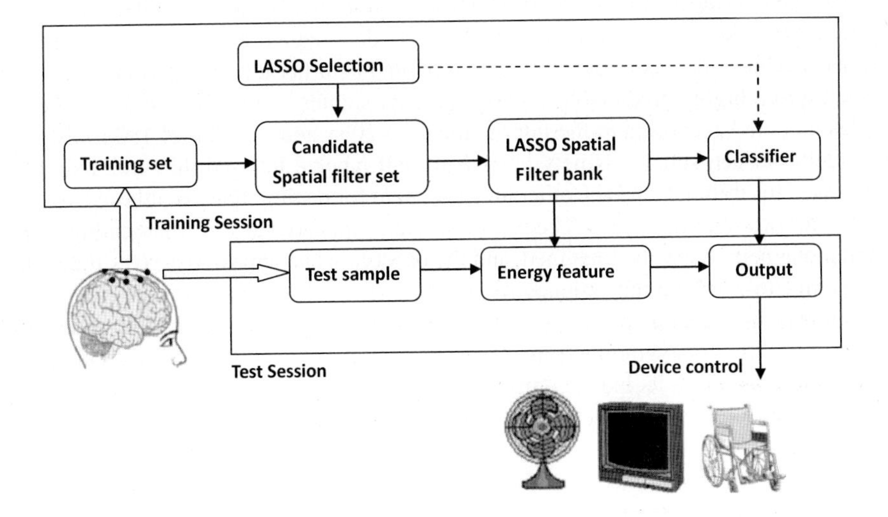

Fig. 1. A structural model for the BCI system with the LASSO spatial filter bank

$$\sum_{i=1}^{n}(Y_i - \beta X_i)^2 + \lambda \sum_{j=1}^{N} |\beta_j| \tag{5}$$

where $\beta = (\beta_1, \cdots, \beta_N)$ and $\lambda \geqslant 0$ is a penalty term. An important property of the L1 penalty is that it can generate exact zero estimated coefficients. Therefore, it can be used for feature selection, and thus we can select the spatial filters corresponding to the selected features to construct the LASSO spatial filter bank.

The relation of λ and the number of spatial filters to be selected should be noted: Briefly speaking, the bigger λ is, the more zero elements of β has. When $\lambda \rightarrow 0$, lots features will be selected. However, since the classifier is too complex, it may have unsatisfactory prediction and be less interpretable. When $\lambda \rightarrow +\infty$, fewer features will be selected. The case of $\lambda = +\infty$ corresponds to the simplest classifier where no input variable is used for classification. As a result, if we select the interval of λ values properly, we can obtain filter bank corresponding any possible number of the spatial filters in it. The optimal number of spatial filters can be determined by cross-validation.

After obtaining the LASSO spatial filter bank as a feature extractor, we can use two strategies to achieve classification task and test the effectiveness of our proposed LASSO spatial filter bank: One is training an independent classifier, and the other is to use the β as a linear classifier to predict the condition of test samples. We denote these two methods as "LASSO spatial filter bank+LDA (LSFB+LDA)" and "LASSO spatial filter bank+LASSO (LSFB+LASSO)", respectively.

4 Experiment

4.1 Data Description and Experimental Setup

The EEG data used in this study were made available by Dr. Allen Osman of University of Pennsylvania during the NIPS 2001 BCI workshop (Sajda, Gerson, Mller, Blankertz, Parra, 2003). There were a total of nine subjects denoted S1, S2,. . ., S9, respectively. For each subject, the task was to imagine moving his or her left or right index finger in response to a highly predictable visual cue. EEG signals were recorded with 59 electrodes mounted according to the international 10-20 system. A total of 180 trials were recorded for each subject. Ninety trials with half labeled left and the other half right were used for training, and the other 90 trials were for testing. Each trial lasted six seconds with two important cues. The preparation cue appeared at 3.75 s indicating which hand movement should be imagined, and the execution cue appeared at 5.0 s indicating it was time to carry out the assigned response.

Signals from 15 electrodes over the sensorimotor area were used in this paper, and for each trial the time window from 4.0 s to 6.0 s was retained for analysis. Other preprocessing operations included common average reference, 8-30 Hz bandpass filtering, and signal normalization to eliminate the energy variation of different recording instants (Müller-Gerking et al., 1999). To fully compare the performances, we changed the number of spatial filters in the filter bank from 2 to 8 and a comparison of classification accuracies of three methods (previous spatial filter bank+LDA which is denoted as "TSFB+LDA", "LSFB+LDA", "LSFB+LASSO") was obtained. Moreover, to test

Table 1. The classification accuracies (%) of three methods with the number of spatial filters in the filter bank changed from 2 to 8

Subject	Method	Number of the spatial filters						
		2	3	4	5	6	7	8
	TSFB+LDA	47.78	—	60.00	—	**67.78**	—	64.44
1	LSFB+LDA	47.78	51.11	**67.78**	66.67	62.22	65.56	64.44
	LSFB+LASSO	48.89	47.78	53.33	53.33	54.44	56.67	48.89
	TSFB+LDA	57.78	—	62.22	—	61.11	—	63.33
2	LSFB+LDA	60.00	71.11	67.78	67.78	61.11	63.33	63.33
	LSFB+LASSO	73.33	73.33	73.33	**75.56**	70.00	68.89	65.56
	TSFB+LDA	67.78	—	66.67	—	70.00	—	75.56
3	LSFB+LDA	67.78	63.33	66.67	**75.56**	74.44	74.44	74.44
	LSFB+LASSO	70.00	70.00	70.00	70.00	72.22	68.89	70.00
	TSFB+LDA	74.44	—	80.00	—	77.78	—	76.67
4	LSFB+LDA	74.44	78.89	80.00	78.89	**81.11**	80.00	80.00
	LSFB+LASSO	66.67	77.78	80.00	80.00	78.89	77.78	77.78
	TSFB+LDA	66.67	—	63.33	—	60.00	—	62.22
5	LSFB+LDA	66.67	67.78	68.89	62.22	64.44	60.00	65.56
	LSFB+LASSO	61.11	61.11	61.11	66.67	67.78	**70.00**	70.00
	TSFB+LDA	44.44	—	61.11	—	67.78	—	64.44
6	LSFB+LDA	44.44	60.00	61.11	67.78	68.89	70.00	**71.11**
	LSFB+LASSO	44.44	46.67	50.00	61.11	63.33	67.78	67.78
	TSFB+LDA	65.56	—	75.56	—	76.67	—	**81.11**
7	LSFB+LDA	65.56	64.44	76.67	76.67	67.78	74.44	75.56
	LSFB+LASSO	65.56	65.56	63.33	63.33	76.67	76.67	78.89
	TSFB+LDA	56.67	—	57.78	—	58.89	—	56.67
8	LSFB+LDA	56.67	60.00	57.78	62.22	60.00	60.00	60.00
	LSFB+LASSO	50.00	51.11	60.00	61.11	**66.67**	64.44	62.22
	TSFB+LDA	53.33	—	61.11	—	58.89	—	56.67
9	LSFB+LDA	53.33	58.89	61.11	62.22	58.89	**63.33**	63.33
	LSFB+LASSO	60.00	60.00	56.67	55.56	57.78	57.78	58.89

the performances in practice BCI application, we compared their classification accuracies with the number of spatial filters in the filter bank was determined by 10-fold cross-validation technology.

4.2 Results and Performance Analysis

Table 1 shows the classification accuracies of three methods (TSFB+LDA, LSFB+LDA, LSFB+LASSO) with different number of spatial filters in the spatial filter bank. The results show the better classification ability of the LSFB+LDA and LSFB+LASSO methods. Moreover, there are some results should be noted: On the dataset of subject 2, when the number of the spatial filters in the filter bank is two, the LASSO filter bank is different to the previous filter bank with their results are 60% and 57%. Note that the spatial filters in the previous filter bank are the first and the last eigenvectors of the eigenvalue spectrum. Thus it proves the fact that the first or the last eigenvectors of the eigenvalue

Table 2. The classification accuracies (%) and the corresponding numbers of spatial filters in the filter bank determined by 10-fold cross-validation technique

Method	Subject									$Mean \pm Std$
	1	2	3	4	5	6	7	8	9	
TSFB+LDA	62.2	61.1	73.3	74.4	66.6	63.3	76.6	56.6	53.3	65.3 ± 8.1
	(14)	(6)	(14)	(2)	(2)	(10)	(6)	(8)	(2)	(7)
LSFB+LDA	**63.3**	67.8	**74.5**	**78.9**	**66.7**	**71.1**	75.6	60.0	**63.3**	$\mathbf{69.0 \pm 6.3}$
	(8)	(5)	(6)	(3)	(2)	(10)	(8)	(3)	(8)	**(6)**
LSFB+LASSO	**63.3**	**73.3**	72.2	**78.9**	**66.7**	66.7	**80.0**	**61.1**	60.0	$\mathbf{69.1 \pm 7.3}$
	(10)	(3)	(11)	(5)	(4)	(8)	(8)	(13)	(5)	(7)

spectrum may not be the best choice to be included in the filter bank. Moreover, many filter banks with odd number of spatial filters are better than the ones with even number of spatial filters. Table 2 lists the performances of three methods with the number of spatial filters in the filter bank is determined by 10-fold cross-validation technique. It also confirms the truth of the stronger generalization ability of our proposed method. Moreover, we can see the two classification strategies have similar performances. However, they have different merits. The "LSFB+LDA" method is more flexible since other classification methods can be used for different scenarios while the "LSFB+LASSO" is much more computationally effectiveness since the β is calculated during the filter bank construction step.

To sum up, these experimental results not only demonstrate the limitations of the previous LASSO filter bank, but also support the effectiveness of our proposed LASSO spatial filter bank.

5 Conclusion

In this paper, we have proposed an improved spatial filter bank of EER method for EEG classification, which is called LASSO spatial filter bank. Specially, we employ the spatial filters corresponding to the energy features selected by a penalized feature selection method called LASSO to construct the filter bank. Moreover, to learn a classifier for performing classification with the LASSO spatial filter bank, we present two strategies which are named "LASSO filter bank+LDA" and "LASSO filter bank+LASSO", respectively. Their excellent results in our experiment demonstrate that the LASSO spatial filter bank can alleviate the limitations of the previous spatial filter bank and obtain a stronger generalization ability.

The LASSO spatial filter bank proposed in this paper is based on the energy features. However, there are a great variety of other features for designing BCI, such as amplitude values of EEG signals [9], Power Spectral Density (PSD) values [10], Auto Regressive (AR) [11], Time-frequency features [12] and so on. As a result, how to construct a filter bank that can combine the advantages of these features and obtain a more outstanding classification ability for EEG classification is worth studying. Moreover, owing to the need of reducing the training session [13], it is significative to extend the method in this paper to construct an optimal filter bank by selecting the spatial filters obtained from other sessions or subjects.

Acknowledgments. This work is supported by the National Natural Science Foundation of China under Projects 60703005 and 61075005.

References

1. Sun, S.: The Extreme Energy Ratio Criterion for EEG Feature Extraction. In: Kůrková, V., Neruda, R., Koutník, J. (eds.) ICANN 2008,, Part II. LNCS, vol. 5164, pp. 919–928. Springer, Heidelberg (2008)
2. Müller-Gerking, J., Pfurtscheller, G., Flyvbjerg, H.: Designing Optimal Spatial Filters for Single-trial EEG Classification in a Movement Task. Clinical Neurophysiology 110, 787–798 (1999)
3. Millán, J.R.: Robust common spatial patterns for EEG signal preprocessing. In: Proceedings of the 30th Annual International Conference of the IEEE Engineering in Medicine and Biology Society, pp. 2087–2090 (2008)
4. Blankertz, B., Tomioka, R., Lemm, S., Kawanabe, M., Müller, K.-R.: Optimizing Spatial Filters for Robust EEG Single-Trial Analysis. IEEE Signal Processing Magazine 25, 41–56 (2008)
5. Tibshirani, R.: Regression Shrinkage and Selection Via the Lasso. Journal of the Royal Statistical Society: Series B (Methodological) 58, 267–288 (1996)
6. Sun, S.: Extreme energy difference for feature extraction of EEG signals. Expert Systems with Applications 37, 4350–4357 (2010)
7. Saeys, Y., Inza, I., Larrañaga, P.: A Review of Feature Selection Techniques in Bioinformatics. Bioinformatics 23, 2507 (2007)
8. Ma, S., Huang, J.: Penalized feature selection and classification in bioinformatics. Briefings in bioinformatics 9, 392–403 (2008)
9. Kaper, M., Meinicke, P., Grossekathoefer, U., Lingner, T., Ritter, H.: BCI competition 2003–data set IIb: Support Vector Machines for the P300 Speller Paradigm. IEEE Transactions on Biomedical Engeneering 51, 1073–1076 (2004)
10. Chiappa, S., Bengio, S.: HMM and Iohmm Modeling of EEG Rhythms for Asynchronous Bci Systems. In: European Symposium on Artificial Neural Networks, pp. 985–992 (2004)
11. Penny, W.D., Roberts, S.J., Curran, E.A., Stokes., M.J.: EEG-based Communication: A Pattern Recognition Approach. IEEE Transactions on Rehabilitation Engeneering 8, 214–215 (2000)
12. Wang, T., Deng, J., He, B.: Classifying EEG-based Motor Imagery Tasks by Means of Time-frequency Synthesized Spatial Patterns. Clinical Neurophysiology 115, 2744–2753 (2004)
13. Schalk, G., Blankertz, B., Chiappa, S., et al.: BCI competition III (2004–2005), `http://ida.first.fraunhofer.de/projects/bci/competitioniii/`

Boolean Algebra and Compression Technique for Association Rule Mining

Somboon Anekritmongkol and M.L. Kulthon Kasamsan

Faculty of information Technology, Rangsit University,
Pathumtani 12000, Thailand
Somboon_a@hotmail.com, kasemsan@rangsit.rsu.ac.th

Abstract. Association Rule represents a promising technique to find hidden patterns in database. The main issue about mining association rule in the large database. One of the most famous association rule learning algorithms is Apriori. Apriori algorithm is one of algorithms for generation of association rules. The drawback of Apriori Rule algorithm is the number of time to read data in the database equally number of each candidate were generated. Many research papers have been published trying to reduce the amount of time to read data from the database. In this paper, we propose a new algorithm that will work rapidly. Boolean Algebra and Compression technique for Association rule Mining (B-Compress) is applied to compress database and reduce the amount of times to scan database tremendously. Boolean Algebra combines, compresses, generates candidate itemset and counts the number of candidates. The construction method of B-Compress has ten times higher mining efficiency in execution time than Apriori Rule.

Keywords: Association rule, Apriori Rule, Boolean algebra, Data mining.

1 Introduction

One of the most popular technique in data mining is Apriori rule [1][2][3][6]. The Data mining is usually involve huge amounts of information. Association rules exhaustively look for hidden patterns, making them suitable for discovering predictive rules involving subsets of data set attributes. Association rule learners are used to discover elements that co-occur frequently within a data set consisting of multiple independent selections of elements (such as purchasing transactions), and to discover rules. Firstly, the discovered association dependent on the data. Secondly, most of information in data set is of the same pattern. Thirdly, the amount of time involve in read the entire database. Fourthly, the pruning candidate in each step of process. This paper proposes the development of algorithm to discover association rules from huge amount of information that is faster than the Apriori rule by using Boolean algebra and compressed technique. The improvement focuses on compressing data and reducing the number of times to read data from the database.

2 Basic in Association Rule

Let D = {T1, T2, . . . ,Tn} [2] be a set of n transactions and let I be a set of items, I = {i1, i2 . . . im}. Each transaction is a set of items, i.e. Ti $\subseteq$ I. An association rule is an

L. Cao, J. Zhong, and Y. Feng (Eds.): ADMA 2010, Part II, LNCS 6441, pp. 150–157, 2010.

implication of the form $X \Rightarrow Y$, where $X, Y \subset I$, and $X \cap Y = \emptyset$; X is called the antecedent and Y is called the consequent of the rule. In general, a set of items, such as X or Y, is called an itemset. In this work, a transaction record is transformed into a binary format where only positive binary values are included as items. This is done for efficiency purposes because transactions represent sparse binary vectors. Let P(X) be the probability of appearance of itemset X in D and let P(Y |X) be the conditional probability of appearance of itemset Y given itemset X appears. For an itemset $X \subseteq I$, support(X) is defined as the fraction of transactions $Ti \in D$ such that $X \subseteq Ti$. That is, P(X) = support(X). The support of a rule $X \Rightarrow Y$ is defined as support($X \Rightarrow Y$) = P(X $\cup$ Y). An association rule $X \Rightarrow Y$ has a measure of reliability called confidence ($X \Rightarrow Y$) defined as P(Y |X) = P(X$\cup$Y)/P (X) = support(X$\cup$Y)/support(X). The standard problem of mining association rules [1] is to find all rules whose metrics are equal to or greater than some specified minimum support and minimum confidence thresholds. A k-itemset with support above the minimum threshold is called frequent. We use a third significance metric for association rules called lift [25]: lift($X \Rightarrow Y$) = P(Y |X)/P (Y) = confidence($X \Rightarrow Y$)/support(Y). Lift quantifies the predictive power of $X \Rightarrow Y$; we are interested in rules such that lift($X \Rightarrow Y$) > 1.

3 Apriori Rule

Finding frequency itemsets using candidate generation. Apriori is a algorithm proposed by R. Agrwal and R. Srikant in 1994. Apriori rule employs an iterative approach know as a level-wise search, where *k*-itemsets are used to explore (k+1)-itemsets. Tthe set of frequency 1-itemsets is found by scanning the database to accumulate the count of each time and collecting those items satisfy minimum support. The resulting set is L_1. Next L_1 used to find the set of frequency 2-itemsets, which is used to find, and so on, until no more frequency *k*-itemsets can be found. The finding of each L_k requires one full scan of database.

Algorithm: Apriori rule.
Find frequent itemsets using an iterative level-wide approach based on candidate generation.
Input: D, a database of transaction;
min_sup, The minimum support count threshold.
Output: L, frequent itemsets in D
Method:

```
L1 = find_frequent_1-itemset(D);
for (k=2;Lk-1!= 0;k++){
   CK = Apriori_gen(Lk-1);
   for each transaction t ∈ D {// scan D for count
        Ct= subset(Ck,t);// Get subset of t that are candidate
              for each candidate c∈ Ct
                 C.count++;
        }
        Lk = {c∈ Ck|c.count ≥ min_sup}
   }
Return L = UkLk;
```

4 Boolean Algebra

Boolean algebra [5], developed in 1854 by George Boole in his book, "An Investigation of the Laws of Thought". Some operations of ordinary algebra, in particular multiplication xy, addition $x+y$, and negation $-x$, have their counterparts in Boolean algebra, respectively the Boolean Operations AND, OR, and NOT also called conjunction $x \wedge y$, disjunction $x \vee y$, and negation or complement $\neg x$ sometime $!x$. Some authors use instead the same arithmetic operations as ordinary algebra reinterpreted for Boolean algebra, treating xy as synonymous with $x \wedge y$ and $x+y$ with $x \vee y$.

Basic Boolean operations can be defined arithmetically as follows.

$$x \wedge y = xy \quad (1)$$

$$x \vee y = x + y - xy \quad (2)$$

$$\neg x = 1 - x \quad (3)$$

Alternatively, the values of $x \wedge y$, $x \vee y$, and $\neg x$ can be expressed without reference to arithmetic operations by tabulating their values with truth tables as follows.

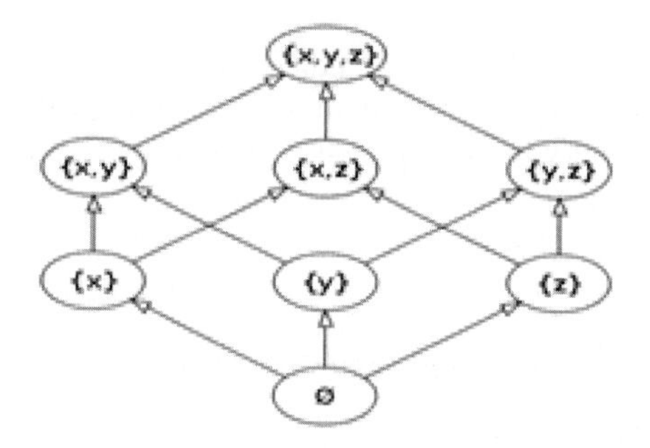

Fig. 1. Boolean algebra structure

5 Boolean Algebra and Compression Data

Psudo-code of Boolean algebra and Compression Data.

Tid_cl = a table of transaction
Tid_cl2 = a table contain candidate
Final Candidate = the result of candidates > Min_sup
T_k = Itemset
Tid = transaction ID
Min_sup = Minimum support count all transactions

Procedure Find L1

Procedure Generate Candidate-2
for each itemset $T_1 \in$ Tid_cl.T_{k-1}
  for each itemset $T_2 \in$ Tid_cl.T_{k-1}
    if($T_2 > T_1$) then
    Add $T_2 \cup T_1$ to Tid_cl2;

Procedure Generate Associate Data
Find L1;
Compress Structure ;
Generate Candidate-2;
add to Final Candidate File;

```
    for each itemset count  Tk ∈ Tid_cl;
    Delete Tk ≤ Min_sup;
     Add to Final Candidate;
Procedure Compress Structure
for each Tid , Tidk ∈ L1 {
    if  same Tidk Then
New_structurek = New_structurek + Tk ;
    else
      New_Structure.countk++;
}
for each New_structurek  Update
Tid_cl.feq;
```

```
for each (k=3; Lk-1 ;k++){
          for each Tid_cl2 Delete
            Tid_cl2.Tk ≤ Min_sup
          for each Tid_cl2 add to final
Candidate
           for each itemset (j=1 ;
              Tid_cl2.eof ; j++)
          add Tid_cl2.Tj ∪ Tid_cl.T;
          Delete Tid_cl2k-1;
 }
```

Example: Minimum support 10% percent = 1.5

Table 1. Transaction Data

Trans#	Item	Trans#	Item	Trans#	Item	Trans#	Item
T001	I1	T004	I4	T009	I1	T012	I3
T001	I2	T005	I1	T009	I2	T013	I1
T001	I5	T005	I3	T009	I3	T013	I2
T002	I2	T006	I2	T010	I1	T013	I4
T002	I4	T006	I3	T010	I2	T014	I1
T003	I2	T007	I2	T010	I4	T014	I2
T003	I3	T007	I3	T011	I1	T014	I3
T004	I1	T008	I1	T011	I2	T015	I2
T004	I2	T008	I2	T011	I3	T015	I3
T004	I3	T008	I3	T012	I2		

Step 1: Find L1 : Find frequency itemset and remove frequency itemset less than minimum support.

Table 2. Eliminate Itemset lower minimum support

Itemsets	Support count	Eliminate {I5} lower Minimum Support	Itemsets	Support count
{I1}	9	Result of candidate-1	{I1}	9
{I2}	14	Itemsets	{I2}	14
{I3}	11		{I3}	11
{I4}	4	→	{I4}	4
{I5}	1			

Step 2: Compression Data.
Create pattern is same structure as T001→ pattern1, T002 → pattern2, {T003, T006, T007, T012, T015} → pattern3. T004→ pattern4, {T010,T013}→ pattern 5, T005 → pattern6 and {T008,T009,T011,T014} → pattern7.

Table 3. Pattern table after compress data

	I1	I2	I3	I4	Count
Pattern-1	X	X			1
Pattern-2		X		X	1
Pattern-3		X	X		5
Pattern-4	X	X	X	X	1
Pattern-5	X	X		X	2
Pattern-6	X		X		1
Pattern-7	X	X	X		4

Step 3: Find Candidate-2 Itemsets.
To discover the set of frequency 2-itemsets, Generate candidate from Pattern table itemsets_1 to itemsets_k and count number of candidate. Starting from candidate-2 itemsets by $\text{Itemset}_1 \cup \text{Itemset}_{k-i}$, scan for all pattern transactions table. Then remove each candidate_{k-1} itemsets < minimum support.

Table 4. Result of Candidate-2 Itemsets

Itemsets (C_2)	Support Count
{I1,I2}	8
{I1,I3}	6
{I1,I4}	3
{I2,I3}	10
{I2,I4}	4

Step 4: Find Candidate-3 Itemsets
Generate Candidate-3 Itemsets by $\text{Itemset}_1 \cup C_2$, scan for all candidate-2 itemsets.

Table 5. Result of Candidate-3 Itemsets

Item sets	Support Count
{I1,I2,I3}	5
{I1,I2,I4}	3

Table 6. Final result

Itemsets	Support Count	Itemsets	Support Count
{I1}	9	{I1,I4}	3
{I2}	14	{I2,I3}	10
{I3}	11	{I2,I4}	4
{I4}	4	{I1,I2,I3}	5
{I1,I2}	8	{I1,I2,I4}	3
{I1,I3}	6		

6 Experiment

In this section, we performed a set of experiments to evaluate the effectiveness of B-Compress. The experiment data set consists of two kinds of data, data from Phanakorn Yontrakarn Co., Ltd. and Generate sampling data. This company sales and offer car services to discover association data. The experiment of three criteria, firstly, increase amount of records from 10,000 to 50,000 records and fixed 10 itemsets. Secondly, increase of itemsets and fixed amount of records = 50,000. Thirdly, increase of minimum support and fixed itemsets and amount of records.

Experiment 1: Increase number of records. Step 10,000 records. Fixed 10 itemsets.

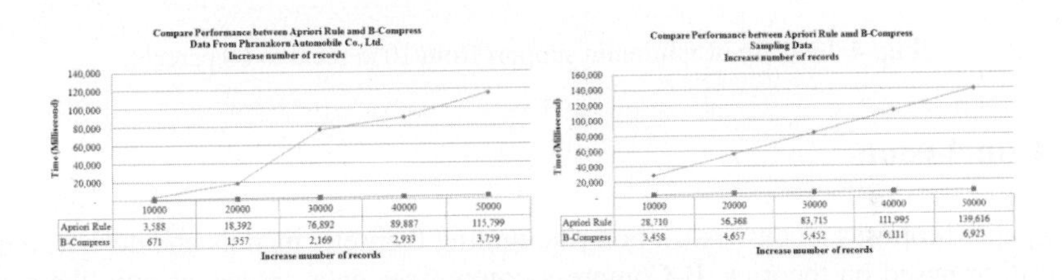

Fig. 2. Increase number of records. Step 10,000 records. Fixed 10 itemsets.

Experiment 2: Increase itemsets. Fixed number of records 10 itemsets

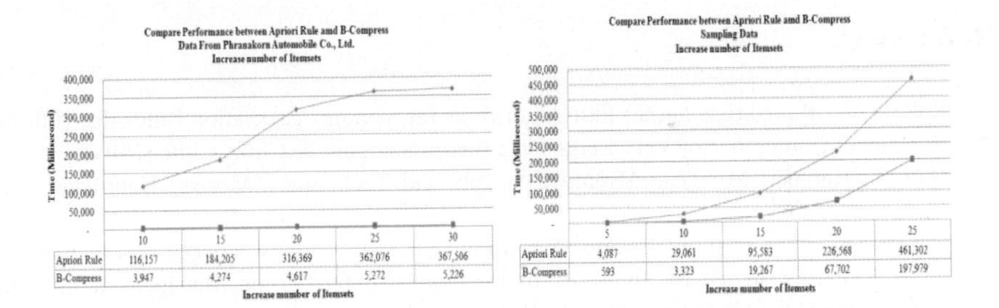

Fig. 3. Increase itemsets and fixed number of records 50,000 records

Experiment 3: Increase of minimum support from 10 percent to 60 percent. The step to change minimum support, Apriori rule low minimum support takes time to process but B-Compress slightly affects the performance because B-Compress compresses data, process only the actual data and reduce to each candidate.

The result of experiments, Boolean Algebra and Compression Technique for Association rule Mining discovers an association has ten times higher mining efficiency in execution time than Apriori rule. If increasing the number of records, Apriori rule will take time to read the whole database. If increasing the number of itemset, Apriori rule will create more candidates depending on the number of itemsets but Boolean Algebra and Compression Technique for Association rule Mining takes shorter time because it will compress data and create candidates in existing data only and delete candidate's data that are lower than minimum support from each steps.

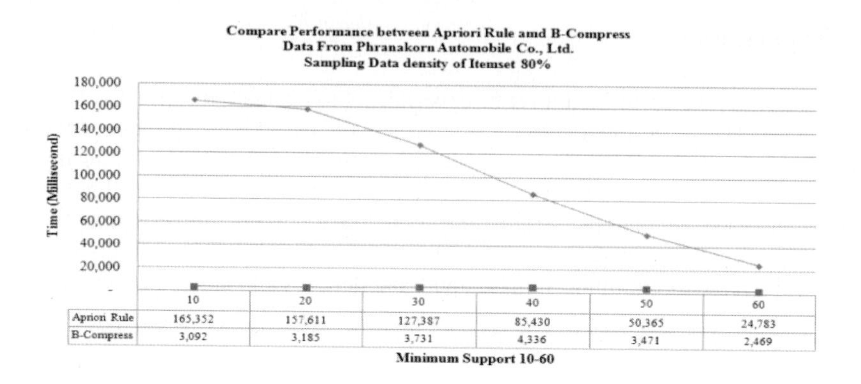

Fig. 4. Increase of minimum support from 10 percent to 60 percent

7 Conclusion

The paper proposes a new association rule mining theoretic models and designs a new algorithm based on theories. B-Compress compresses data, processes only the actual data and reduces the data of each candidate. The Boolean Algebra and Compression Technique for Association rule Mining is able to discover data more than ten times faster than Apriori rule.

References

1. Rakesh, A., Ramakrishnan, S.: Fast Algorithms for Mining Association Rules. In: 20th International Conference of Very Large Data Bases, Santiago de Chile, Chile (1994)
2. Han, J., Kamber, M.: Data Mining Concepts and Techniques. Morgan Kaufmann, San Francisco (2006)
3. Rakesh, A., Imielinski, T., Swami, A.: Mining association rules between sets of items in large databases. In: ACM SIGMOD International Conference on Management of Data, Washington D.C., USA, pp. 207–216 (1993)
4. Ramakrishnan, S., Rakesh, A.: Mining Generalized Association Rules. In: 21th VLDB Conference, Zurich, Swizerland (1995)
5. Yaichroen, U.: Applied Boolean Algebra, Bangkok, Thailand (1989)
6. Savasere, A., Omiecinski, E., Navathe, S.: An Efficient Algorithm for Mining Association Rules in Large Database. In: 21st VLDB Conference, Zurich, Swizerland, pp. 432–444 (1995)
7. Margahny, M.: Fast Algorithm for Mining Assiciation Rules. In: AIML 2005 Conference, Cairo, Egypt (2005)
8. Fukuda, T., Morimoto, Y., Morishita, S.: Data Mining using Two Dimensional Optimized Association Rules, Scheme, Algorithms and Visualization. In: ACM SIGMOD International Conference on the Management of Data, New York, pp. 13–23 (1996)
9. Goethals, B.: A Priori Versus A Posteriori Filtering of Association Rules. In: ACM SIGMOD Workshop on Research Issues in Data Mining and Knowledge Discovery, Philadelphia, USA (1999)

10. Hipp, J., Guntzer, U., Nakheizadeh, G.: Algorithms for Association Rule Mining A General Survey and Comparison. In: ACM SIGKDD, New York, pp. 58–64 (2000)
11. Ramakrishnan, S., Rakesh, A.: Mining Quanyitative Association Rules in Large Relational Tables. In: SIGMOD, Montreal, Cannada, pp. 1–12 (1996)
12. Shenoy, P., Jayant, R., Sudarshan, S., Bhalotia, G., Bawa, M., Shah, D.: Turbo-changing Vertical Mining of Large Database. In: MOD 2002, Dallas, USA, pp. 22–33 (2000)
13. Zaki, M., Parthasarathy, S., Ogihara, M., Li, W.: Algorithms for Fast Discovery of Association Rules. In: 3rd International Conference on Knowledge Discovery and Data Mining, CA, USA, pp. 283–296 (1997)
14. Borgelt, C.: Simple Algorithms for Frequent Item Set Mining, pp. 351–369. Springer, Heidelberg (2010)

Cluster Based Symbolic Representation and Feature Selection for Text Classification

B.S. Harish[1], D.S. Guru[1], S. Manjunath[1], and R. Dinesh[2]

[1] Department of Studies in Computer Science,
University of Mysore, Mysore 570 006, India
[2] Honeywell Technologies Ltd
Bangalore, India
bsharish@ymail.com, dsg@compsci.uni-mysore.ac.in,
manju_uom@yahoo.co.in, dinesh.ramegowda@honeywell.com

Abstract. In this paper, we propose a new method of representing documents based on clustering of term frequency vectors. For each class of documents we propose to create multiple clusters to preserve the intraclass variations. Term frequency vectors of each cluster are used to form a symbolic representation by the use of interval valued features. Subsequently we propose a novel symbolic method for feature selection. The corresponding symbolic text classification is also presented. To corroborate the efficacy of the proposed model we conducted an experimentation on various datasets. Experimental results reveal that the proposed method gives better results when compared to the state of the art techniques. In addition, as the method is based on a simple matching scheme, it requires a negligible time.

Keywords: Text Document, Term Frequency Vector, Fuzzy C Means, Symbolic Representation, Interval Valued Features, Symbolic Feature Selection, Text Classification.

1 Introduction

In automatic text classification, it has been proved that the term is the best unit for text representation and classification [1]. Though a text document expresses vast range of information, unfortunately, it lacks the imposed structure of a traditional database. Therefore, unstructured data, particularly free running text data has to be transformed into a structured data. To do this, many preprocessing techniques are proposed in literature [2]. After converting an unstructured data into a structured data, we need to have an effective representation model to build an efficient classification system. Many representation schemes can be found in the literature [3].

Although many text document representation models are available in literature, frequency based Bag of Word (BOW) model gives effective results in text classification task. Unfortunately, BOW representation scheme has its own limitations. Some of them are: high dimensionality, loss of correlation and loss of semantic relationship that exists among the terms in a document. Also, in conventional supervised classification an inductive learner is first trained on a training set, and then it is used to classify a

L. Cao, J. Zhong, and Y. Feng (Eds.): ADMA 2010, Part II, LNCS 6441, pp. 158–166, 2010.

testing set, about which it has no prior knowledge. However, for the classifier it would be ideal to have the information about the distribution of the testing samples before it classifies them. Thus in this paper, to deal with the problem of learning from training sets of different sizes, we exploited the information derived from clusters of the term frequency vectors of documents.

Clustering has been used in the literature of text classification as an alternative representation scheme for text documents. Several approaches of clustering have been proposed. Given a classification problem, the training and testing documents are both clustered before the classification step. Further, these clusters are used to exploit the association between index terms and documents [4]. In [5], words are clustered into groups based on distribution of class labels associated with each word. Information bottle neck method is used to find a word cluster that preserves the information about the categories. These clusters are used to represent the documents in a lower dimensional feature space and naïve bayes classifiers is applied [6]. Also, in [7] information bottleneck is used to generate a document representation in a word cluster space instead of word space, where words are viewed as distributions over document categories. Dhillon et al., (2002) in [7] proposed an information theoretic divisive algorithm for word clustering and applied it to text classification. Classification is done using word clusters instead of simple words for document representation. Two dimensional clustering algorithms are used to classify text documents in [8]. In this method, words/terms are clustered in order to avoid the data sparseness problem. In [9] clustering algorithm is applied on labeled and unlabeled data, and introduces new features extracted from those clusters to the patterns in the labeled and unlabeled data. The clustering based text classification approach in [10] first clusters the labeled and unlabeled data. Some of the unlabeled data are then labeled based on the clusters obtained.

All in all, the above mentioned clustering based classification algorithms work on conventional word frequency vector. Conventionally the feature vectors of term document matrix (very sparse and very high dimensional feature vector describing a document) are used to represent the class. Later, this matrix is used to train the system using different classifiers for classification. Generally, the term document matrix contains the frequency of occurrences of terms and the values of the term frequency vary from document to document in the same class. Hence to preserve these variations, we propose a new interval representation for each document. Thus, the variations of term frequencies of document within the class are assimilated in the form of interval representation. Moreover conventional data analysis may not be able to preserve intraclass variations but unconventional data analysis such as symbolic data analysis will provide methods for effective representations preserving intraclass variations. The recent developments in the area of symbolic data analysis have proven that the real life objects can be better described by the use of symbolic data, which are extensions of classical crisp data [11].

Thus these issues motivated us to use symbolic data rather than using a conventional classical crisp data to represent a document. To preserve the intraclass variations we create multiple clusters for each class. Term frequency vectors of documents of each cluster are used to form an interval valued feature vector. On the other way, in interval type representation there are chances of overlap between the features which leads to low classification rate and hence we need to select only those interval features which will have less overlap between the features. To the best of our knowledge no work has been reported in the literature which uses symbolic feature selection method for text document classification. Feature selection are of two types i.e., filters and wrappers method. There is evidence that wrapper methods often perform better on small scale problems, but on large scale problems, such as text classification, wrapper methods are shown to be impractical because of its high computational cost [12]. Hence in this paper we propose a filter method to select the best features.

With this backdrop, in our previous work (Guru et al., 2010) in [13], we made an initial attempt towards application of symbolic data concepts for text document representation. In this paper the same work is extended towards creating multiple representatives per class using clustering before symbolic representation. In order to select features with less overlap we propose a novel filter based feature selection method. To the best of our knowledge no work has been reported in the literature which uses symbolic representation and symbolic feature selection method for text document classification.

The rest of the paper is organized as follows: A detailed literature survey and the limitations of the existing models are presented in section 1. The working principle of the proposed method is presented in section 2. Details of dataset used, experimental settings and results are presented in section 3. The paper is concluded in section 4.

2 Proposed Method

The proposed method has 3 stages: i). Cluster based representation of documents ii). Symbolic feature selection and iii) Document classification.

2.1 Cluster Based Representation

In the proposed method, documents are represented by a set of term frequency vectors. Term frequency vector of sample documents of an individual (class) have considerable intra class variations. Thus, we propose to have an effective representation by capturing these variations through clustering and representing each cluster by an interval valued feature vector called symbolic feature vector as follows:

Let there be S number of classes each containing N number of documents, where each document is described by a t dimensional term frequency vector. The term document matrix [14], say X of size $(SN \times t)$ is constructed such that each row represents a document of a class and each column represents a term. We recommend applying any existing dimensionality reduction techniques [15] on X to obtain the transformed term document matrix Y of size $(SN \times m)$, where m is the number of features chosen out of t which is not necessarily optimum. Now, the training documents of each class are first clustered based on the reduced term frequency vector. Let $[D_1, D_2, D_3, ..., D_n]$ be a set of n samples of a document cluster of lth class say C_j^l; $j = 1, 2, 3, ..., P$ (P denotes number of clusters) and $l = 1, 2, 3, ..., S$. Let $F_i = [f_{i1}, f_{i2}, ..., f_{im}]$ be a set of m features characterizing the document sample D_i of a cluster C_j^l. Further, we recommend capturing intra class variations in each kth feature values of the jth cluster in the form of an interval valued feature $\left[f_{jk}^{-}, f_{jk}^{+}\right]$. The interval $\left[f_{jk}^{-}, f_{jk}^{+}\right]$ represents the upper and lower limits of a feature value of a document cluster in the knowledge base. Now, the reference document for a cluster C_j^l is formed by representing each feature $(k=1,2,3,...,m)$ in the form of an interval and is given by

$$RF_j^l = \left\{\left[f_{j1}^{-}, f_{j1}^{+}\right], \left[f_{j2}^{-}, f_{j2}^{+}\right], ..., \left[f_{jm}^{-}, f_{jm}^{+}\right]\right\}, \tag{1}$$

where, $j=1,2,...,P$ represents the number of clusters of documents of class l. It shall be noted that unlike conventional feature vector, this is a vector of interval valued feature and this symbolic feature vector is stored in the knowledge base as a representative of the jth cluster. Thus, the knowledge base has P number of symbolic vectors representing clusters corresponding to a class. In total there will be $(S \times P)$ representative vectors in the database to represent all S number of classes.

2.2 Symbolic Feature Selection

As the term document matrix (which is of huge dimension) is represented by interval type data, there may be chances of overlapping in features among classes and this may reduce the classification accuracy. Hence there is a need to select the best interval features which have less overlapping among the classes. Further the dimension of the term document matrix is also reduced thereby reducing the classification time. In order to select the best features from the class representative matrix F we need to study the variance present among the individual features of each class. The features which have maximum variance shall be selected as the best features to represent the classes. Since F is an interval matrix we compute a proximity matrix of size $(S * P) \times (S * P)$ with each element being of type multivalued of dimension m by computing the similarity among the features. The similarity from class i to class j with respect to k^{th} feature is given by [16]

$$S_{i \to j}^{k} = \left(\frac{|I_{ik} \cap I_{jk}|}{|I_{jk}|}\right) \quad . \tag{2}$$

where, $I_{ik} = [f_{ik}^{-}, f_{ik}^{+}] \ \forall k = 1,2,...,m$ are the interval type features of the clusters C_i and $I_{jk} = [f_{jk}^{-}, f_{jk}^{+}] \ \forall k = 1,2,...,m$ are the interval type features of the clusters C_j.

From the obtained proximity matrix, the matrix M of size $(S \times P)^2 \times m$ is constructed by listing out each multivalued type element one by one in the form of rows. In order to select the features we study the correlation among the features and the features which have highest correlation will be selected as the best features.

Let $TCorr_k$ be the total correlation of the kth column with all other columns of the matrix M and let $AvgTCorr$ be the average of all total correlation obtained due to all columns. i.e.,

$$TCorr_k = \sum_{q=0}^{m} Corr(k^{th}\ Column, q^{th}\ Column) \text{ and} \tag{3}$$

$$AvgTCorr = \frac{\sum_{k=0}^{m} TCorr_k}{m}. \quad (4)$$

We are interested in those features which have high discriminating capability, we recommend here to select those features, $TCorr_l$ of which is higher than the average of correlation $AvgTCorr$.

2.3 Document Classification

The document classification proposed in this work considers a test document, which is described by a set of m feature values of type crisp and compares it with the corresponding interval type feature values of the respective cluster stored in the knowledge base. Let, $F_t = [f_{t1}, f_{t1}, ..., f_{tm}]$ be a m dimensional feature vector describing a test document.

Let RF_j^l be the interval valued symbolic feature vector of jth cluster of lth class. Now, each mth feature value of the test document is compared with the corresponding interval in RF_j^l to examine whether the feature value of the test document lies within the corresponding interval. The number of features of a test document, which fall inside the corresponding interval, is defined to be the degree of belongingness. We make use of Belongingness Count B_c as a measure of degree of belongingness for the test document to decide its class label.

$$B_c = \sum_{k=1}^{m} C\left(f_{tk}, \left[f_{jk}^-, f_{jk}^+\right]\right) \text{ and} \quad (5)$$

$$C\left(f_{tk}, \left[f_{jk}^-, f_{jk}^+\right]\right) = \begin{cases} 1; & if\ (f_{tk} \geq f_{jk}^-\ and\ f_{tk} \leq f_{jk}^+). \\ 0; & Otherwise \end{cases} \quad (6)$$

The crisp value of a test document falling into its respective feature interval of the reference class contributes a value 1 towards B_c and there will be no contribution from other features which fall outside the interval. Similarly, we compute the B_c value for all clusters of remaining classes and the class label of the cluster which has highest B_c value will be assigned to the test document as its label.

3 Experimental Setup

3.1 Dataset

To test the efficacy of the proposed model, we have used the following four datasets. The first dataset consists of vehicle characteristics extracted from wikipedia pages (vehicles- wikipedia) [17]. The dataset contains 4 categories that have low degrees of similarity. The dataset contains four categories of vehicles: Aircraft, Boats, Cars and Trains. All the four categories are easily differentiated and every category has a set of unique key words. The second dataset is a standard 20 mini newsgroup dataset [18] which contains about 2000 documents evenly divided among 20 Usenet discussion groups. This dataset is a subset of 20 newsgroups which contains 20,000 documents. In 20 MiniNewsgroup, each class contains 100 documents in 20 classes which are randomly picked from original dataset. The third dataset is constructed by a text corpus of 1000 documents that are downloaded from Google-Newsgroup [19]. Each class contains 100 documents belonging to 10 different classes (Business, Cricket, Music, Electronics, Biofuels, Biometrics, Astronomy, Health, Video Processing and Text Mining). The fourth dataset is a collection of research article abstracts. All these research articles are downloaded from the scientific web portals. We have collected 1000 documents from 10 different classes. Each class contains 100 documents.

3.2 Experimentation

In this section, the experimental results of the proposed method are presented. Initially the term document matrix X of size $(SN \times t)$ is constructed such that each row represents a document of a class and each column represents a term. We used Regularized Locality Preserving Indexing (RLPI) [20] on X to obtain the transformed term document matrix Y of size $(SN \times m)$. The reason behind choosing RLPI for our experiment is that it has a capability of discovering discriminating structure of the document space. More details on theoretical and algorithmic analysis of RLPI can be found in [20]. In the first set of experiments, 50% of documents in the corpus are considered to train the system and remaining 50% are considered for testing. On the other hand, in the second set of experiments 60% of documents are considered for training and 40% of documents are considered for testing. For both the experiments, we have randomly selected the training documents to create the symbolic feature vectors for each class. While conducting the experimentation we have varied the number of features m selected through RLPI from 1 to 30 dimensions. For each obtained dimension we create cluster based interval representation as explained in section 2.1. At this juncture, we used Fuzzy C Means (FCM) clustering algorithm to create a cluster based symbolic representation. The reason behind using FCM is its ability to discover cluster among the data, even when the boundaries among the data are overlapping. Also FCM based techniques has the advantage over the conventional statistical techniques like NN classifier, maximum likelihood estimate etc, because its distribution is free and no knowledge about the distribution of data is required [21]. After obtaining the symbolic feature vectors for the documents, we employ the symbolic feature selection methods to obtain the best subset features d as explained in section 2.2 and subsequently the classification of testing documents is done in the way that is explained in section 2.3. The experiments are repeated 3 times and for each iteration the training set is randomly selected. The average classification accuracy of the proposed model of all the 3 trials is presented in the result Table 1. Also, the selection of number of clusters in Table 1 is empirically studied. A comparative analysis of the proposed method with other state of the art techniques on benchmark dataset viz., 20 MiniNewsgroup, Wikipedia dataset, Google dataset and Research article dataset is given in Table 2. From the Table 2 it is analyzed that the proposed method achieves better classification accuracy than the state of the art techniques.

Table 1. Classification accuracy of the proposed method on different data sets

Dataset	Training vs Testing	Number of Clusters	Minimum Accuracy	Maximum Accuracy	Average Accuracy
Wikipedia Dataset	50 vs 50	1	80.56	81.21	80.91
	60 vs 40	1	81.23	81.99	81.62
	50 vs 50	2	82.19	83.90	83.11
	60 vs 40	2	84.12	84.86	84.31
	50 vs 50	3	86.25	95.00	91.87
	60 vs 40	3	94.00	98.00	95.50
	50 vs 50	4	82.00	85.00	83.33
	60 vs 40	4	85.00	93.75	90.83
20 MiniNews Group	50 vs 50	1	73.26	74.81	74.10
	60 vs 40	1	74.38	75.12	74.70
	50 vs 50	2	74.38	75.39	74.99
	60 vs 40	2	75.31	76.22	75.79
	50 vs 50	3	77.50	86.50	82.06
	60 vs 40	3	87.33	89.75	88.41
	50 vs 50	4	70.12	87.62	78.12
	60 vs 40	4	83.50	89.25	86.66
Google Dataset	50 vs 50	1	63.83	65.86	64.65
	60 vs 40	1	64.81	65.38	65.05
	50 vs 50	2	66.22	66.91	66.45
	60 vs 40	2	66.86	67.57	67.25
	50 vs 50	3	67.00	84.80	73.33
	60 vs 40	3	74.00	91.75	81.58
	50 vs 50	4	55.60	66.60	62.20
	60 vs 40	4	87.50	90.50	88.83
Research Dataset	50 vs 50	1	89.38	90.16	89.81
	60 vs 40	1	90.27	91.49	91.04
	50 vs 50	2	90.38	91.23	90.85
	60 vs 40	2	91.39	91.86	91.56
	50 vs 50	3	96.00	98.25	96.83
	60 vs 40	3	97.20	99.00	97.86
	50 vs 50	4	92.50	96.50	94.25
	60 vs 40	4	95.60	97.00	96.20

Table 2. Comparative analysis of the proposed method with other state of the art techniques

<table>
<tr><th colspan="3" rowspan="2">Method</th><th colspan="2">Dataset</th></tr>
<tr><th>Wikipedia Dataset</th><th>Mini 20 News Group</th></tr>
<tr><td colspan="3">Status Matrix Representation [19]</td><td>76.00</td><td>71.12</td></tr>
<tr><td colspan="3">Symbolic Representation [13]</td><td>81.99</td><td>NA</td></tr>
<tr><td rowspan="5">Probability based representation [17]</td><td colspan="2">Naive Bayes Classifier with flat ranking</td><td>81.25</td><td>NA</td></tr>
<tr><td rowspan="4">Naïve Bayes Classifier + SVM</td><td>Linear</td><td>85.42</td><td>NA</td></tr>
<tr><td>RBF</td><td>85.42</td><td>NA</td></tr>
<tr><td>Sigmoid</td><td>84.58</td><td>NA</td></tr>
<tr><td>Polynomial</td><td>81.66</td><td>NA</td></tr>
<tr><td colspan="2" rowspan="3">Bag of word representation [18]</td><td>Naive Bayes Classifier</td><td>NA</td><td>66.22</td></tr>
<tr><td>KNN</td><td>NA</td><td>38.73</td></tr>
<tr><td>SVM</td><td>NA</td><td>51.02</td></tr>
<tr><td colspan="3">Proposed Method</td><td>95.50
(3 Clusters)</td><td>88.41
(3 Clusters)</td></tr>
</table>

*NA: Not available

4 Conclusions

The main finding of this work is that the document classification using symbolic representation of clusters achieves considerable increase in classification accuracy when compared to the other existing works. We have made a successful attempt to explore the applicability of symbolic data for document classification. Overall the following are the contributions of this paper:

i. The new method of cluster based representation for document classification
ii. Introduction of a novel symbolic feature selection method
iii. Targeting a good classification accuracy on different datasets.

References

1. Rigutini, L.: Automatic Text Processing: Machine Learning Techniques. PhD thesis, University of Siena (2004)
2. Hotho, A., Nurnberger, A., Paab, G.: A brief survey of text mining. Journal for Computational Linguistics and Language Technology 20, 19–62 (2005)
3. Harish, B.S., Guru, D.S., Manjunath, S.: Representation and classification of text documents: A brief review. International Journal of Computer Applications Special Issue on Recent Trends in Image Processing and Pattern Recognition, 110–119 (2010)
4. Kyriakopoulou, A., Kalamboukis, T.: Text classification using clustering. In: Proceedings of ECML-PKDD Discovery Challenge Workshop (2006)

5. Pereira, F., Tishby, N., Lee, L.: Distributional clustering of english words. In: Proceedings of the 31st Annual Meeting of the Association for Computational Linguistics, pp. 183–190 (1993)
6. Slonim, N., Tishby, N.: The power of word clustering for text classification. In: Proceedings of the European Colloquium on IR Research, ECIR (2001)
7. Dhillon, I., Mallela, S., Kumar, R.: Enhanced word clustering for hierarchical text classification. In: Proceedings of the 8th ACM SIGKDD International Conference on Knowledge Discovery and Data Mining, Canada, pp. 191–200 (2002)
8. Takamura, H., Matsumoto, Y.: Two-dimensional clustering for text categorization. In: Proceedings of the Sixth Conference on Natural Language Learning (CoNLL- 2002), Taiwan, pp. 29–35 (2002)
9. Raskutti, B., Ferr, H., Kowalczyk, A.: Using unlabeled data for text classification through addition of cluster parameters. In: Proceedings of the 19th International Conference on Machine Learning ICML, Australia (2002)
10. Zeng, H.J., Wang, X.H., Chen, Z., Lu, H., Ma, W.Y.: Cbc: Clustering based text classification requiring minimal labeled data. In: Proceedings of the 3rd IEEE International Conference on Data Mining, USA (2003)
11. Bock, H.H., Diday, E.: Analysis of symbolic Data. Springer, Heidelberg (1999)
12. John, G.H., Kohavi, R., Pfleger, K.: Irrelevant features and the subset selection problem. In: Proceedings of International Conference on Machine Learning, pp. 121–129 (1994)
13. Guru, D.S., Harish, B.S., Manjunath, S.: Symbolic representation of text documents. In: Proceedings of Third Annual ACM Bangalore Conference (2010)
14. Zeimpekis, D., Gallopoulos, E.: Tmg: A matlab toolbox for generating term document matrices from text collections. In: Grouping Multidimensional Data: Recent Advances in Clustering, pp. 187–210. Springer, Heidelberg (2006)
15. Tang, B., Shepherd, M., Heywood, M.I., Luo, H.: Comparing dimension reduction techniques for document clustering. In: Kégl, B., Lapalme, G. (eds.) Canadian AI 2005. LNCS (LNAI), vol. 3501, pp. 292–296. Springer, Heidelberg (2005)
16. Guru, D.S., Kiranagi, B.B., Nagabhushan, P.: Multivalued type proximity measure and concept of mutual similarity value useful for clustering symbolic patterns. Journal of Pattern Recognition Letters 25, 1003–1013 (2004)
17. Isa, D., Lee, L.H., Kallimani, V.P., Rajkumar, R.: Text document preprocessing with the bayes formula for classification using the support vector machine. IEEE Transactions on Knowledge and Data Engineering 20, 23–31 (2008)
18. Elnahrawy, E.M.: Log-based chat room monitoring using text categorization: A comparative study. In: International Conference on Information and Knowledge Sharing. Acta Press Series, pp. 111–115 (2002)
19. Dinesh, R., Harish, B.S., Guru, D.S., Manjunath, S.: Concept of status matrix in classification of text documents. In: Proceedings of Indian International Conference on Artificial Intelligence, India, pp. 2071–2079 (2009)
20. Cai, D., He, X., Zhang, W.V., Han, J.: Regularized locality preserving indexing via spectral regression. In: ACM International Conference on Information and Knowledge Management (CIKM 2007), Portugal, pp. 741–750 (2007)
21. Bezdek, J.C.: Pattern Recognition with Fuzzy Objective Algorithms. Kluwer Academic Publishers, Dordrecht (1981)

SimRate: Improve Collaborative Recommendation Based on Rating Graph for Sparsity

Li Yu, Zhaoxin Shu, and Xiaoping Yang

School of Information, Renmin University of China,
Beijing 100872, P.R. China
buaayuli@ruc.edu.cn, innocent.vivi@gmail.com, yang@ruc.edu.cn

Abstract. Collaborative filtering is a widely used recommending method. But its sparsity problem often happens and makes it defeat when rate data is too few to compute the similarity of users. Sparsity problem also could result into error recommendation. In this paper, the notion of SimRank is used to overcome the problem. Especially, a novel weighted SimRank for rate bi-partite graph, SimRate, is proposed to compute similarity between users and to determine the neighbor users. SimRate still work well for very sparse rate data. The experiments show that SimRate has advantage over state-of-the-art method.

Keywords: Collaborative Filtering, SimRank, Similarity, Sparsity.

1 Introduction

A promising technology to overcome such an information overload is recommender systems. One of the most successful recommendation techniques is *Collaborative Filtering* (CF) which identifies customers whose tastes are similar to those of a given customer and it recommends products those customers have liked in the past [1,2]. Although widely being used, there are a lot of issues, such as sparsity, cold-starting etc. [2]. In this paper, we focus on the sparsity of collaborative recommendation. Specially, the notion of famous *SimRank* is exploited to measure the similarity of users. A novel improved *SimRank*, *SimRate*, is proposed for collaborative filtering. The experiments show that proposed method makes better performance than state-of-the-art when rating data is sparse.

There are two primary research contributions made in this paper. (1) Proposing an effective method to overcome the sparsity of collaborative filtering by using *SimRank* intuition, and proposed *SimRate* also extend *SimRank* and *SimRank*++ for collaborative recommendation based on rating. (2) Making varied experiments to verify that proposed method has advantage over the state-of-the art approaches.

The rest of the paper is organized as follows. The next section provides a brief review on collaborative filtering and similarity computing. In section 3, collaborative recommender and its sparsity are introduced. In section 4, famous *Simrank* and *SimRank*++ is analysed and explained why they cannot be used to collaborative filtering based on rate. In section 5, a novel weighted *SimRank* for rate graph, *SimRate*, is designed for sparsity of collaborative filtering. Section 6 describes our experimental work, including experiment datasets, procedure and results of different experiments. In final, some concluding remarks and directions for future research are provided.

L. Cao, J. Zhong, and Y. Feng (Eds.): ADMA 2010, Part II, LNCS 6441, pp. 167–174, 2010.

2 Related Works

Collaborative filtering is a widely used recommendation techniques in many fields, such as news, Movie, book, CD, video, joke, etc [3,4]. A survey on collaborative recommender can be found in [1]. Although being widely used, collaborative filtering has two key limitation, respectively sparsity and cold-starting. In this paper, the sparsity issue is focused. The problem can be resulted for too few rating data which make it hard to compute the similarity between users and make impossible to determine the neighbour users of active user. In practical application, rate data is very sparse, and sparsity often happen [1].

In essence, the sparsity issue in collaborative recommendation is related to similarity computation of users. There are a lot of works dedicate to the problem. Traditionally, collaborative recommender is modelled into a user-item rating matrix, and a user is profiled as a vector. According the notion, widely used measures for the similarity between users include *Cosine*, *Pearson* etc[3,7]. For these methods, it is assumed that the preference of user can be structurally described though their rated items. Recently, being inspired by *PageRank*, the similarity measure based linkage is proposed, such as *SimRank*[5]. Using the measure, the similarity between two nodes is computed by averaging the similarity of their neighbour node. But *SimRank* work for no-weighted graph. When being used in collaborative recommender, rating information will be lost. Most recently, *SimRank++*, a weighted *SimRank* [6], is proposed to compute the similarity of two queries in a weighted bi-partite click graph, where the weight of edge is click times from a query to an ad. Clicking times as weight in *SimRank++* has different means with rate in collaborative recommender systems. *SimRank++* could be used in collaborative recommender. In fact, the experiments in section 6 verify the conclusion. Other works on *SimRank* include *MatchSim*[9], *PageSim*[10] etc. But these methods are not adapted to collaborative filtering for sparsity.

3 Collaborative Recommendation and Its Sparsity

Collaborative filtering is a widely used recommendation technique to date [1,3]. The schemes rely on the fact that each person belongs to a larger group of similar behaving individuals. Formally, for a recommender systems, let R be an $n \times m$ user-item matrix containing rating information of n customers on m items, where $R_{i,j}$ is rate of the i customer for the jth item, as shown in Table 1.

Table 1. User-Item Rating Matrix for Collaborative Recommendation

User/Item	I_1	I_2	…	I_j	…	I_n
U_1	R_{11}	R_{12}	…	R_{1j}	…	R_{1n}
U_2	R_{21}	R_{22}	…	R_{2j}	…	R_{2n}
…	…	…	…	…	…	…
U_i	R_{i1}	R_{i2}	…	$R_{ij}=?$	…	R_{in}
…	…	…	…	…	…	…
U_m	R_{m1}	R_{m2}	…	R_{mj}	…	R_{mn}

CF work as following procedures. First, identify the k most similar users for active user. After the k most similar users *Neighbor(a)* (also called as neighbor users) for active user a have been discovered, their rating information is aggregated to predict the preference of active user for un-rating items. Finally, the predicted item is order by predicting score, and the items with high score will be recommended.

In above algorithm, the similarity between users is the most key step. If the data is too sparse, it is possible that the similarity cannot be computed by widely used similarity measures include *Cosine*, *Pearson*, etc. because there is no common rated item. For example, as shown in Table 2, the similarity between user U_1 and U_2 could not be computed because they have not common rated item. Sometimes, although two users have common rated items (for example, U_1 and U_3 have one common item), the similarity of them could be not accurate because the number of common rated items is too few. But recently proposed *SimRank* is still effective to compute the similarity for sparse data, as shown in Table 3, using *SimRank* metric, the similarity between user U1 and U2, $s(U_1,U_2)=0.537$.

Table 2. An example of Sparse Rating matrix

User \ Item	I_1	I_2	I_3
U_1	3	null	5
U_2	null	5	null
U_3	2	3	null

Table 3. Similarity Matrix by SimRank Matrix (for C=0.8)

User \ User	U_1	U_2	U_3
U_1	1	0.537	0.667
U_2	0.537	1	0.481
U_3	0.667	0.481	1

4 Analysis on SimRank for Collaborative Recommendation

4.1 Basic SimRank

SimRank is a method for computing object similarities [7], used to measure the similarity of the structural context in which objects occur based on their relationships with other objects. Let $S(A, B)$ denote the similarity between user A and user B, and $S(X, Y)$ denote the similarity between item X and Y. For $A \neq B$, $S(A, B)$ be computed as the following,

$$S(A,B) = \frac{C_1}{|E(A)||E(B)|} \sum_{i \in E(A)} \sum_{j \in E(B)} S(i,j) \ . \tag{1}$$

For $X \neq Y$, $S(X, Y)$ be computed as the following,

$$S(X,Y) = \frac{C_2}{|E(X)||E(Y)|} \sum_{i \in E(X)} \sum_{j \in E(Y)} S(i,j) \ , \qquad (2)$$

Where C_1 and C_2 is a constant between 0 and 1.

Although the *SimRank* is an effective method used to compute the similarity of users when rating data is very sparse[7], the *SimRank* is only used for no-weighted bipartite graph. If the algorithm is directly used in collaborative recommender, all rating information will be lost. Although the similarity between two users can be computed, the accuracy of the similarity is doubtful. For example, as shown in the Fig.1, both user *A* and *B* rate 1 for item *X* in Fig.1.(a), both user *A* and *B* rate 5 for item *X* in Fig.1.(b), user *A* rate 1 for item *X* while user B rate 5 for item *X* in Fig.1.(c). If *SimRank* is used, the similarity of user A and B in the three figure case are equal for reflecting rating information. It is obviously unreasonable.

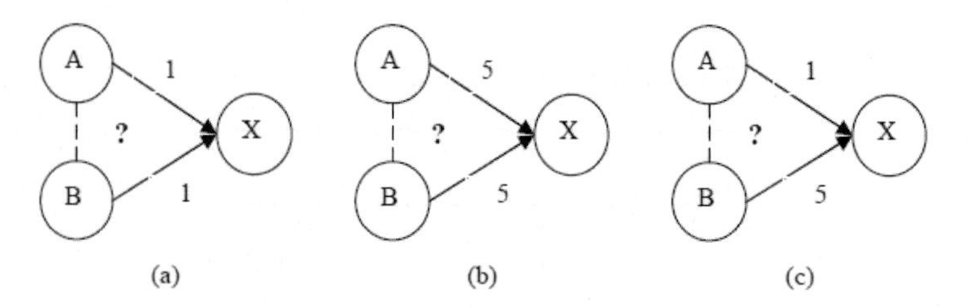

Fig. 1. User A and User B make different rate for the item X

4.2 SimRank++

Recently, a weighted *SimRank* algorithm, *SimRank*++[9], is designed to compute the similarity of the two queries though clicking ad, where weight means the number of clicks that an ad received as a result of being displayed for a query. Weight means relation strength between the ad and the query.

There exist two premises on the weight of *SimRank*++. The first one is that a query strongly related to an ad means the query has lower relationship with other ads. The premise is not true for collaborative recommendation, where a user could rate high for many items. For example, Tom could rate high for a '*football*' book and '*English*' book because he likes '*football*' and '*English*'. Another premise in *SimRank*++ is that smaller weight means lower relationship of the query with the ad and make smaller role when the similarity is computed. The *SimRank* score of the two queries q1 and q2 is consistent with the weights on the click graph. This premise also could not true. In collaborative recommendation, if two users give high rate for common item, as shown in Fig.1.(b), they have similar interest. But if two users give low rate for common item, as shown in Fig.1.(a), then both of the two users dislike the common item, they also have similar preference to some degree. That is to say, in collaborative recommendation, the similarity between two users should be not consistent with their rates. In the next section, a new weighted *SimRank* for rate graph is proposed to improve the two premises.

5 SimRate: A Similarity Measure for Rate Graph

First, in order to use proposed method, rating matrix is firstly translated into bi-partite rate graph. Formally, if there is a rating matrix $R_{m \times n}$ for collaborative filtering, it will be translated into bi-partite rate graph $G = (U, I, E)$, where E is a set of edges that connect user with item. G has an edge (user, item) if the user have rated the item. For any node v in rate graph, we denote by E(v) the set of neighbors of v. If the node v is a user, E(v) denote the set of items rated by user v. If the node v is an item, then E(v) denote the set of users who rated the item v. As shown in Fig.2, rate matrix in Table 2 is modeled into weighted rate bi-partite digraph.

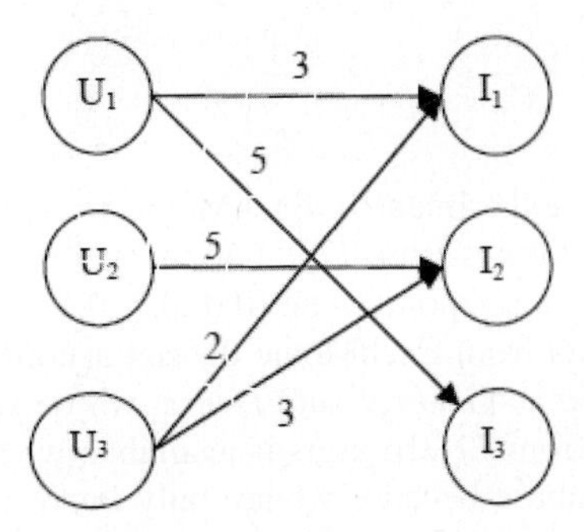

Fig. 2. Bi-partite rate graph

According to above the observation, proposed weighted *SimRank* for rate graph, called as *SimRate*, has following key idea.

Supposing item $X=(X_1,X_2,\ldots,X_m)$ is the rated items set by user A, and item $Y=(Y_1,Y_2,\ldots,Y_n)$ is the rated items set by user B.

1) The similarity between user A and B is got by integrating the similarity between element of set X and the element of set Y, but not by averaging.

2) The contribution of similarity between element pair of items X and Y depend on agreement that user A and B respectively rate for item X and Y. That is to say, R_{AX} is more closer to R_{BY}, the $S(X_i, Y_j)$ has greater contribution on the $S(A,B)$, and higher weight for $S(X_i, Y_j)$will attributed when the $S(A,B)$ is computed. On the contrary, little weight is attributed.

So, the similarity of user A and B is computed as the following,

For $A \neq B$,

$$S(A,B) = \frac{C_1}{\sum_{i \in E(A)} \sum_{j \in E(B)} \frac{1}{|R_{A,i} - R_{B,j}| + 1}} \sum_{i \in E(A)} \sum_{j \in E(B)} S(i,j) \frac{1}{|R_{A,i} - R_{B,j}| + 1}, \quad (3)$$

otherwise, $S(A,B)=0$.

For $i \neq j$,

$$S(i,j) = \frac{C_2}{\sum_{A \in E(i)} \sum_{B \in E(j)} \frac{1}{|R_{A,i} - R_{B,j}| + 1}} \sum_{A \in E(i)} \sum_{B \in E(j)} S(A,B) \frac{1}{|R_{A,i} - R_{B,j}| + 1}, \quad (4)$$

Where C_1 and C_2 is a constant between 0 and 1.

If $i=j$, $S(i,j)=1$.

As done in *SimRank*, an iterative computing method is exploited to compute *SimRate*.

6 Experiments

6.1 Datasets

The datasets used in our experiment is EachMovie provided by DEC Systems Research Center. It contains ratings from 72,916 users on 1,628 movies[11]. User ratings were recorded on a numeric six-point scale (0.0, 0.2, 0.4, 0.6, 0.8, 1.0). We extracted three different data subsets from EachMovie dataset according to the voting density, respectively called as *Sparse*, *General*, and *Dense*, whose statistics are shown in Table 4. Although the data from 72,916 users is available, we restrict the analysis to the first 400 users in the database because we are only interested in the performance of the algorithm under conditions where the number of users and items is low. For each data subset, 400 movies were used as training data, while the other 100 movies were used as testing data in our experiments.

Table 4. Statistics of Experiment Datasets

Dataset	Profile	Ratings of per user	Sparsity
Sparse	400users/500movies/8000ratings	20 movies per user	96%
General	400users/500movies/37795ratings	90-100 movies per user	81.1%
Dense	400users/500movies/73663ratings	150-300 movies per user	63.22%

Mean Absolute Error (MAE) widely used is chose to measure the accuracy of recommendation based on *SimRate*.

In this section, two experiments are made for verify our conclusions. The first experiment is used to verify whether *SimRate* has better performance for sparse data than traditional similarity measure. In the experiment, *SimRate* is compared with *Cosine* and *Pearson*. The second experiment is used to compare the accuracy of *SimRate* with the accuracy *SimRank* and *SimRank++*.

6.2 Experiment Results

All experiment results are shown in Fig.3. First, as shown in Fig.3.(a), apparently *SimRate* has an overwhelming advantage over traditional similarity measure, *Cosine* and *Pearson*, for all data subsets. It is shown the similarity measure based on link for bi-partite rate graph is better than the similarity measure based on characteristic for rate matrix. With the increasing of the sparsity of data subset, *SimRate* has greater advantage over *Cosine* and *Pearson*.

Secondly, as shown in Fig.3.(b), although all of *SimRate*, *SimRank*, *SimRank++* are based on linkage and can work for sparse data, *SimRate* has higher accuracy than *SimRank* and *SimRank++*. It is explained that the *SimRank* with no-weight lose detail rating information. Although both *SimRate* and *SimRank++* focus on weighted bi-partite graph, rate in *SimRate* and clicking in *SimRank++* have different meanings, *Simrank++* will result heavy error when used in collaborative recommend based on rate.

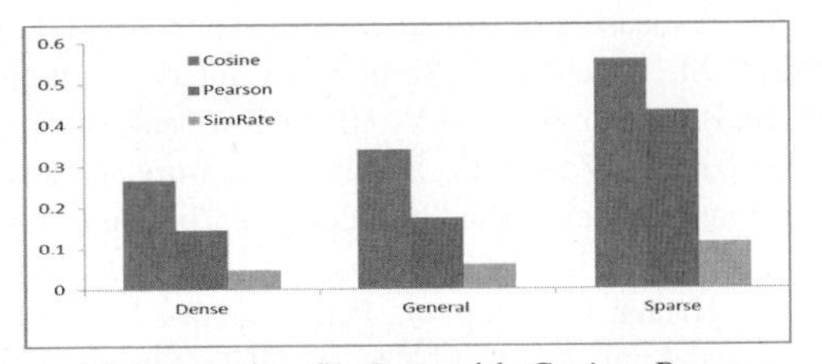

(a) Comparing *SimRate* with *Cosine*, *Pearson*

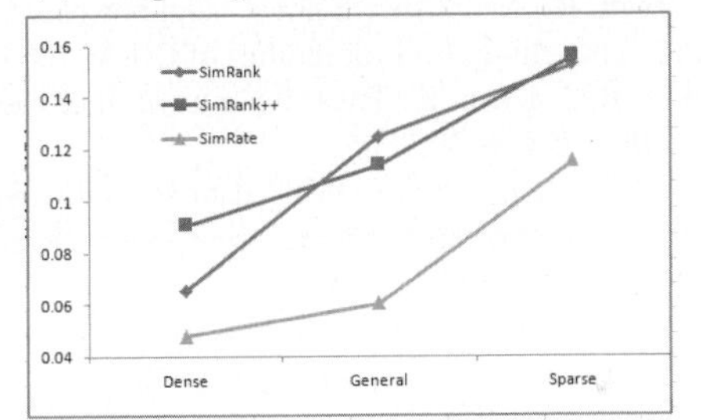

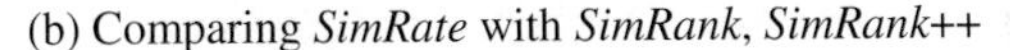

(b) Comparing *SimRate* with *SimRank*, *SimRank++*

Fig. 3. Experiment Results

7 Conclusion

Collaborative filtering is a key recommendation method. Being inspired by the notion of *SimRank*, a novel similarity measure, *SimRate*, is proposed to compute the similarity of users. The method is effective to overcome the sparsity of collaborative filtering, and has advantage over *SimRank*, *SimRank++*. In the future, more dataset would be tested for *SimRate*. It is also interesting to research the performance of *SimRate* for cold-starting in recommender systems.

Acknowledgments. This research was partly supported by National Science Foundation of China under grants No.70871115, RUC Planning on Publishing in International Journal under grants No.10XNK090. Special thank is given to the System Research Center of Digital Equipment Corporation for providing EachMovie database available for research. We also like to thank anonymous reviewers for their valuable comments.

References

1. Adomavicius, G., Tuzhilin, A.: Toward the next generation of recommender systems: a survey of the tate-of-the-art and possible extensions. IEEE Transactions on Knowledge & Data Engineering 17, 734–749 (2005)
2. Hyung, J.A.: A new similarity measure for collaborative filtering to alleviate the new user cold-starting problem. Information Sciences 178, 37–51 (2008)
3. Linden, G., Smith, B., York, J.: Amazon.com recommendations: item-to-item collaborative filtering. IEEE Internet Computing 7, 76–80 (2003)
4. MovieLens, MovieLens dataset (2006), http://www.grouplens.org/ (accessed on April 2006)
5. Glen, J., Jennifer, W.: A measure of structural-context similarity. In: KDD (2002)
6. Ioannis, A., Hector, G.M., Chichao, C.: SimRank++: query rewriting through link analysis of the click graph. In: Proceedings of the VLDB Endowment, vol. 1 (2008)
7. Zeng, C., Xing, C., Lizhu, Z., Zheng, C.: Similarity measure and instance selection for collaborative filtering. International Journal of Electronic Commerce 8, 115–129 (2004)
8. Xu, J., Yuanzhe, C., Hongyan, L., Jun, H., Xiaoyong, D.: Calculating Similarity Efficiently in a Small World. In: Huang, R., Yang, Q., Pei, J., Gama, J., Meng, X., Li, X. (eds.) Advanced Data Mining and Applications. LNCS, vol. 5678, Springer, Heidelberg (2009)
9. Lin, Z., Lyu, M.R., Irwin, K.: Matchsim: a novel neighbor-based similarity measure with maximum neighborhood matching. In: Proceedings of CIKM 2009, pp. 1613–1616 (2009)
10. Zhenjiang, L., Michael, R.L., Irwin, K.: PageSim: a novel link-based measure of web page aimilarity. In: Proceedings of WWW (2006)
11. McJones, P.: EachMovie collaborative filtering data set. DEC Systems Research Center (2002), http://www.research.digital.com/SRC/eachmovie/

Logistic Regression for Transductive Transfer Learning from Multiple Sources

Yuhong Zhang, Xuegang Hu, and Yucheng Fang

School of Computer and Information, Hefei University of Technology, Hefei 230009, China
yuhong.hfut@gmail.com, jsjxhuxg@hfut.edu.cn, ycheng_fang@163.com

Abstract. Recent years have witnessed the increasing interest in transfer learning. And transdactive transfer learning from multiple source domains is one of the important topics in transfer learning. In this paper, we also address this issue. However, a new method, namely TTLRM (Transductive Transfer based on Logistic Regression from Multi-sources) is proposed to address transductive transfer learning from multiple sources to one target domain. In term of logistic regression, TTLRM estimates the data distribution difference in different domains to adjust the weights of instances, and then builds a model using these re-weighted data. This is beneficial to adapt to the target domain. Experimental results demonstrate that our method outperforms the traditional supervised learning methods and some transfer learning methods.

Keywords: transductive transfer learning, classification, multiple sources, logistic regression.

1 Introduction

As a new field of machine learning research, transfer learning is proposed to solve the fundamental problem of mismatched distribution between the training and testing data. And it consists of three categories [1], that is, (1) inductive transfer learning[2,3], a few labeled data in target domain are available; (2) transductive transfer learning[4,5], a lot of labeled data in the source domain are available while no labeled data in the target domain are available; (3) unsupervised transfer learning[6,7], labeled data in both source and target domains are unavailable.

In many applications, labeling data in the target domain is tough. Even though labeling a few data, it would be costly. In addition, when multiple similar sources are available in applications, it is the problem that how to transfer the knowledge from the multi sources to one target domain. In this case, transductive transfer learning is feasible to address these problems. Therefore, we propose a method to address the problem, namely transductive transfer learning from multiple sources.

In fact, many efforts have been focused on this field. Mansour et al.[8] estimated the data distribution of each source to reweight the patterns from different sources. And it is proved that hypothesis combinations weighted by the source distributions benefit from favorable theoretical guarantees. Crammer et al.[9] also addressed the problem based on the assumption that the data distribution of all source domains are

L. Cao, J. Zhong, and Y. Feng (Eds.): ADMA 2010, Part II, LNCS 6441, pp. 175–182, 2010.

the same, and the labels change only due to the varying amounts of noise. Obviously, the assumption usually is not valid in real world.

Bliter et al. proposed Structual Correspondence Learning(SCL)[10], in which a set of pivot features is defined firstly from the unlabeled data, then these pivot features are treated as a new label vector. This model can be adaptive to the target domain by solving the vector. Experimental results show that SCL can reduce the difference between domains, but how to select pivot features is difficult and domain-dependent. As the follow-up work, MI-SCL introduces the Mutual Information (MI) to choose the pivot features[11], it tries to find the pivot features that is dependent highly on the labels in the source domain. Mark introduced the Confidence-Weighted (CW) learning[12] to maintain the probabilistic measure of confidence in each parameter, it aims to improve the parameter estimates and reduce the distributuion's variance.

In contrast to the works mentioned above, in this paper, a new method namely TTLRM is proposed. It depends on an effective tool, logistic regression for transfer learning. Our method aims to handle the issue that all the data in source domains are labeled. Experimental studies present that our method is effective and performs better than the traditional supervised learning methods and some transfer learning methods.

2 Transductive Transfer Learning Based on Logistic Regression from Multiple Sources (TTLRM)

In this paper, there are K source domains and one target domain and they are similar but not the same. A lot of labeled data in the source domain are available while only unlabeled data in the target domain are available. Our goal is to train a high-quality classification model for the target domain.

More formally, let X^{S_k} (k=1...K) be the source domain instance space, X^T be the target domain instance space, and Y={0,1} be the set of labels. According to the definition of the problem, the source domains are denoted as $D_{S_k} = \{(x_1^{S_k}, y_1^{S_k}),(x_2^{S_k}, y_2^{S_k}),...,(x_{N_{S_k}}^{S_k}, y_{N_{S_k}}^{S_k})\}$, where $x_i^{S_k} \in X^{S_k}$, $y_i^{S_k} \in Y$ is the corresponding label and N_{S_k} is the size of the source domain D_{S_k} $(k=1,2,...,K)$.

And the target domain is denoted as $D_T = \{(x_1^T, ?),(x_2^T, ?),...,(x_{N_T}^T, ?)\}$, where $x_i^T \in X^T$ and N_T is the size of D_T, and the symbol'?'means the label is unknown.

Here, there is an assumption that the conditional probability $P_{D_{S_k}}(y \mid x) = P_{D_T}(y \mid x)$, though $P(x)$, the probability of an instance could vary. The assumption indicates that if the identical instances appear in both source and target domains, the labels should be the same.

TTLRM consists of 3 steps: First, the logistic regression model is extended to adapt to transfer learning; Second, the data distribution difference between source domain and target domain is exploited, and the weight of each instance is adjusted; Finally, the model is built on the re-weighted data, which make the classifier more adaptive to the target domain. The description of algorithm is shown as follows.

Algorithm: TTLRM

Input: D_S, D_T and parameter of termination λ.

Output: the classification model of target domain.

(1) Mix the source and target data, and the label of data in D_{mix} is set as the j-th domain where the data come from;

(2)On the labeled D_{mix}, we can estimate the value of distribution difference with the approach of logistic regression;

(3) $w = w^{old} = 0$;

(4)Based on the distribution difference, the weight of data (the vector w) will be updated;

(5) $\forall i = 0,1,...,n$, if $abs(w_i - w_i^{old}) \leq \lambda$ is satisfied, then goto step 6, else goto step 4;

(6)The parameter vector $\{w_j^k\}_{j=0}^n$ ($k = 1,2,...,K$) gotten, we can train the classification model for the target domain based on logistic regression.

2.1 Logistic Regression

Logistic regression is one of the well known and effective tools for classification[13,14]. As the set of instances X is given, conditional probability $P(Y \mid X)$ denotes the probability that X belongs to Y, where Y is discrete-valued and X is any vector containing discrete or continuous variables. Logistic regression estimates the parameter model from the training data set through optimizing the objective function. When Y is a Boolean value, the classification model is as follows,

$$P(y_i = 1 \mid x_i) = \frac{\exp(w^T x_i)}{1 + \exp(w^T x_i)}, \tag{1}$$

$$P(y_i = 0 \mid x_i) = 1 - P(y_i = 1 \mid x_i), \tag{2}$$

where $x_i = \left(1, x_{i1}, ..., x_{ip}\right)^T$, x_{ij} is the j-th attribute of instance x_i , p is the number of attributes, $w = \left(w_0, w_1, ..., w_n\right)^T$ is parameter vector and w_j follows the normal distribution, donated as $w_j \sim N(\mu_j, \sigma_j)$. Parameter vector w can be obtained from the training data set $D = \{x_i, y_i\} |_{i=1}^N$, in term of maximizing log-likelihood function as formula 3.

$$\sum_{i=1}^{N} \log P(y_i \mid x_i) - \sum_{j=1}^{n} \left(\frac{w_j - \mu_j}{\sigma_j} \right). \tag{3}$$

In general, $\mu_j = 0$, $\sigma_j = \sigma$ (σ is a constant) are set in advance. Formula 3 is a concave function for w , therefore, the global optimum can be obtained by using nonlinear numerical optimization methods. When the parameter vector w is obtained, formula 1 and 2 can be used to calculate the probability that the testing data belongs to the positive (or negative) category.

2.2 Logistic Regression for Transfer Learning Model

According to [15], logistic regression is applied in K different source domains to obtain their respective parameter vectors $\{w_j^k\}_{j=0}^n, k=1,2,...,K$. Then, the parameter vector of the target domain is denoted as $w \sim N(\mu,\sigma)$, where the parameter vector w in the target domain can be obtained by maximizing log-likelihood function as formula 4.

$$\sum_{i=1}^{N_T} \log P(y_i^T \mid x_i^T) - \sum_{j=0}^{n} (\frac{w_j - \mu_j}{2\sigma_j})^2 . \tag{4}$$

It should be mentioned that we need labeled data in the target domain during maximizing formula 4 to obtain parameter vector of the target domain. However, in transductive transfer learning setting, no labeled data are available in the target domain. Meanwhile, as the source domains and the target domain follow different distributions, the labeled data in the source domains can not be utilized directly. To solve this problem, the maximizing log-likelihood function is rewritten as formula 5 so that the labeled data in source domains can be used. And $f(x,y,w) = \log P(y \mid x)$ is supposed.

$$\begin{aligned} &\sum_{i=1}^{N_T} \log P(y_i^T \mid x_i^T) - \sum_{j=0}^{n} (\frac{w_j - \mu_j}{2\sigma_j})^2 \\ &= \sum_{i=1}^{N_T} f(x_i^T, y_i^T, w) - \sum_{j=0}^{n} (\frac{w_j - \mu_j}{2\sigma_j})^2 \\ &= \sum_{(x,y)\in X\times Y} \frac{P_{D_T}(x,y)}{P_{D_{S_k}}(x,y)} P_{D_{S_k}}(x,y) f(x,y,w) - \sum_{j=0}^{n} (\frac{w_j - \mu_j}{2\sigma_j})^2 \\ &\approx \sum_{i=1}^{N_{S_k}} \frac{P_{D_T}(x_i^{S_k}, y_i^{S_k})}{P_{D_{S_k}}(x_i^{S_k}, y_i^{S_k})} f(x_i^{S_k}, y_i^{S_k}, w) - \sum_{j=0}^{n} (\frac{w_j - \mu_j}{2\sigma_j})^2 \end{aligned} \tag{5}$$

According to formula 5, the data in the source domains can be utilized by given the corresponding weight $\frac{P_{D_T}(x_i^S, y_i^S)}{P_{D_S}(x_i^S, y_i^S)}$ for each instance (x_i^S, y_i^S). Thus, a precise parameter vector can be learned through maximizing the objective function.

2.3 Estimate the Weight of the Source Data

The weight in formula 5 reflects the difference of the data distribution between the source domain and target domain, and it is irrelative to the label, but is caused by the difference between $P_{D_T}(x_i^S)$ and $P_{D_{S_k}}(x_i^{S_k})$ (seen form formula 6). Thus, our goal is to estimate the weight of each instance.

$$\frac{P_{D_T}(x_i^S, y_i^S)}{P_{D_{S_k}}(x_i^{S_k}, y_i^{S_k})} = \frac{P_{D_T}(y_i^S \mid x_i^S) P_{D_T}(x_i^S)}{P_{D_{S_k}}(y_i^{S_k} \mid x_i^{S_k}) P_{D_{S_k}}(x_i^{S_k})} = \frac{P_{D_T}(x_i^S)}{P_{D_{S_k}}(x_i^{S_k})} \tag{6}$$

Here, we introduce a method to estimate the value of the difference in data distribution between source and target domain. Firstly, merging all the data in source domain and target domain, the instance set $X^{mix} = \bigcup_{k=1}^{K} x_i^{S_k} \bigcup x_i^T$ of the mixed domain D_{mix} is obtained. Then, ε is set to the label of the mixed domain, i.e., the difference of instance come from, where $\varepsilon = j$ infers that instance *x* in the mixed domain comes from the *j*-th source domain, $\varepsilon = 0$ specifies that *x* comes from the target domain. Therefore it can also be denoted as $D_{mix} = \left\{(x_i^{S_1},1)|_{i=1}^{N_{S_1}},...,(x_i^{S_K},K)|_{i=1}^{N_{S_K}},(x_i^T,0)|_{i=1}^{N_T}\right\}$.

Distinctly, the distribution probabilities of source and target domains were rewritten in the mixed domain as the formula 7 shows.

$$\frac{P_{D_T}(x)}{P_{D_{S_k}}(x)} = \frac{P_{D_{mix}}(x \mid \varepsilon=0)}{P_{D_{mix}}(x \mid \varepsilon=k)} = \frac{P_{D_{mix}}(\varepsilon=k)}{P_{D_{mix}}(\varepsilon=0)} \frac{P_{D_{mix}}(\varepsilon=0 \mid x)P_{D_{mix}}(x)}{P_{D_{mix}}(\varepsilon=k \mid x)P_{D_{mix}}(x)} = \frac{P_{D_{mix}}(\varepsilon=k)}{P_{D_{mix}}(\varepsilon=0)} \frac{P_{D_{mix}}(\varepsilon=0 \mid x)}{P_{D_{mix}}(\varepsilon=k \mid x)} \tag{7}$$

To evaluate the value of $\frac{P_{D_{mix}}(\varepsilon=0 \mid x)}{P_{D_{mix}}(\varepsilon=k \mid x)}$ in formula 7, the strategy of multinomial logistic regression is used. It is assumed that the data in mixture domain satisfies the following formula.

$$\log \frac{P_{D_{mix}}(\varepsilon=k \mid x)}{P_{D_{mix}}(\varepsilon=0 \mid x)} = \beta_k^T x, \tag{8}$$

where $\beta_k = (\beta_{k0}, \beta_{k1}, ..., \beta_{kn})^T$, $k = 1, 2, ..., K$. The parameter vector $\boldsymbol{\beta} = (\beta_1, \beta_2, ..., \beta_K)^T$ can be obtained through maximizing the log-likelihood function. As $\boldsymbol{\beta}$ is obtained, we can calculate the value of $\frac{P_{D_{mix}}(\varepsilon=0 \mid x)}{P_{D_{mix}}(\varepsilon=k \mid x)}$ according to formula 8 and estimate the weight value of $\frac{P_{D_T}(x_i^S)}{P_{D_{S_k}}(x_i^{S_k})}$ according to formula 7.

The difference of the data distribution between the source domains and the target domain, $\frac{P_{D_T}(x_i^S)}{P_{D_{S_k}}(x_i^{S_k})}$ is obtained, the weight vector *w* can be solved by maximizing the formula 6. Lastly, the model is trained from the re-weighted data. It is conductive to make the classifier more adaptive to the target domain.

3 Experiments and Result Analysis

3.1 Data Sets

In order to evaluate TTLRM, we perform the experiments on the sentiment classification data set, including the product reviews of Amazon.com. It was first used in [10]. Corpus is derived from four domains: Books, DVDs, Electronics, Kitchens. Each domain contains 2000 texts respectively and there are two labels {positive, negative}. Different domain distributions are similar but not identical. We randomly select 1000 texts from each domain, as a labeled data set. The remaining ones are regarded as an unlabeled data set. In this section, we compare our algorithm against the traditional supervised classification learning (such as NB, SVM and LR) and transductive transfer learning (includes SCL, SCL-ML and CW). In addition, we set the parameter $\lambda = 0.05$ [16] in the experiments.

3.2 Experiment Results

Table 1 reports the experiment results. In this table, first column is the data set. The first three letters denote the source domains and the last letter denotes the target domain. For example, "BEK-D" indicates that the classifiers are trained in books, electronics, kitchens and tested in DVDs.

In the observation form table 1, we can find that all of the transfer learning methods outperform the baselines, and TTLRM performs better than traditional supervised machine learning algorithm (NB, SVM and LR) on all of the data sets. More specifically, the accuracy could be improved by [7.45%,14.15%] on average, which indicates that our algorithm is more adaptive to the target domain when data distributions are different. Meanwhile, as compared with the transfer learning methods, MIILR does not perform as well as CW-PA, but there is little deviation. However, in comparison with SCL and SCL-MI, TTLRM can achieve higher accuracy. The improvement is in the range of [2.58%, 7.48%] for all of the data sets except BDE-K. On the BDE-K dataset, the accuracy in TTLRM is improved by 1.06% compared to SCL, while it is lower than SCL-MI by 0.17%. It is still comparative. In sum, TTLRM can build an effective model for the target domain where no labels are available from multiple source domains.

Table 1. Comparisons of MTTLRM and Other Method in Classification

Source-Target	Baselines			TL Methods			
	NB	SVM	LR	SCL	SCL-MI	CW-PA	TTLRM
BEK-D	70.85	70.15	71.15	74.57	76.30	80.29	79.35
DEK-B	68.35	69.40	68.85	72.77	74.57	80.45	80.25
BDE-K	69.80	71.55	73.50	80.83	82.06	82.60	81.89
BDK-E	67.35	73.05	72.30	78.43	78.92	84.36	81.50

4 Conclusion

We proposed a new method TTLRM based on logistic regression for transductive transfer learning in this paper. Our method aims to handle the problem of knowledge transfer from multiple sources to one target domain. It adjusts the weight of each instance by exploiting the distribution difference between the source data and the target data, and trains a model on the re-weighted data which make the classifier more adaptive to the target domain. Experimental results show that TTLRM is effective.

However, the proposed algorithm assumes that the source domain and the target domain are similar. When they are dissimilar, the brute-force transfer may fail, even may destroy the performance of the target task, namely negative transfer. Thus, how to judge the similarity among different domains to avoid negative transfer is our future work.

Acknowledgments. The work was supported by the 973 program of china under the grant No. 2009CB326203 and the National Nature Science Foundation of China (NSFC) under the grant No. 60828005.

References

1. Yang, Q.: An Introduction to Transfer Learning. In: Tang, C., Ling, C.X., Zhou, X., Cercone, N.J., Li, X. (eds.) ADMA 2008. LNCS (LNAI), vol. 5139, Springer, Heidelberg (2008)
2. Dai, W., Yang, Q., Xue, G., Yu, Y.: Boosting for Transfer learning. In: 24th International Conference on Machine Learning, Corvallis, Oregon USA, pp. 193–200 (2007)
3. Lee, S., Chatalbashev, V., Vickrey, D.: Learning a meta-level prior for feature relevance from multiple related tasks. In: 24th International Conference on Machine Learning, pp. 489–496. ACM, Corvalis (2007)
4. Sugiyama, M., Nakajima, S., Kashima, H.: Direct importance estimation with model selection and its application to covariate shift adaptation, `http://books.nips.cc/cgi-bin/archer_query.cgi?first=1&scope=all&hits_per_page=10&description=long&keywords=importance+estimation&submit=Search+NIPS`
5. Huang, J., Smola, A., Gretton, A.: Correcting sample selection bias by unlabeled data. In: 19th Annual Conference on Neural Information Processing Systems, `http://books.nips.cc/cgi-bin/archer_query.cgi?first=1&scope=all&hits_per_page=10&description=long&keywords=Correcting+sample&submit=Search+NIPS`
6. Dai, W., Yang, Q., Xue, G.: Self-taught clustering. In: 25th International Conference of Machine Learning, Helsinki, Finland, pp. 200–207 (2008)
7. Wang, Z., Song, Y., Zhang, C.: Transferred dimensionality reduction. In: Daelemans, W., Goethals, B., Morik, K. (eds.) ECML PKDD 2008, Part II. LNCS (LNAI), vol. 5212, pp. 550–565. Springer, Heidelberg (2008)
8. Mansour, Y., Mohri, M., Rostamizadeh, A.: Domain adaptation with multiple sources. Advances in Neural Information Processing Systems 21, 1041–1048 (2009)
9. Crammer, K., Kearns, M., Wortman, J.: Learning from multiple sources. Journal of Machine Learning Research 9, 1757–1774 (2008)

10. Blitzer, J., Dredze, M., Pereira, F.: Biographies, Bollywood, Boom-boxes and Blenders: Domain Adaptation for Sentiment Classification. In: The 45th Annual Meeting of the Association for Computational Linguistics, Prague, Czech Republic, pp. 440–447 (2007)
11. Blitzer, J., McDonald, R., Pereira, F.: Domain adaptation with structural correspondence learning. In: the Conference on Empirical Methods in Natural Language, Sydney, Australia, pp. 120–128 (2006)
12. Dredze, M., Kulesza, A., Crammer, K.: Multi-domain learning by confidence-weighted parameter combination. Machine Learning 79(1-2), 123–149 (2010)
13. Liao, X., Xue, Y., Carin, L.: Logistic regression with an auxiliary data source. In: 21st International Conference on Machine Learning, Bonn, Germany, pp. 505–512 (2005)
14. Glasmachers, T., Igel, C.: Maximum Likelihood Model Selection for 1-Norm Soft Margin SVMs with Multiple Parameters. IEEE Transactions Pattern Analysis and Machine Intelligence 32(8), 1522–1528 (2010)
15. Marx, Z., Rosenstein, M., Kaelbling, L.: Transfer learning with an ensemble of background tasks. In: NIPS Workshop on Transfer Learning, http://iitrl.acadiau.ca/itws05/papers.htm
16. Fang, Y.: Logistic Regression for Ttansductive Transfer Learning. The dissertation of Hefei University of Techonology. 19-25 (2010)

Double Table Switch: An Efficient Partitioning Algorithm for Bottom-Up Computation of Data Cubes

Jinguo You[1], Lianying Jia[2], Jianhua Hu[1], Qingsong Huang[1], and Jianqing Xi[2]

[1] School of Information Engineering and Automation,
Kunming University of Science and Technology, Kunming, Yunnan 650051
{jgyou,jhhu,qshuang}@kmust.com

[2] School of Computer Science and Engineering, South China University of Technology,
Guangzhou, Guangdong 510000
{jia.ly,csjqxi}@scut.edu.cn

Abstract. Bottom-up computation of data cubes is an efficient approach which is adopted and developed by many other cubing algorithms such as H-Cubing, Quotient Cube and Closed Cube, etc. The main cost of bottom-up computation is recursively sorting and partitioning the base table in a worse way where large amount of auxiliary spaces are frequently allocated and released. This paper proposed a new partitioning algorithm, called Double Table Switch (DTS). It sets up two table spaces in the memory at the beginning, where the partitioned results in one table are copied into another table alternatively during the bottom-up computation. Thus DTS avoids the costly space management and achieves the constant memory usage. Further, we improve the DTS algorithm by adjusting the dimension order, etc. The experimental results demonstrate the efficiency of DTS.

Keywords: Data warehouse, Data cube, Bottom-up computation, Partitioning algorithm, Double Table Switch.

1 Introduction

Data cube is an important model and operator in data warehouses and OLAP. It improves query performance by pre-aggregating some group-bys instead of computing them on the fly. However, the data cube computation consumes large amount of CPU time and disk storage. To address this issue, a lot of works are studied. Most of previous data cubing algorithms such as Multi-Way Array Cube [1] compute from the base cuboid (the most aggregated group-by) to less aggregated group-bys, which is called top-bottom computation. On the contrary, BUC [2] first proposed bottom-up cubing approach by computing from the ALL cuboid to others. BUC shares fast sorting and partitioning and facilitates Apriori pruning, which makes it very efficient, especially for iceberg and sparse cubes. Due to its importance and usability, BUC is adopted by extensive other cubing algorithms including H-Cubing [3], Quotient Cube [4,5] and Closed Cube [6,7], etc. Star-Cubing [8] integrates top-down and bottom-up computation. In spite of such lots of works, BUC still warrants a thorough study. Any improvement on it may have wide effect.

L. Cao, J. Zhong, and Y. Feng (Eds.): ADMA 2010, Part II, LNCS 6441, pp. 183–190, 2010.

Essentially, BUC recursively partitions the tuples in the base table by depth first searching the next dimension. As pointed out in [2], the majority of the time in BUC is spent partitioning and sorting the data. Thus partitioning optimization plays a key role for improving the performance of BUC. By far, none of cubing technologies in literatures, to our knowledge, study efficient partitioning of BUC. This paper proposed a new partitioning algorithm called Double Table Switch, DTS. It optimizes the memory management by avoiding lots of auxiliary space used by partitioning frequently allocated and released. It is further improved by adjusting the dimension order.

The rest of the paper is organized as follows. Section 2 describes the problem of original BUC algorithm. In section 3, we present the DTS mechanism and develop its algorithm. Further, the refinements of DTS are discussed in section 4. The efficiency of DTS is demonstrated in Section 5. Finally, section 6 concludes our study.

2 Original Bottom-Up Cubing Algorithm

Unlike top-down computation, BUC starts at the ALL cuboid and moves upward to the base cuboid. As shown in Fig. 1(a), BUC reads the first dimension and partitions the tuples in the base table P_0 based on its cardinality, namely the distinct values in dimension A. In general, sorting or partitioning data in a dimension needs an auxiliary space to store the sorted result or exchange the keys. Therefore, a new space P_1 is allocated to accommodate the return results of partitioning, a1 partition and a2 partition. For each partition in dimension A, BUC partitions the next dimension. Then P_2 is allocated, and then P_3, until the partition doesn't meet the minimum support (for full cube computation, it is one, i.e. the partition contains only one tuple). For now, we have the maximum space usage, P_0+P_1+P_2+P_3. Next, we don't need P_3 any more and it is released and retrieved. Then (a1, b2) partition of P_2 is processed. The recursive partitioning procedure of BUC can be expressed by a processing tree shown in Fig. 1(b), where the node denotes the partition and the edge the space allocated. If we depth first search the tree, we have the space allocation and release sequence, P_0, P_1, P_2, P_3, P_3, P_3'...

The main problem of BUC is that the auxiliary space is allocated and released much frequently. Among efficient sort algorithms, the space complexity of quick sort

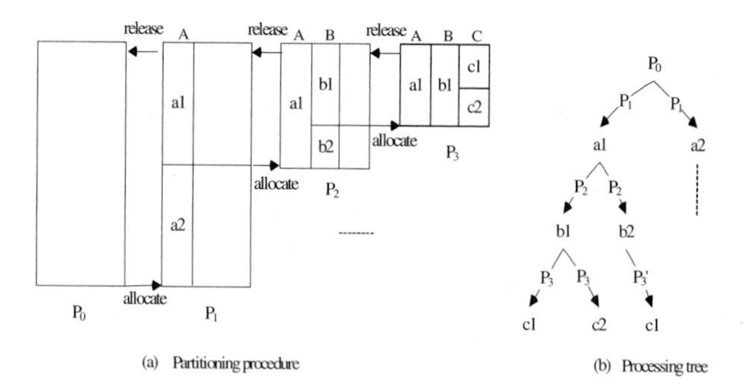

Fig. 1. BUC partitioning

is O(log$n$), whereas merge sort O(n). CountingSort is often used in the BUC algorithm when a dimension of a partition is packed, in the sense that the sort key in that dimension should be integer value between zero and its cardinality, and that the cardinality is known in advance [2]. CountingSort needs the O(n+k) space. Although the space cost may be trivial and insignificant for one way sorting, the recursive partitioning results in multi-way sorting and thus the heavy space management. Quite a lot of partitioning time is unnecessarily spent on allocating and releasing the space. When the dimension is high, the situation becomes even worse due to more partitions.

3 Double Table Switch Technology

To overcome the difficulty raised by bottom-up computation, this section describe a new partitioning approach, called Double Table Switch, DTS. Also the algorithm implementation of DTS is given and the complexity is analyzed.

3.1 Mechanism

Initially, Double Table Switch sets up two spaces in the memory. One space is used to store the basic table. The other space is the copy of the basic table, which only stores the tuple address of the basic table rather than the entire tuple data.

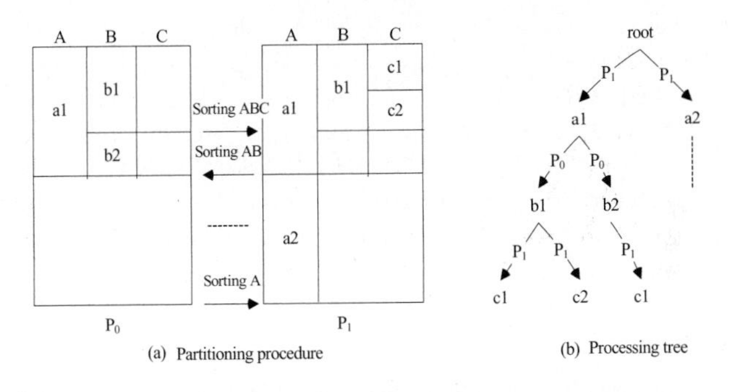

Fig. 2. DTS partitioning

Fig. 2(a) shows the partitioning computation with DTS. There are two spaces: P_0 and P_1. First, the data of the base table is loaded into the P_0. The partitioning begins at the first dimension in P_0. The tuples in P_0 are sorted in this dimension. There are two distinct values in dimension A, so two partitions are generated. To accommodate the results, the addresses of tuples in all partitions are copied into P_1. Thus the next partitioning can proceed in P_1. Next, the tuples in a1 partition of P_1 are sorted in dimension B. The partitions, (a1, b1) and (a1, b2) are formed and then wrote back to P_0 in turn, just overriding the a1 partition. The dimension C is sorted and partitioned in (a1, b1) partition of P_0. Then (a1, b1, c1) partition and (a1, b1, c2) partition are copied into P_1, ... It switches P_0, P_1, then P_0, ..., until the partition contains only one single tuple or the minsup is not met. When a partition is searched, we proceed to next partition in the previous dimension. The processing tree is formed as shown in Fig. 2(b). If we explore the processing tree in depth first searching way, we have the space access sequence, P_0, P_1, P_0, P_1, P_1, P_0, P_1,…It's delicate that copying the partition results in P_0 to P_1 can guarantee not influencing other partitions in P_1 and vice versa.

3.2 Algorithm

Fig. 3 presents the partitioning algorithm of DTS. Initially, the memory allocates two spaces, the base table $B[0]$ and its copy $B[1]$. The global variable *sw* indicates which space is currently used between $B[0]$ and $B[1]$. The variables *bPos* and *ePos* are the boundary of the partition that will be sorted.

```
Algorithm 1. DTS
Global: base table B[0], the copy B[1], the switch sw
Input: int bPos, int ePos, int d, int card, int *freq
Output: void
Method:  //using the counting sort approach
1:        for int i = 0; i < card; i++ do
2:          freq[i] = 0;
3:        end for
4:        for i = bPos; i < ePos; i++ do
5:          freq[ B[sw][i][d] ] ++;
6:        end for
          //calculate the begin position per element.
7:        let freq[0] = bPos;
8:        for i = 1; i < card; i++ do
9:          freq[i] = freq[i] + freq[i - 1];
10:       end for
11:       for i = bPos; i < ePos; i++ do
            //copy the data of B[sw] to B[!sw]
12:         B[!sw][ freq[ B[sw][i][d] ] - 1 ] = B[sw][i];
13:         freq[ B[sw][i][d] ] --;
14:       end for
15:       sw = !sw;  //make B[!sw] the current space
16:       return;
```

CountingSort approach is adopted since it excels most other sort methods and attains linear running time cost (from line 1 to line 14). The array variable *freq* obtains the start position of each partition result (line 9). Then the tuples of the current space $B[sw]$ are copied to another space $B[!sw]$ according to the start position (line 12). Finally, line 15 makes $B[!sw]$ the current space.

3.3 Complexity

Let the base table contains T tuples, N dimensions with a cardinality of C per dimension. From Fig. 1(a) and Fig. 2(a), the maximum space occupied by BUC partitioning is $P_0 + \sum_{i=0}^{N-1} P_i = O(2T + T(C^{N-1} - 1)/(C^N - C^{N-1}))$, whereas DTS has the even less and constant space, $O(2T)$. In practice, the copy table only stores the tuple pointers of the base table, so the space of DTS is less than $O(2T)$. For instance, if the base table contains 1,000,000 tuples and 10 dimensions with 4 byte size per dimension, the copy table will occupy 4 MB memory in a 32-bit PC machine, while the size of the base table is 40 MB. Once the two table spaces are allocated in memory, the memory usage is not varied, fixed at a constant value, say 44MB.

From Fig. 1(b) and Fig. 2(b), let every node has C branches except the leaf nodes, so the total number of partitioning times is C^N at most. Although the two algorithms have the same partitioning times, BUC partitioning needs extra C^N times space management increasing exponentially. DTS doesn't need it and thus saves the computation time greatly.

4 Refinements

This section discusses some technologies that refine and improve the partitioning algorithm by DTS. We prove the efficiency of dimension ordering by the cardinality theoretically. Integer mapping is also discussed here.

4.1 Dimension Ordering

BUC recursively partitions in the dimension of the input partition by depth first searching until the partition contains only one tuple or doesn't meet the minimum support. To reduce the recursion, BUC should be stopped as early as it can. The influence of the dimension ordering was referred in [2]. The higher the cardinality in a dimension, the closer BUC is to prune some partition for saving computation. Here we get the same result but for multiple dimensions and further prove it.

Let the base table contains T tuples, N dimensions with the cardinality $C_0, C_1, \ldots, C_{N-1}$ respectively per dimension. Also assume that $T = C_0 \times C_1 \times \ldots \times C_{N-1}$. BUC sorts the tuples in the first dimension in the base table, generating C_0 partitions with the average size T/C_0. Then the partitions are re-partitioned in the second dimension, generating C_1 partitions with the average size $(T/C_0)/ C_1$. In this way, we can induce the following:

$$P_N = \frac{P_{N-1}}{C_{N-1}}, \tag{1}$$

P_N denotes the average size of the partitions, $P_0 = T$, and BUC stops if P_N =1or P_N=minsup.

From (1), it is easy to conclude that if we sort the dimensions according to the cardinality each dimension in descending order, P_N firstly reaches the limitation, while P_N stops recursion at last by sorting the dimensions in ascending order. Therefore, if the cardinality is known in advance, the recursive partitioning time by DTS will be reduced further by adjusting dimension order.

4.2 Integer Mapping

Mapping an attribute value with the string or date type to an integer is a common technology since the integer values are small and sort quickly. When a value is searched in an attribute, the value is firstly looked up in a hashed symbol table. If it exists in the table, the mapped integer is returned. If not, it is assigned a new integer. For integer attribute values originally, should they be mapped into integer values again? It's wise to do so. Because the integer attribute values may not continuous, mapping them to continuous and dense integer values saves the space and scanning time for CountingSort. For instance, the maximum value in an attribute is 1,000, but the cardinality is 50. Therefore, assigning a 1,000 size array is not appropriate. The values should be mapped to 50 distinct integer values with the maximum value 50.

5 Performance Analysis

To check the efficiency and scalability of the proposed algorithm, we conducted two main experiments on Intel dual core 1.83GHz system with 1 GB RAM running Windows 2003. Test 1 is to compare the computation performance of BUC using DTS with the original BUC algorithm. Test 2 is to examine the effect of dimension ordering. The two algorithms are all coded using C++. The minsup is 1 and the closed check is added. Accordingly, we generated two synthetic datasets using the same data generating tools as in [7,8]. The parameter configurations of the dataset in test 1 are as follows: the number of tuples is 1,000,000, and the number of dimensions increases from 5 to 13 with a cardinality of 95 per dimension. The parameter configurations of the dataset in test 2 are as follows: 1,000,000 tuples, 7 dimensions with the cardinality 160, 110, 70, 40, 20, 10, 2 respectively. We adjust the dimension order by re-arranging the cardinalities and generate 4 dimension orders in the following table 1.

Table 1. Dimension ordering by cardinality per dimension

No.	Cardinality
1	160, 110, 70, 40, 20, 10, 2
2	110, 160, 70, 40, 20, 10, 2
3	110, 2, 70, 20, 40, 160, 10
4	2, 10, 20, 40, 70, 110, 160

5.1 Performance of DTS and BUC

Fig. 3(a) shows that DTS is much faster than BUC. When the base table contains 5 dimensions, DTS and BUC have comparable performance. However, as the number

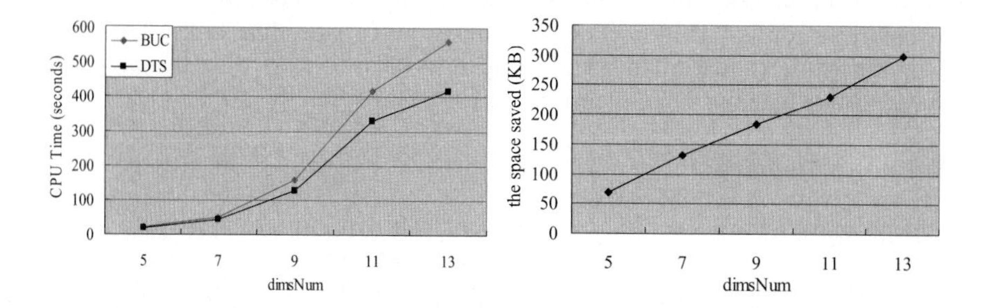

Fig. 3. (a) CPU time w.r.t. dimensions; (b) Memory space saved by DTS w.r.t. dimensions

of dimensions increases, DTS outperforms BUC totally. With 11 dimensions, DTS saves 80 seconds and improves the performance by 20%. The reason is that the more the dimensions, the more partitions and the more recursions are called. For BUC, it needs to spend plenty of time on dealing with the memory management.

5.2 Memory Space of DTS and BUC

Also we observed that the memory space used by BUC varies continually because the partition space is created and released frequently, while DTS keeps on a constant memory usage all the running time. Fig. 3(b) shows that the memory space is fairly saved by DTS compared with BUC as the number of dimensions is increased. Although the saved space is not very obvious in this test, the complexity analysis suggests that DTS saves the C^N times space management by avoiding space allocation and release. The above performance test further demonstrates it. Thus it needs to be emphasized that DTS is a time efficient and space fairly efficient partitioning algorithm.

5.3 Effect of Dimension Ordering

Fig. 4 examines the effect of dimension ordering. We run DTS and BUC respectively with 4 different dimension orders as shown in Table 1. The No. 1 arranges the cardinalities in descending order, while the No. 4 in ascending order. The No. 2 and No. 3 arranges the cardinalities arbitrarily and differs slightly each other.

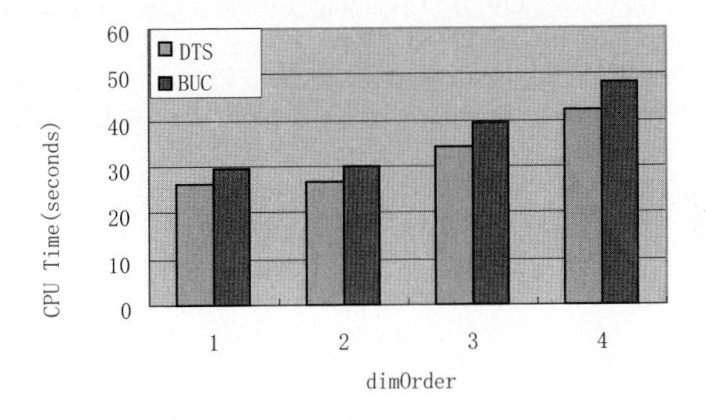

Fig. 4. CPU time w.r.t. different dimension orders

The experimental results show that No. 1 achieves the least CPU time, 26 seconds, while No. 4 needs the maximum time, 42 seconds. It's interesting that No. 1 improves the performance by 38% only by adjusting the dimension ordering. The CPU time of No. 2 and No. 3 are about the same and are between those of No. 1 and No. 4.

6 Conclusions

Double Table Switch is an interesting and efficient approach for recursively partitioning and sorting. We demonstrated that DTS outperforms BUC partitioning by complexity analysis and experimental evaluation. DTS is also important for other BUC family algorithms. Furthermore, we note that DTS is not only limited to improving BUC, but also it can be generalized and applied to other scenarios. Sorting on a composite key or grouping by multiple attributes using DTS are issues for future research.

Acknowledgments. The research was supported by Yunnan Educational Foundation (09C0109) and the Nature Science Foundation of Yunnan Province of China (2006PT06, 14118226).

References

1. Zhao, Y., Deshpande, P., Naughton, J.F.: An Array-Based Algorithm for Simultaneous Multidimensional Aggregates. In: SIGMOD, pp. 159–170 (1997)
2. Beyer, K., Ramakrishnan, R.: Bottom-Up Computation of Sparse and Iceberg CUBEs. In: SIGMOD, pp. 359–370 (1999)
3. Han, J.W., Pei, J., Dong, G., Wang, K.: Efficient computation of iceberg cubes with complex measures. In: SIGMOD, pp. 441–448 (2001)
4. Lakshmanan, L., Pei, J., Han, J.W.: Quotient Cubes: How to Summarize the Semantics of a Data Cube. In: VLDB, pp. 778–789 (2002)
5. Lakshmanan, L., Pei, J., Zhao, Y.: QCTrees: An Efficient Summary Structure for Semantic OLAP. In: SIGMOD, pp. 64–75 (2003)
6. Li, S.E., Wang, S.: Research on Closed Data Cube Technology. Journal of Software 15(8), 1165–1171 (2004)
7. Xin, D., Shao, Z., Han, J.W., Liu, H.: C-Cubing: Efficient Computation of Closed Cubes by Aggregation-Based Checking. In: ICDE (2006)
8. Xin, D., Han, J.W., Li, X., et al.: Computing Iceberg Cubes by Top-Down and Bottom-Up Integration: The StarCubing Approach. IEEE Trans. Knowl. Data Eng. 19(1), 111–126 (2007)

Tag Recommendation Based on Bayesian Principle

Zhonghui Wang[2] and Zhihong Deng[1,2]

[1] The State Key Lab of Computer Science, Institute of Software,
Chinese Academy of Sciences, Beijing 100190, China
[2] Key Laboratory of Machine Perception (Ministry of Education),
School of Electronics Engineering and Computer Science, Peking University
wangzh@cis.pku.edu.cn, zhdeng@cis.pku.edu.cn

Abstract. Social tagging systems have become increasingly a popular way to organize online heterogeneous resources. Tag recommendation is a key feature of social tagging systems. Many works has been done to solve this hard tag recommendation problem and has got same good results these years. Taking into account the complexity of the tagging actions, there still exist many limitations. In this paper, we propose a probabilistic model to solve this tag recommendation problem. The model is based on Bayesian principle, and it's very robust and efficient. For evaluating our proposed method, we have conducted experiments on a real dataset extracted from BibSonomy, an online social bookmark and publication sharing system. Our performance study shows that our method achieves good performance when compared with classical approaches.

Keywords: Tag recommendation, Bayesian principle, Social tagging Algorithm.

1 Introduction

With the development of Web2.0, web users now are encouraged to create multimedia contents by themselves, and share those contents on line. One kind of user-created contents is tag. Tag is keywords or phrases used to label a resource, and can help one to organize resources by personal interest. Tagging can be useful in many areas, such as query expansion [1], web search [2], personalized search [3, 4], web resource classification [5] and clustering [6]. Meanwhile, social tagging is the progressing that people share their tags to each others, and can help user to find resources he/she would like to view, or to find other users who share similar habits with him/her.

Tag recommendation can assist users by suggesting a set of tags that users are likely to use to bookmark a web resource. Personalized tag recommendation takes a user's previous tagging behaviors into account when making suggestions, and usually obtains better performance compared with general tag recommendation. Because tag-recommendation technical could successfully meet personally requirements, nowadays more and more social tagging websites have added their tag recommendation model. According to what kind of resources are supported, there are BibSonomy, Flickr[1] (sharing photos), del.icio.us[2] (sharing bookmarks), Movielens[3] (sharing movies), etc.

---

[1] http://www.flickr.com/
[2] http://delicious.com/
[3] http://movielens.umn.edu/

L. Cao, J. Zhong, and Y. Feng (Eds.): ADMA 2010, Part II, LNCS 6441, pp. 191–201, 2010.

Nowadays, many researches are focus on this problem. A detailed account of different types of tagging systems can be found in [7, 8]. Generally speaking, there exist two types of methods which are content-based methods and graph-based methods. Content-based methods [9, 10] use content of resources to expand the recommendation tag-collection, therefore could provide recommendation for new users and new resources. Graph-based approaches make full use of the correlation between data to give recommendations, and usually perform better than content-based methods.

One of the most dominant methods used in recommender systems is Collaborative filtering (CF) [11, 12 and 13]. The idea of CF is that user may tag a resource just like other users who have similar tagging habit to him. Due to its simplicity and its promising results, CF has been used wildly, and various approaches have been developed. FolkRank [14] is an adaptation of PageRank that can generate high quality recommendations which are shown empirically to be better than other previous proposed collaborative filtering models. Another effective algorithm is Tensor Reduction algorithm [15]. It performs a 3-dimensional analysis on the usage data, using the Higher Order Singular Value Decomposition (HOSVD) technique.

In this paper, we propose a new method called Bayesian-based Tag Recommendation algorithm (BTR). Using Bayesian principle, we can quickly find out the tags which user may want. For evaluating our method, we have conducted experiments on a real dataset extracted from BibSonomy[4]. Experiment results show that our method achieves good performance when compared with classical existing approaches.

The organization of the rest of the paper is as follows. Section 2 introduces the formal statement of the problem. Section 3 introduces the Tag Recommendation Algorithm based on Bayesian principle. Section 4 presents our performance study. In section 5, we conclude with a summary and point out some future research issues.

2 Problem Statement and Preliminaries

In a social tagging system, users can bookmark web resources by assigning tags to them. Tag recommendation can recommend to the user some personalized tags which the user will use, that is, given a user and a resource without tags, the algorithm should predict which tags the user will use to tag this resource.

Here, Let $U = \{u_1, u_2... u_k\}$ be the set of users, $R= \{r_1, r_2... r_m\}$ be the set of resources and $T = \{t_1, t_2... t_n\}$ be the set of tags. A tagging action, that a user u tags a resource r with a tag t, can be denoted as a post $s= \{u, r, t\}$. Let $S= \{s_1, s_2... s_l\}$ be the set of tagging actions. So a tagging system $F = <U, R, T, S>$.

When user u tags a resource r, which tags in T would likely be used according to his personal habit? If we know this, we can recommend those tags to him before his tagging. It will be a big benefit. Let $p(t \mid u, r)$ be probability that user u will use tag t to annotate resource r. When $<u, r>$ is known, intuitively, the strategy is to recommend those tags which has the biggest $p(t \mid u, r)$.

[4] http://bibsonomy.org

3 The Tag Recommendation Algorithm Based on Bayesian Principle

As pointed out in Section 2, when user *u* look at resource *r*, we should estimate all the probability $p(t \mid u, r)$ ($t \in T$) based on the tagging system *F*. After that, we pick the top-k tag *t* with the biggest $p(t \mid u, r)$ to recommend.

3.1 Recommendation Framework

According to Bayesian principle, we have the following Probability relation:

$$p(t \mid u, r) = \frac{p(u, r \mid t) * p(t)}{p(u, r)} \tag{1}$$

In (1), $p(t \mid u, r)$ means the probability of user *u* uses tag *t* to bookmark resource *r*. $p(u, r \mid t)$ means the Posterior probability of user *u* and resource *r* given a tag *t*. $p(t)$ is the prior probability of tag *t*, and $p(t)$ is the prior probability of tag *t*.

So when estimate $p(t \mid u, r)$, we first need to get $p(u, r \mid t)$, $p(t)$ and $p(u, r)$. But the problem is that data in an actual tagging system *F* is always very sparse. Like in Amazon, there are millions of books, and most users could only tag 0.01% or less. If we using the exact estimate of those probabilities to compute $p(t \mid u, r)$, we may only get an over fitting result. For the purpose of robust and efficiency of the algorithm, we assume that *U* and *R* are independent and identically distributed. According to this assumption, (1) can transform to:

$$p(t \mid u, r) = \frac{p(u, r \mid t) * p(t)}{p(u, r)} = \frac{p(u \mid t) * p(r \mid t) * p(t)}{p(u, r)} \tag{2}$$

Here, $p(u \mid t)$ is the conditional prior probability of user *u* given the tag *t*, $p(r \mid t)$ is the conditional prior probability of resource *r* given the tag *t*, $p(t)$ is the prior probability of tag *t*. $p(u \mid t)$, $p(r \mid t)$ and $p(t)$ respectively represent the effect of *U*, *R* and *T* when generating a recommendation. Because *U* and *R* can't be independent distributed, and the effects of *U*, *R* and *T* for recommending can't be equal, we rewritten (2) as follows:

$$p(t \mid u, r) \propto \frac{p(u \mid t)^{\partial} * p(r \mid t)^{\beta} * p(t)^{\gamma}}{p(u, r)} \tag{3}$$

Here, α, β, γ are parameters that respectively represent the effect of *U*, *R* and *T*. These parameters could be adjusted by manual, and also can be learn from the training data automatically.

Given a user *u* and an resource *r*, for all tag *t*, $p(u,r)$ is the same. So, this part of (3) has nothing to do with the comparing operation and can be removed.

$$p(t \mid u, r) \propto p(u \mid t)^{\partial} * p(r \mid t)^{\beta} * p(t)^{\gamma} \tag{4}$$

After all, as logarithmic computer can keep the original ordering relation, we put a logarithmic computer on (4):

$$y(t,u,r) = \alpha \ln p(u \mid t) + \beta \ln p(r \mid t) + \gamma \ln p(t) \tag{5}$$

Then, given a user *u* and a resource *r*, our algorithm will rank the tags by $y(t,u,r)$.

Given tagging system *F*, all the $\ln p(u \mid t)$, $\ln(r \mid t)$ *and* $\ln p(t)$ can be computed before recommendation. Here we give some definitions:

S_t : the set of post *s*, $s \in F \wedge t \in s$.

$S_{t,u}$: the set of post *s*, $s \in F \wedge t,u \in s$.

$S_{t,r}$: the set of post *s*, $s \in F \wedge t,r \in s$.

Based on these definitions and Principles of Probability and Statistics, we can estimate:

$$\begin{cases} p(t) = \dfrac{|S_t|}{|S|} \\ p(u \mid t) = \dfrac{|S_{t,u}|}{|S_t|} \\ p(r \mid t) = \dfrac{|S_{t,r}|}{|S_t|} \end{cases} \tag{6}$$

Combine (5) and (6):

$$\begin{aligned} y(t,u,r) &= \alpha \ln p(u \mid t) + \beta \ln p(r \mid t) + \gamma \ln p(t) \\ &= \alpha \ln \frac{|S_{t,u}|}{|S_t|} + \beta \ln \frac{|S_{t,r}|}{|S_t|} + \gamma \ln \frac{|S_t|}{|S|} \\ &= \alpha \ln |S_{t,u}| + \beta \ln |S_{t,r}| + (\gamma - \alpha - \beta) \ln |S_t| - \gamma \ln |S| \end{aligned} \tag{7}$$

For the same reason of ignoring *p(u,r)* in (3), here we remove *ln|S|*:

$$\begin{aligned} y'(t,u,r) &= y(t,u,r) - \gamma \ln |S| \\ &= \alpha \ln |S_{t,u}| + \beta \ln |S_{t,r}| + (\gamma - \alpha - \beta) \ln |S_t| \end{aligned} \tag{8}$$

When $|S_t| = 0$, we have $\ln |S_t| = -\infty$, and then (8) will come to invalid. The same situation will occurred when $|S_{t,u}| = 0$, or $|S_{t,u}| = 0$. To avoid these situations, we adopt an optimization strategy as following:

$$l(t,u,r) = \alpha \ln |S_{t,u} + 0.5| + \beta \ln |S_{t,r} + 0.5| + (\gamma - \alpha - \beta) \ln |S_t + 1| \tag{9}$$

Using (9), we can easily estimate *l(t,u,r)*. So Instead of comparing *p(t|u,r)*, we can compare *l(t,u,r)* to get the recommendation.

3.2 Example and Implement

For easier understanding, let us give a simple example. If we have a tagging system as following:

Table 1. An example

user	resource	tag
u_1	r_1	t_1, t_2,t_5
u_1	r_3	t_3, t_4, t_5
u_2	r_2	t_3, t_4
u_2	r_3	t_2, t_5

Now, which tags should be recommended to <u_1, r_2>?

If we set $\alpha=1$, $\beta=1$, $\gamma=1$, thus the effect *U*, *R* and *T* are equal. Then (9) comes to be:

$$l(t,u,r)=\ln|S_{t,u}+0.5|+\ln|S_{t,r}+0.5|-\ln|S_t+1| \quad (10)$$

The negative sign before $\ln|S_t+1|$ represents the meaning of information coverage. Let's explain it. Tags that appear only in a specific group of resources carry more information than those tags that appear in a wide range of resources. Thus when recommending, the former tags should have higher score than the latter tags.

We can use this Formula to estimate $l(t_5,u_1,r_2)$:

$$\begin{aligned} l(t_5,u_1,r_2) &= \ln(|S_{t5,u1}|+0.5)+\ln(|S_{t5,r2}|+0.5)-\ln(|S_{t5}|+1) \\ &= \ln(2+0.5)+\ln(0+0.5)-\ln(3+1) \\ &= -0.5051 \end{aligned}$$

After a similar computer progress, we can get:

$$l(t_1,u_1,r_2)=-0.4259,\ l(t_2,u_1,r_2)=-0.602,$$
$$l(t_3,u_1,r_2)=-0.1249,\ l(t_4,u_1,r_2)=-0.1249$$

So the recommending order will be: t_3,t_4,t_1,t_5,t_2. The result can be explained as follow: t_3,t_4 were used by u_1 before, and they also were used to tag r_2. So u_1 have a big chance to choose t_3,t_4 when bookmarking r_2. For t_1,t_5,t_2, we can see that they are all used by u_1, but t_5,t_2 are also used by u_2 too, so that t_1 have bigger information coverage than t_5, t_2. After all, when comparing t_5, t_2, t_5 is used more than t_2 by u_1. When α, β, γ are different from assumption, the result will be different too.

For the same reason, when we need recommendation to <*u*$_2$, *r*$_1$>, the result is:

$$l(t_1,u_2,r_1) = -0.4259, l(t_2,u_2,r_1) = -0.1249, l(t_3,u_2,r_1) = -0.602,$$
$$l(t_4,u_2,r_1) = -0.602, l(t_5,u_2,r_1) = -0.2498$$

So the recommending order of <*u*$_2$, *r*$_1$> will be: t_2, t_5, t_1, t_3, t_4. It can be understood well like the recommending order of <*u*$_1$, *r*$_2$>.

Let's set *k* be the number of recommendation tags for a query <*u*, *r*>. Then we give the pseudo code as follows:

```
Program GetPro {

  Input: F = <U, R, T, S>

  for all t in T{
     v(t) = ln(|S_t| + 1);
     for all u in U
        v(t,u) = ln(|S_t,u| + 0.5);
     for all r in R
        v(t,r) = ln(|S_t,r| + 0.5);
  }//end for

  store all v(t), v(t,u) and v(t,r)
}
Processes Recommendation {

  Input: query <u, r>
              α, β, γ
              Recommendation number k

  for all t in T
        l(t,u,r) = αv(t,u) + βv(t,r)
                          + (γ-α-β)v(t)

  sort all l(t,u,r)

  output top k tags
}
```

4 Experimental Evaluation

In this experiment, we choose the dataset from BibSonomy, a web-based social bookmarking system that enables users to tag web documents as well as bibtex entries of scientific publications. We get the dataset from ECML PKDD'09 Challenge[5]. Brief statistics of the data can be found in the web-site of ECML PKDD'09 Challenge, and the details of the dataset can be found in [16]. We also describe it in Table. 2.

Table 2. Statistics of dataset

|User|	|resource|	|tag|	|post|
1185	22389	13276	253615

Test data include 778 query, every query gives an <*u, r*> pair, and tags are missing.

We choose three algorithms for comparison. One is Most Popular Tags by Resource (MPT). For a given resource we counted for all tags in how many posts they occur together with this resource. Those tags that occurred most frequently with this resource then will be recommended.

Another comparison algorithm is user-user Collaborative Filtering (uCF). Collaborative Filtering is the most famous methods to solve tag recommendation problem. CF has several variants, one of which is user-user CF (uCF). Here we give a briefly describe of this variant. Firstly, uCF computes similarity between users by their tagging history behavior (user tagging history behavior is represent as a Vector Space Model of tags). When <*u, r*> is given, uCF finds similar users of u, and combines the tags they tag r weighted by user similarity. At last uCF chooses the top-k tags to recommend. In our experiment, we set $k = 10$.

The last one is mixed matrix Collaborative Filtering model (mCF). Like uCF, another effective variant of CF is resource-resource Collaborative Filtering (rCF). The process of rCF is very similar to uCF, but based on resource. Many researches show that rCF is worse than uCF in almost all situations (like [7]), so in this paper this variant is ignored. But we can merge uCF and rCF to form a mixed matrix Collaborative Filtering model (mCF). This is very effective. In our implement, we choose the parameters carefully to reach its best represent.

To measure the performance of those algorithms, we use F1-score as standard. F1-score is used as standard in almost every tag recommendation tasks. The definition of F1-score is as follows:

$$F1-score = \frac{2 * precision * recall}{precision + recall} \tag{11}$$

[5] http://www.kde.cs.uni-kassel.de/ws/dc09/

4.1 Performance Comparison

By setting $\alpha = 0.1$, $\beta = 0.9$, and $\gamma = 1$, we get the best performance of BTR. The results of those four algorithms are shown in Figure2.

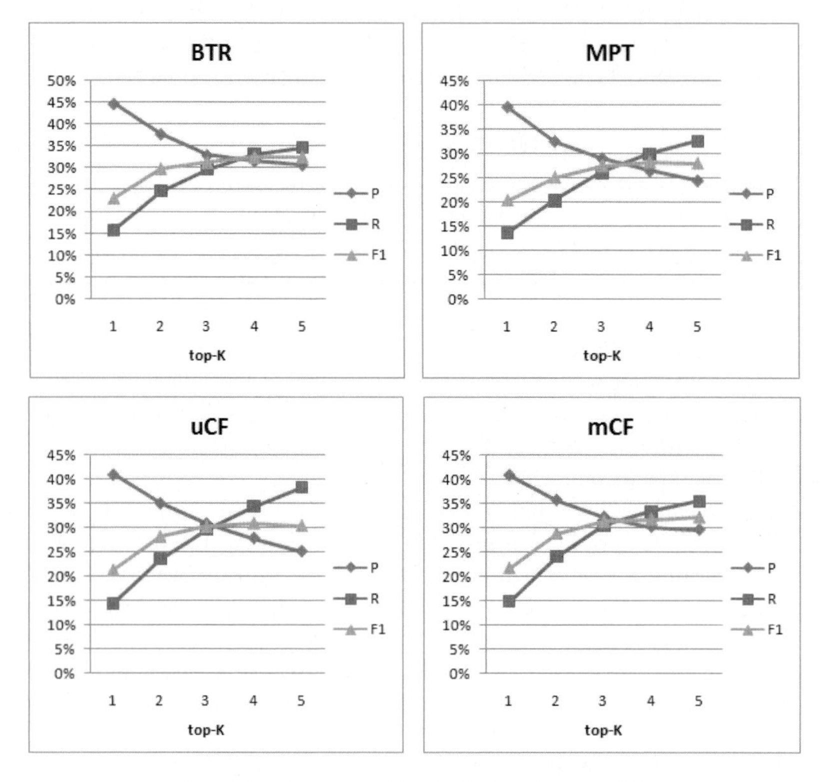

Fig. 1. Performance of BTR, MPT, uCF and mCF

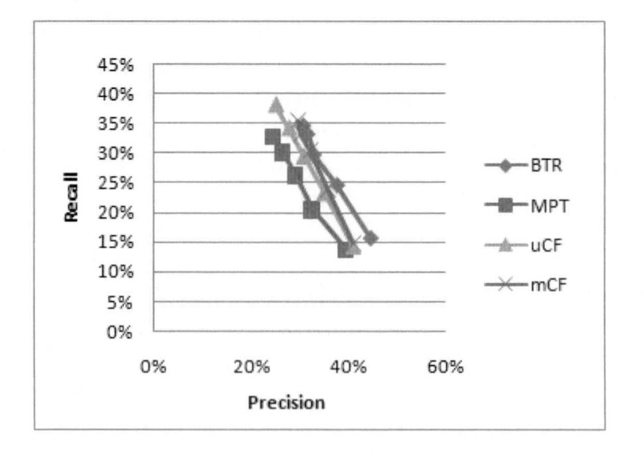

Fig. 2. The precision and recall of these four methods

Let us look at Fig. 1. Although only little deference they performed, we can also say that these are an order. Our BTR outperforms its three competitors with the highest f1-score 32.6%. The second one is mCF with the f1-score 32.2%, very closely to BTR. The next is uCF, and the loser is MPT, whose f1-score cannot over 30%.

Fig. 2 shows the precision and recall of the four methods. The slopes of these four curves are almost the same. The top-rightmost curve depicts the performance of BTR and it can be clearly seen that BTR outperforms the other methods in both precision and recall.

4.2 Parameters Optimization

Here we study the effect of *U*, *R* and *T*. Firstly we compare the effect of *U* and *R*. So we ignore the effect of *T*, thus in experiments we set $\gamma = 1$ and $\gamma = \alpha+\beta$. After this assumption, (9) turns to:

$$l(t,u,r) = \alpha v(t,u) + (1-\alpha) v(t,r) \tag{12}$$

Then we adjust α step by step. Fig. 3 a) shows the result. We can see from figure 4 that resources are more important than users when generating a recommend. In our experiments, when $\alpha / \beta = 9$, the recommendation gets its best F1-measure. This result tells us that the characters of resources are more important than of users in s tag action and the personality of user are also useful to make the prediction better.

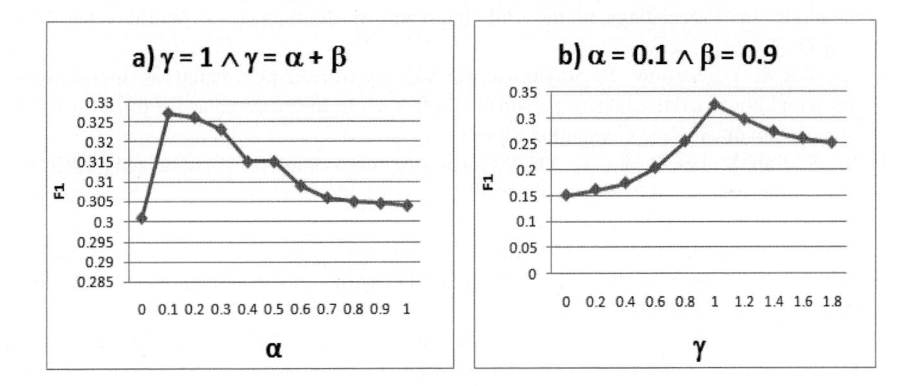

Fig. 3. a)Effectiveness of parameters when $\gamma = 1 \wedge \gamma = \alpha + \beta$; b) Effectiveness of parameters when $\alpha = 0.1$ and $\beta = 0.9$

Next we consider the effect of *T*. We set $\alpha = 0.1$ and $\beta = 0.9$, then adjust the parameter γ. Fig. 3 b) shows the results.

When $\gamma = 1$, we get the best result. It means that the effectiveness is best when γ is equal to the sum of α andβ. It is not a coincidence. In fact, after some carefully examination, a conclusion come to us that if α and β are fixed, the best recommendation always came when $\gamma = \alpha+\beta$.

5 Conclusions

In this paper, we briefly describe a new algorithm BTR that using Bayesian principle to solve tag recommendation problem. Firstly we turned the tag recommendation task into a probability predict problem so that a classical Bayesian method could be easily applied. Based on this we got a simple but effective algorithm named BTR. The experimental results show that our method performs very well.

In future work we want to improve and expand this method to a new level, meanwhile investigate new kinds of techniques for automatically tag recommendation.

Acknowledgments. This work is partially supported by Supported by the National High Technology Research and Development Program of China (863 Program) under Grant No. 2009AA01Z136 and the National Natural Science Foundation of China under Grant No.90812001. The authors also gratefully acknowledge the helpful comments and suggestions of the reviewers, which have improved the presentation.

References

1. Biancalana, C., Micarelli, A., Squarcella, C.: Nereau: a social approach to query expansion. In: Proceedings of the 10th ACM Workshop on Web Information and Data Management (2008)
2. Bao, S., Xue, G., Wu, X., Yu, Y., Fei, B., Su, Z.: Optimizing web search using social annotations. In: Proceedings of the 16th International Conference on World Wide Web (2007)
3. Schmidt, K.-U., Sarnow, T., Stojanovic, L.: Socially filtered web search: an approach using social bookmarking tags to personalize web search. In: Proceedings of the 24th ACM Symposium on Applied Computing (2009)
4. Xu, S., Bao, S., Fei, B., Su, Z., Yu, Y.: Exploring folksonomy for personalized search. In: Proceedings of the 31st Annual International ACM SIGIR Conference on Research and Development in Information Retrieval (2008)
5. Yin, Z., Li, R., Mei, Q., Han, J.: Exploring social tagging graph for web object classification. In: Proceedings of the 15th ACM SIGKDD International Conference on Knowledge Discovery and Data Mining (2009)
6. Ramage, D., Heymann, P., Manning, C.D., Garcia-Molina, H.: Clustering the tagged web. In: Proceedings of the Second ACM International Conference on Web Search and Data Mining (2009)
7. Golder, S., Huberman, B.A.: The structure of collaborative tagging systems. In: Computing Research Repository (2005)
8. Marlow, C., Naaman, M., Davis, M., Boyd, D.: Tagging paper, taxonomy, Flickr, academic article, toread. In: Proceedings of the seventeenth ACM conference on Hypertext and hypermedia (2006)
9. Tatu, M., Srikanth, M., Silva, T.D.: Tag Recommendations using Bookmark Content. In: Proceeding of ECML PKDD Discovery Challenge (2008)
10. Lu, Y.-T., Yu, S.-I., Chang, T.-C., Hsu, J.Y.-J.: A Content-Based Method to Enhance Tag Recommendation. In: Proceedings of the Twenty-First International Joint Conference on Artificial Intelligence (2009)

11. Resnick, P., Iacovou, N.: GroupLens: an open architecture for collaborative filtering of netnews. In: Proceedings of the ACM conference on Computer supported cooperative work (1994)
12. Herlocker, J.L., Konstan, J.A., Terveen, L.G.: Konstan Evaluating collaborative filtering recommender systems. ACM Transactions on Information Systems (2004)
13. Su, X.Y., Khoshgoftaar, T.M.: A survey of collaborative filtering techniques. In: Advances in Artificial Intelligence (2009)
14. Schke, R.J., Marinho, L., Hotho, A.: Tag Recommendations in Folksonomies. In: Proceeding of ECML PKDD Discovery Challenge (2007)
15. Symeonidis, P., Nanopoulos, A., Manolopoulos, Y.: Tag Recommendations Based on Tensor Dimensionality Reduction. In: Proceedings of the ACM conference on Recommender system (2008)
16. Hotho, A., J"aschke, R., Schmitz, C., Stumme, G.: BibSonomy: A Social Bookmark and Publication Sharing System. In: Kok, J.N., Koronacki, J., Lopez de Mantaras, R., Matwin, S., Mladenič, D., Skowron, A. (eds.) ECML 2007. LNCS (LNAI), vol. 4701, Springer, Heidelberg (2007)

Comparison of Different Methods to Fuse Theos Images

Silong Zhang[1,2] and Guojin He[1]

[1] Center for Earth Observation and Digital Earth, Chinese Academy of Sciences, China
[2] Graduate University of Chinese Academy of Science, China, 10086
{slzhang,gjhe}@ceode.ac.cn

Abstract. Along with the development of the remote sensing, an increasing number of remote sensing applications such as land cover classification, feature detection, urban analysis, require both high spatial and high spectral resolution. On the other hand, the Satellite can't get high spatial and high spectral resolution at the same time because of the incoming radiation energy to the sensor and the data volume collected by the sensor. Image fusion is an effective approach to integrate disparate and complementary information of multi-source image. As a new type of Remote Sensing data source, the lately launched Theos can be widely used in many applications. So the fusion of its high spatial resolution image and multi-spectral image is important. This paper selects several widely used methods for the fusion of data of high spatial resolution and high spectral resolution. The result of each approach is evaluated by qualitative and quantitative comparison and analysis.

Keywords: IHS Transform, Image Fusion, Remote Sensing, Spectral Response, Color Distortion.

1 Introduction

Theos was launched on 1 October 2008. It provides two different resolution images: the Pan (panchromatic) image with high spatial resolution (2m) and low spectral resolution and the multi-spectral image with high spectral resolution and low spatial resolution (15m). Image fusion is the technique to combine those two classes of images to form a new image with high spatial resolution and high spectral resolution [1]. Numerous image fusion techniques have been developed over the last two decades. The most commonly approaches are IHS (Intensity-Hue-Saturation) transform, PCA (Principal Component Analysis) transform, Brovey transform, and SFIM (Smoothing Filter-based Intensity Modulation). We select those methods because they are widely used in popular Remote Sensing software, they are easy and simple to use. Each method has its own merits and is been used in different applications. In our work, we will find the most suitable approach for Theos image fusion.

Before our work, the simultaneously acquired Theos multi-spectral and Pan images should been preprocessed. In other words, those images should been corrected and registrated precisely. When the fusion finished, the assessment of the image fusion quality are presented by correlation coefficient, mean, standard deviation and other indexes [2].

L. Cao, J. Zhong, and Y. Feng (Eds.): ADMA 2010, Part II, LNCS 6441, pp. 202–209, 2010.

The goal of this paper is to select the appropriate approach to fuse the new remote sensing image. In our work, we assume that the Pan image and multispectral images are co-registered, no noise in images.

2 Fusion Methods

2.1 IHS Fusion

In general, the IHS method converts a remote sensing image from RGB(red, green blue) space into IHS(intensity-hue-saturation) space, then use the high resolution Pan image replace the intensity(I) band before converts back to RGB space. This can be expressed as follows [3]:

(1) Converts the image from RGB space to IHS space.

$$\begin{bmatrix} I \\ V_1 \\ V_2 \end{bmatrix} = \begin{bmatrix} \frac{1}{3} & \frac{1}{3} & \frac{1}{3} \\ \frac{-\sqrt{2}}{6} & \frac{-\sqrt{2}}{6} & \frac{2\sqrt{2}}{6} \\ \frac{1}{\sqrt{2}} & \frac{-1}{\sqrt{2}} & 0 \end{bmatrix} \begin{bmatrix} R \\ G \\ B \end{bmatrix}, \tag{1}$$

$$H = tan^{-1}(\frac{v_2}{v_1}), \tag{2}$$

$$S = \sqrt{v_1^2 + v_2^2}\,, \tag{3}$$

(2) The intensity component I is replaced by the Pan image.

(3) Converts back to RGB space.

$$\begin{bmatrix} F(R) \\ F(G) \\ F(B) \end{bmatrix} = \begin{bmatrix} 1 & \frac{-1}{\sqrt{2}} & \frac{1}{\sqrt{2}} \\ 1 & \frac{-1}{\sqrt{2}} & \frac{-1}{\sqrt{2}} \\ 1 & \sqrt{2} & 0 \end{bmatrix} \begin{bmatrix} Pan \\ v_1 \\ v_2 \end{bmatrix}. \tag{4}$$

2.2 PCA Method

PCA is another popular approach in image fusion. At first, the original multi-spectral images were transformed into new and uncorrelated images which called components.

$$\begin{bmatrix} PC_1 \\ PC_2 \\ PC_3 \end{bmatrix} = \begin{bmatrix} \varphi_{11} & \varphi_{12} & \varphi_{13} \\ \varphi_{21} & \varphi_{22} & \varphi_{23} \\ \varphi_{31} & \varphi_{32} & \varphi_{33} \end{bmatrix} \begin{bmatrix} R \\ G \\ \langle \end{bmatrix} = A \begin{bmatrix} R \\ G \\ B \end{bmatrix}, \tag{5}$$

Where, A is the covariance matrix.

The dimensionality of the original data will be reduced in this way, the first several components contain majority of information. The first component PC_1 will be substituted with the Pan image.

Then the data converts back to the original image space, forms the fused multispectral image with high spatial resolution.

$$\begin{bmatrix} F(R) \\ F(G) \\ F(B) \end{bmatrix} = \begin{bmatrix} \varphi_{11} & \varphi_{12} & \varphi_{13} \\ \varphi_{21} & \varphi_{22} & \varphi_{23} \\ \varphi_{31} & \varphi_{32} & \varphi_{33} \end{bmatrix}^T \begin{bmatrix} Pan \\ \%_0 C_2 \\ PC_3 \end{bmatrix} = A^T \begin{bmatrix} Pan \\ PC_2 \\ PC_3 \end{bmatrix} \tag{6}$$

Where A^T is inverse matrix of covariance matrix A. Pan is high resolution image.

2.3 Brovey Transform

The Brovey is a color normalized method which normalizes multispectral bands combination of high spatial data. The formula is:

$$\begin{bmatrix} F(R) \\ F(G) \\ F(B) \end{bmatrix} = \frac{Pan}{R+G+B} \begin{bmatrix} R \\ G \\ B \end{bmatrix}, \tag{7}$$

Where Pan is the high resolution Pan image, R, G and B are multispectral images. F(R), F (G), F (B) is bands of fused image.

2.4 SFIM Method

The SFIM method defined by equation[4]:

$$\mathrm{IMAGE}_{SFIM} = \frac{\mathrm{IMAGE}_{low}\mathrm{IMAGE}_{high}}{\mathrm{IMAGE}_{mean}}, \tag{8}$$

Where $\mathrm{IMAGE}_{low}$is the multispectral image, $\mathrm{IMAGE}_{high}$is Pan image, $\mathrm{IMAGE}_{mean}$is a smoothed pixel of IMAGE_{high}using averaging filter over a neighborhood equivalent to the actual resolution of IMAGE_{low}. The filter kernel size is decided based on the resolution ratio between the higher and lower resolution images. In this case, the smoothing filter kernel size for calculating the local mean of the Pan image is7×7.

3 Evaluation Criterions for Fusion Results

There are several image fusion methods; each has its own merits. There are several evaluation criterions to evaluate the image fusion technique. Image fusion aim at integrate the spatial information and the spectral information from different data sources. The assessment of effect of the fusion methods should be in enhancement of spatial information and spectral information preservation. The following parameters were selected to assess the quality of the results [5][6].

3.1 Mean Value (μ) and Standard Deviation (σ)

Mean is defined as the average gray value of the image. Standard deviation means varies from its mean value. They can be gotten below:

$$\mu = \frac{1}{MN} \sum_{i=1}^{M} \sum_{j=1}^{N} f(i,j), \tag{9}$$

$$\sigma = \sqrt{\frac{1}{MN-1}\sum_{i=1}^{M}\sum_{j=1}^{N}(f(i,j)-\mu)^2}\,, \tag{10}$$

Where $f(i,j)$ means gray value of pixel (i,j). MN means the total number of pixels in image. If the value of μ of original image and fused image is difference. That means the spectral information was changed.

3.2 Entropy

Entropy is an important indicator of the degree of information abundance of an image. It was correlated with the amount of information. The ability to describe details of different images can be known by comparing their entropy. The entropy of an image is:

$$\mathrm{H(x)} = -\sum_{i=0}^{255} p_i log_2 p_i\,, \tag{11}$$

Where p_i is the probability of pixels with gray value i.

3.3 Correlation Coefficient

Correlation coefficient (cc) measures the degree of similarity between original image and fused image. High value means more correlate between two images.

$$\mathrm{cc} = \frac{\sum_{i=1}^{M}(X_i-\bar{X})(Y_i-\bar{Y})}{\sqrt{\sum_{i=1}^{M}(X_i-\bar{X})^2(Y_i-\bar{Y})^2}}\,, \tag{12}$$

Where cc is correlation coefficient, X_i and Y_i are gray values of original image and fused image. $\bar{X}$ and $\bar{Y}$ are their mean gray value.

3.4 Average Gradient

Average gradient if Denotation of image's spatial details. So it can be used to assess spatial quality. The average gradient is greater, the image is more legible.

$$\mathrm{G} = \frac{1}{n}\sum\sqrt{(\Delta I_x^2 + \Delta I_x^2)/2}\,, \tag{13}$$

Where $\Delta I_x$and ΔI_ymeans the difference of x-direction and y-difference respectively. n is the total number of pixels in the image.

4 Experimental Results

The image used in this research was acquired on November 10, 2009. Landcover in this image includes water, vegetation highway and buildings. Some specifications of Theos data are given in table 1:

Table 1. Some Specifications of Theos

THEOS	Pan	Multispectral
Spatial resolution/m	2	15
Spectral range / μm	0.45-0.90	0.45-0.52
		0.53-0.60
		0.62-0.69
		0.77-0.90
SNR	>110	>>117

Before our work, those two images have been co-registered with accuracy less than 1 pixel. Band 4, 3, 2 and Pan bands were selected, and the multispectral image was resembled to the size of Pan image. The original images and the fusion results are showed in figure 1.

4.1 Qualitative Evaluation

Visually, all fused images' spatial resolution has been improved apparently. The details of fused images have been significantly enhanced, so it's easier for interpretation. The fused image obtained by Brovey and IHS are particularly clearer compare with other images. The edge information of fused image gained by PCA and SFIM is partly lost. The color of fused image obtained by IHS transform and PCA transform approaches is distorted. The vegetation is apparently darker compare with original multispectral image. The overall brightness of the result of Brovey technique is obviously lower than original image. The SFIM approach can preserve spectral information of the multispectral image well[7].

4.2 Quantitative Evaluation

We select mean value, standard deviation, entropy, correlation coefficient, and average gradient to assess the performance of each fusion technique. Where mean value, correlation coefficient are the measures of the ability of preserve of spectral information, while the rest indexes are descriptors of spatial information. The values of those indexes can be seen in table 2 and table 3.

Table 2 shows mean values and standard deviation of original image and fused images; it clearly indicates that the mean value of fused images gained by SFIM is similar with original images. PCA has similarity compare with other methods. SFIM and IHS have greater standard deviate than Brovey and PCA. According to the correlation coefficient, SFIM-fused image's value is much larger compare with other methods. So the SFIM method is the best way to preserve spectral information of original multispectral image. Brovey is the second one.

The measures entropy and average gradient are descriptors of spatial information. The goal of image fusion is not only preserve spectral information but also spatial information. Through table 3, we can get: all methods can improve original image spatial resolution. The value of entropy of SFIM-fused image is the largest one, and then is IHS-fused image. SFIM-fused image has largest average gradient value too; the PCA-fused image has the second large average gradient value.

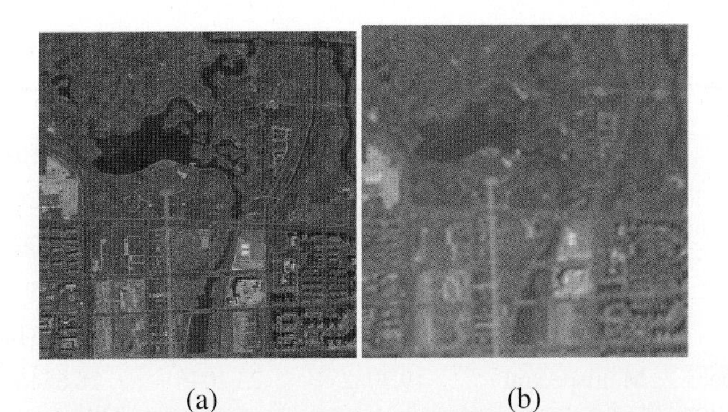

(a) (b)

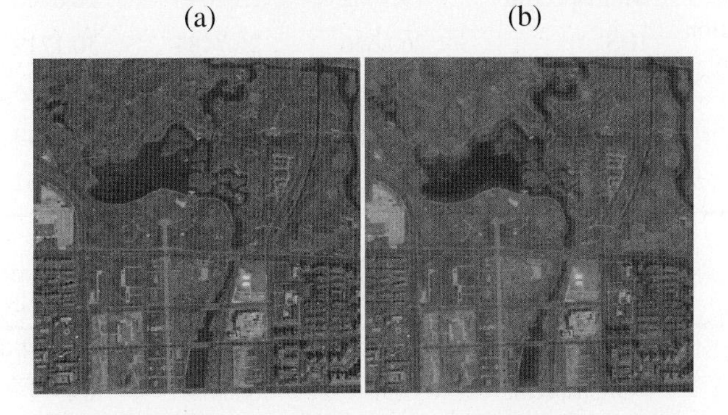

(c) (d)

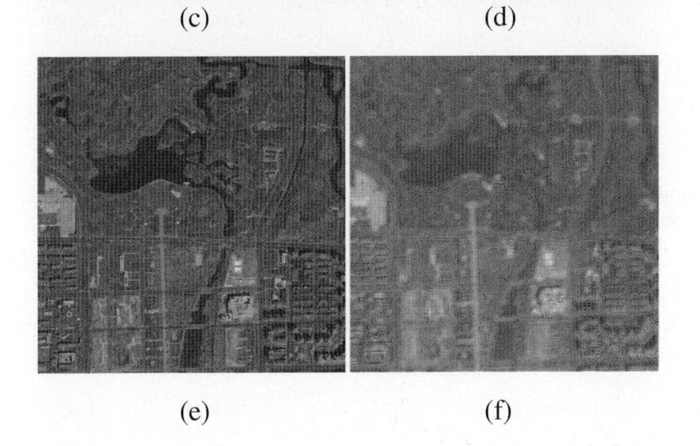

(e) (f)

Fig. 1. Compare of result of different fusion algorithm on Theos image.(a) original Pan image,(b) original multispectral image, (c) result of IHS approach, (d) result of PCA approach (e) result of Brovey approach, (f) result of SFIM approach.

From the above, the SFIM method not only gets the best spatial resolution but also preserve spectral information. So we can say SFIM is the most appropriate method for Theos image fusion.

Table 2. Comparative mean and standard deviation of original and fused images

index	method	Band4	Band3	Band2
Mean μ	Multispectral	66.5465	118.6717	105.0866
	IHS	45.2664	96.5449	82.9628
	PCA	66.5465	95.1399	89.4509
	Brovey	51.6329	91.5103	80.7739
	SFIM	66.6373	118.7315	105.1314
Standard deviation σ	Multispectral	19.3082	23.2046	28.6546
	IHS	26.8656	26.7484	30.1719
	PCA	19.3082	27.1407	31.2374
	Brovey	20.4203	29.9689	30.0547
	SFIM	22.5477	31.0850	34.0662

Table 3. Comparative measures of original and fused images

index	method	Band4	Band3	Band2
entropy H	Multispectral	6.1966	6.0819	6.4879
	IHS	6.4728	6.6463	6.7866
	PCA	6.1966	6.6309	6.8755
	Brovey	6.3512	6.8025	6.7402
	SFIM	6.4906	6.8085	6.8752
correlation coefficient cc	Multispectral	1	1	1
	IHS	0.6100	0.6006	0.7069
	PCA	0.6214	0.4870	0.3593
	Brovey	0.7241	0.4599	0.6053
	SFIM	0.8559	0.7345	0.8297
average gradient G	Multispectral	1.0775	1.0628	1.3277
	IHS	8.9990	9.2744	9.2886
	PCA	10.0775	10.0270	12.4225
	Brovey	6.4060	11.3700	10.0282
	SFIM	8.3006	14.6382	13.0042

5 Conclusion

We fused Theos Pan image and multispectral images through IHS, PCA, Brovey and SFIM methods. And we assessed the fused image qualitatively and quantitatively. We select mean value, standard deviation, entropy, correlation coefficient and average gradient as measures to evaluate different fusion techniques. The SFIM method is very strong in preserving spectral information and spatial information.

But the spectral information is lost or distorted in all fused images, because the values of Pan and multispectral are difference. So this method should be improved to fuse Theos images better.

Acknowledgments. The research has been supported by the National Natural Science Foundation of China (NSFC) Funding: under grant number 60972142.

References

1. Yun, Z.: Understanding Image Fusion. Photo Grammetric Engineering & Remote Sensing 7, 657–661 (2004)
2. Wald, L.: Some Terms of Reference in Data Fusion. IEEE Trans. Geosci. Remote Sens., 1190–1193 (1999)
3. Pohl, C., Genderen, J.L.: Multisensor Image Fusion in Remote Sensing: Concepts, Methods and Applications. International Journal of Remote Sensing 19(5), 823–854 (1998)
4. Liu, J.G.: Smoothing filter-based intensity modulation: a spectral preserve image fusion technique for improving spatial details International J. Journal of Remote Sensing 18(21), 3461–3472 (2000)
5. Wald, L., Ranchin, T., Mangolini, M.: Fusion of Satellite Images of Different Spatial Resolutions: Assessing The Quality of Resulting Images. Photogrammetric Engineering and Remote Sensing 63, 691–699 (1997)
6. Chavez, P.S., Sildes, S.C., Anderson, J.A.: Comparison of Three Different Methods to Merge Multiresolution and Multispectral Data: Landsat TM and SPOT Panchromatic. Photogrammetric Engineering and Remote Sensing 57, 295–303 (1991)
7. Liu, J.G.: Evaluation of Landsat-7 ETM+ Panchromatic Band for Image Fusion with Multispectral bands. Natural Resources Research 9(4), 269–276 (2000)

Using Genetic K-Means Algorithm for PCA Regression Data in Customer Churn Prediction

Bingquan Huang, T. Satoh, Y. Huang, M.-T. Kechadi, and B. Buckley

School of Computer Science and Informatics, University College Dublin, Belfield, Dublin 4, Ireland
bingquan.huang@ucd.ie, takeshi.sato@ucdconnect.ie

Abstract. Imbalance distribution of samples between churners and non-churners can hugely affect churn prediction results in telecommunication services field. One method to solve this is over-sampling approach by PCA regression. However, PCA regression may not generate good churn samples if a dataset is nonlinear discriminant. We employed Genetic K-means Algorithm to cluster a dataset to find locally optimum small dataset to overcome the problem. The experiments were carried out on a real-world telecommunication dataset and assessed on a churn prediction task. The experiments showed that Genetic K-means Algorithm can improve prediction results for PCA regression and performed as good as SMOTE.

Keywords: Genetic K-means Algorithm, PCA Regression, Nonlinear Discriminant, Churn Prediction, Imbalanced Distribution of Classes.

1 Introduction

Customer Churn has become a serious problem for companies mainly in telecommunication industry. This is as a result of recent changes in the telecommunications industry, such as, new services, new technologies, and the liberalisation of the market. In recent years, Data Mining techniques have emerged as one of the method to tackle the Customer Churn problem[8,7,16].

Usually speaking, the study of customer churn can be seen as a classification problem, with two classes: Churners and Non-Churners. The goal is to build a classifier from training dataset to predict potential churn customers. However, the performance of a classifier is subject to the distribution of class samples in a training dataset. Imbalanced distribution of class samples is an issue in data mining as it leads to lower classification performances[11]. In telecommunication sector, failure to identify potential churner can result in a huge financial loss. The methods solve such problem can be categorised into: 1) Data Sampling and 2) Cost Sensitive Learning[12]. In this paper, we are interested in increasing the size of churners by data sampling approach.

The aim of sampling approaches is to correctly set the distribution samples to build an optimal classifier by adding minority class samples known as over-sampling. There have been various sampling approaches proposed to counter

L. Cao, J. Zhong, and Y. Feng (Eds.): ADMA 2010, Part II, LNCS 6441, pp. 210–220, 2010.

non-heuristic sampling problems. Synthetic Minority Over-sampling Technique (SMOTE)[2] generates artificial data along the line between minor class samples and K minority class nearest neighbours. This causes the decision boundaries for the minor class space to spread further into majority class space. Alternative method of minority class addition is data generation by regression. In particular regression by PCA is our focus. PCA[14] reveals the internal structure of a dataset by extracting uncorrelated variables known as Principal Components (PC) which best explains the variance in the dataset. PC is capable of reconstructing the data back linearly with loss information[6]. However, an imbalanced telecommunication dataset is usually non-linear (see Figure 2). We believe an approach which straightly uses PCA on whole data set without any reasonable separation to estimate the missing samples, might be ineffective for the churn prediction.

The main idea of our approach is to form number of small minority datasets from original minority class dataset. The datasets can be formed by traditional K-means algorithm but we are interested if Genetic Algorithm(GA) combined with K-means algorithm could be useful in forming locally optimal clustered sets. We apply PCA regression on each minority class clustered datasets to generate new minority class samples.

This paper is organized as follows: the next section outlines the proposed approach on churn prediction task and then Section 3 explains the process of the experiments and its evaluation criteria. The Evaluations of the proposed approach are presented as well. We conclude and highlight with limitations in Section 4.

2 Approach

Our proposed approach basically applies PCA technique with the Genetic Algorithm K-means algorithm to generate new data for minority class. First and foremost, we form minority class data,d_{churn} by extracting minority class samples from an original raw data d_{raw}. Assume that d_{raw} is formed after data sampling and data pre-processing. After the minority class data is formed, GA K-means clustering technique is applied on the data to form K clusters. This step is important in order to apply PCA regression locally rather on whole minority data. We believe that clustering dataset formed by GA K-means avoids the inclusion of redundant information because of lower variance when applying PCA regression. The next step is PCA regression on each cluster. The main objective of this step is to transform each data inside a cluster back to original feature space in terms of selected Principal Components. These transformed data are then added to d_{raw} to boost the distribution of minority class samples. Finally, we make use of this data to build a churn prediction model for a classification purpose. Figure 1 describes the approach in picture. We explain the main steps in our approach in the following section.

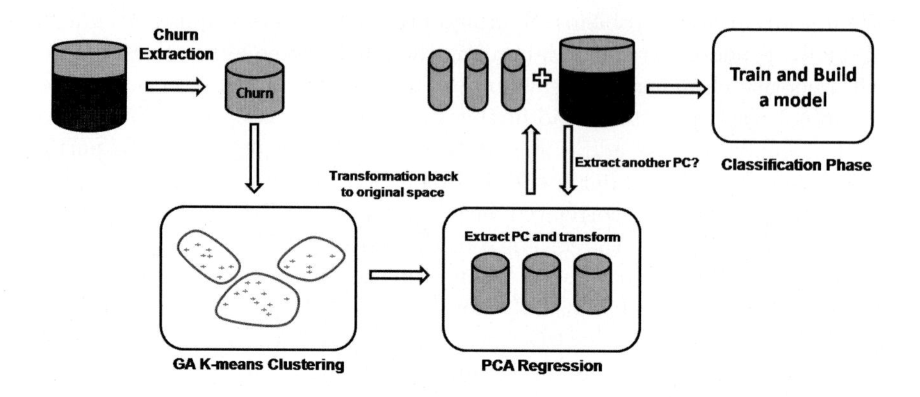

Fig. 1. The proposed approach with the four main steps to generate new samples of minority class

2.1 GA K-Means Clustering Algorithm

The main step of the approach after the formation of d_{chur} is to apply GA K-means clustering algorithm on the data. The standard K-means algorithm is sensitive to the initial centroid and poor initial cluster centres would lead to poor cluster formation. Recently, Genetic Algorithm (GA)[13] is employed to avoid sensitivity problem in centre point selection that is locally optimum and then apply the K-means algorithm.

In GA K-means algorithm, a gene represents a cluster centre of n attributes and a chromosome of K genes represents a set of K cluster centres. The squared error function is employed as the GA fitness function. The GA K-means clustering algorithm works as follows:

1. **Initialization:** Randomly select K data points as cluster centres for W times from original data set as chromosomes, apply k-means with selected centroid, the initial population is composed of each of the resultant partition. Compute fitness value for each chromosome.
2. **Selection:** Using roulette-wheel[10] to select chromosomes from initial population for mating, the roulette-wheel would be spun W times to establish the same size of population.
3. **Crossover:** The selected chromosomes are randomely paired with other parents for reproduction. The bits of chromosomes are swapped by the probability P_c.
4. **Mutation:** Compute average fitness value F_{avg} for each chromosome. If the fitness value for one chromosome is greater than F_{avg}, mutation operation is applied to insure diversity in the population.
5. **Elitism:** In order to avoid missing the best chromosome at the end, a list, L_{best} is created to store the best chromosome that is best generated by each generation, with evolution the previous best will be replaces by the current best one.

6. **Iteration:** Go to step 2, until the termination criterion is satisfied. The iteration process terminates when the variation of fitness value within the best chromosomes is less than a specific threshold.

2.2 Linear PCA and Data Generation

We apply the PCA regression technique on each cluster to generate a new dataset in original feature space in terms of selected principal components (PC). PCA has a property of searching for PC that accounts for large part of total variance in the data and projecting data linearly onto new orthogonal bases using PC.

Consider a dataset $X=\{x_i, i = 1, 2, \ldots, N, x_i \in \Re^N\}$ with attribute size of d and N samples. The data is standardised so that the standard deviation and the mean of each column are 1 and 0, respectively. PC can be extracted by solving the following Eigenvalue Decomposition Problem[14]:

$$\lambda\alpha = \mathbf{C}\alpha, \quad subject\ to\ ||\alpha||_2 = \frac{1}{\lambda}. \tag{1}$$

where α is the eigenvectors and $\mathbf{C}$ is the covariance matrix. After solving the equation (1), sort the eigenvalues in descending order as larger eigenvalue gives significant PC. Assume that matrix α contains *only* a selected number of eigenvectors (PC). The transformed data is computed by

$$X_{tr} = \alpha^T X^T. \tag{2}$$

From equation (2), matrix X^T can be obtained by $X^T = \alpha^{T^{-1}} X_{tr}$. Finally, the matrix X^T is transposed again to get the matrix X^{new}. Since we standardised the data in the first step, the original standard deviation and the mean of each column must be included in each X_{ij}^{new}. The newly generated data X^{new} is then added to d_{raw} to adjust the distribution of the samples. We continue this process until all clusters are transformed.

We run two data generation approaches on PCA regression. The first approach utilises all clusters on PCA regression (Local1). The second approach only uses the centroid of each cluster to form a dataset of centre points (Local2) and use this data to extract principal components.

3 Experiments

3.1 Data Description

139,000 customers were randomly selected from a real world database provided by Eircom for the experiments. The distribution of churner and non-churners is very imbalanced in both the training data and the testing data. These data contain respectively 6,000, resp. 2000, churners and 94,000, resp. 37000, non-churners. These data contain 122 features which describe individual customer characteristics, see [15] for more detailed descriptions.

3.2 Evaluation Criteria

According to the literature and previous experiments in [15], the Decision Tree C4.5 (DT) [5] and the SVM [1] outperformed the Neural Networks(ANN) classifier when applied on telecommunication data. In this paper, 4 classifier techniques, DT C4.5, SVM, Logistic Regression (LR)[3] and Naive Bayes (NB) were employed to build prediction models.

The performance of the predictive churn models has to be evaluated. If a_{11}, resp. a_{22} is the number of the correctly predicted churners, resp. non-churners, and a_{12}, resp. a_{21} is the number of the incorrectly predicted churners, resp. non-churners, the following evaluation criteria are used in the experiments:

- the accuracy of true churn (TP) is defined as the proportion of churn cases that were classified correctly: $\text{TP} = \frac{a_{11}}{a_{11}+a_{12}}$.
- the false churn rate (FP) is the proportion of non churn cases that were incorrectly classified as churn: $\text{FP} = \frac{a_{21}}{a_{21}+a_{22}}$.

A good solution should have both a high TP with a low FP. When no solution is dominant, the evaluation depends on the expert strategy, i.e. to favour TP or FP.

The Receiver Operating Curve technique (ROC) is recommended to evaluate learning algorithms. It shows how the number of correctly classified positive samples varies with the number of incorrectly classified negative samples. However, there is no clear dominating relation between multiple ROC curves and it is often difficult to make a decision on which curve is better. The Area under ROC curve(AUC)[9] provides single number summary for the performance of learning algorithms. AUC is a statistically consistent and more discriminating measure. We calculate the AUC threshold on FP as 0.5 as telecom companies are generally not interested in FP above 50%.

3.3 Experimental Setup

We first visualize the dataset pattern using the Self Organizing Map(SOM)[4] to observe whether the churners are dispersed or concentrated in certain area before the experiments. If the churners are dispersed(i.e. non-discriminant), there is a possibility that traditional K-means algorithm would be hard to cluster optimally.

The aim of the experiments is to observe if additional churner samples generated by Local PCA regression would improve churn prediction results. The following 3 experiments are conducted. The first experiment examines the most optimal cluster size K of the GA K-means clustering algorithm for PCA regression. For this experiment, we set the range of K to be in $\{4, 8, 16, 36, 40, \ldots, 72\}$ and then applly the PCA regression on each cluster. When the churner samples are added and a new data is formed, churn prediction models are built. The second experiment compares the prediction results of each classifier by PCA regression from experiment 1 to Linear regression(LiR), standard PCA based data generation and SMOTE. The cluster size for this experiment is chosen from experiment 1. Finally the third experiment examines whether or not increasing the size of churner samples improves the accuracy of churn prediction. We increase the number of churners from original size of 6000 up to 30000(6000, 12000...30000) by setting the PC threshold to 0.9, 0.8...0.6. A new datasdet is generated based on 2 types of local PCA generation (Local1 & Local2) in all experiments.

3.4 Results and Discussion

Figure 2 is a SOM map of 39 by 70 nodes. The maximum 9,000,000 training cycles and 3 folds cross-validation were used to train the SOM. The gray color cells and purple color cells represent churn and non-churn, respectively. The figure shows that there are small portion of churners and dispersed in the map. Therefore, we believe it is hard to obtain optimal cluster datasets using traditional K-means algorithm.

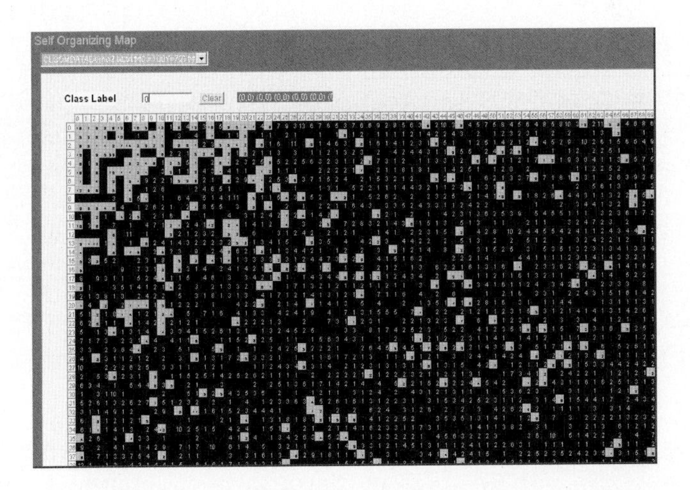

Fig. 2. Data Visualization, which shows that churn samples are in different locations in the map

Figure 3 presents the graph of AUC against the size of cluster. The narrow line at AUC 0.2 mark is when FP threshold was set to 20%. The blue and red lines represent AUC of GA-based local PCA. The former and latter lines represent the AUC results following *Local*1 and *Local*2 data generation, respectively. The green line represents the AUC when traditional K-means clustering was applied to local PCA. The GA K-means clustering produced higher AUC than traditional K-means in general. However, the difference between them was at most 0.15 with the exception of the red line, therefore, one can say there is no significant difference in GA and traditional K-means algorithm. The cluster size ranging from 36 to 72 produced better results for Local PCA. On the contrary, *Local*2 tended to have opposite behavior as the smaller the cluster size is the better AUC.

The FP and TP rates of 3 data regression methods, GA-based local PCA, standard PCA and LiR ,were compared in Figure 4. For local PCA we selected 2 best cluster size from Figure 3 for each classifier. The standard PCA operates similar to local PCA but clustering technique is not applied on churn data which means larger variance within the data. In logistic regression, both types of local PCA performed better than Standard PCA and LiR. This behaviour can also be seen in other classifiers but C4.5. For C4.5 it is hard to conclude which data generation method is better because some lines lie on top of the other line. The LiR did perform generally better than Standard PCA but the Local PCA performed the best prediction results.

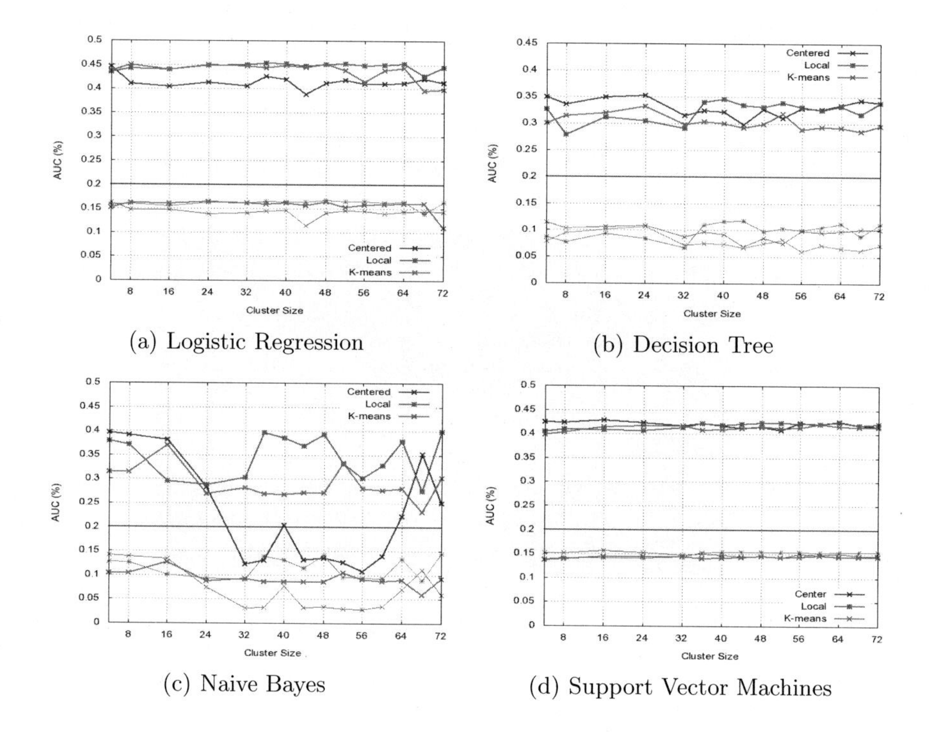

(a) Logistic Regression

(b) Decision Tree

(c) Naive Bayes

(d) Support Vector Machines

Fig. 3. AUC vs Cluster Size, based on the 4 different modelling techniques (Logistic Regression, Decision Tree, Naive Bayes and Support Vector Machines)

Our next experiment is to examine the classification results after adjusting the distribution of class samples. Figure 5(a) to 5(d) present the ROC graphs for each classifier using *Local*1 data generation as it gave the best prediction results among other data regression method. Churn size of 600 to 6000 was selected from original raw data. From 6000 onward, additional Churn samples generated by PCA were added to raw data. The SVM, NB and LR performed well with

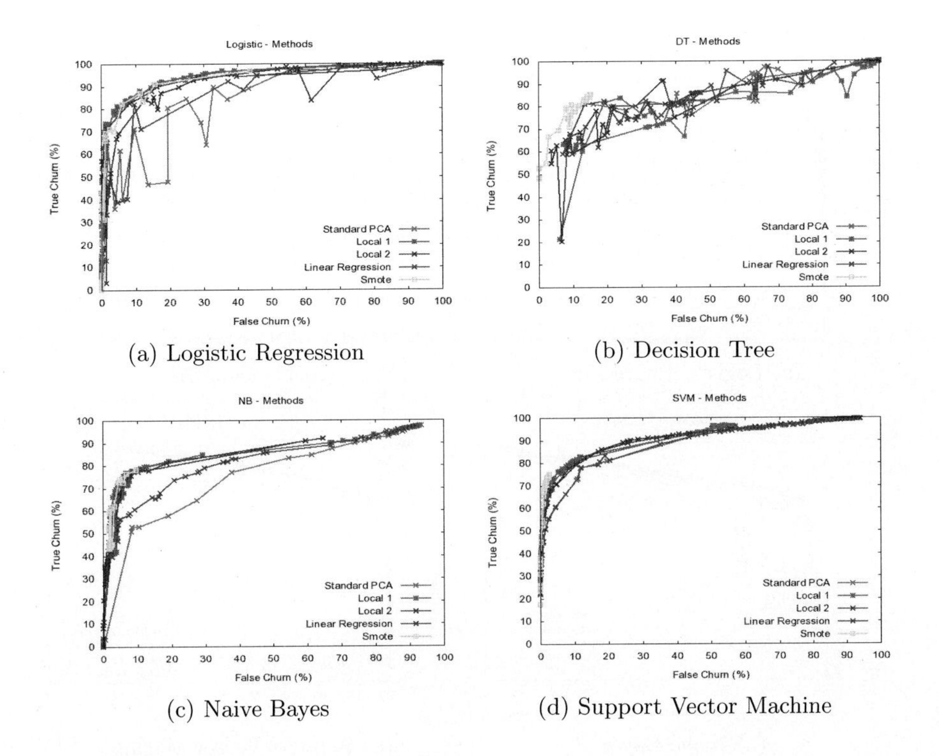

Fig. 4. ROC curves vs various data generation methods, based on the 4 different modelling techniques (Logistic Regression, Decision Tree, Naive Bayes and Support Vector Machines)

churn size 6000 to 12000 but churn size afterwards did not produce acceptable TP and FP rates as this can be seen from Figure 6 clearly which illustrates the AUC rate against churn size. The C4.5 classifier showed that the more samples are added, the higher TP and FP.

In summary, the experiments showed that 1) Clustering size K did produce different AUC results according to the size as larger K tend to perform well with minor improvements 2) Local PCA data regression performed better than Standard PCA and LiR and finally 3) Adding similar churn samples to original data did improve the TP rate in all classifiers in contrast to the FP where it reached over 50% as the churn size reached over 12000. One of the reasons for the high FP rate is due to the change in decision boundaries. As the churn samples are added, the churn space enters further into non-churn space. This leads to more non-churn samples inside churn space and hence, high number of incorrectly classified non-churn. Classifiers such as DT, SVM and LR require defining decision regions. NB classifier relies on Bayes' theorem. Changes in the number of churn samples would likely to affect prior probability and hence the possibility of increase in FP.

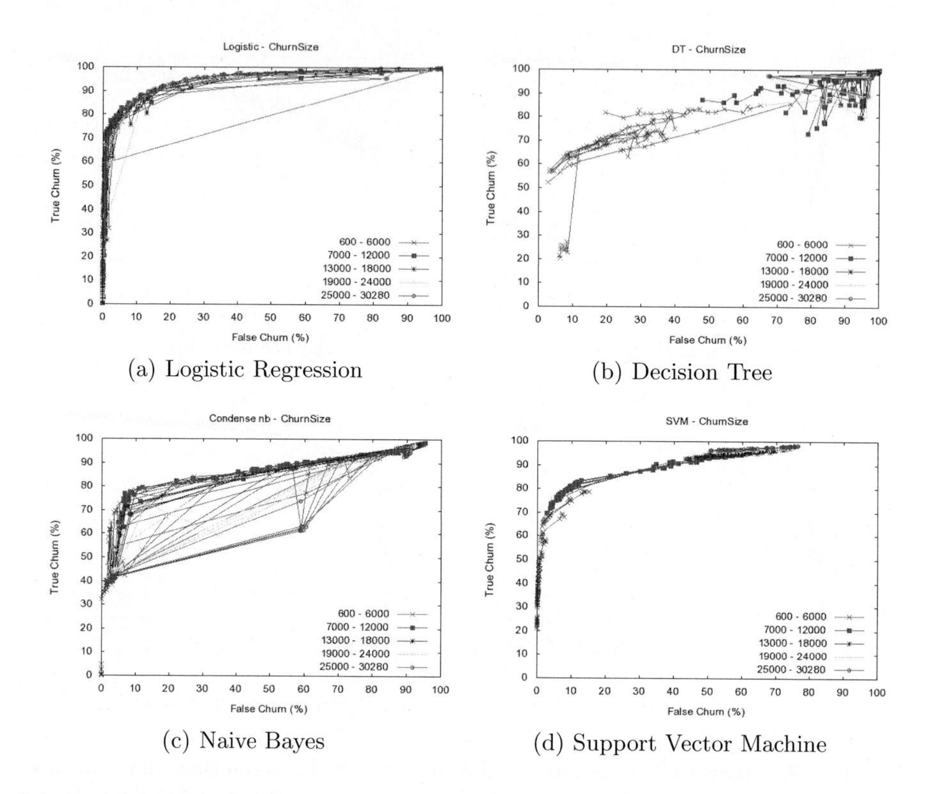

Fig. 5. ROC graph of different churn size, based on the 4 different modelling techniques (Logistic Regression, Decision Tree, Naive Bayes and Support Vector Machines)

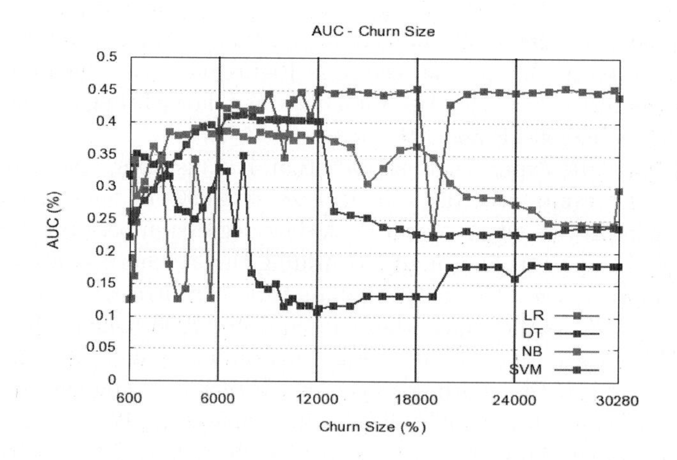

Fig. 6. A graph of AUC against the size churn samples, based on the 4 different modelling techniques (Logistic Regression, Decision Tree, Naive Bayes and Support Vector Machines)

4 Conclusion and Future Works

In this paper, we have applied GA on PCA regression to generate churn class samples to solve classification problem lead by imbalance distribution. We employed GA to search optimal cluster dataset that is discriminant so that linear PCA can find relevant PC for regression. The newly generated data were added to original raw data afterwards.

The approach was tested on a telecommunication data on churn prediction task. The results showed that PCA regression in combination with GA performed better than PCA regression without GA, linear regression in general and it performed as good as SMOTE. The optimal clustering size in GA K-means for PCA regression was within the range of 36-72. The GA K-means clustering performed better that traditional K-means algorithm. Additional samples would improve TP rate for churn size 6000-12000 but the FP rate would increase over 50% for churn size afterwards in each classifier. Since we are more interested in identifying customer who will likely to churn as losing a client causes greater loss for the Telecom company, improvement in TP is a good results. Nevertheless, FP rate must be limited as the cost of high FP can be expensive for future marketing campaign.

It is up to experts strategy whether to focus more on TP than FP and decide maximum acceptable FP rate. Some telecommunication companies might be interested in recognising potential churners more than non-churners as it is less expensive to provide extra services to potential churners, which is much less than non-churners in size, to retain them than providing extra services to all customers.

The difficulty of selecting locally optimum size K of clustering is a problem with the K-means clustering for Local regression. In this work, we had to spend time on running numerous experiments to find best cluster size K based on AUC rate. Additional problem of the approach is the prediction results of churn size after 12000 as it produced high FP rate. We are interested in understanding as to why additional churn samples would give high FP. There is a possibility that new churn data generated by various PC threshold can lead to poor classification of FP. For example, adding new churn data generated by threshold 0.9 might give different learning behaviour as opposed to new churn data generated by threshold 0.6. We are interested in learning if classifier can be improved through ensemble learning in the case of poor learning classifier due to new churn samples.

References

1. Burges, C.J.C.: A tutorial on support vector machines for pattern recognition. Data Mining and Knowledge Discovery 2(2), 121–167 (1998)
2. Chawla, N.V., Bowyer, K.W., Hall, L.O., Kergelmeyer, W.P.: SMOTE: Synthetic Minority Over-sampling Technique. JAIR 16, 321–357 (2002)
3. Hosmer, D.W., Lemeshow, S.: Applied Logistic Regression. Wiley, New York (1989)
4. Kohonen, T.: Self-Organizing Maps. Series in Information Sciences, vol. 30. Springer, Heidelberg (2001)

5. Quinlan, J.R.: Improved use of continuous attributes in c4. 5. Journal of Artificial Intelligence Research 4, 77–90 (1996)
6. Zhang, J., Yang, Y., Lades, M.: Face Recognition: Eigenface, Elastic Matching, and Neural Nets. In: The IEEE, pp. 1423–1435 (1997)
7. Luo, B., Peiji, S., Juan, L.: Customer Churn Prediction Based on the Decision Tree in Personal Handyphone System Service. In: International Conference on Service Systems and Service Management, pp. 1–5 (2007)
8. Au, W., Chan, C.C., Yao, X.: A novel evolutionary data mining algorithm with applications to churn prediction. IEEE Transactions on Evolutionary Computation 7, 532–545 (2003)
9. Bradley, A.P.: The Use of the area under the ROC curve in the evaluation of machine learning algorithms. Pattern Recognition 30, 1145–1159 (1997)
10. Bäck, T.: Evolutionary Algorithms in Theory and Practice. ch. 2. Oxford Univeristy Press, Oxford (1996)
11. Chawla, N.V., Japkowicz, N., Kotcz, A.: Editorial: special issue on learning from imbalanced data sets. SIGKDD Explor. Newsl. 6, 1–6 (2004)
12. Domingos, P.: MetaCost: A general method for making classifiers cost sensitive. In: The 5th International Conference on Knowledge Discovery and Data Mining, pp. 155–164 (1999)
13. Goldberg, D.E.: Genetic Algorithms in Search, Optimization and Machine Learning. Kluwer Academic Publishers, Dordrecht (1989)
14. Jolliffe, I.T.: Principal Components Analysis. Springer, New York (1986)
15. Huang, B.Q., Kechadi, M.T., Buckley, B.: Customer Churn Prediction for Broadband Internet Services. In: Pedersen, T.B., Mohania, M.K., Tjoa, A.M. (eds.) Data Warehousing and Knowledge Discovery. LNCS, vol. 5691, pp. 229–243. Springer, Heidelberg (2009)
16. Wei, C., Chiu, I.: Turning telecommunications call details to churn prediction: a data mining approach. Expert Systems with Applications 23, 103–112 (2002)

Time-Constrained Test Selection for Regression Testing

Lian Yu, Lei Xu, and Wei-Tek Tsai

School of Software and Microelectronics, Peking University, P.R. China
Department of Computer Science and Engineering, Arizona State University, USA

Abstract. The strategy of regression test selection is critical to a new version of software product. Although several strategies have been proposed, the issue, how to select test cases that not only can detect faults with high probability but also can be executed within a limited period of test time, remains open. This paper proposes to utilize data-mining approach to select test cases, and dynamic programming approach to find the optimal test case set from the selected test cases such that they can detect most faults and meet testing deadline. The models have been applied to a large financial management system with a history of 11 releases over 5 years.

Keywords: Test case selection, regression testing, test case classification, Time-constrained, P-measure.

1 Introduction

Testers face two challenges when performing regression testing: (1) they have a large volume of test cases to select; and (2) they have limited test resources to utilize in terms of time and testers. Currently, many of them manually select various test cases based on their experiences, resulting in time-consuming and low capability of defect detections.

This paper uses Chi-square to extract the features of a test case, associates them with a certain type of defects that the test case can detect, and uses SVM (Support Vector Machine) to classify test cases according to their capabilities to detect pertinent defects, and applies P-measure and dynamic programming approaches to select test cases from each of the categories with the aim to discover more defects within a limited time span. The process of test case classification and selection is described as follows:

1. Test Case Classification: We collect historical test-case dataset and classify them according different defect types using Support Vector Machine (SVM) model, such that we can choose specific sets of test cases to detect the target types of defects.
2. Test Case Selection: With a limited period of testing time, we use an optimal resource allocation model to fully utilize the test resources with the highest ability of detecting defects. In this paper, test resources refer to the test execution time. We use the test case dataset to build an optimal allocation model and estimate the model parameters to make sure the model can achieve the desired optimal goals and meet the time resource constraints.

L. Cao, J. Zhong, and Y. Feng (Eds.): ADMA 2010, Part II, LNCS 6441, pp. 221–232, 2010.

The paper is organized as follows. Section 2 presents the approach to classify existing test cases according to defect types that they are likely to reveal. Section 3 depicts an optimal model to select test cases with the limitation of test execution time. Section 4 gives a survey on related work. Finally, Section 5 concludes the paper.

2 Test Case Classification

The test case classification consists of two steps: 1) collect historical test cases and perform pre-processing on the data; 2) build an SVM (Support Vector Machine) 1 model to classify test cases according to defect types that they are associated with.

2.1 Collecting Historical Test Cases and Pre-processing the Test Dataset

This paper collects the historical test data (test cases and defect reports) of a software product from a multinational financial corporation. The product consists of 4 components (User Interface (UI), Domain Functions (FUN), Data Management (DM), Interface (IF)), and there are 11 release versions over 5 years. This paper defines 5 types of defects: four defect types are associated with the four components, and one for system (SYS) related defects. As the raw data are vulnerable to the impacts of noise data, data preprocessing is needed to improve the data quality and the classification accuracy. The process has the following main steps:

1. Extract useful information from each test case, including the description and test execution time, and then manually label each test case according to a defect type in terms of modules that the test case can reveal during the testing.
2. Filter the punctuation using regular expressions and common words using stop-words dictionary. It aims at increasing the classification accuracy.
3. This paper uses chi-square algorithm as the feature extraction algorithm to select the features, represented as terms/words which are most comprehensive text contents to portray the characteristic of a test case regarding a particular defect type.

We collect 1521 test cases from a large financial management system with a history of 11 releases over 5 years. We pre-process on the collected data using the 1) and 2) steps, and using the step 3) to extract the features. For example, in Table 1, the highest chi-values (in bold) of features "Add-in" and "browse" are associated with the categories of Interface and UI, respectively. A feature belongs to a defect type where the feature has the highest chi-value. Therefore feature "Add-in" belongs to Interface, and "browse" to UI. As a result, 1260 individual features are taken out from the historical test cases.

Table 1. Feature Chi-Values and Categories

Features \Defect Types	UI	DM	FUN	Interface	SYS
Add-in	5.6	2.3	9.7	**22.6**	7.9
Browse	**4.9**	0.3	1.4	0.3	2.1

The average chi-value is used to select the most comprehensive features and we observed that the features with chi-values lower than 0.20 contain lots of invalid information. For example, as shown in Table 2, the three features, "Colon", "TABS", and "PREP" have low average chi-values. With the literal meanings of these features, they are unhelpful for the test-case classification and even decline the classification accuracy; therefore we define them as invalid features for the classification, and scrub out such words. On the other hand, the three features "PARTYTYPEDESC", "DROP", and "METADATA" have relatively high average chi-values. By the literal meanings of the features, they are helpful for the test-case classification and increase the classification accuracy.

Table 2. Invalid Features and Valid Features

Feature	Category	Average chi-values
Colon	SYS	0.00181343607657669
TABS	FUN	0.01481480448331962
PREP	IF	0.04428857965618085
PARTYTYPEDESC	SYS	0.2592934583067385
DROP	DM	0.3871442755803598
METADATA	DM	0.4264642415551175

Consequently we set the threshold of average chi-value as 0.20. Retain the features with chi-values above the threshold, and obtain 1076 features with 14.6% of features removed.

2.2 Classifying Test Cases Using SVM Model

The paper selects SVM classification algorithm for the two reasons: (1) SVM supports text classification and multi-label classification; and (2) SVM can solve high-dimensional and non-linear problems. We train the model to classify the test cases and improve the classify accuracy from the following three aspects:

1. Handling stop-words and punctuations in training and testing data: In the early experiment, we find there are still some stop-words in the defect summaries, which worsen the prediction accuracy. Accordingly, we append more stop-words such as the uppercase word to the dictionary.
2. Managing data unbalance: Initially, some type of defects accounts for large proportions in the sample data while the others are very small even no data. For example, the proportion of defects belonging to interface module (IF) is relatively small, 9.6%, while the proportion of defects belonging to financial function module (FUN) reaches 32.4%, more than three times as shown in Table 3. To manage the problem, we increase the amount of the defect data whose types account for small proportions. In addition we define a weight for each bug type during the feature extraction, and use the weight to reduce the dominance of the bug types which account for large proportions.
3. Tuning parameters of SVM tool: During modeling test-case classification, we can obtain different prediction accuracy with the same data by varying the threshold of the feature value, or changing the proportion between training data and testing data.

Table 3. Proportions of defect types

Defect type	DM	FUN	IF	SYS	UI
Proportion	15.3%	32.4%	9.6%	14.5%	29.3%

At the beginning of the experiments, we have totally 299 test cases as sample data, which are divided into three ratios of training data to testing data: 2:1, 3:1 and 4:1 while keeping the thresholds and costs cross the three ratios the same. The averages of accuracies (the number of defects correctly classified vs. the total number of defect data) for the ratios (2:1), (3:1), (4:1) are 59.9%, 56.1% and 54.3% as shown in Table 4. It is observed that ratio 2:1 gets relatively high accuracy where among the test cases of testing data, i.e., more than half of test cases are classified into the correct defect types.

Table 4. Accuracy varying with ratios of training data to testing data

Feature #	Threshold	2:1	3:1	4:1
88	0.1	59.6%	51.3%	54.2%
85	0.3	64.6%	59.6%	54.2%
66	0.5	57.6%	62.1%	51.3%
63	0.7	57.6%	51.3%	57.3%
Averages of accuracy		59.9%	56.1%	54.3%.

Because the number of test cases in the initial experiments is not enough, the accuracy is not as high as expected. When we adjusted the feature threshold during the experiments, the accuracy did not change much. We consider that the main reasons are due to the small size of sample data. Accordingly, we collect more historical test data from the software vendor to improve the accuracy. Now, we have 1521 test cases increasing 1222 sample data. We try also three ratios of training data to testing data: 2:1, 3:1 and 4:1 while keeping the threshold and cost the same. The averages of accuracies for the ratios (2:1), (3:1), (4:1) are 84.0%, 80.0% and 81.6% as shown in Table 5. It is observed that ratio 2:1 gets relatively high accuracy as well.

Table 5. Second round experiments

Feature #	Threshold	2:1	3:1	4:1
47	0.05	79.1%	80.4%	81.5%
47	0.10	83.9%	75.9%	79.9%
46	0.15	85.3%	80.1%	82.2%
41	0.20	87.6%	83.7%	82.9%
Averages of accuracy		84.0%	80.0%	81.6%

3 Test Case Selection

In most of situations, we are left with a huge number of test cases in each category after the test case classification per relevant defect types discussed in the previous section. How to select test cases from different categories that maximize the defect detection capability with the limited resource constraints is the focus of this section. The process of test case selection includes establishing a test case selection optimal model, estimating parameters in the model, and running the model by taking in the classified test cases and outputting optimized test case selection.

3.1 Test Case Selection Model

3.1.1 Establishing the Objective Function

Test case selection model is based on P-measure algorithm, which is to quantify the defect detecting ability of a testing strategy. The formula is described as follows:

$$P = 1 - \prod_{i=1}^{k} (1-\theta_i)^{n_i} \tag{1}$$

P is the probability of incurring at least one failure. θ_i is the failure rate or error rate and is predicted using the historical data. Test cases are divided into k subsets corresponding to the k defect types. n_i is the number of test cases selected in the ith subset. The objective function in the test selection model is to maximize the P function under certain constrains as discussed below.

3.1.2 Identifying Constraints

The constraints of the optimal model are shown as follows:

$$\sum_{i=1}^{k} c_i n_i \le w_i C, n_i \ge 1, \textit{ where } \mathrm{c}_i \text{ and } n_i \text{ are integers, } 0 < w_i < 1, \mathrm{i} = 1, \cdots, \mathrm{k} \tag{2}$$

C is the total test resources, referring to the test execution time (hours), w_i is the weight of time resource allocated to detect defect type i, and c_i is the resource for defect type i, where the resource c_i is estimated as the mean value of total hours in each defect type.

3.2 Solving the Optimal Model

This section derives the solution of the optimal model stated in equations (1) and (2). There are two relevant tasks: one is to estimate the parameters in the model, and the other is to develop the model solution.

3.2.1 Parameter estimation

The main job in this subsection is to estimate the failure rate θ_i. We use two approaches to fulfilling the task: data smoothing and linear regression analysis.

The exponential smoothing is a technique applied to time series data to produce smoothed data or to make forecasts. We tried 8 versions of historical data, and each has five categories of test cases. Take the first category as an example. The exponential smoothing approach has higher relative errors and average error up to 0.3391. On the other hand, linear regression approach gets smaller relative errors and average error up to 0.059. Therefore, we choose the linear regression approach to predict the error rate.

Let ci,j be the hours that a test case j in category i needs to execute, and Ni the number of test cases in category *i*. The c_i is estimated in the equation (3), where $c_{i,j}$ and N_i can be collected from historical data.

$$c_i = \sum_{j=1}^{N_i} c_{i,j} / N_i \tag{3}$$

Let θ_i be the dependent variable and c_i the independent variable. The formula is shown as follows:

$$\theta_i = f(c_i) = a \times c_i + b \tag{4}$$

The historical error rate θ_i is calculated as the ratio of the number of failed test cases to the total number of test cases. Parameters *a* and *b* in equation (4) can be determined using historical data of θ_i and the estimated c_i. Hence, θ_i in the new version release can be predicted using formula (4).

3.2.2 Optimal Model solution

We use dynamic programming to solve the optimization problem as described below.

1. Derive a recursion formula for the P-measure formula in equation (1) in the optional model.

Let $P_i(S)$ be the maximum probability to detect defect types *i*, *i*+1,…,*k*. *S* refers to the hours allocated to the defect types *i*, *i*+1,…,*k*. From equation (1), we can get the following formulas:

$$P_i(S) = 1 - \prod_{l=i}^{k} (1-\theta_l)^{n_l} \tag{5}$$

Let $(S-c_i n_i)$ refer to the hours allocated to the defect types *i*+1,…,*k*. From (5), we obtain the following formula conversion:

$$P_{i+1}(S - c_i n_i) = 1 - (1-\theta_{i+1})^{n_{i+1}} \cdots (1-\theta_k)^{n_k} \tag{6}$$

Putting (6) into (5), we get a recursion formula:

$$P_i(S) = \max\{1-(1-\theta_i)^{n_i} \cdot [1 - P_{i+1}(S - c_i n_i)]\} \tag{7}$$

Where n_i satisfies the condition:

$$c_i n_i \le S - \sum_{l=i+1}^{k} c_l n_l \le S - \sum_{l=i+1}^{k} c_l, 1 \le n_i \le \left\lfloor \left(S - \sum_{l=i+1}^{k} c_l n_l \right) / c_i \right\rfloor \tag{8}$$

2. Determine integer n_k based on inequality (8).

Let $i = k$, then $P_{i+1}(S-c_i n_i)=0$. Therefore we obtain the basic equation of dynamic programming from equation (7):

$$P_k(S) = \max[1-(1-\theta_k)^{n_k}] \tag{9}$$

The range of n_k in equation (9) satisfies the condition:

$$1 \leq n_k \leq \lfloor S / c_k \rfloor \tag{10}$$

where n_k is a natural number. We choose at least one test case in each category, i.e., $n_i \geq 1$. As the equation (9) is a monotonic increasing function, taking the upper-bound of integer n_k in inequality (10) can maximize $P_k(S)$.

3. Determine integer n_i (i=1,...,k-1) based on equation (7) and constraint (8).

According to (8), we obtain the following recursion constraint on n_{k-1}:

$$1 \leq n_{k-1} \leq \lfloor (S - c_k n_k) / c_{k-1} \rfloor \leq \left\lfloor (C - \sum_{l=1}^{k-2} c_l - c_k n_k) / c_{k-1} \right\rfloor, where \sum_{l=i}^{k} c_l \leq \sum_{l=i}^{k} c_l n_l \leq S \leq C - \sum_{l=1}^{i-1} c_l n_l \leq C - \sum_{l=1}^{i-1} c_l \tag{11}$$

Likewise, we can get n_{k-2}, ..., n_1. Through the above three steps, we get the optimal sizes for test cases selected from each category: $n^{opt} = (n_1^{opt}, n_2^{opt}, \cdots n_k^{opt})$.

3.3 Evaluating the Test Case Selection

We collect the historical data on error rates and average test times of test cases in the five categories across 8 versions from 1.0 through 1.7 from a financial management system, and run the optimal model using the collected data, where in each version, θ_i and c_i are associated with the different five types of defects. To evaluate the validity of the test case selection model, this section compares the optimal model with three selection approaches: random selection, average selection and proportional selection.

Using the collected data, we estimate the parameters a and b in equation (4), and get the results as 0.005 and 0.215, respectively. When there is a new release, the average time c_i can be calculated based on existing test case set. With the equation (4), for example, the error rate θ_i of each category can be estimated for new version 1.9 as shown in Table 6.

Table 6. Selection Model Version 1.9

Defect Type	UI	DM	FUN	IF	SYS
θ_i	35%	25%	38%	30%	30%
c_i	3.95	6.75	3.50	5.10	5.20

3.3.1 Optimal Model based Selection Strategy

Taking as inputs the estimated average time c_i and the estimated error rate θ_i of each category of test cases, and using the dynamic programming model described in Section 3.2, we attain the optimal number of test cases selected from each category as shown below: $n^{opt} = (n_1^{opt}, n_2^{opt}, n_3^{opt}, n_4^{opt}, n_5^{opt}) = (1, 16, 1, 1, 1)$.

It is observed that although we acquire the optimal results based the model (1) and (2), the numbers of test cases from each category are out of balance. Category "FUN" has the largest number of test cases up to 16, while the rest of categories are only 1. Although we can concentrate on one type of defect detection, there is no guarantee that other categories have no defects.

We propose two ways to improve the balance that is different from what paper 4 does. One is to allow testers to specify the number proportion of each category. The other is to allow testers to specify lower bound of the number of test cases in each category. We use weight w_i to adjust the upper limit on n_i, and modify the constraints as follows:

$$1 \le n_i \le \min\{w_i \cdot C / c_i, \left\lfloor \left(S - \sum_{l=i+1}^{k} c_l n_l \right) / c_l \right\rfloor\} \tag{12}$$

The new optimal result is shown as follows, which is more balanceable among different test case categories. $n^{opt} = (n_1^{opt}, n_2^{opt}, n_3^{opt}, n_4^{opt}, n_5^{opt}) = (1, 5, 3, 3, 3)$.

It is observed above that original dynamic programming model just guarantees the optimal time utilization and defect detection capability, but does not address the test case balances among different categories. By adjusting the weight w_i in the constraints, the balance can be improved and still maintain the overall time utilization and defect detection capability.

3.3.2 Random Selection Strategy

The first assessment is to compare with the random results of test case selection:

1. Randomly select a test case from the whole test set.
2. Subtract the time of test case execution from the total available time C.
3. Repeat steps 1 and 2, till the available time is not enough to execute a test case.

Using the same sample data, we get the selection results for the new release as follows. $n^{ran} = (n_1^{ran}, n_2^{ran}, n_3^{ran}, n_4^{ran}, n_5^{ran}) = (2, 7, 2, 5, 1)$.

3.3.3 Average Selection Strategy

The average selection schema emphasizes the time allocation balance among different categories. It apportions time to each category. Equation (13) shows the average selection schema, where each category has at least one test case selected.

$$n_i^{ave} = \begin{cases} \left\lceil (C - \sum_{i=1}^{k} c_i) / c_i \cdot k \right\rceil & (C - \sum_{i=1}^{k} c_i > 0) \\ 1 & (C - \sum_{i=1}^{k} c_i = 0) \end{cases} \tag{13}$$

Using the same sample data, we get the selection results for the new release as follows. $n^{ave} = (n_1^{ave}, n_2^{ave}, n_3^{ave}, n_4^{ave}, n_5^{ave}) = (2, 4, 3, 3, 3)$.

3.3.4 Proportional Selection Strategy

The proportional selection schema calls attention to the error rates associated with test case categories. The higher the error rate is, the more test cases it will select. Equation

(14) shows the proportional selection schema, where each category has at least one test case selected.

$$n_i^{prop} = \begin{cases} \left\lceil \theta_i / \sum_{j=1}^{k} \theta_j \cdot (C - \sum_{i=1}^{k} c_j) / c_i \right\rceil & (C - \sum_{i=1}^{k} c_j > 0) \\ 1 & (C - \sum_{i=1}^{k} c_j = 0) \end{cases} \tag{14}$$

Using the same sample data, we get the selection results for the new release as follows. $n^{pro} = (n_1^{pro}, n_2^{pro}, n_3^{pro}, n_4^{pro}, n_5^{pro}) = (2, 4, 3, 3, 4)$.

3.3.5 Comparison among Different Strategies

We compare different strategies in terms of resource utilization and defect detecting capability. The resource utilization η is defined as follows:

$$\eta = \sum_{i=1}^{k} c_i n_i / C, \text{ where } 0 \le \eta \le 1 \tag{15}$$

The higher η is, the better the test resource is utilized; the lower η is, the resource is wasted. Figure 1 shows the comparison of resource utilizations among different strategies, where the abscissa stands for 10 different versions of data sets and ordinate stands for the resource utilization rate that represents the test time used vs. the total test time available.

It is observed that average and proportional strategies get lower resource utilization rates, the random strategy has a large fluctuation, and the optimal allocation strategy has stable and high resource utilization rate.

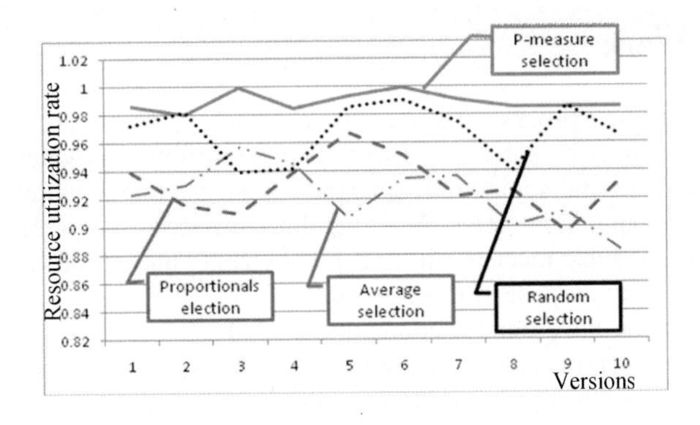

Fig. 1. Comparison of resource utilization

This paper uses P-measure (see formula (1) in Section 3.1) to evaluate the detection capabilities of different test case selection strategy. Figure 2 shows the defect detection capabilities of the four strategies. It is observed that random strategy get the lowest detection capability; the proportional strategy is better than the average, as it selects test cases based on the error rate θ_i and its pertinence is obvious; the average strategy is inferior to the proportional strategy and with high vacillation; among them the optimal allocation strategy has the most stable and highest detection capability.

On the other hand, we notice that there are no much differences in terms of average detection capability among the optimal model, the average strategy and the proportional strategy as they are all reaching up to 0.9999 although the average strategy and the proportional strategy have large fluctuation. We consider that this may be due to the small numbers of selected test cases from different categories, varying from 1 to 7. Our future work in this regard will apply the proposed to other systems which need more test cases, also will utilize simulation approach to observe the outcomes when increasing the sizes of selected test cases.

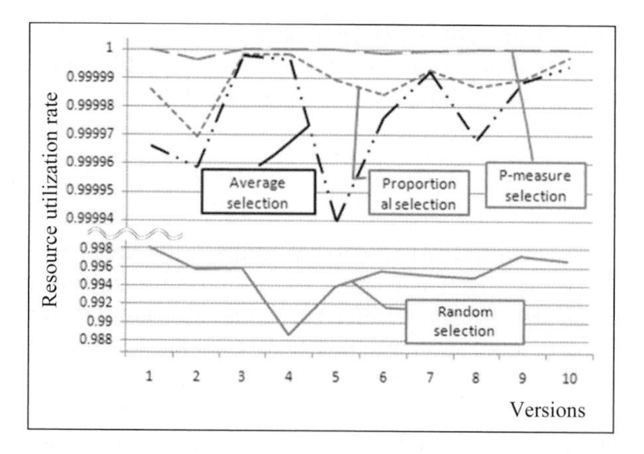

Fig. 2. Comparison of defect detecting capability

4 Related Work

In the software testing process, the quantity and quality of test cases determines the cost and effectiveness of software testing. Current software products have a large number of test cases and some are redundancy. It means that a subset of a test suite can also meet testing requirements. Selecting those test cases out of a large number of a test suite, which fully meet the testing requirements and as far as possible non-redundant is the main purpose of this research.

Yoo and Harman have done a comprehensive survey on regression testing techniques 3. Various approaches of test case selection procedure have been proposed using different techniques and criteria including heuristics for the minimal hitting set problem, integer programming, data-flow analysis, symbolic execution, dynamic slicing, CFG (control flow graph) graph-walking, textual difference in source code, SDG (system dependency graph) slicing, path analysis, modification detection, firewall, CFG cluster identification and design-based testing.

Optimal approaches, such as greedy algorithms, heuristic algorithms and integer programming, can be used to solve this problem 4, selecting proper test cases with resource constraints. We extend Zhang's work by adding weighs to improve balances among different categories. Orso et al. presented two meta-content based techniques of regression test selection for component based software 5. W.T. Tsai et al. 6 proposed a Model-based Adaptive Test (MAT) case selection and ranking technique to eliminate redundant test cases, and rank the test cases according to their potency and coverage.

Data mining has been applied to software reliability 7, software engineering, and software testing. Mark Last et al. 8 used data mining approach in automating software testing. They demonstrated the potential use of data mining algorithms for automated induction of functional requirements from execution data and applied a novel data mining algorithm called Info-Fuzzy Network (IFN) to execution data of a general-purpose code for solving partial differential equations. Mockus et al. 9 demonstrated that historical change information can support to build reliable software systems by predicting bugs and effort. In software testing, module complexity can be evaluated using several measures, but the relationships between these measures and the likelihood of a module failure are still not fully understood. As shown by Dick and Kandel 10, these relationships can be automatically induced by data mining methods from software metrics datasets that are becoming increasingly available in the private and public domain.

Predictions of number and locations of faults have been carried by many researchers. Ostrand et al. 11 developed a negative binomial regression model to predict the expected number of faults in each file of the next release of a system. Sergiy A. Vilkomir' paper 12 presented an approach to using Markov chain techniques for combinatorial test case selection. Ing-Xiang Chen 13 presented an implicit social net work model using Page Rank to establish a social net work graph with the extracted links. When a new bug report arrives, the prediction model provides users with likely bug locations according to the implicit social network graph constructed from the co-cited source files.

5 Conclusion

This paper uses the classification techniques to create test case classification model in regarding to defect types those test cases can detect; creates an optimal test case selection model with P-measure as the objective function and time allowance as the constraints; and uses dynamic programming approach to obtain the optimal solution. We collected historical data from a financial management system produced by an international software company. The Chi-square algorithm is used to extract features from test case descriptions and defect reports, and SVM is used to model the classifiers. These experiments used 1521 test cases and identified 1571 defects. The accuracy of the model is affected by the number and the pre-processing of the training data. When the quantity of the historical data is more than 1000 in the experiment, classification accuracy is able to achieve 75% to 87%. By using this model, the efficiency of the test process can be improved by 20% on average without changing the underlying technology or tools. In particular, the efficiency can improve more than 25% for those peoples who are new in testing and/or the product.

Our future work will collect more datasets on integration testing, increasing the accuracy of predicting results. We will gather new test-case data, classify them with different defect types and then give priorities to those serious faults so that testers can focus their energy on the high priority items first.

Acknowledgements. This work is partially supported by the National Science Foundation of China (No.60973001), IBM China Research Lab (No.20090101), and U.S. Department of Education FIPSE project. The authors would thank Jingtao zhao, Hui Lv, Ting Xu for working on the empirical study described in this paper.

References

1. Han, J.W., Micheline, K.: Data Mining: Concepts and Techniques, 2nd edn. Morgan Kaufmann, San Francisco (2006)
2. Yang, Y., Pedersen, J.O.: A comparative study on feature selection in text categorization. In: Machine Learning: Proceedings of the Fourteenth International Conference, pp. 412–420 (1997)
3. Yoo, S., Harman, M.: Regression testing minimization, selection and prioritization: a survey. Software Testing, Verification and Reliability (2010)
4. Zhang, D.P., Nie, C.G., Xu, B.W.: Optimal Allocation of Test Case Considering Testing-Resource in Partition Testing. The journal of NanJing university, Natural Sciences, 554–555 (2005)
5. Orso, A., Harrold, M.j.: Using component meta-content to support the regression testing of component based software. In: Proc. of IEEE Int. Conf on Software Maintenance, pp. 716–725 (2001)
6. Tsai, W.T., Zhou, X.Y., Paul, R.A., Chen, Y.N., Bai, X.Y.: A Coverage Relationship Model for Test Case Selection and Ranking for Multi-version Software. In: 10th IEEE High Assurance System Engineering Symposium, pp. 105–112 (2005)
7. Cai, K.F.: A Critical Review on Software Reliability Modeling Reliability Engineering and System Safety, vol. 1(32) (1991)
8. Last, M.: The Data Mining Approach to Automated software Testing, Conference on Knowledge Discovery in Data. In: Proceedings of the Ninth ACM SIGKDD International Conference on Knowledge Discovery and Data Mining, pp. 24–27 (2003)
9. Mockus, A., Weiss, D.M., Zhang, P.: Understanding and predicting effort in software projects. In: Proceedings of the 25th International Conference on Software Engineering, pp. 274–284 (2003)
10. Dick, S., Meeks, A., Last, M., Bunke, H., Kandel, A.: Mining in Software Metric Databases. Fuzzy Sets and Systems 145(1), 81–110 (1998)
11. Ostrand, T., Weyuker, E., Bell, R.M.: Predicting the location and number of faults in large software system. IEEE Transactions in Software Engineering 31(4), 340–355 (2005)
12. Sergiy, A., Vilkomir, S.W.T., Jesse Poore, H.: Combinatorial test case selection with Markovian usage models. Fifth International Conference on Information Technology: New Generations, 3–8 (2008)
13. Xiang, I., Yang, C.Z., Lu, T.K., Jaygarl, H.J.: Implicit Social Network Model for Predicting and Tracking the Location of Faults. In: The 32nd Annual IEEE International Computer Software and Applications Conference, pp. 136–143. IEEE Computer Society, Los Alamitos (2008)

Chinese New Word Detection from Query Logs

Yan Zhang[1], Maosong Sun[1], and Yang Zhang[2]

[1] State Key Laboratory on Intelligent Technology and Systems Technology
Deptment of Computer Science and Technology,
Tsinghua University, Beijing 100084, China
[2] Sohu Inc. R&D center, Beijing 100084, China
zhang-y-05@mails.tsinghua.edu.cn, sms@mail.tsinghua.edu.cn,
zhangyang@sohu-rd.com

Abstract. Existing works in literature mostly resort to the web pages or other author-centric resources to detect new words, which require highly complex text processing. This paper exploits the visitor-centric resources, specifically, query logs from the commercial search engine, to detect new words. Since query logs are generated by the search engine users, and are segmented naturally, the complex text processing work can be avoided. By dynamic time warping, a new word detection algorithm based on the trajectory similarity is proposed to distinguish new words from the query logs. Experiments based on real world data sets show the effectiveness and efficiency of the proposed algorithm.

Keywords: new word detection, dynamic time warping, query logs, search engine.

1 Introduction

Thanks to the rapid development of the internet, Chinese language changes over a wide range of areas, and more and more new words are brought into birth. These new words, usually corresponding to real world events, are so informative that can reflect the social interests. On the other hand, these new words, which have not been collected by any vocabulary in time, bring many difficulties into text processing. Thus, research on new word detection becomes very essential and deserves tense attentions. Meanwhile, due to the rapid development and large quantities of new words, manually recognizing the new words requires a huge amount of human efforts. Therefore, people desire to detect new words automatically by introducing machine learning methods. Former works on new word detection all resort to the web pages to distinguish new words from the texts, which requires complex text processing, e.g., word segmentation. Taking the wisdom of crowd into account, new words are concerned even mainly created by web users and user behaviors on the internet. Specifically, the query log from search engines, can serve as a major resource of new word detection. Our work is based on visitor-centric resource, query logs, to shed light on the automatical detection of Chinese new words.

For the ease of clearance, we give several examples of the real query logs collected from a commercial search engine, as described in Table 1.

The examples described in Table 1 include the query terms,“猪流感(Swine Flu)”,“电影(movie)” and “优酷(the website named Youku)”, the IP information which recorded

L. Cao, J. Zhong, and Y. Feng (Eds.): ADMA 2010, Part II, LNCS 6441, pp. 233–243, 2010.

Table 1. Examples of query log data

IP	Query	URL	Time
..*	猪流感	http://sohu.com	20090502
..*	电影	http://google.com	20090101
..*	优酷	http://youku.com	20091231

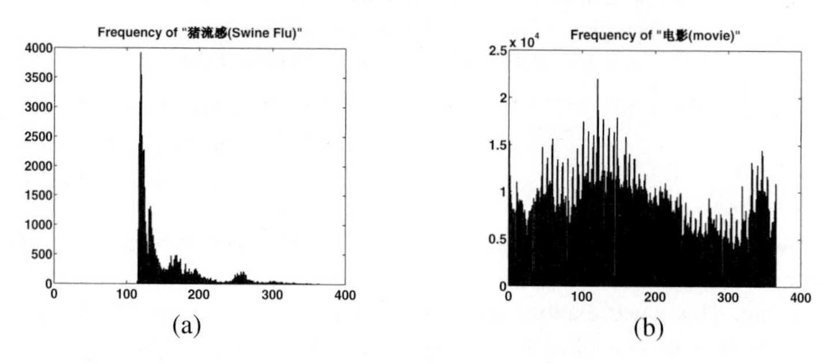

Fig. 1. Word Frequency. (a) "猪流感(Swine Flu)". (b) "电影(movie)".

the IP of users who issued these word terms, the URL and the time stamps when users issued the query terms.

In Figure 1(a), we note that the frequency trajectory of the word in temporal dimension reaches its maximum around the occurrence of Swine Flu expansion, which due to the burstiness of the word term "猪流感(Swine Flu)". Meanwhile, the frequency trajectory of the word "电影(movie)" in temporal dimension do not have a significant spike which may due to the fact that the word term "电影(movie)" is not a relatively bursty word.

Based on this observation, we exploit the frequency trajectory to detect new word since it can reflect the dynamics of public interests, so as the probable birth of new words.

The rest of our paper is organized as follows. Section 2 introduces the related work. Section 3 gives the details of the dataset and representation. Section 4 presents our similarity measures of new words and section 5 describes the new word detection algorithm. Section 6 gives the evaluation metric and discuss the results. In the last section, we conclude the paper.

2 Related Work

In the literature, there are two understandings of new word: One corresponds to the word out of vocabulary and the other means the popular word that has been newly created or old word with new meaning. For the work of detecting words out of vocabulary, Ref. [1] is the first to exploit visitor-centric data, e.g., user behaviors on Chinese input method, to detect new and representative words to enlarge the corpus for word segmentation and expert knowledge base. For the new word of the second category, there have been few works yet except for these works [2, 3, 4]. In this paper, we also focus to the second class of new words.

However, almost of the previous works detect new words from the author-centric resource without taking the visitor-centric resources into account. Meanwhile, there has been a field of extracting information from the visitor-centric data resource [5,6,7,8,9]. Our work focuses on detecting new words from the visitor-centric resources, which has not been mentioned in existing works.

Our work differs from the above works in following ways: First, we focus on detecting new words from the visitor-centric resources, which has not been exploited in former works. Second, to our best knowledge, we are the first to take the dynamics of word term frequency into accounts by dynamic time warping similarity measures. We will give detailed statement in the later sections.

3 Data Representation

As mentioned in the first section, query log has the advantage for detecting new words: Naturally segmented, user-centric and recorded with time stamps. In this section we take a real world query log as an example to describe the characteristics for detecting new words.

Given the normalized query logs, we first group each query term with its frequency into sequence. Let T denote the duration or period of the collected data, and f_w be the normalized frequency sequence of a word term w, organized chronologically as $f_w = \{f_w(1), f_w(2), ..., f_w(T)\}$, where $f_w(t)$ represents the number of appearances in time index t. Once data is segmented into chronological sequence of query log, similarity comparison between time series can be accomplished.

To smooth out the short-term fluctuations and highlight long-term trends, We exploit moving average method to avoid the effects of fluctuates of data, since moving average is the dominate method for smoothing data in the literature. For example, given the normalized frequency sequence f_w, the smoothed sequence f'_w, of which the ith component is denoted by $f'_w(i)$, is calculated as the following:

$$f'_w(i) = \frac{1}{l} \sum_{k=i-l+1}^{i} f_w(k) \tag{1}$$

Because queries on weekdays are typically more than that on weekends and holidays, the step length l was set to be 7 to reduce the effect of weekly fluctuations of the issued queries.

4 Words Filtering

Before diving into the details of our experimental setup, we first introduce our dataset used for our work.

4.1 Dataset

For our experiments, the query logs are collected from a commercial search engine[1] within 365 day units, from January 1, 2009 through December 31, 2009 and includes

[1] http://www.sogou.com/

72,021,172 different queries that appears during the time. As shown in the Table 1, the IP information has also been collected by the search engine. However, to protect the privacy of users, we eliminate the IP information. Given all the query logs from the search engine, we observe that the frequencies of words in the dataset roughly yield the Zipf's Law, which are illustrated in Figure 2.

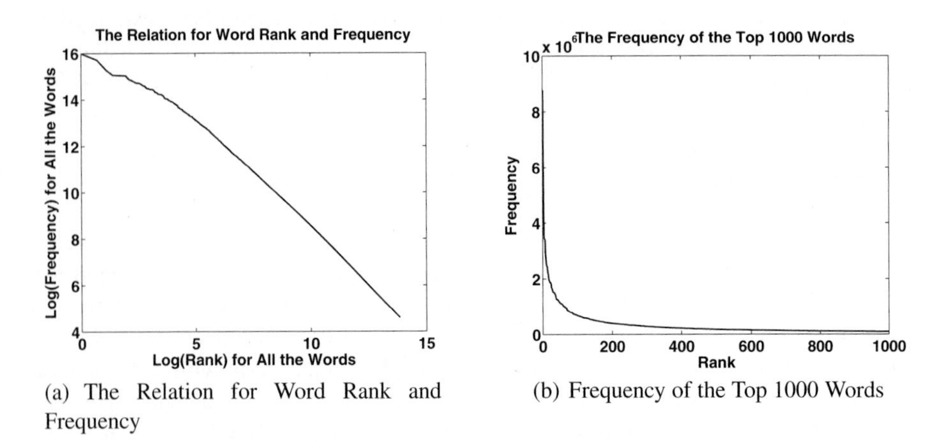

(a) The Relation for Word Rank and Frequency

(b) Frequency of the Top 1000 Words

Fig. 2. Distribution for the Words in Dataset

Among the 72,021,172 distinct queries, there are 71,155,114 (occupy about 98.8%) queries with low frequency less than 100 appearances within this period. We discard the non-GBK decode queries and the queries with low frequency less than 100 and get 866,058 queries sequences with the length of 365 day units which is detailed in the Table 2 with different lengths of queries.

Given the word frequency, we first remark that detecting words by using the time series similarity algorithm to cluster or classify the words. However, after the removal of low frequency queries, the number of queries remain too large for the similarity calculation of time series. Thus, we attempt to filter words roughly by the trajectories classification, which can be accomplished with much lower complexity.

Table 2. The distribution of word frequency with different length after word filtering

Length	2	3	4	5	6	7
Frequency	75,752	109,386	234,880	154,630	172,790	118,620

4.2 Trajectory Category Based on Burstiness

In order to distinguish the bursty words from the query logs, we perform a user study on a labeled data set. We collect the new words from the website[2] which is provided from January 1, 2009 to December 31, 2009, and select the new words appear in our dataset

[2] http://pinyin.sogou.com

as our labled set of new words, denoted as LNW (labeled new words). Meanwhile, we also randomly select 1000 common words from the a general Chinese as our corpus of non-new words. However, there are only 66.5% common words from this corpus appear in our dataset based on our observation of 10 times of random selection. Hence, we choose one group of common words randomly selected from the corpus which also appear in our dataset as our labeled set of non-new words, denoted as $LNNW$ (labeled non-new words). All the sets are described as in Table 3 and the labeled data set for new words and non-new words mentioned in the latter sections all mean this.

Table 3. Description of label set, listed by the length of word term

Length	2	3	4	5	6	7	Total
LNW	137	525	216	66	16	24	984
$LNNW$	611	27	12	8	3	1	662

According to the observation on the characteristics of these labeled sets, we formulate frequency trajectories into four categories, which are described in the Table 4.

Table 4. Description for trajectory categories of queries

Category	Description
HB	High-energy and bursty words
LB	Low-energy and bursty words
HU	High-energy and unbursty words
LU	Low-energy and unbursty words

By using a user study for the trajectories of manually labeled new words and non-new words, we formulate the trajectories of query logs from the Chinese search engine into four types, HB (High-energy and bursty words), LB (Low-energy and bursty words), HU (High-energy and unbursty words) and LU (Low-energy and unbursty words), where the energy correspond to the frequency. Here, the burstiness of word sequence is calculated by the algorithm using the discrepancy theory [10]. Our approach is the one dimensional special case of the method proposed in [10], which can be decribed as the following formula

$$f_w^b = \frac{max(f_w(i)) - \mu(f_w(i))}{\sigma(f_w(i))} \tag{2}$$

, where, $\mu(f_w(i))$ and $\sigma(f_w(i))$ represents the average value and the variance of the word frequency sequence $f_w(i)$, respectively.

Here, the threshold between high and low energy is set to be 10000, correspond to 4 in the logarithm scale and the threshold of burstiness is set to be 7. For the four types of the frequency trajectories, the type HB occupies 0.05%, the type HU occupies 1.33%, the type LB occupies 5.24% and the type LU occupies 93.38%, which is detailed in the Table 5.

Table 5. The proportion(%) of the four types based on energy and burstiness, listed by the length of words

Length	2	3	4	5	6	7
HB(%)	0.07	0.06	0.02	0.02	0.02	0.03
HU(%)	3.54	1.50	1.26	0.80	0.55	0.49
LB(%)	4.17	6.40	4.71	5.30	5.25	5.57
LU(%)	92.22	92.05	94.01	93.88	94.18	93.91

4.3 Trajectory Category Based on Length

In addition to the types of the trajectory, we also explore the performance of classification based on the length of the trajectory.

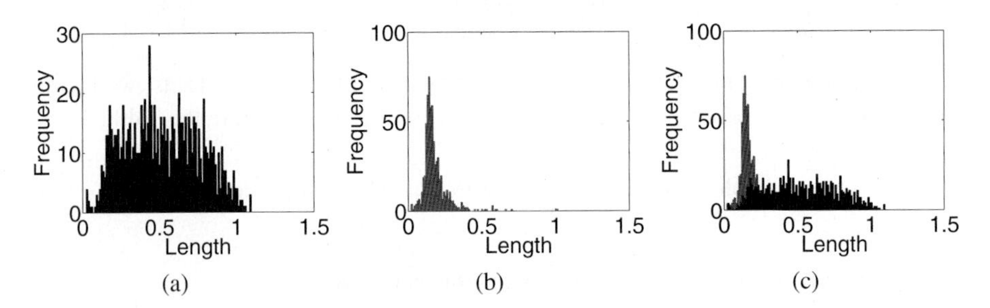

Fig. 3. The histogram of trajectories length of words.: (a) new words with normalization; (b) non-new words with normalization; (c) correlation of both the two types of words with normalization

From Figure 3 we notice that the two types of words can be distinguished roughly when the threshold was set to be 0.2. Thus, as the threshold was determined, we can calculate all the lengths of smoothed sequence frequency of query logs, and get the distribution of the length of query logs shown in the Figure 4, separated by the length of words, ranging from 2 to 7. Note that in this figure, the number of the words below the threshold accounts for more than 80%, which will be eliminated. Hence, after the threshold of the length of trajectories, the number of all words candidate for new word detection will be reduced effectively.

5 New Word Detection Based on Trajectory Similarity Measures

After word filtering, we get 45,367 and 77,945 words by the two methods as described in the former section. Thus, the space for new word detection has been reduced largely and the similarity measures can be executed now. In this section we motivate the similarity measures and propose our new word detection algorithms. Given two sequence P and Q, with length of M, N, represented as $P = (p_1, p_2, ..., p_M)$ and $Q = (q_1, q_2, ..., q_N)$, respectively, we utilize two similarity measures to evaluate the relationship between words and aim to detect new words based on the labeled data using a general machine learning algorithm.

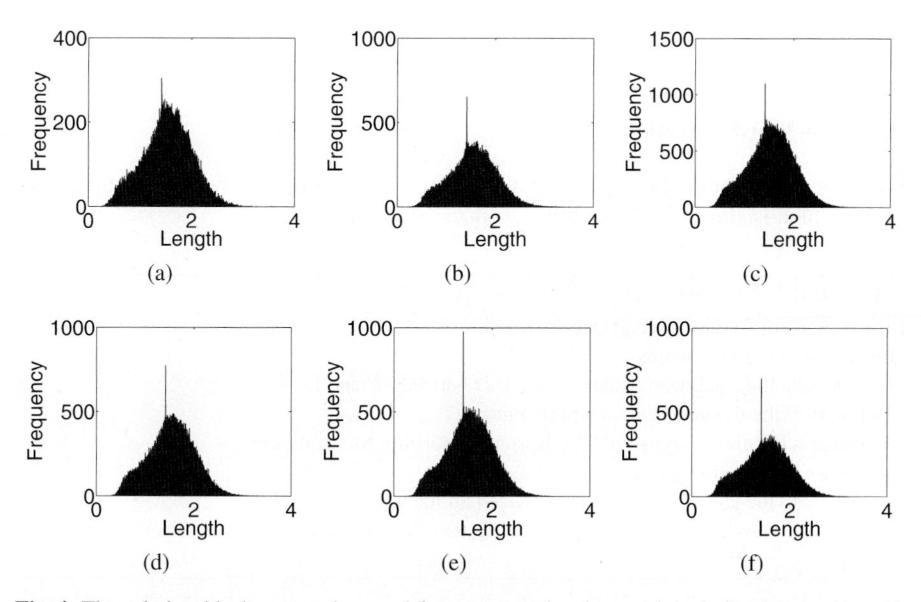

Fig. 4. The relationship between the word frequency and trajectory length for the queries with length n: (a) $n = 2$; (b) $n = 3$; (c) $n = 4$; (d) $n = 5$;(e) $n = 6$; (f) $n = 7$

5.1 Similarity Measures Based on Correlation

It is common to think of the similarity measure with correlation between two sequences. We introduce the similarity measures based on sequence correlation coefficient as our baseline algorithm. Given the two sequences, the correlation value is calculated according to the following equation,

$$Corr\_similarity = \frac{n\sum_n (PQ) - \sum_n P \sum_n Q}{\sqrt{\left[n\sum_n P^2 - (\sum_n P)^2\right]\left[n\sum_n Q^2 - (\sum_n Q)^2\right]}}. \tag{3}$$

5.2 Similarity Measures Based on Dynamic Time Warping

Though the correlation similarity between sequences is used widely, it can only catch the linear relationship between sequences. To evaluate the similarity in addition to the linear one, we adopt the time warping algorithm [11], which can avoid the warping of time scale, to measure the similarity between sequences. Firstly, a $M * N$ matrix is built, where the element $(d_{i,j})$ denotes the distance between P and Q (Euclidean distance is used here without loss of generality). A warping path, $W = (w_1, w_2, ...w_K)$, is used to record a mapping between sequence P and Q, which aims to minimize the distance on the warping path, where, K denotes the length of the warping path and the element $w_k = (p_i, q_j)_k$. Once the warping path is constructed, the similarity measures based on the dynamic time warping, can be calculated as follows:

$$DTW\_similarity(P, Q) = \exp^{-\sqrt{\sum_{i=1}^{K} w_i}}. \tag{4}$$

We observe that when the two sequences P and Q are the same sequence, then w_i equal to 0 and the similarity measure $DTW\_similarity$ reaches to 1.

5.3 New Word Detection

Based on the dynamic time warping algorithm, we give our new word detection algorithm, which is shown in Algorithm 1.

Algorithm 1. New Word Detection from Query Logs

Input: A set of frequency sequences of queries
Output: A set of new words
1: Calculate the Euclidean distance matrix D between P and Q.
2: Initialize the dynamic time warping matrix D'.
3: Perform the classic dynamic time warping algorithm for sequences.
4: Return the warping paths $\{w_k\}$.
5: Calculate the similarity measures between all the word sequences.
6: Group all the candidate words into two clusters using k-means algorithm to get the new word set

---

6 Experimental Setup

In this section, we demonstrate the experiment setup of our approach to new word detection. Firstly, we detail the results after the filtering processing. Secondly, we analyze the performance of new word detection with the two different similarity metric by evaluating the proposed metric with user experience. Then, we describe our proposed evaluation metric. Finally, some examples of detected new words are listed and analyzed. All the experiments have been conducted on an 8Ghz Intel Pentium 4, with 4GB of RAM.

6.1 Evaluation Metric

To evaluate our experiment, we investigate the performance of new word detection described in the Algorithm 1. Here, we give description of evaluation metric. For the clustering experiment, we adopt the number of detected new words from the dataset and the precision as our performance metric. To measure the global quality of the new word detection results, five measures are used, including *miss (m)*, *false alarm (f)*, *recall (r)*, *precision (p)* and the F_1, which are defined as follows based on the contingency table.

- $p = a/(a+b)$, *if* $(a+b) > 0$, *otherwise undefined*;
- $r = a/(a+c)$, *if* $(a+c) > 0$, *otherwise undefined*;
- $m = c/(a+c)$, *if* $(a+c) > 0$, *otherwise undefined*;
- $f = b/(b+d)$, *if* $(a+c) > 0$, *otherwise undefined*;
- $F_1 = 2rp/(r+p)$, *if* $(r+p) > 0$, *otherwise undefined*;

Table 6. The cluster-term contingency table

	new word	not new word
in cluster	a	b
not in cluster	c	d

6.2 Results

According to the new word detection algorithm described as Algorithm 1, we perform the new word detection and get 8,973 new words. To evaluate the performance of our approach to new word detection, we randomly select 200 detected new words from each of the four groups of results as our test set. Note that there are four group of results with the combination of two filtering methods and two similarity measures, so we finally get totally 800 candidates of new words. Because there is no benchmark dataset for event detection from click-through data. We find 3 experts to evaluate the results and got the average precision of 63.5% for detected new words.

Each new word candidate was rated by 3 human raters using the 5-point Likert scale defined in [12], from irrelevant to highly relevant. The agreement between all the three assessors on the results of random selected test set as computed by Cohen's Kappa is 0.609. Moreover, we also give the result new words detection on our created test set with manually labeled new words and non-new words, as shown in Table 7, where the – mean undefined evaluation metric. We do 5-fold cross validation in this experiment and finally achieve the average evaluation results, in which the F-measure and precision is 0.765 and 0.799.

Table 7. The results of event detection

	p	r	m	f	F_1
LNW	0.799	0.755	0.244	0.059	0.765
$LNNW$	0.651	–	–	0.464	–

6.3 Quality of Detected Words

To illustrate the results of word filtering, for each of the four types, we give a representative trajectory for example, which is shown in the Figure 5. Here, to maintain the original properties of the query trajectories, all the frequency values are the appearances of these queries without smoothing.

In Figure 6, we give another illustrative example of detected word with its 5-nearest words trajectories of frequency in week unit and its corresponding smoothing frequency. The concerned word term in the figure is "猪流感(Swine Flu)", which is a new word since it did not appear before the Swine Flu emerged. The frequency trajectories by the two similarity measures are plotted. The pictures in each column are the frequency trajectory of the top k (here is fixed to be 5) nearest neighbors by the two similarity measures, which are "快乐女声(A TV show named 'Happy Girls')", "陈绍基(A person named Chen Shaoji)", "五月(May)", "高兴(Happy)", "猪流感症

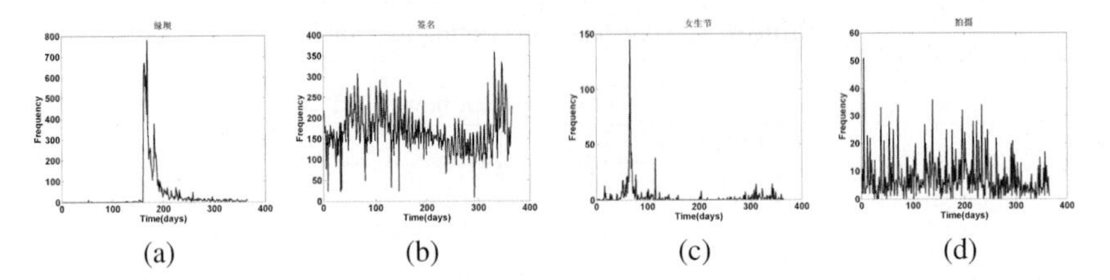

Fig. 5. Four representative word frequency trajectories: (a) HB: "绿坝(Green Dam)"; (b) HU: "签名(signature)"; (c) LB: "女生节(Girls' Day)"; (d) LU: "拍摄(shoot)"

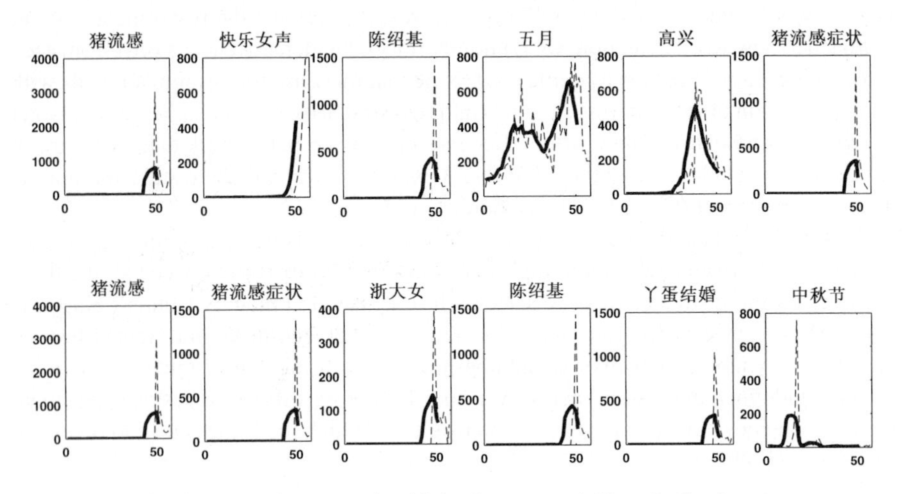

Fig. 6. Detected new words with its 5 nearest neighbors in the dataset

状(Symptom of Swine Flu)", respectively. Remained word terms in this figure include "浙大女(The girl from Zhejiang University)", "丫蛋结婚(A person named Yadan get married)" and "中秋节(Mid-autumn Holiday)". From this figure, we observe that the word feature "猪流感症状(symptoms of Swine Flu)" is a most similar word both in trajectory, bursty and energy-related aspects, which matches to our intuition.

7 Conclusions and Future Work

This paper analyzed the dynamics of word features collected from the query logs of a commercial search engine. By considering the query as a temporal signal from the energy and bursty perspectives, we reveal many new characteristics of the visitor-centric data and detect new words. Because there is no benchmark for our work to compare with other works, to validate our proposed methods, we evaluated our experimental results using manually-labeled new words assessed by 3 experts. Future work will devote to proposing a simple and effective evaluation method or benchmark.

Acknowledgments. This work is supported by the National Natural Science Foundation of China under Grant No. 60873174.

References

1. Zheng, Y., Liu, Z., Sun, M., Ru, L., Zhang, Y.: Incorporating user behaviors in new word detection. In: IJCAI 2009: Proceedings of the 21st International Joint Conference on Artificial Intelligence (July 2009)
2. Liu, H.: A noval method for fast new word detection. Journal of Chinese Information Processing 20, 17–23 (2006) (in Chinese)
3. Cui, S., Liu, Q., Meng, Y., Hao, Y., Nishino, F.: New word detection based on large-scale corpus. Jounral of Computer Research and Development 43, 927–932 (2006) (in Chinese)
4. Jia, Z., Shi, Z.: Probability techniques and rule methods for new word detection. In: Computer Engineering, vol. 30 (October 2004) (in Chinese)
5. Zhao, Q., Liu, T.Y., Bhowmick, S.S., Ma, W.Y.: Event detection from evolution of click-through data. In: KDD 2006: Proceedings of the 12th ACM SIGKDD International Conference on Knowledge Discovery and Data Mining, pp. 484–493. ACM, New York (2006)
6. Chen, L., Hu, Y., Nejdl, W.: Using subspace analysis for event detection from web click-through data. In: WWW 2008: Proceeding of the 17th International Conference on World Wide Web, pp. 1067–1068. ACM, New York (2008)
7. Chen, L., Hu, Y., Nejdl, W.: Deck: Detecting events from web click-through data. In: ICDM 2008: Proceedings of the 2008 Eighth IEEE International Conference on Data Mining, pp. 123–132. IEEE Computer Society, Los Alamitos (2008)
8. Wang, C., Zhang, M., Ru, L., Ma, S.: Automatic online news topic ranking using media focus and user attention based on aging theory. In: Proc. of CIKM, pp. 1033–1042 (2008)
9. Wang, C., Zhang, M., Ru, L., Ma, S.: Automatic online news topic ranking using media focus and user attention based on aging theory. In: CIKM 2008: Proceeding of the 17th ACM Conference on Information and Knowledge Management, pp. 1033–1042. ACM, New York (2008)
10. Lappas, T., Arai, B., Platakis, M., Kotsakos, D., Gunopulos, D.: On burstiness-aware search for document sequences. In: KDD 2009: Proceedings of the 15th ACM SIGKDD International Conference on Knowledge Discovery and Data Mining, pp. 477–486. ACM, New York (2009)
11. Keogh, E.: Exact indexing of dynamic time warping. In: VLDB 2002 : Proceedings of the 28th International Conference on Very Large Data Bases, VLDB Endowment (2002) pp. 406–417 (2002)
12. Sahami, M., Heilman, T.D.: A web-based kernel function for measuring the similarity of short text snippets. In: WWW 2006: Proceedings of the 15th International Conference on World Wide Web, pp. 377–386. ACM, New York (2006)

Exploiting Concept Clumping for Efficient Incremental E-Mail Categorization

Alfred Krzywicki and Wayne Wobcke

School of Computer Science and Engineering
University of New South Wales
Sydney NSW 2052, Australia
{alfredk,wobcke}@cse.unsw.edu.au

Abstract. We introduce a novel approach to incremental e-mail categorization based on identifying and exploiting "clumps" of messages that are classified similarly. Clumping reflects the local coherence of a classification scheme and is particularly important in a setting where the classification scheme is dynamically changing, such as in e-mail categorization. We propose a number of metrics to quantify the degree of clumping in a series of messages. We then present a number of fast, incremental methods to categorize messages and compare the performance of these methods with measures of the clumping in the datasets to show how clumping is being exploited by these methods. The methods are tested on 7 large real-world e-mail datasets of 7 users from the Enron corpus, where each message is classified into one folder. We show that our methods perform well and provide accuracy comparable to several common machine learning algorithms, but with much greater computational efficiency.

Keywords: concept drift, e-mail classification.

1 Introduction

Incremental, accurate and fast automatic document categorization is important for online applications supporting user interface agents, such as intelligent e-mail assistants, news article recommenders and Helpdesk request sorters. Because of their interaction with the user in real time, these applications have specific requirements. Firstly, the document categorizer must have a short response time, typically less than a couple of seconds. Secondly, it must be sufficiently accurate. This is often achieved by trading higher accuracy for lower coverage, but from a user point of view, high accuracy is preferred, so that the categorizer must either provide a fairly accurate category prediction or no prediction at all. The problem of e-mail categorization is difficult because of the large volume of messages needed to be categorized, especially in large organizations, the requirement for consistency of classification across a group of users, and the dynamically changing classification scheme as the set of messages evolves over time.

In our earlier work (Wobcke *et al.* [19]), we pointed out that the classification patterns of messages into folders for the e-mail data set studied typically changed abruptly as new folders or topics were introduced, and also more gradually as the meaning of the classification scheme (the user's intuitive understanding of the contents of the folders)

L. Cao, J. Zhong, and Y. Feng (Eds.): ADMA 2010, Part II, LNCS 6441, pp. 244–258, 2010.

evolved. We suggest that it is particularly important for classification algorithms in the e-mail domain to be able to handle these abrupt changes, and that many existing algorithms are unable to cope well with such changes. On the other hand, we observe that the e-mail classification exhibits a kind of "local coherence" which we here term *clumping*, where over short time periods, the classification scheme is highly consistent. An extreme example of this is when there are many messages in a thread of e-mail that are categorized into the same folder. However, much existing work on e-mail classification fails to address these complex temporal aspects to the problem.

What we call "clumping" is related to, but is different from, concept drift, which is usually taken to be a gradual shift in the meaning of the classification scheme over time. There is much research specifically on detecting and measuring various types of concept drift. A popular method is to use a fixed or adaptive shifting time window [7], [6], [17]. Nishida and Yamauchi [12] use statistical methods to detect concept drift by comparing two accuracy results, one recent and one overall. Vorburger and Bernstein [15] compare the distribution of features and target values between old and new data by applying an entropy measure: the entropy is 1 if no change occurs and 0 in the case of an extreme change. Gama *et al.* [5] measure concept drift by analysing the training error. When the error increases beyond certain level, it is assumed that concept drift has occurred. Our approach to changing context is different in that we address unexpected local shifts in the context, rather than gradual temporal changes. Our methods are also suitable for randomly sequenced documents as long as the sequence shows some sort of clumping.

In our previous work (Krzywicki and Wobcke [9]) we introduced a range of methods based on Simple Term Statistics (STS) that addresses computational requirements for e-mail classification. We showed that, in comparison to other algorithms, these simple methods give relatively high accuracy for a fraction of the processing cost. For further research in this paper we selected one of the overall best performing STS methods, referred to as M_{b2}, further explained in Section 3. In this paper we aim to undertake a more rigorous analysis of clumping behaviour and propose two new methods, *Local Term Boosting* and *Weighted Simple Term Statistics*, which can take advantage of clumping to improve classification accuracy. These methods weight each term according to its contribution to successful category prediction, which effectively selects the best predicting terms for each category and adjusts this selection locally. The weight adjusting factor is also learned in the process of incremental classification. *Weighted Simple Term Statistics* is also based on term statistics but weighted according to the local trend. Our expectation was that this method would benefit from using the global statistical properties of documents calculated by STS and local boosting provided by Local Term Boosting.

All methods were tested on the 7 largest sets of e-mails from the Enron corpus. The accuracy of these methods was compared with two incremental machine learning algorithms provided in the Weka toolkit [18]: Naive Bayes and k-Nearest Neighbour (k-NN), and one non-incremental algorithm, Decision Tree (called J48 in Weka). We also attempted to use other well known algorithms, such as SVM, but due to its high computation time in the incremental mode of training and testing it was not feasible for datasets of the size we used.

Our categorization results on the Enron corpus are directly comparable with those of Bekkerman *et al.* [1], who evaluate a number of machine learning methods on the same Enron e-mail datasets as we use in this research. These methods were Maximum Entropy (MaxEnt), Naive Bayes (NB), Support Vector Machine (SVM), and Winnow. In these experiments, methods are evaluated separately for 7 users over all major folders for those users. Messages are processed in batches of 100. For each batch, a model built from all previous messages is tested on the current batch of 100 messages. Accuracy over the 100 messages in the batch is calculated, however if a message in the batch occurs in a new category (i.e. not seen in the training data), it is not counted in the results. The paper does not mention how terms for the classification methods are selected, however it is reasonable to assume that all terms are used. The most accurate methods were shown to be MaxEnt (an algorithm based on entropy calculation) and SVM. It was noted that the algorithms differed in terms of processing speed, with Winnow being the fastest (1.5 minutes for the largest message set) and MaxEnt the slowest (2 hours for one message set). Despite a reasonably high accuracy, the long processing time makes SVM and MaxEnt unsuitable for online applications. Only Winnow, which was not the best overall, can be used as an incremental method, while the other two require retraining after each step. The temporal aspects of e-mail classification, including concept drift, are mentioned in the paper, but no further attention was given to these issues in the evaluation.

The rest of the paper is structured as follows. In the next section, we describe concept clumping and introduce clumping metrics. In Section 3, we present Simple Term Statistics in the context of concept clumping and introduce the Local Term Boosting and Weighted Local Term Boosting methods. Section 4 discusses the datasets, evaluation methodology and experimental results. In Section 5, we discuss related research on document categorization, concept drift and various boosting methods. Finally, Section 6 provides concluding points of the paper and summarizes future research ideas.

2 Concept Clumping

In this section we provide definitions of two types of clumping: *category* and *term-category clumping*, and introduce a number of metrics to quantify their occurrence.

2.1 Definitions

Our definition of *category clumping* is similar to that of *permanence* given in [2]. A concept (a mapping function from the feature space to categories) is required to remain unchanged for a number of subsequent instances. The definitions of concept clumping are given below.

Let $D = \{d_1, ...\}$ be a sequence of documents presented in that order, $F = \{f_1, ...\}$ a set of categories, and $T_d = \{t_{d1}, ...\}$ a set of terms in each document d. Each document is classified into a single category and each category f is assigned a set of documents $D_f = \{d_{f1}, ...\}$ so that $\bigcup_f D_f = D$ and $\sum_f |D_f| = |D|$.

Definition 1. *A* ***category clump*** $\{d_{f1}, ...\}$ *for category* f *in* D *is a maximal (contiguous) subsequence of* D *for which* $d_{f1-1} \in D_f$ *and each* $d_{fi} \in D_f$. *Let us call this type of clumping* CL_f.

That is, a category clump is a (contiguous) sequence of documents (excluding the first one) classified in the same folder. Since each document is in a single category, a document can be in only one category clump.

Definition 2. *Let D_t be the possibly non-contiguous subsequence of D consisting of all documents in D that contain t. Then a* ***term-category clump*** *for term t and category f in D is defined as a category clump for f in D_t. That is, a* ***term-category clump*** *$D_{tf} = \{d_{tf1}, ...\}$ for term t and category f in D is a maximal contiguous subsequence of D_t such that $d' \in D_f$ where d' is the document preceding d_{tf1} in D_t, and each $d_{tfi} \in D_f$. Let us call this type of clumping CL_t.*

Less formally, a term-category clump is a (possibly non-contiguous) sequence D' of documents from D having a common term t that are classified in the same folder f, for which there is no intermediate document from D not in D' in a different folder that also contains t and so "breaks" the clump. A document may be in more than one term-category clump, depending on how many terms are shared between the documents.

Notice that in the above definitions, the first document in a sequence of documents with the same classification does not belong to the clump. The reason for defining clumps in this way is so that the clump corresponds to the potential for the categorizer to improve its accuracy based on the clump, which is from the second document in the sequence onwards.

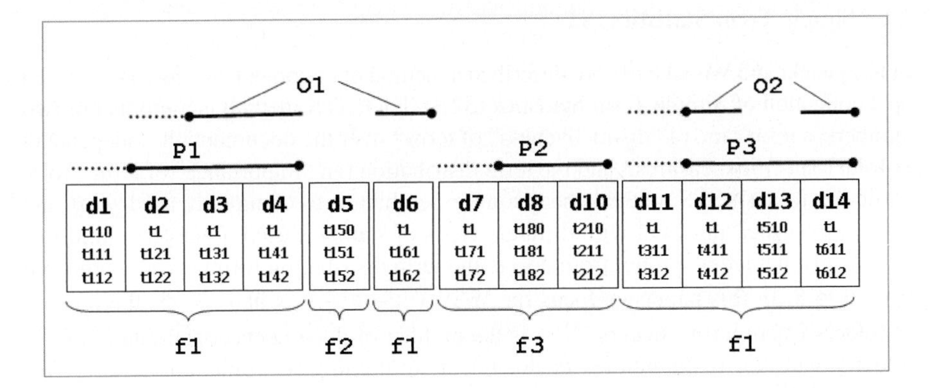

Fig. 1. Clumping Types

Figure 1 illustrates the idea of clumping, where category clumps P1, P2 and P3 span over a number of adjacent instances in the same category, whereas term-category clump O1 for category f1 continues for the same term t1 and skips instance d5 of category f2, which does not contain t1. There are documents of category f1 with term t1 later again, but they cannot belong to clump O1, because document d7 earlier in the sequence contains term t1, but is in a different category. Therefore instances d12 and d14 in category f1 form another term-category clump O2. Document d13 is not in O2 because it does not contain t1. Broken lines denote documents that start the sequence but, according to the definition, do not belong to clumps.

To measure how clumping affects the accuracy of the methods, we define a metric, the *clumping ratio*. Note that this metric is not used to detect clumps, as this is not required by our categorization methods.

Let M be a set of n clumps $M = \{M_1, ..., M_n\}$, each consisting of one or more documents from a sequence.

Definition 3. *The* ***theoretical maximal number of clumps*** $c_{max}(d)$ *for a document d in a document sequence is defined by* $c_{max}(d) = 1$ *for category clumping, and* $c_{max}(d) = |T_d|$ *for term-category clumping.*

Definition 4. *The* ***clumping ratio*** r *of a non-empty document sequence* D *(for both category clumping and term-category clumping) is defined as* $r = \frac{\sum_{i=1}^{n} |M_i|}{\sum_{j=1}^{|D|} c_{max}(d_j)}$.

Note that this ratio is always between 0 and 1.

3 E-Mail Categorization Methods

In this section, we provide a summary of Simple Term Statistics (STS) and introduce two new incremental methods: Local Term Boosting (LTB) and Weighted Simple Term Statistics (WSTS).

3.1 Simple Term Statistics (STS)

In Krzywicki and Wobcke [9], we described a method of document categorization based on a collection of Simple Term Statistics (STS). Each STS method is a product of two numbers: a term ratio (a "distinctiveness" of term t over the document set, independent of folder-specific measures), and the term distribution (an "importance" of term t for a particular folder). STS methods bear some resemblance to commonly used *tf-idf* and *Naive Bayes* measures, but with different types of statistics used. Each term ratio formula is given a letter symbol from a to d and term distribution formulas are numbered from 0 to 8. In this paper we focus on $M_{b2} = \frac{1}{N_{ft}} * \frac{N_{dtf}}{N_{dt}}$, where N_{ft} is the number of folders where term t occurs, N_{dtf} is the number of documents containing term t in folder f and N_{dt} is the number of documents in training set containing term t. This method performs well across most of the datasets used in this research, therefore it is used as a representative of STS methods in the experimental evaluation section.

The predicted category f_p for document d is defined as follows:

$$f_p = argmax_f(w_f) \text{ when } max_f(w_f) \geq \theta, \quad (1)$$

where $w_f = \sum_{t \in d} w_{t,f}$. In the previous work, we were particularly interested in datasets where not all documents needed to be classified, hence a threshold θ was used to determine when a document was classified. In the datasets in this paper, every document is classified, so θ is set to 0.

3.2 Local Term Boosting (LTB)

In online applications, documents are presented in real time and the categorizer learns a general model. Such a model may cover all regularities across the whole set of documents, for example the fact that some keywords specifically relate to some categories. However, the training data may contain sequences of documents having the same local models that differ among themselves. These differences are not captured by general models of the data, so it is advantageous if a learner can adapt to them. For example, a sequence of documents may have a single category, or may contain sets of similar term-to-category mappings. In Section 2 these local regularities have been defined as *category clumping* and *term-category clumping* respectively. We show that they can be exploited to increase the prediction accuracy by a method called Local Term Boosting (LTB), which we introduce in this section. In short, Local Term Boosting works by adjusting the term-category mapping values to follow the trend in the data. LTB is also sensitive to same-category sequences (*category clumping*). The method can be used for text categorization by itself or can be mixed with other methods, as shown in the next section.

Similar to STS, an array of weights $w_{t,f}(|T| \times |F|)$ is maintained for each term t and each folder f. Note that in online applications the array would need to be expanded from time to time to cover new terms and folders. The difference is in the way the term weights for each category $w_{t,f}$ are calculated. Initially, all weights are initialized to a constant value of 1.0. After processing the current document d, if the predicted folder f_p is incorrect, weights are modified as follows, where f_t is the target category. As above, weights are calculated for both the predicted folder f_p and the target folder f_t using the formula $w_f = \sum_{t \in d} w_{t,f}$, where b is a boosting factor and represents the speed of weight adjustment, and ϵ is a small constant (here set to 0.01).

Algorithm 1. (Adjusting Weights for Local Term Boosting)
1 $\delta_b = |w_{fp} - w_{ft}|/2 + \epsilon$
2 $b = (b * N + \delta_b)/(N + 1)$, where N is current number of instances
3 **forall** terms t in document d
4   **if** $t \in f_t$ **then** $w_{t,ft} := w_{t,ft} * (1 + b/(w_{ft} + b))$
5   **if** $t \in f_p$ **then** $w_{t,fp} := w_{t,fp} * (1 - b/(w_{fp} + b))$
6 **endfor**

Essentially, the term-folder weight is increased for the target category f_t and decreased for the incorrectly predicted folder f_p.

The predicted category f_p for document d is calculated in the same way as for STS as defined in Equation 1 but where the weights are calculated as above.

It seems intuitive that the boosting factor b should be smaller for slow and gradual changes and larger for abrupt changes and should also follow the trend in the data. Experiments using a sliding window with varying values for b showed, however, that this was not the case. It was possible to find an optimal *constant* boosting factor, different for each dataset, that provided the best accuracy for that entire data set. For this reason the following strategy was adopted in LTB to automatically converge to such a constant b, based on the running average of the differences in weights δ_b. After classifying an instance, according to Equation 1, an array of sums of weights is available for each category. If the category prediction is incorrect, a minimal adjustment δ_b is calculated in such a way that, if added to the target category weight and subtracted from the predicted category weight, the prediction according to Equation 1 would be correct. This adjustment is then added to the running average of b. In the general case where $\theta \neq 0$, instead of δ_b, the absolute value $|\delta_b|$ is used to address the following problem. If a message should be classified into a folder, but is not classified due to the threshold θ being higher than the maximum sum of term weights w_{max}, δ_b may take a large negative value. This would decrease the running average of b, while in fact it should be increased, therefore a positive value of δ_b is used.

Since the goal is to adjust the sum of weights, we need to apply only a normalized portion of b to each individual term weight, hence the $b/(w_{ft} + b)$ expression in Equation 1. As N becomes large, b converges to some constant value and does not need further adjustments (although periodic adjustment may still be required when dealing with a stream of documents rather than a limited number of messages in the datasets). We observed that for all of the datasets, after around 1000 iterations of Algorithm 1, b had converged to a value that produced a high classification accuracy.

The success of the Local Term Boosting method depends on how many clumps there are in the data. In Section 2 we defined two types of clumping: category clumping CL_f and term-category clumping CL_t. We will discuss below how Local Term Boosting adapts to both types of clumping. For CL_t, this is apparent from Algorithm 1. After a number of examples, the term-folder weights are adjusted sufficiently so that same terms (or group of terms) indicate the same folders. For CL_f, this is not so obvious, as the category clumping does not seem to depend on term-folder weights at all. Let us assume that there is a sequence of documents D_σ such that the target folder is always f, but all the documents are classified into some other folders, different from f. Any term t common to this sequence of documents will have its weight $w_{t,f}$ increased. If the term t occurs m times in this sequence of documents, its weight will become roughly $w_{t,f} * (1 + b_n)^m$, where b_n is the normalized boosting factor used in Algorithm 1. Therefore, as long as the same terms occur regularly in the sequence of documents, so that their weights all continue to increase, it becomes more likely that f is predicted. Since it is likely that at least some terms occur frequently in all documents in such a sequence, this condition is highly likely. The results in Section 4.3 confirm this expectation.

3.3 Weighted Simple Term Statistics (WSTS)

STS bases the prediction on the assumption that the statistical properties of terms present in a document and their distribution over categories indicate the likelihood of the document class. As the term significance may change over time and new terms are introduced that initially do not have sufficienly reliable statistics, by combining these statistics with locally boosted weights, we would expect that the classifier would follow the local trend even more closely. This, in summary, is the main idea of Weighted Simple Term Statistics (WSTS) introduced in this section, which is a combination of STS and LTB.

The WSTS method maintains two types of term-category weights: one for STS, which we will denote by $\rho_{t,f}$, and one for LTB, denoted as before by $w_{t,f}$. Again the predicted category f_p for document d is calculated according to Equation 1, except that the total term-folder weight is obtained by multiplying STS weights by LTB weights, i.e. $w_f = \sum_{t \in d}(\rho_{t,f} * w_{t,f})$. As in LTB, the boosting factor b is used to adjust the weights $w_{t,f}$ after the prediction is made, according to Algorithm 1.

3.4 Comparison of Methods

Based on the above definitions of STS, LTB and WSTS, it is interesting to compare all three methods in terms of their potential ability to classify documents in the presence on concept drift and clumping. The STS method M_{b2} should have its best prediction if there are terms unique to each category. This is because if such a term t exists for folder f, the STS weight $w_{t,f}$ would be set to its maximal value of 1. Similarly, STS is able to efficiently utilize new terms appearing later in the sequence of documents when they are unique to a single folder. If the same term is in more than one folder, then the weight would become gradually smaller. In circumstances where no new folders are introduced, as the number of documents N_{dtf} in the folder increases, the term-folder weight also increases, and tends to stabilize at some value determined by the number of folders and the distribution of the term across folders. Therefore STS can sometimes positively respond to both category and term-category clumping.

The main advantage of LTB is its ability to modify weights to improve folder prediction. Since the prediction is based on a maximal sum of weights of terms for a folder, by adjusting the weights, LTB should be able to identify groups of terms that best indicate each category. Unlike STS, however, LTB is able to increase the weights theoretically without limit, therefore is better able to adapt to both types of clumping.

WSTS is expected to take advantage of both LTB and STS, therefore should be able to utilize new terms for a folder and to increase the weights beyond the limit of STS when required. WSTS weights are expected to be generally smaller than LTB weights as they are multiplied by the STS component, which is in the 0–1 range.

4 Experimental Evaluation

In this section, we present and discuss the results of testing all methods described in Section 3. For comparison with our methods, we also include three additional algorithms available in the Weka toolkit: two updatable algorithms: Naive Bayes and k-NN (called IBk1 in Weka), and Decision Tree (J48 in Weka).

4.1 Statistics on Enron Datasets

We present evaluation results on 7 large e-mail sets from the Enron corpus that contain e-mail from the folder directories of 7 Enron e-mail users. We call these e-mail sets by the last name of their users: *beck, farmer, kaminski, kitchen, lokay, sanders* and *williams*. A summary of dataset features is provided in Table 1.

There is a great variation in the average number of messages per folder, ranging from 19 for *beck* to over 200 for *lokay*. Also, some message sets have specific features that affect prediction accuracy, for example *williams* has two dominant folders, which explains the high accuracy for all methods. For all users, the distribution of messages

Table 1. Statistics of Enron Datasets

	beck	*farmer*	*kaminski*	*kitchen*	*lokay*	*sanders*	*williams*
#e-mails	1971	3672	4477	4015	2489	1188	2769
#folders	101	25	41	47	11	30	18
% e-mails in largest folder	8.4	32.4	12.2	17.8	46.9	35.2	50.2
Category Clumping Ratio	0.151	0.28	0.162	0.235	0.38	0.387	0.891
Term-Category Clumping Ratio	0.25	0.435	0.292	0.333	0.404	0.459	0.806

across folders is highly irregular, which is typical for the e-mail domain. Some more details and peculiarities of the 7 Enron datasets can be found in Bekkerman *et al.* [1].

The last two rows of the table show the values of the earlier introduced clumping measures applied to each dataset. It is noticeable that datasets with dominating folders (e.g. *williams* and *lokay*) also have higher clumping measures. This is not surprising, since messages in these folders tend to occur in sequences, especially for the *williams* dataset, which contains two very large folders. The percentage of messages in the largest folder, although seeming to agree with the measures, does not fully explain the value of clumping ratios as they also depend on the term and folder distribution across the entire set.

4.2 Evaluation Methodology

Training/testing was done incrementally, by updating statistics (for STS and WSTS) and weights (for LTB and WSTS) for all terms in the current message d_i before testing on the next message d_{i+1}. STS and WSTS statistics on terms, folders and messages are incrementally updated, and the term-folder weights based on these statistics are recalculated for all terms in a document being tested. LTB and WSTS weights are initialized to 1 for each new term and then updated before the prediction step. Accuracy was calculated as a micro-average, that is, the number e-mails classified correctly in all folders divided by the number of all messages. For comparison with Bekkerman *et al.* [1], if a test message occurred in a newly introduced folder, the result was not counted in the accuracy calculation.

The Naive Bayes and k-NN methods are implemented in Weka as updatable methods, therefore, similar to our algorithms, the prediction models for these methods were updated incrementally without re-training. The Decision Tree method, however, had to be re-trained for all past instances at each incremental step.

The execution times for all methods were taken on a PC with Quad Core 4GHz processor and 4GB memory.

4.3 Evaluation Results

Table 2 provides a summary of the performance of the methods on all message sets. The table shows the accuracy of the methods in the experiments for predicting the classification over all major folders (i.e. except Inbox, Sent, etc.), and the total execution time

Table 2. Accuracy and Execution Times on All Enron Datasets

	beck	*farmer*	*kaminski*	*kitchen*	*lokay*	*sanders*	*williams*
LTB	0.542	0.759	0.605	0.55	0.79	0.75	0.939
STS	0.534	0.719	0.609	0.475	0.750	0.701	0.917
WSTS	**0.587**	**0.783**	**0.632**	**0.58**	**0.805**	**0.795**	**0.933**
WSTS/STS	1.098	1.089	1.037	1.221	1.073	1.134	1.017
WSTS/LTB	1.086	1.029	1.039	1.045	1.020	1.056	0.993
Execution times	8s	9s	19s	23s	10s	7s	6s
Naive Bayes	0.16	0.58	0.273	0.271	0.624	0.454	0.88
Execution times	20s	14s	24s	29s	7s	5s	7s
k-NN (IBK1)	0.316	0.654	0.37	0.307	0.663	0.569	0.87
Execution times	47s	192s	336s	391s	105s	13s	114s
Decision Tree	0.286	0.619	0.334	0.304	0.616	0.517	0.888
Execution times	2h	14h	20h	38h	4.5h	32min	3h
Bekkerman *et al.*	0.564	0.775	0.574	0.591	0.836	0.73	0.946
Best Method	SVM	SVM	SVM	MaxEnt	SVM	SVM	SVM

for updating statistics and classifying all messages in the folder. For WSTS the table also shows the relative increase in accuracy over the basic methods (STS and LTB).

The highlighted results are for the method that produces the highest accuracy for the given dataset. Out of the presented methods, WSTS is the best method overall. On average, WSTS is 9.6% better than STS and 2.9% better than LTB. Moreover, using WSTS does not increase the execution time compared to STS, which is of high importance for online applications. Out of the three general machine learning methods, k-NN provided the best accuracy, but is still much below STS.

The last row in the table shows the most accurate method obtained by Bekkerman *et al.* [1] chosen from a selection of machine learning methods. Note that these results are obtained by training on previous messages and testing on a window of size 100. While the most accurate method is typically SVM, it is impractical to retrain SVM after every message. The results of Bekkerman *et al.* on all Enron sets were compared with WSTS using the microaverage ($\sum_{allsets}(accuracy \times \#messages)/\sum_{allsets}(\#messages)$) and the macroaverage ($\sum_{allsets}(accuracy)/\#\ sets$). The WSTS accuracy is 1.8% higher when using the first measure, and 2% with the second measure.

One of the important requirements for online applications is processing speed. Comparing the execution times, all LTB and STS methods are much faster than the updatable Naive Bayes and k-NN methods, with Naive Bayes being the faster but less accurate of the two. The execution time for STS and LTB is proportional to $\#terms\ in\ e\text{-}mail \times \#e\text{-}mails$, which explains the differences between the datasets. Bekkerman *et al.* [1] does not give runtimes, however notes that training the SVM, which is the best performing method, can take up to half an hour on some datasets, but the machine details are not provided. Given that our incremental step is one document, as opposed to 100 documents for Bekkerman *et al.*, STS and Local Term Boosting are much faster.

4.4 Discussion: Accuracy vs. Clumping

In this section we discuss the dependency between the accuracy of our STS, LTB and WSTS methods and term-category clumping. Figure 2 shows the WSTS accuracy (upper line) and term-category clumping (lower line) for selected Enron users. The other datasets, with the exception of *williams*, exhibit similar trends; the *williams* dataset is highly abnormal, with most messages contained in just two large folders. Clumping numbers were obtained by recording the number of terms that are in clumps at each point in the dataset. For this figure, as well as for the correlation calculations presented later, the data was smoothed by the running average in a window of 20 data points.

Visual examination of the graph suggests that the accuracy is well aligned with the term-category clumping for all datasets. To confirm this observation, we calculated Pearson correlation coefficients between the accuracy and clumping measures for all

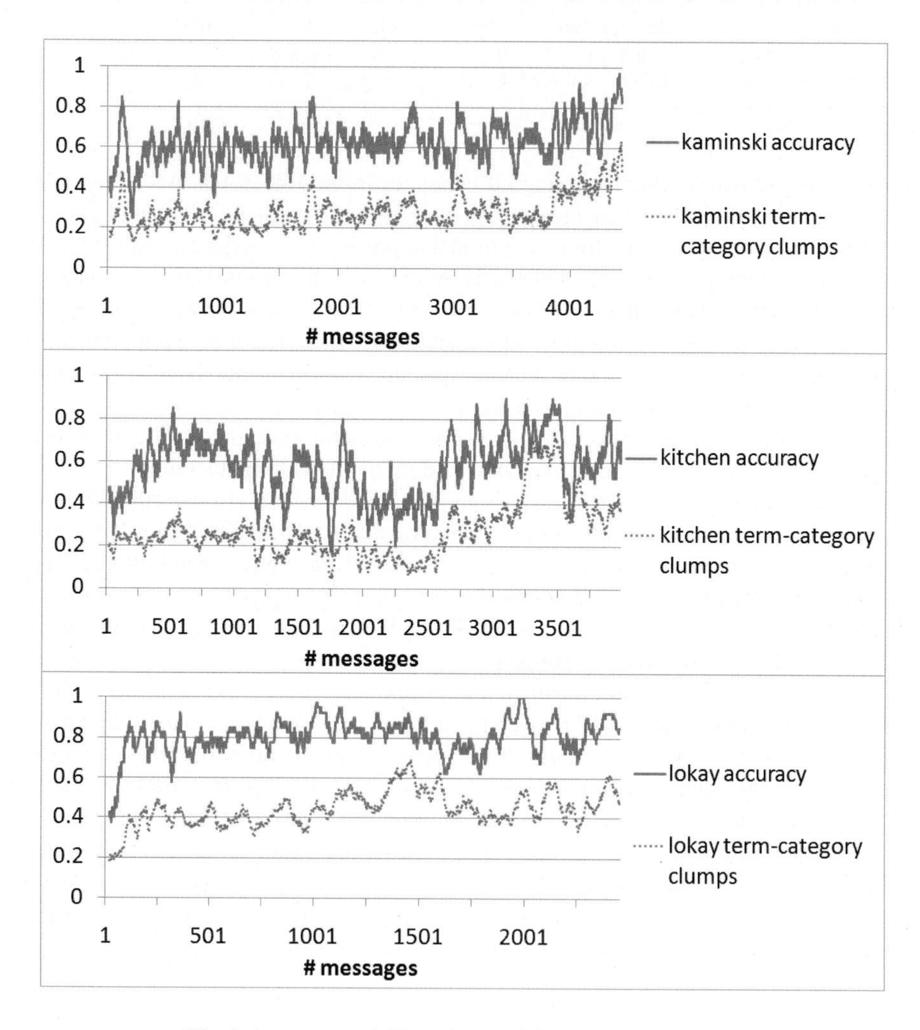

Fig. 2. Accuracy and Clumping on Selected Enron Datasets

Table 3. Correlation of Accuracy and Term-Category Clumping for All Methods

	beck	*farmer*	*kaminski*	*kitchen*	*lokay*	*sanders*	*williams*
LTB CL_f	0.504	0.21	0.204	0.547	0.207	0.289	0.696
LTB CL_t	0.754	0.768	0.695	0.711	0.509	0.674	0.822
STS CL_f	0.452	0.217	0.193	-0.083	0.344	0.145	0.648
STS CL_t	0.771	0.597	0.738	0.044	0.425	0.426	0.778
WSTS CL_f	0.443	0.190	0.231	0.552	0.362	0.201	0.718
WSTS CL_t	0.784	0.737	0.737	0.688	0.588	0.662	0.847
Naive Bayes CL_f	0.336	0.078	0.098	0.451	-0.017	0.199	0.706
Naive Bayes CL_t	0.365	0.341	0.292	0.489	0.264	0.447	0.817
k-NN CL_f	0.245	0.083	0.116	0.154	0.084	0.321	0.663
k-NN CL_t	0.379	0.408	0.413	0.210	0.390	0.434	0.825
Decision Tree CL_f	0.266	0.186	0.150	0.320	-0.005	0.158	0.647
Decision Tree CL_t	0.371	0.341	0.363	0.354	0.309	0.399	0.807

methods and all datasets, shown in Table 3. Although we mainly consider term-category clumping (CL_t), the category clumping (CL_f) is also shown for completeness. The correlations of LTB and WSTS accuracy with term-category clumping is above 0.5 for all datasets, which indicates that these methods are able to detect and exploit local coherencies in the data to increase accuracy. The average correlation for WSTS on all datasets is slightly higher than LTB (about 2%), which to some degree explains its higher accuracy. STS also shows a correlation above 0.5 for 4 out of 7 datasets, although much lower than WSTS (about 25% on average). Other methods (Naive Bayes, k-NN and Decision Tree) show some degree of correlation of accuracy with term-category clumping, but about 40% lower on average. This suggests that these commonly used machine learning methods, even when used in an incremental (Naive Bayes and k-NN) or pseudo-incremental (Decision Tree) mode, are not able to track changes in the data and local coherence to the same degree as LTB and WSTS.

It is apparent in the graphs in Figure 2 that the degree of clumping itself varies over time for all the datasets, which is again a property of the e-mail domain. There are several sections in the datasets that exhibit greater than normal fluctuations in clumping, and here the WSTS method is still able to track these changes with high accuracy. We now look at these more closely. For example, there are abrupt changes in clumping in the *kaminski* dataset between messages 1 and 190 which is aligned extremely closely with WSTS accuracy (correlation 0.96), the *kitchen* dataset between messages 1600 and 1850 (correlation 0.96), the *kitchen* dataset between messages 3250 and 3600 (correlation 0.93), and the *lokay* dataset between messages 1970 and 2100 (correlation 0.87). It is interesting that these high spikes in clumping are aligned with much higher accuracy (reflected in these extremely high correlations) than the averages over the rest of the datasets.

5 Related Work

A variety of methods have been researched for document categorization in general and e-mail foldering in particular. Dredze *et al.* [3] use Latent Semantic Analysis (LSA) and

Latent Dirichlet Allocation (LDA) to pre-select a number of terms for further processing by a perceptron algorithm. The paper is focused on selecting terms for e-mail topics, rather than machine learning for document categorization. In contrast to the work of Bekkerman *et al.* [1], where all major folders are used for each of the 7 Enron users, Dredze *et al.* [3] use only the 10 largest folders for each user (the same 7 users). We used both types of data in our evaluation and, as expected, found that it is much easier to provide accurate classifications for the 10 folder version of the experiment. For this reason we focused only on the data sets used by Bekkerman *et al.*

Our methods show some similarities to methods presented in Littlestone [10], Littlestone and Warmuth [11] and Widmer [16]. Littlestone [10] described a binary algorithm called Winnow2, which is similar to Local Term Boosting in that it adjusts weights up or down by a constant α. In Winnow2, however, weights are multiplied or divided by α, while Local Term Boosting uses additive weight modification and the adjustment factor is learned incrementally. We chose an additive adjustment because it could be easily learned by accumulating differences after each step. Another difference is that Winnow2 uses a threshold θ to decide if an instance should be classified into a given category or not, whereas Local Term Boosting uses a threshold to determine if a message should be classified into a category or remain unclassified. If a document term can be treated as an expert, then the idea used in Weighted Simple Term Statistics is also similar to the Weighted Majority Algorithm of Littlestone and Warmuth [11], and the Dynamic Weighted Majority Algorithm of Kolter and Maloof [8]. One important difference, however, is that each expert in our method maintains a set of weights, one for each category, which allows for faster specialization of experts in categories. Another difference is that the weight adjusting factor itself is dynamically modified for Weighted Simple Term Statistics, which is not the case for the above algorithms. Simple Term Statistics methods also show some similarity to the incremental implementation of Naive Bayes of Widmer [16], except that we use a much larger collection of term statistics and their combinations.

We use Local Term Boosting as a way to increase the accuracy in the presence of local context shifts in a sequence of documents. Classical boosting (Freund and Schapire [4]), adjusts the weights of the training examples, forcing weak learners to focus on specific examples in order to reduce the training error. Schapire and Singer [13] adapted AdaBoost to the text categorization domain by changing not only the weights for documents but also weights associated with folder labels. Instead of weighting documents and labels, we use weights that connect terms with folder labels. In this sense, our method is closer to incremental text categorization with perceptrons as used by Schütze *et al.* [14]. As in Local Term Boosting, the weights remain unchanged if the document is classified correctly, increased if a term predicts the correct folder (which can be regarded as a positive example), and decreased if the term prediction is incorrect (corresponding to a negative example). The difference is, however, that a perceptron is a binary classifier and calculates the weight adjustment Δ in a different way from our methods.

6 Conclusion

In this research, we introduce a novel approach to e-mail classification based on exploitation of local, often abrupt but coherent changes in data, which we called clumping.

We evaluated two new methods that use concept clumping for e-mail classification: Local Term Boosting (LTB) based on dynamic weight updating that associates terms with categories, and Weighted Simple Term Statistics (WSTS), being a combination of Local Term Boosting and a Simple Term Statistics (STS) method introduced previously. We showed that these methods have very high accuracy and are viable alternatives to more complex and resource demanding machine learning methods commonly used in text categorization, such as Naive Bayes, k-NN, SVM, LDA, LSI and MaxEnt. Both STS and LTB based methods require processing only the terms occurring in each step, which makes them truly incremental and sufficiently fast to support online applications. In fact, we discovered that these methods are much faster, while showing similar performance in terms of accuracy, than a range of other methods.

Our experiments showed that Local Term Boosting and Weighted Local Term Boosting are able to effectively track clumping in all datasets used to evaluate the methods. We also devised metrics to measure the clumping in the data and showed that the degree of clumping is highly correlated with the accuracy obtained using LTB and WSTS.

WSTS, which is a combination of STS and LTB, is generally more accurate than STS or LTB alone. We believe that the combined method works better on some data sets with specific characteristics, for example, if different types of clumping occur many times in the data, but the data always returns to its general model rather than drifting away from it. In this case the general model alone would treat these sequences as noise, but combined with Local Term Boosting it has a potential to increase the accuracy by exploiting these local changes. The combined method may accommodate to the changes faster in those anomalous sequences, because the statistics collected by STS would preset the weights closer to the required level.

If a single method was to be selected to support an online application, the choice would be WSTS, since it combines a global model (STS) with utilization of local, short but stable sequences in the data (LTB), has very good computational efficiency and works best on all datasets we have tested so far. The definition and experimentation with more complex metrics, or even a general framework including the application of all presented methods to multi-label documents, is left to future research.

References

1. Bekkerman, R., McCallum, A., Huang, G.: Automatic Categorization of Email into Folders: Benchmark Experiments on Enron and SRI Corpora. Technical Report IR-418, Center for Intelligent Information Retrieval, University of Massachusetts, Amherst (2004)
2. Case, J., Jain, S., Kaufmann, S., Sharma, A., Stephan, F.: Predictive Learning Models for Concept Drift. In: Proceedings of the Ninth International Conference on Algorithmic Learning Theory, pp. 276–290 (1998)
3. Dredze, M., Wallach, H.M., Puller, D., Pereira, F.: Generating Summary Keywords for Emails Using Topics. In: Proceedings of the 13th International Conference on Intelligent User Interfaces, pp. 199–206 (2008)
4. Freund, Y., Schapire, R.E.: A Short Introduction to Boosting. Journal of the Japanese Society for Artificial Intelligence 14, 771–780 (1999)
5. Gama, J., Medas, P., Castillo, G., Rodrigues, P.: Learning with Drift Detection. In: Bazzan, A.L.C., Labidi, S. (eds.) SBIA 2004. LNCS (LNAI), vol. 3171, pp. 286–295. Springer, Heidelberg (2004)

6. Hulten, G., Spencer, L., Domingos, P.: Mining Time-Changing Data Streams. In: Proceedings of the 7th ACM SIGKDD International Conference on Knowledge Discovery and Data Mining (KDD 2001), pp. 97–106 (2001)
7. Klinkenberg, R., Renz, I.: Adaptive Information Filtering: Learning in the Presence of Concept Drifts. In: Workshop Notes of the ICML/AAAI 1998 Workshop Learning for Text Categorization, pp. 33–40 (1998)
8. Kolter, J., Maloof, M.: Dynamic Weighted Majority: A New Ensemble Method for Tracking Concept Drift. In: Proceedings of the Third International IEEE Conference on Data Mining, pp. 123–130 (2003)
9. Krzywicki, A., Wobcke, W.: Incremental E-Mail Classification and Rule Suggestion Using Simple Term Statistics. In: Nicholson, A., Li, X. (eds.) AI 2009. LNCS, vol. 5866, pp. 250–259. Springer, Heidelberg (2009)
10. Littlestone, N.: Learning Quickly When Irrelevant Attributes Abound: A New Linear-Threshold Algorithm. Machine Learning 2, 285–318 (1988)
11. Littlestone, N., Warmuth, M.: The Weighted Majority Algorithm. Information and Computation 108, 212–261 (1994)
12. Nishida, K., Yamauchi, K.: Detecting Concept Drift Using Statistical Testing. In: Corruble, V., Takeda, M., Suzuki, E. (eds.) DS 2007. LNCS (LNAI), vol. 4755, pp. 264–269. Springer, Heidelberg (2007)
13. Schapire, R.E., Singer, Y.: BoosTexter: A Boosting-Based System for Text Categorization. Machine Learning 39, 135–168 (2000)
14. Schütze, H., Hall, D.A., Pedersen, J.O.: A Comparison of Classifiers and Document Representations for the Routing Problem. In: Proceedings of the 18th International Conference on Research and Development in Information Retrieval, pp. 229–237 (1995)
15. Vorburger, P., Bernstein, A.: Entropy-Based Concept Shift Detection. In: Perner, P. (ed.) ICDM 2006. LNCS (LNAI), vol. 4065, pp. 1113–1118. Springer, Heidelberg (2006)
16. Widmer, G.: Tracking Context Changes through Meta-Learning. Machine Learning 27, 259–286 (1997)
17. Widmer, G., Kubat, M.: Learning in the Presence of Concept Drift and Hidden Contexts. Machine Learning 23, 69–101 (1996)
18. Witten, I.H., Frank, E.: Data Mining. Morgan Kaufmann, San Francisco (2005)
19. Wobcke, W., Krzywicki, A., Chan, Y.-W.: A Large-Scale Evaluation of an E-Mail Management Assistant. In: Proceedings of the 2008 IEEE/WIC/ACM International Conference on Web Intelligence and Intelligent Agent Technology, pp. 438–442 (2008)

Topic-Based User Segmentation for Online Advertising with Latent Dirichlet Allocation

Songgao Tu and Chaojun Lu

Deptartment of Computer Science and Engineering, Shanghai Jiao Tong University,
No. 800 Dongchuan Road, Shanghai 200240, China
tusonggao@sjtu.edu.cn, lu-cj@cs.sjtu.edu.cn

Abstract. Behavioral Targeting (BT), as a useful technique to deliver the most appropriate advertisements to the most interested users by analyzing the user behaviors pattern, has gained considerable attention in online advertising market in recent year. A main task of BT is how to automatically segment web users for ads delivery, and good user segmentation may greatly improve the effectiveness of their campaigns and increase the ad click-through rate (CTR). Classical user segmentation methods, however, rarely take the semantics of user behaviors into consideration and can not mine the user behavioral pattern as properly as should be expected. In this paper, we propose an innovative approach based on the effective semantic analysis algorithm Latent Dirichlet Allocation (LDA) to attack this problem. Comparisons with other three baseline algorithms through experiments have confirmed that the proposed approach can increase effectiveness of user segmentation significantly.

Keywords: Latent Dirichlet Allocation, Behavioral targeting, User segmentation.

1 Introduction

The World Wide Web (WWW) has been rapidly and continuingly growing for nearly two decades and is playing an increasingly important role as an access to useful information, which has brought with it a fast developing field known as online advertising science. Sponsored search [1] and contextual advertising [2] are two of the most widely explored online advertising business models. Besides, behavioral targeting has been validated to be helpful in delivering the more appropriate ads to potentials consumers. A key challenge of BT is user segmentation, which tries to divides users into different groups, with users of similar interest placed into the same group. Advertisers are expected to gain more profits from their online campaigns provided that the ads are targeted to the more interested users. As advertisers can select the segment most relevant to their ads, the quality of user segmentation has dominant impact on the effectiveness of behavioral targeted advertising.

In this paper, we mainly address the problem of user segmentation for BT in the context of commercial search engine. The problem can be stated as follows.

L. Cao, J. Zhong, and Y. Feng (Eds.): ADMA 2010, Part II, LNCS 6441, pp. 259–269, 2010.

Suppose there is a set of online users. For each user, we take the queries issued by the users as the their behavior. Some ads have been displayed to each of these users according to the queries through sponsored search. With every ad displayed, whether or not it is clicked by the users is recorded. Our goal is to segment all the users into appropriate segments by the analysis of the quires so that we can hopefully help the online advertisers achieve their marketing objective by displaying the ads only to the more relevant users.

Conventional user segmentation methods are mainly based on classical clustering such as k-means, hierarchical clustering, etc. Most traditional algorithm, however, generally make use of keywords as features with Vector Space Model [3] and fail to exploit the semantic relativeness between the queries. As such, two users who have relevant but not exactly the same query keywords shall never be grouped into the same segment. To overcome the shortness of these methods and to mine the semantics underlying the queries, we propose in this paper an approach based on Latent Dirichlet Allocation (LDA) [4]. LDA is a powerful topic-based semantic modeling algorithm and has been intensively explored in academia these years. To the best of our knowledge, however, no previous work has attempted to do user segmentation through LDA. Experiments have demonstrated that our proposed methods can boost the CTR compared significantly with other traditional algorithms.

The rest of paper is structured as follows. In Section 2, we gives a short overview of the related work about behavioral targeting and user segmentation. Section 3 formulates the problem we are to solve and describes in detail our proposed user segmentation approach based on LDA. Performance evaluation of our proposed method compared with three other baseline algorithms is illustrated in Section 4. Finally, we conclude our work in Section 5 and point out some promising directions for further study.

2 Related Work

In recent years, many general web services, such as search engines, websites, etc. have devoted to analyzing the users' online behaviors to provide a more customized web service and advertising campaign. As is defined in Wikipedia [5], "Behavioral targeting is a technique used by online publishers and advertisers to increase the effectiveness of their campaigns. Behavioral targeting uses information collected on an individual's web-browsing behavior, such as the pages they have visited or the searches they have made, to select which advertisements to display to that individual. Practitioners believe this helps them deliver their online advertisements to the users who are most likely to be interested."

BT is now playing an increasingly important role as useful online advertising scheme in industry and many commercial systems involving behavioral targeting emerged: DoubleClick [6], which utilizes some extra features such as browse type and the operating system of users; Adlink [7], which takes the short user session into consideration for BT; Specificmedia [8], which enables advertisers to target audiences through advanced proprietary demographic, behavioral, contextual, geographic and retargeting technologies; and the Yahoo! Smart ads [9], which converts creative campaign elements and offerings into highly customized and relevant display ads by delivering ads according to the Web sufer's age, gender, location and online activities. Though an increasing number of commercial BT system appeared, there are no public work in academia to demonstrate how much BT can really help online advertising in search engine until [10], the evaluation metrics introduced in which is also adopted in this work to test the effectiveness of our proposed approach.

Demographic prediction and user segmentation are two of mostly used techniques for behavioral targeting. Neither of these two issues, however, are reasonably fully studied within the computer science research community. [11] is one of the very few works come into our view about demographic prediction. It makes endeavor to predict internet users' gender and age information based on their browsing behaviors by treating the webpage view record as a hidden variable to propagate demographic information between different users. As to user segmentation, we only find two related work [12] and [13]. [12] gives a method for web user segmentation, which integrates finite mixture models and factor analysis, resulting in a statistical method which concurrently performs clustering and local dimensionality reduction within each cluster. [13] applies the semantic analysis algorithm probabilistic Latent Sematic Analysis(pLSA) to solve user segmentation and pronounces to boost performance obviously with respect to the evaluation metrics given by [10].

3 Topic-Based User Segmentation with LDA

3.1 Problem Formulation

Before introducing our proposed approach to User Segmentation, we formulate the problem as follows: Let $U = \{u_1, u_2, ..., u_n\}$ be a web user id set, where each $u_i, i = 1, 2, ..., n$, is a web uer id and can be represented as $u_i = \{q_{i,1}, q_{i,2}, ..., q_{i,m_i}\}$, where $q_{i,j}, j = 1, ..., m_i$, is the jth query in all the m_i queries issued by user u_i to the search engine. With the simplified assumption that the interest of each user u_i can be fully derived from all the queries issued by u_i, we define the interest of u_i by $I_i = Interest(q_{i,1}, q_{i,2}, ..., q_{i,m_i})$. And we further denote the difference and similarity between I_i and I_j as function $dist(i,j)$ $sim(i,j)$, respectively. It is evident that there are many different ways to partition the user id set U into d subsets $g_1, g_2, ..., g_d$ such that $g_i \cap g_j = \emptyset$ and $\cup_{i=1}^{d} g_i = U$. Among all the ways of partition $\mathscr{P}$, our task is to find the optimal one P^* which suffices

$$P^* = argmin_P \left[\sum_{k=1}^{d} \sum_{i,j \in g_k} dist(i,j) - \sum_{1 \le k_1 < k_2 \le d} \sum_{i \in g_{k_1}, j \in g_{k_2}} sim(i,j) \right]. \quad (1)$$

3.2 Topic-Based User Segmentation with LDA

First we briefly describe Latent Dirichlet Allocation, which our proposed approach is based on. A more detailed elaboration can be found in [4]. LDA is a fully generative graphical model for describing the latent topics of documents. LDA takes every topic as a distribution over the words of the vocabulary, and every document as a distribution over the topics. All of these are sampled from Dirichlet distributions. There are several methods proposed for learning in LDA such as variational expectatin maximization [4], expectation propagation [14], and Gibbs sampling [15]. We take the Gibbs sampling as the learning algorithm in this paper.

Assume we have a vocabulary V consisting of words, a set T of k topics and n documents of arbitrary length. For every topic z a distribution ϕ_z on V is sampled from $Dir(\beta)$, where $\beta \in \mathbf{R}_+^V$ is a smoothing parameter. Similarly, for every document d a distribution θ_d on T is sampled from $Dir(\alpha)$, where $\alpha \in \mathbf{R}_+^V$ is also a smoothing parameter.

The words of the documents are drawn as follows: for every word position of document d a topic z is drawn from θ_d, and then a word is drawn from ϕ_z and filled into the position.

LDA can be thought of as a Bayesian network as Fig. 1.

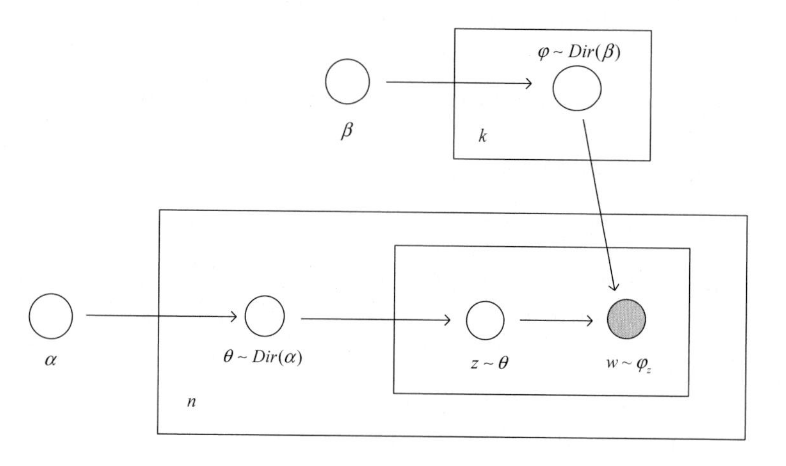

Fig. 1. LDA as a Bayesian network

Gibbs sampling is one of the often used method for making inference for LDA. It is a Markov chain Monte Carlo algorithm for sampling from a joint distribution $p(x), x \in \mathbf{R}^n$, if all conditional distributions $p(x_i|x_{-i})$ are known $(x_{-i} = (x_1, x_2, ..., x_{i-1}, x_{i+1}, ..., x_n))$. The kth transition $x^k \rightarrow x^{k+1}$ of the Markov chain is generated as follows: Choose an index $1 \leq i \leq n$ (usually $i = k \bmod n$), and let $x^{k+1} = x^k$ everywhere except at index i where x_i^{k+1} is sampled from $p(x_i|x_{-i}^k)$.

In LDA the goal is to estimate the distribution $p(z|w)$ for $z \in T^P, w \in V^P$ where P denotes the set of word positions in the documents. Thus for Gibbs sampling one has to caculate $p(z_i|z_{-i}, w)$ for $i \in P$. This has an efficiently computable closed form

$$p(z_i|z_{-i}, w) = \frac{n_{z_i}^{t_i} - 1 + \beta_{t_i}}{n_{z_i} - 1 + \sum_t \beta_t} \cdot \frac{n_d^{z_i} - 1 + \alpha_{z_i}}{n_d - 1 + \sum_z \alpha_z}. \tag{2}$$

Here d is the document of position i, t_i is the actual word in position $i, n_{z_i}^{t_i}$ is the number of positions with topic z_i and word t_i, n_{z_i} is the number of positions with topic $z_i, n_d^{z_i}$ is the number of topics z_i in document d, and n_d is the length of document d. After a sufficient number of iterations we arrive at a topic assignment sample z. Knowing z, we can estimate ϕ and θ as

$$\phi_{z,t} = \frac{n_z^t + \beta_t}{n_z + \sum_t \beta_t} \tag{3}$$

and

$$\theta_{d,z} = \frac{n_d^z + \alpha_z}{n_d + \sum_z \alpha_z}. \tag{4}$$

For an unseen document d the θ topic distribution can be estimated exactly as in (4) once we have a sample from its word-topic assignment z. Sampling z can be performed with a similar method as before, but now only for the positions i in d:

$$p(z_i|z_{-i}, w) = \frac{\tilde{n}_{z_i}^{t_i} - 1 + \beta_{t_i}}{\tilde{n}_{z_i} - 1 + \sum_t \beta_t} \cdot \frac{n_d^{z_i} - 1 + \alpha_{z_i}}{n_d - 1 + \sum_z \alpha_z}. \tag{5}$$

The notation $\tilde{n}$ refers to the union of the whole corpus and the unseen document d.

Specifically in the context of user segmentation as specified by the above subsection, each document d_i represents a user u_i, and each query $q_{i,j} (j = 1, 2, ..., m_i)$ issued by the user u_i can be taken as a word in the correspond document. Thus each topic z_i can be treated as an interest group, all the words t with the highest $\phi_{z,t}$ can show what the interest group z is actually about. There can be many ways to decide which interest group a user should belong to. Here we just take the most intuitive and reasonably effective one: user u_i (corresponding to document d_i) is set into interest group z_t if $\theta_{i,t} = max\{\theta_{i,j}, j = 1, 2, ..., m_i\}$.

4 Experiments

We have done extensive experiments to evaluate our proposed method's effectiveness. We performed our experiment on a data set prepared from the log file of a commercial online shopping website and compared the results with three other baseline algorithms.

Table 1. Typical format of the data set

UserId	+HllBJtiQ20CAQTW0N7zOr3x
Query_list	sports clothes Lining Adidas Nike
Clicked_adid_list	4b16ddfe5bb1d45b917266f23bb71d61
	0636aa4d2dd938f277a7c97b81e4dce8
Unclicked_adid_list	af381958673452d7db83f0069693e2f4
	0636aa4d2dd938f277a7c97b81e4dce8
	0636aa4d2dd938f277a7c97b81e4dce8

4.1 Data Set

The data set prepared for the empirical study in this section originate from one day's(May 1st, 2010) pageview and click-through log data by one of China's most popular commercial online shopping website[1]. As not everyone who visits the website will log in, using only the actual user id will throw away a great deal of useful information. So we take *cookie id* as an equivalence to the actual user id. The first time a user visits the website by a browser, the website will plant a cookie in the browser, which will be maintained until the user clear it away manually, in which case a new cookie will be planted the next time the user visit the website. We can imagine that in most cases one cookie can really correspond to one online user, whether or not he or she has logged in. From the pageview log file we can obtain the information: the cookie id, the queries submitted and the id of the ads displayed to that user. From the click-through log file, the information about cookie id and the id of the ads that have been clicked through can be extracted. We join the two part of log information by the key *cookie id* and obtain information about more than 6.5 million cookie id, which is too many for the experiments. For the convenience of the experiment, we sample about 300,000 users out of them, preferably selecting those having issued more than three queries and having clicked at least one ad displayed for them. With all these preprocessing operations done, the final format of the data set which we arrive at can be illustrated in Table 1.

4.2 Evaluation Metrics

As user segmentation is an under-explored problem in academia, there is no universally accepted evaluation metrics for it yet. In this paper, we take the evaluation metrics proposed in [10], which we think can properly serve the purpose of testing the effectiveness of different user segmentation methods. Before introducing the evaluation metrics, first we define some mathematical symbols that will be used. Let $A = \{a_i, a_i, ..., a_n\}$ be a set of the n advertisement in our data set, $U_i = \{u_{i1}, u_{i2}, ..., u_{im_i}\}$ be the group of users who have been displayed ad a_i. A Boolean function

[1] This data set was collected for research use when the first author was doing his internship in Alibaba Research and Development Center.

$$\delta(u_{ij}) = \begin{cases} 1 & \text{if } u_{ij} \text{ clicked } a_i \\ 0 & \text{otherwise} \end{cases} \tag{6}$$

is defined to show whether the user u_{ij} has clicked ad a_i. We use the notation

$$G(U_i) = \{g_1(U_i), g_2(u_i), ..., g_K(U_i)\}, i = 1, 2, ..., m \tag{7}$$

to represent the distribution of U_i under a given user segmentation results, where $g_k(U_i)$ stands for all the users in U_i who are grouped into the kth user segment. Obviously, the kth user segment can be defined as

$$g_k = \bigcup_{i=1,2,...,n} g_k(U_i). \tag{8}$$

Ads Click Through Rate Improvement. With all the mathematical symbols given above, we define the Click Through Rate (CTR) of ad a_i as

$$CTR(a_i) = \frac{1}{m_i} \sum_{j=1}^{m_i} \delta(u_{ij}) \tag{9}$$

and CTR of a_i over user segment g_k as

$$CTR(a_i|g_k) = \frac{1}{|g_k(U_i)|} \sum_{u_{ij} \in g_k(U_i)} \delta(u_{ij}), \tag{10}$$

where $|g_k(U_i)|$ is the number of the users placed in $g_k(U_i)$. In order to measure the improvement of CTR by user segmentation, we define

$$\Delta(a_i) = \frac{CTR(a_i|g^*(a_i)) - CTR(a_i)}{CTR(a_i)}, \tag{11}$$

where $g_*(a_i) = argmax\{CTR(a_i|g_k), k = 1, 2, ..., d\}$

Ads Click Entropy. As another evaluation metric, we define

$$Enp(a_i) = -\sum_{k=1}^{d} P(g_k|a_i) log P(g_k|a_i) \tag{12}$$

as the ads click entropy of ad a_i, where $P(g_k|a_i) = \frac{1}{m_i} \sum_{u_{ij} \in g_k(U_i)} \delta(u_{ij})$ is used to estimate the probability of the users in user segment g_k will click the ad a_i. According to the intuitive meaning behind the entropy, the larger the ads click entropy is, the more uniformly the users who have clicked ad a_i distribute among all the user segments and the worse result we achieve. To put it in a converse way, smaller the click Entropy illustrates more effective user segment outcome.

4.3 Results

In this part, we compare our proposed approach with three other baseline algorithms using the evaluation metrics given above. Two of the three baseline algorithms are classical vector-distance based clustering algorithms: k-means and hierarchical clustering. Another baseline algorithm is probabilistic Latent Semantic Analysis (pLSA), another semantic text analysis algorithm like LDA.

Firstly, we compare the four methods using the evaluation metric CTR improvement defined by (11) on different user segments. To make the results more convincing, we have carried out the experiments independently five times, and taken the averaged results as the final outcome, which is shown in Fig. 2. From Fig. 2 we can generally arrive at the following three points: First, the semantic analysis methods, pLSA and LDA are always doing better than the other two traditional distance-based clustering algorithm. In particular, when the segment number is set to be 30, the LDA-based method exceeds k-means method and hierarchical clustering method by as much as 21.6% and 23.2%, respectively; the PLSA-based approach exceeds the same two by 20.2% and 21.7%, respectively. Second, the performances of the four methods improve stably with the user segment number increasing while the differences between the four tend to be smaller at the same time. This phenomenon is understandable because when the segment number approaches to infinity and every web user is placed into distinct segment, the ad CTR of the any segment method will be the same. Third, while they are both semantic analysis methods, our proposed LDA-based method consistently outperforms the pLSA-based one. The main advantage of LDA over pLSA is that LDA assumes a Dirichlet prior with the topic distribution and reasonably derives a more accurate mixtures of topics.

Secondly, we compute the ad click entropy given by (12) and the averaged results over all ads can be seen in Fig. 3 with increasing the number of segments increasing the same way as above experiment. From Fig. 3 we can apparently observe the following two phenomena: first, on the whole, k-means based approach gets the lowest click entropy, followed by our proposed LDA-based method. However, when we closely study the the segmentation results of k-means we find that almost one third of the segment given by k-means contains only one user, which is the main reason for its low click entropy. Obviously, in practice these segment are trivial and should be removed. Taking this factor into consideration, we can say that our proposed still works as well as k-means based method, if not better. Second, unlike CTR improvement, ad click entropy curves of all of the four approaches exhibit no strict pattern of growth with the segment number increasing but fluctuate irregularly by the various segment number.

Finally, we intend to find out the relation of iteration number with the effectiveness of our proposed approach, which is of practical significance to decide when to terminate the algorithm properly. From Fig. 2 we can see usually our proposed method obtain the best performance compared with other approaches when segment number is set to be 30. So we calculate the CTR improvement and click entropy of different iteration numbers on the condition that the segment number is 30, with the results listed in Table 2. From the results we can find

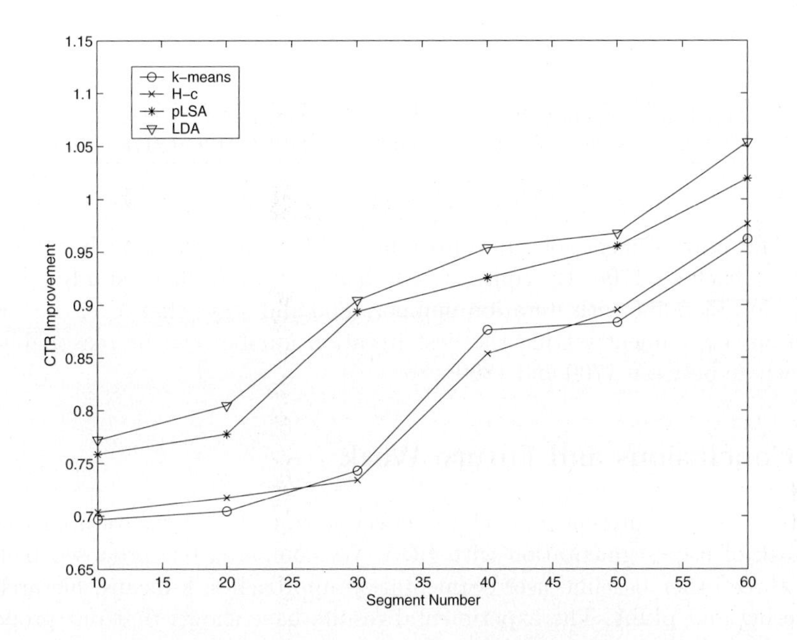

Fig. 2. CTR improvement of the four methods with different segment numbers

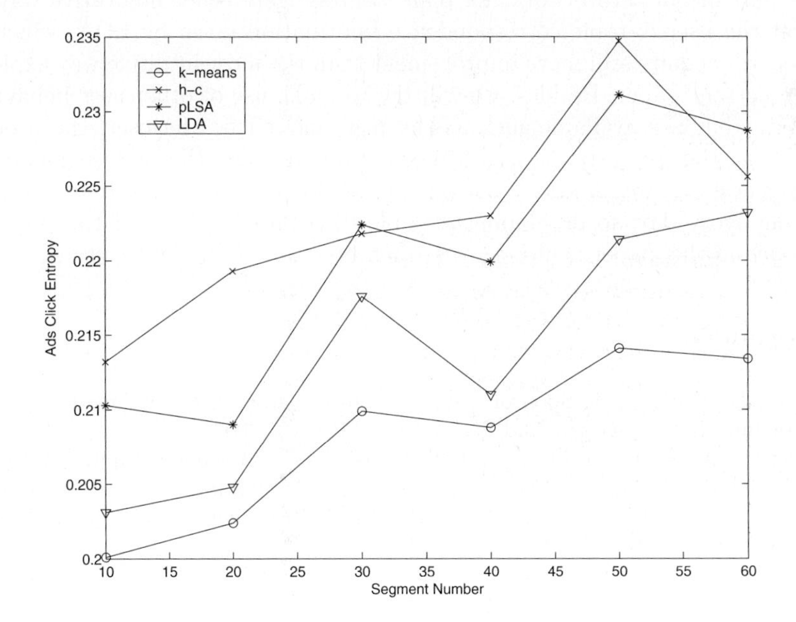

Fig. 3. Ads Click Entropy of the four methods with different segment numbers

Table 2. CTR improvement and click entropy with different iteration numbers(segment number is set to 30)

Iteration Num	1000	1200	1500	1700	1800	1900	2000
CTR Improvement	0.7876	0.8123	0.8340	0.9042	0.9042	0.9042	0.9042
Click Entropy	0.2337	0.2249	0.2249	0.2231	0.2176	0.2176	0.2176

that CTR improvement goes up consistently and arrive a plateau when iteration number reaching 1700. By contrast, ad click entropy declines stably , arrives at the lowest point with iteration number 1800 and stays that level thereafter. So in our experiment setting the best iteration number can be reasonably set somewhere between 1700 and 1800.

5 Conclusions and Future Work

In this paper, we present a novel and effective topic-based way of approaching the task of user segmentation with LDA. We compared our proposed method with three other baseline user segmentation approaches: k-means, hierarchical clustering and pLSA. The experimental results have shown that our proposed method exhibits apparent better performance for user segmentation in contrast to the other baseline user clustering algorithms, both in terms of evaluation metrics CTR improvement and click entropy.

As part of our future work, we plan to investigate other innovative ways to exploit the user-to-topic correspondence information given by LDA. which we expect will acquire still more improvement than the most intuitive way explored in the current paper. Besides, we will try to make use of other user behavioral log data that are available such as the page viewed by the user, the product bought by the user, etc. as extra information to boost the user segmentation effectiveness. As we all know, the size of web log data is quite large and keeps growing every day, so developing scalable distributed edition of the proposed approach in this paper is also a rewarding direction of our future study.

References

1. Fain, D., Pedersen, J.: Sponsored search: A brief history. Bulletin-American Society For Information Science And Technology 32(2), 12 (2006)
2. Broder, A., Fontoura, M., Josifovski, V., Riedel, L.: A semantic approach to contextual advertising. In: Proceedings of the 30th Annual International ACM SIGIR Conference on Research and Development in Information Retrieval, p. 566. ACM, New York (2007)
3. Salton, G., Wong, A., Yang, C.S.: A vector space model for automatic indexing. Commun. ACM 18(11), 613–620 (1975)
4. Blei, D., Ng, A., Jordan, M.: Latent dirichlet allocation. The Journal of Machine Learning Research 3, 993–1022 (2003)
5. http://en.wikipedia.org/wiki/Behavioral_targeting

6. http://www.doubleclick.com/
7. https://www.google.com/adsense/login/en_US/?gsessionid=Dc28hZShnCI
8. http://www.specificmedia.co.uk/
9. http://advertising.yahoo.com/central/marketing/smartads.html
10. Yan, J., Liu, N., Wang, G., Zhang, W., Jiang, Y., Chen, Z.: How much can behavioral targeting help online advertising? In: Proceedings of the 18th International Conference on World Wide Web, pp. 261–270. ACM, New York (2009)
11. Hu, J., Zeng, H., Li, H., Niu, C., Chen, Z.: Demographic prediction based on user's browsing behavior. In: Proceedings of the 16th International Conference on World Wide Web, p. 160. ACM, New York (2007)
12. Zhou, Y., Mobasher, B.: Web user segmentation based on a mixture of factor analyzers. E-Commerce and Web Technologies, 11–20 (2006)
13. Wu, X., Yan, J., Liu, N., Yan, S., Chen, Y., Chen, Z.: Probabilistic latent semantic user segmentation for behavioral targeted advertising. In: Proceedings of the Third International Workshop on Data Mining and Audience Intelligence for Advertising, pp. 10–17. ACM, New York (2009)
14. Minka, T., Lafferty, J.: Expectation-propagation for the generative aspect model. In: Proceedings of the 18th Conference on Uncertainty in Artificial Intelligence, Citeseer, pp. 352–359 (2002)
15. Griffiths, T., Steyvers, M.: Finding scientific topics. Proceedings of the National Academy of Sciences 101(Suppl. 1), 5228–5235 (2004)

Applying Multi-Objective Evolutionary Algorithms to QoS-Aware Web Service Composition

Li Li, Peng Cheng, Ling Ou, and Zili Zhang

Southwest University, Chongqing, P.R. China
{lily,chengp,ouling,zhangzl}@swu.edu.cn

Abstract. Finding optimal solutions for QoS-aware Web service composition with conflicting objectives and various restrictions on quality matrices is a NP-hard problem. This paper proposes the use of multi-objective evolutionary algorithms (MOEAs for short) for QoS-aware service composition optimisation. More specifically, SPEA2 is introduced to achieve the goal. The algorithm is good at dealing with multi-objective combinational optimisation problems. Experimental results reveal that SPEA2 is able to approach the Pareto-optimal front with well spread distribution. The Pareto front approximations provide different trade-offs, from which the end-users may select the better one based on their preference.

Keywords: Multi-objective evolutionary algorithms, Service composition, QoS, Pareto front.

1 Introduction

Service composition is an important part of the whole life-cycle of services innovation research. It is evident from the fact that Web service composition has attracted increasing attention [1]. QoS-based composition for meeting non-functional requirements has been widely studied recently [2,3,4]. However, most of the existing QoS-aware compositions are simply based on the assumption that multiple criteria, no matter whether they are competing or not, can be combined into a single criterion to be optimised, according to some utility functions. In practice, this can be very difficult as utility functions or weights are not well known a priori [5]. In most cases, Pareto-optimal solutions[1], which are produced by applying multi-objective evolutionary algorithms (MOEAs), are often preferred to single (criterion) solutions because the final solution is always a trade-off in practice. Pareto-optimal sets are also known as efficient, or non-inferior sets. QoS-aware service composition is a multi-objective optimisation problem, which requires simultaneous optimisation of multiple and often competing criteria.

[1] Pareto-optimal solutions are sets of solutions that are non-dominated with respect to each other. In other words, Pareto-optimal solutions are those for which any improvement in one objective can only occur through the worsening of at least one other objectives [6].

L. Cao, J. Zhong, and Y. Feng (Eds.): ADMA 2010, Part II, LNCS 6441, pp. 270–281, 2010.

The popular travel scenario probably best illustrates the above situation. For example, we may think that four service components are involved. They are *hotel Booking*, *air-ticket Booking*, *car Rental*, and *theme Parks Pass Booking*. Suppose a large number of candidate Web services with different quality criteria are available for each of the above service components, our task is to provide a solution to find the optimal candidate services for each of them.

As a special class of evolutionary algorithms, the MOEAs are a good fit in finding the optimal sets. They are well-suited for solving multi-objective problems [7,8,9] such as QoS sensitive service composition. Particularly, we use SPEA2 [10] in this paper. Note that terms such as criteria and objectives, qualities and characteristics are used interchangeably unless otherwise specified.

Our main contributions are summarised as follows:

- We propose the use of SPEA2 to handle QoS-aware Web service composition.
- SPEA2 for QoS-aware Web service composition is studied experimentally.

The rest of the paper is organised as follows. Section 2 presents the problem and QoS model followed by the problem formulation. Section 3 explains how to customise SPEA2 to govern QoS composition with detailed experimental results. Section 4 provides an overview of the related work. Finally, Section 5 concludes the paper.

2 Problem Description

Web services have the potential to offer enterprises the capability to integrate in-house business services with external Web services in order to conduct complex business transactions. At the centre of Web services is service composition. The main purpose of QoS composition is to compose independently developed applications at a high level of abstraction in order to have the required complex service. One critical issue of service composition is the selection of the best services to fulfill the role specified by the business logic. It is very important to differentiate between a number of functionally equivalent services and to select the optimal ones. However, it is not an easy job yet to make a decision when a large number of services are presented [11]. Moreover, selecting optimal concrete services becomes complex if a client wants to make sure of receiving a service which meets a specific performance, for example, within a given cost level and a minimum time delay, but a higher availability. This is because different dimensional qualities may conflict with one another in the real world. A typical example is the *time* and *cost* pair. Usually, quicker response and cheaper price is highly demanded, but in practice they are often in conflict. When more quality criteria are required, the aforementioned question becomes more complex. It is clear that service composition is a typical combinational optimisation problem.

Qualities play an important role in decisively identifying the best set of services available at runtime. Obviously, it is important to incorporate QoS into service composition. Unfortunately, finding an optimal solution for QoS-aware Web service composition with conflicting objectives and global constraints between quality dimensions is a NP-hard problem [12].

Evolutionary algorithms are suitable to solve multi-objective optimising problems because they are able to produce a set of solutions in parallel. A growing attention to the multi-objective optimisation problems of Web service composition in recent years is evident. SPEA2 [10], an improved version of SPEA (Strength Pareto Evolutionary Algorithm) is a relatively recent technique for finding or approximating the Pareto-optimal set for multi-objective optimisation problems. The promising results yielded from comparison of SPEA2 with SPEA and NSGA-II [13], on different test problems indicates that SPEA2 is one of the most suitable algorithms in dealing with the problem presented in this paper. Also based on our experience in using evolutionary algorithms in optimisation problems [14], we propose the use of SPEA2 to cope with QoS-aware Web service composition issues.

Below we first introduce the QoS model and then the description of the optimisation problem as a multi-objective problem.

2.1 QoS Model

QoS is an integral part of Web services. It is not uncommon that more than one concrete service realising a particular feature is available. Basically, these concrete services are functionally equivalent therefore they can be interchanged. As different concrete services may operate at different QoS measures, these QoS attributes can be used to differentiate a number of functionally equivalent concrete services. In practice, the choice between them is dictated by QoS criteria.

Although Web services may have quantitative and qualitative characteristics, basically *response time*, invoking *cost* and service *availability* are included.

- *Availability. Availability* [15] is defined as the ratio of time period in which a Web service exists or is ready for use. Usually it is expressed in percentage so a service availability belongs to $[0, 1]$.
- *Cost. Cost* is the amount of money that a service consumer has to pay in order to use this service. In this paper, it is normalised to $[0, 1]$.
- *Response time. Response time* [15] is measured at an actual Web service call. It is normalised to $[0, 1]$ in this paper.

Currently three quality dimensions are discussed in the travel scenario, however, there is no limit to the number of characteristics to be handled by the algorithm.

2.2 Problem Formulation

The main purpose of service composition is to select a set of best fitted concrete services that guarantee the success of composition. Without loss of generality, we assume that all objectives are to be minimised and all equally important. In other words, no additional knowledge about the problem itself is available (e.g., no prioritising, scaling and weighing of the objectives). Suppose there are k objectives (i.e., k Web service quality dimensions). The minimisation[2] of a multi-objective problem with k objectives is defined as follows:

[2] The maximisation of a multi-objective problem can be implemented as a reverse of minimisation functions.

$$minimise\ [f_1(\overline{x}), f_2(\overline{x}), ..., f_k(\overline{x})],$$

where $\overline{x} = [x_1, x_2, ..., x_n]^T$ $(n, k \in \mathbb{N})$ is a vector of decision variables which satisfies a series of constraints. It specifies which concrete service to select for each abstract service group. Each x_i is an integer. $[f_1, f_2, ..., f_k]^T$ is a vector of the objective space.

For example, if we have a vector of concrete services with $\overline{x} = [2, 5, 7, 6]^T$, it means $\mathbf{x}_{[1][2]} = 1$, $\mathbf{x}_{[2][5]} = 1$, $\mathbf{x}_{[3][7]} = 1$ and $\mathbf{x}_{[4][6]} = 1$. In other words, it indicates that the second concrete service in the first group, the fifth concrete service in the second group, the seventh concrete service in the third group, and the sixth concrete service in the fourth group have been selected.

In our example, each concrete service has three characteristics. They are: `availability`, `cost` and `response time`. For the selected concrete services, the objective vector is the aggregation of the corresponding attributes in the decision space. How to calculate these objectives from a decision vector is a question we are facing now. In order to do it, we need to know the transformation rules from the decision vector to the corresponding objective vector. The search process is guided by Pareto-optimal dominance on the objective vector. Table 1 shows the transformation by means of aggregation functions.

Table 1. Aggregation functions

Criteria (Objectives)	Aggregation function
aggregated availability (f_1)	$\Pi_{i=1}^{4} availability\_matrix[i][x_i]$
aggregated cost (f_2)	$\Sigma_{i=1}^{4} cost\_matrix[i][x_i]$
aggregated response time (f_3)	$\Sigma_{i=1}^{4} response\_time\_matrix[i][x_i]$

In this paper, there are four groups of concrete services and three objectives to be optimised, i.e., $n = 4$ and $k = 3$. The objective vector is $[f_1, f_2, f_3]^T$.

Accordingly, there are three matrixes: `availability_matrix`, `cost_matrix` and `respons_time_matrix` to save the values of `availability`, `cost` and `response time` of each concrete service. For example, if the selected solution is $\overline{x} = [2, 5, 7, 6]^T$, the aggregated cost (i.e., f_2) will be:

$$f_2 = cost\_matrix[1][2] + cost\_matrix[2][5] + cost\_matrix[3][7] + cost\_matrix[4][6].$$

Similarly, we can obtain f_1 and f_3, respectively.

3 Experimental Evaluation

We consider four abstract services (i.e. the typical travel scenario) in the experiment. We assume there are 90 concrete services available for each abstract service. The detailed task is to choose the optimal concrete services to achieve better composition results that satisfy three objectives aggregation functions best.

3.1 Experimental Parameters

The detailed MOEA parameters depend on the nature of the problem. Regarding the chromosome coding, an integer string of length m is used to encode the solution $x \in \{1, \ldots, n\}^m$. In other words, there are m digits in the string and each digit is between 1 and n. Individually, each solution is calculated from its chromosome based on the provided attribute matrices.

In order to evaluate the fitness of each individual in the population, we first calculate the objective functions of the individual from its chromosome, i.e., to map a solution from the decision space to the objective space. It is implemented based on the formulae introduced in Section 2.2. Other parameters in the experimentation are shown in Table 2. We follow the given parameters for all tests in the rest of the paper.

Table 2. SPEA2 parameters

Parameter	Value
parent population size	100
child parent population size	100
archive population size	200
chromosome size	4
chromosome encoding scheme	integer encoding
selector	binary tournament selection
crossover	single point (0.95)
mutation	independent bit mutation (0.01)
termination condition	fixed number of generation

3.2 Dataset and Experimental Results

The test data are generated according to some empirical studies of QoS in this domain. Each attribute is normalised between 0 and 1 in this paper. Totally three tests are presented in this section. In the first two tests, each abstract service has 90 candidate services. Consequently, it creates an 4×90 matrix for each attribute. The initial individuals in the first generation are generated randomly.

The proposed algorithm is implemented and tested with the generated dataset and parameters given above. Fig. 1 and Fig. 2 show the initial population and the population at generation 100, but with different angles of view. The experimental results indicate that SPEA2 is capable of guiding the search towards the Pareto-optimal front efficiently. It is shown that SPEA2 already converges to the Pareto-optimal front at generation 100. In other words, evolving with more generation makes no difference in terms of moving the solutions to the true Pareto-optimal front. The results shown in Fig. 3 and Fig. 4 strongly support the above statement.

As the initial attribute matrix data and the initial selected solution in the first generation are created randomly, we have no idea where the true Pareto-optimal front is. However, we understand that better solutions would be the ones

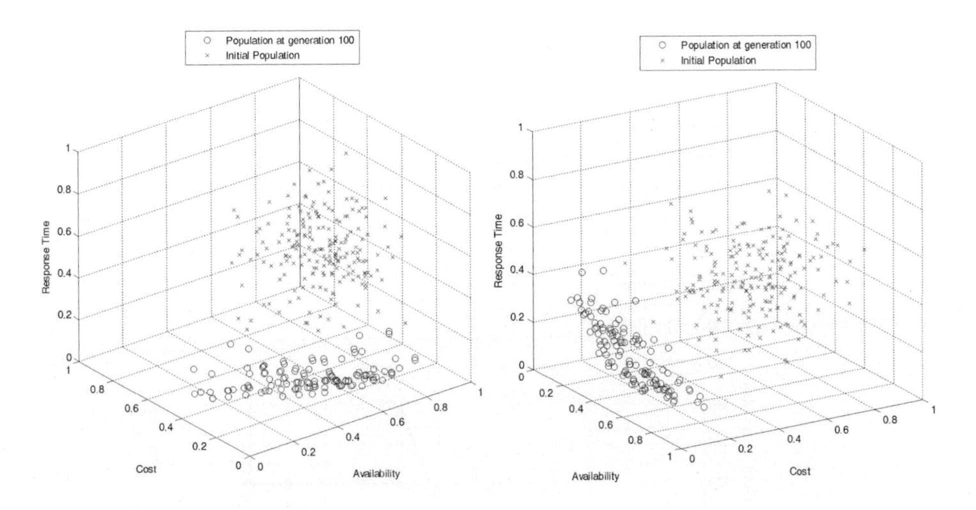

Fig. 1. View1:Initial population VS. population at generation 100

Fig. 2. View2:Initial population VS. population at generation 100

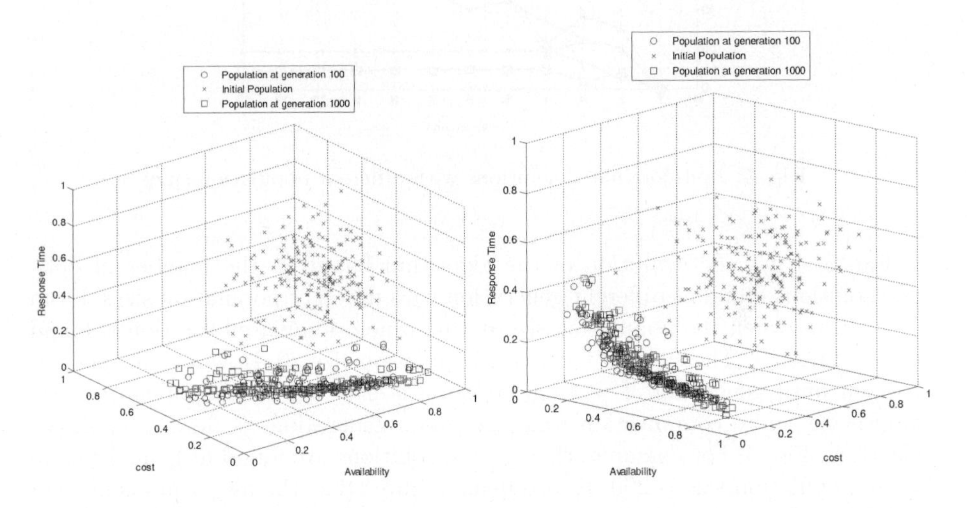

Fig. 3. View1: focusing on population at generation 100 and 1000

Fig. 4. View2: focusing on population at generation 100 and 1000

with lower cost, lower response time, but higher availability. The search process should converge towards to the direction. It has been evident (i.e. the above experimental results). Because the algorithm already converged at generation 100, below we mainly focus on analysing and interpreting findings of Fig. 1 and Fig. 2.

Fig. 1 and Fig. 2 reveal something interesting regarding Web service composition. For example, at generation 100, Fig. 1 clearly shows that the optimal solutions have achieved lower cost and response time, but greater availability, which are centred between 0.5 and 0.8. In order to have a clear view, Fig. 2 is created by rotating the `availability` and `cost` these two dimensions of Fig. 1. Apparently, it shows that the solutions at generation 100 do have lower cost and response time centred around 0.2.

The next two tests are performed to display the convergence property with the presence of different population sizes and various concrete services.

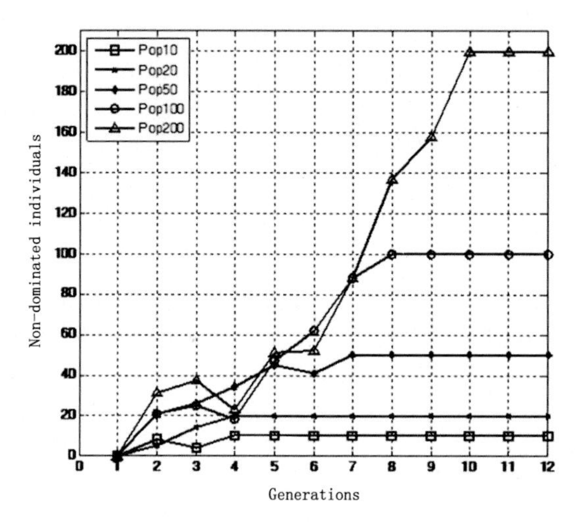

Fig. 5. Non-dominated solutions with different population sizes

Fig. 5 shows the evolution of the algorithm based on the number of optimal solutions found at different generation with different population sizes. From bottom-up, when the population size is 10, no non-dominated solution is found at generation 4. The same is true with population size 20. However, it does not happen until generation 7 when population size is 50. When population size is growing, more computation time (i.e. more generation) is needed as shown clearly in Fig. 5. For example, the optimal solutions are found at generation 10 when population size is 200. It also demonstrates that the algorithm converges quickly with a small population size compared with a large population size.

The next test is different from the previous two tests in which the number of concrete services is not fixed. In Fig. 6, we extend the experiment to show the evolution of the algorithm with various concrete services given that the archive population size is 200. Still, four abstract services are considered. We discuss five different cases with the number of concrete services varying from 30 to 500 for each abstract service. Generally, it takes longer to find a set of optimal solutions with the increase of services. For instance, in the case of 30 services, the algorithm converges at generation 12, while for the cases of 90 services and 150 services, the algorithm finds the non-dominated solutions at nearly the same generation, i.e. at generation 15. We anticipate the same tendency will continue for any other bigger population sizes. As a matter of fact, there is an exception.

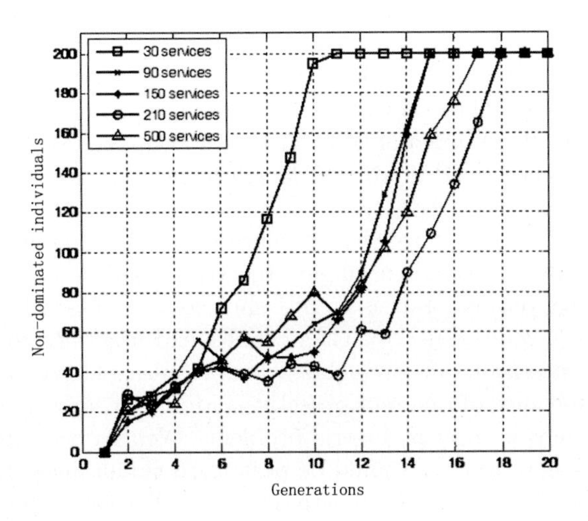

Fig. 6. Non-dominated solutions with various concrete services

In the case of 500 services, the algorithm converges a bit quicker than that in the case of 210 services, given other settings are the same. Currently, we do not know the reason yet.

In short, SPEA2 ensures to find the optimal solutions for service composition problems with different cases. It is able to provide a set of Pareto-optimal concrete services for service composition optimisation problems.

4 Related Work

Recently, some promising results have been reported in the area of Web service and service composition. For example, Multi-objective Evolutionary Algorithms (MOEA) [16,17], Integer Programming (IP) [18,19], Mixed Integer Programming (MIP) [2,20], and Constraint Programming [21]. The following is a brief overview of some recent work in terms of QoS Web service composition.

Zeng et al. [19] introduced a QoS model in which aggregation functions are defined in order to aggregate and measure constraints into a single objective function. The major issues of the QoS-driven service selection approach presented in [19] are scaling (amongst objectives) and weighting. Its weighting phrase requires the selection of proper weights to characterise the user's preferences, which can be very difficult in practice. Furthermore, the method from [19] cannot always guarantee the fulfillment of global constraints, since Web service composition is not separable.

Research on QoS-based composition by using evolutionary computations in order to find an optimal solution has been more active and productive. However, as pointed out by Berbner et al. in [20], the IP approach is hardly feasible in dynamic real-time scenarios when a large number of potential Web services were concerned. In contrast, GAs are able to handle this problem better [22]. GAs are well-suited for Web service selection when multiple QoS attributes are presented.

Canfora et al. [12] proposed the use of Genetic Algorithms (GAs) for the problem mentioned above. It has shown that GAs outperform integer programming used in [19] when a large number of services are available. Moreover, GAs are more flexible than the mixed IP since GAs allow the consideration of nonlinear composition rules [2].

While the selection of the weights of characteristics is required in order to aggregate multi-objectives into a single objective function in GAs, in multi-objective GAs, by contrast, this is totally unnecessary. The advantage of not selecting weights in multi-objective GAs enables end-users not to worry about the accuracy and precision of the weights, which require additional knowledge of the different importance of different objectives. Moreover, instead of optimising objectives separately, in most real world problems, Web service composition, for instance, optimising objectives should be performed simultaneously to get optimal solutions whether or not the objectives are 'conflicting' with to each other. For example, it is always the case to pursue the lowest cost with the highest performance in the minimum time. When the objectives contradict to each other, it is unlikely to find an optimal result in one dimension of the objectives without causing unnecessary suffering to another. In this case, returning a single solution will hardly take place, rather a trade-off between these objectives is much more likely. In addition, the weighted sum approach in GAs largely depends on the formulation (i.e., the weighted formula of the objectives), which has to be readjusted and computed again when the scenario changes.

Apparently, traditional GAs have some inherent limitations in solving QoS-aware composition problems. Several heuristics [6,23,24,25] have been proposed in order to improve traditional GAs in finding optimal or semi-optimal solutions. Multi-objective evolutionary algorithms [26] leverage the QoS-aware service composition problem to provide a set of optimal solutions with different levels of trade-offs. Moreover, reformulating the solutions is not required if there is a change (e.g., the change of the user's preferences). Furthermore, most GA-based multi-objective optimisation algorithms do not require the user to prioritise, scale, or weigh objectives in advance, which is more realistic and reliable.

Using multi-objective GAs is also taken in [27] by Claro et al. The authors discussed the advantages of multi-objective GAs in selecting optimal sets in service selection and a popular Multi-objective algorithm, NSGA-II [13], is used to find sets of services. However, the discussion in [27] is based on NSGA-II, which according to [10] has some limits compared with SPEA2, especially with higher dimensional objective space. In other words, SPEA2 provides a broader distribution and hence better performance.

Apart from GAs, there are some other efforts with different formalisms to in order to target QoS-aware service composition. Yu and Lin [28] studied multiple QoS constraints. The composition problem is modelled as a multi-dimension multi-choice 0-1 knapsack problem (MMKP). A multi-constraint optimal path (MCOP) algorithm with heuristics is presented in [28]. However, the aggregation of parameters using the *Min* function is neglected. Furthermore, the evaluation lacks a metric describing the tightness of the used constraints.

Maximilien and Singh [11] describe the Web Service Agent Framework (WSAF) to achieve service selection by considering the preferences of service consumers as well as the trustworthiness of providers.

At the time of writing, particle swarm optimisation (PSO) [29,30] is becoming popular. Sharing many similarities with GAs, the PSO has the potential to search for optima by updating generations. However, unlike GAs, the PSO has no evolution operators such as crossover and mutation. Deployment of the PSO in order to achieve better QoS composition will be studied in a separate paper.

5 Conclusion

We have proposed the use of multi-objective evolutionary algorithms (MOEAs) in order to optimise multi-objective optimisation problems. Particularly, we discussed the use of SPEA2 to solve QoS-aware Web service composition problems. The experimental results of using SPEA2 has illustrated that SPEA2 is able to converge to the true Pareto-optimal solutions with a wide diversity among solutions. The experimental results also revealed some other important characteristics of MOEAs.

Although there are only four abstract services in our example, there is no limit to the number of abstract services or concrete services to be considered by our implementation of SPEA2, but it requires extra work, especially when fine visual analysis is required. Other future challenges include the discussion of constraints between objectives. We also envision a comprehensive evaluation of SPEA2 and other MOEAs with empirical QoS data in the future.

Acknowledgments

The work is supported in part by Natural Science Foundation Project of CQ CSTC (No. CSTC,2010BB2006) and the Fundamental Research Funds for the Central Universities of P. R. China (No. XDJK2009C015, No. XDJK2009C030).

References

1. Fan, W., Geerts, F., Gelade, W., Neven, F., Poggi, A.: Complexity and composition of synthesized web services. In: PODS 2008, pp. 231–240. ACM, New York (2008)
2. Ardagna, D., Pernici, B.: Adaptive service composition in flexible processes. IEEE Trans. on Software Engineering 33, 369–384 (2007)
3. Jaeger, M.C., Rojec-Goldmann, G., Mühl, G.: Qos aggregation in web service compositions. In: EEE 2005, pp. 181–185. IEEE Computer Society, Los Alamitos (March 2005)
4. Jiang, Z.Y., Han, J.H., Zhao, W.: Optimization model for dynamic qos-aware web services selection and composition. Chinese Journal of Computers 32, 1014–1025 (2009)

5. Fonseca, C.M., Fleming, P.J.: Genetic algorithms for multiobjective optimization: Formulation, discussion and generalization. In: Forrest, S. (ed.) ICGA 1993, pp. 416–423. Morgan Kaufmann, San Francisco (June 1993)
6. Konak, A., Coit, D.W., Smith, A.E.: Multi-objective optimization using genetic algorithms: A tutorial. Reliability Engineering & System Safety 91, 992–1007 (2006)
7. Deb, K.: Multi-Objective Optimization using Evolutionary Algorithms. Wiley, Chichester (2001)
8. Carlos, A., Coello Coello, D.A.V.V., Lamont, G.B.: Evolutionary Algorithms for Solving Multi-Objective Problems. Springer, Heidelberg (2007)
9. Das, S., Panigrahi, B.: Multi-objective Evolutionary Algorithms, vol. 3, pp. 1145–1151. Idea Group Publishing, USA (2008)
10. Zitzler, E., Laumanns, M., Thiele, L.: Spea2: Improving the strength pareto evolutionary algorithm. Technical report, ETH, Zürich (2001)
11. Maximilien, E.M., Singh, M.P.: A framework and ontology for dynamic web services selection. IEEE Internet Computing 8, 84–93 (2004)
12. Canfora, G., Penta, M.D., Esposito, R., Villani, M.L.: An approach for qos-aware service composition based on genetic algorithms. In: Beyer, H.G., O'Reilly, U.M. (eds.) Genetic and Evolutionary Computation Conference, GECCO 2005, pp. 1069–1075. ACM, New York (June 2005)
13. Deb, K., Agrawal, S., Pratap, A., Meyarivan, T.: A fast and elitist multiobjective genetic algorithm: Nsga-II. IEEE Trans. Evolutionary Computation 6, 182–197 (2002)
14. Zhang, Z., Yang, P., Wu, X., Zhang, C.: An agent-based hybrid system for microarray data analysis. IEEE IS 24, 53–63 (2009)
15. Kim, E., Lee, Y.: Quality model for web service. Technical report, OASIS Open Consortium (2005)
16. Coello, C.A.C.: A comprehensive survey of evolutionary-based multiobjective optimization techniques. Knowledge and Information Systems 1, 269–308 (1999)
17. Deb, K.: Multi-objective genetic algorithms: Problem difficulties and construction of test problems. Evolutionary Computation 7, 205–230 (1999)
18. Aggarwal, R., Verma, K., Miller, J.A., Milnor, W.: Constraint driven web service composition in meteor-s. In: SCC 2004, pp. 23–30. IEEE Computer Society, Los Alamitos (September 2004)
19. Zeng, L., Benatallah, B., Ngu, A.H.H., Dumas, M., Kalagnanam, J., Chang, H.: Qos-aware middleware for web services composition. IEEE Trans. Software Eng. 30, 311–327 (2004)
20. Berbner, R., Spahn, M., Repp, N., Heckmann, O., Steinmetz, R.: Heuristics for qos-aware web service composition. In: ICWS 2006, pp. 72–82 (September 2006)
21. Hentenryck, P.V.: Constraint satisfaction in logic programming. MIT Press, Cambridge (1989)
22. Canfora, G., Penta, M.D., Esposito, R., Villani, M.L.: Qos-aware replanning of composite web services. In: ICWS 2005, pp. 121–129. IEEE Computer Society, Los Alamitos (2005)
23. Coello, C.A.C., Lamont, G.B., Veldhuizen, D.A.V.: Evolutionary Algorithms for Solving Multi-Objective Problems (Genetic and Evolutionary Computation). Springer, New York (2006)
24. Veldhuizen, D.A.V., Lamont, G.B.: Multiobjective evolutionary algorithms: Analyzing the state-of-the-art. Evol. Comput. 8, 125–147 (2000)

25. Ai, L., Tang, M.: Qos-aware web service composition accommodating inter-service dependencies using minimal-conflict hill-climbing repair genetic algorithm. In: IEEE International Conference on eScience (eScience 2008), pp. 119–126. IEEE Computer Society Press, Los Alamitos (December 2008)
26. Taboada, H.A., Espiritu, J.F., Coit, D.W.: Moms-ga: A multi-objective multi-state genetic algorithm for system reliability optimization design problems. IEEE Transactions on Reliability 57, 182–191 (2008)
27. Claro, D.B., Albers, P., Hao, J.K.: Selecting web services for optimal composition. In: ICWS 2005 Workshop, pp. 32–45 (July 2005)
28. Yu, T., Lin, K.J.: Service selection algorithms for composing complex services with multiple qos constraints. In: Benatallah, B., Casati, F., Traverso, P. (eds.) ICSOC 2005. LNCS, vol. 3826, pp. 130–143. Springer, Heidelberg (2005)
29. Eberhart, R., Kennedy, J.: A new optimizer using particle swarm theory. In: Proceedings of the Sixth International Symposium on Micromachine and Human Science, pp. 39–43 (1995)
30. Kennedy, J., Eberhart, R.: Particle swarm optimization. In: Proceedings of IEEE International Conference on Neural Networks (ICNN 1995), pp. 1942–1948 (1995)

Real-Time Hand Detection and Tracking Using LBP Features

Bin Xiao, Xiang-min Xu, and Qian-pei Mai

School of Electronic and Information Engineering, South China University of Technology
Guangdong, Guangzhou 510640 China
xiaobin0725@gmail.com,
xmxu@scut.edu.cn

Abstract. In this paper a robust and real-time method for hand detection and tracking is proposed. The method is based on AdaBoost learning algorithm and local binary pattern (LBP) features. The hand is detected by the cascade of classifiers with LBP features. A detailed study was developed to select the parameters for the hand detection classifiers. When tracking the hand, a region of interest (ROI) is defined based on the hand region detected in the last frame, and in order to improve robustness on rotation affine transformation is applied to the ROI. The experimental result demonstrates that this method can successfully detect the hand and track it in real-time.

Keywords: hand detection, hand tracking, boosting, LBP, HCI.

1 Introduction

Hand detection and hand tracking play important roles in Human-Computer Interfaces (HCI). To achieve natural HCI for virtual environment applications, human hand could be considered as an input device. Gesture is a powerful human-to-machine communication method. Two primary problems of gesture recognition are hand detection and hand tracking. The tasks of hand detection and tracking are challenging because the hand is a non-rigid object. Considering the global hand pose and each finger joint, the human hand motion has roughly with 27 degrees of freedom (DOFs) [1].

Several hand detection and tracking systems had been proposed. The first generation approaches require glove-based devices to help recognize the hand. However, the gloves and their attached wires are still quite cumbersome and awkward for users. Moreover, the cost of the glove is often too expensive for regular users. The second generation approaches use skin color or shape feature [3][4][5]. However, those methods are lack of robustness when dealing with dynamic environments and various kinds of lighting. The third generation approaches are based on a cascade architecture using boosting algorithm, which was first introduced by Viola and Jones [2] to for face detection and tracking problems. That approach allows robust and fast detection of hands [6][7].

L. Cao, J. Zhong, and Y. Feng (Eds.): ADMA 2010, Part II, LNCS 6441, pp. 282–289, 2010.

In this paper, we introduce a robust and real-time method for hand detection and tracking, which is based on the boosting architecture combining with local binary pattern (LBP) features.

The rest of the paper is organized as follows. In Section 2 and Section 3, we present the AdaBoost algorithm with LBP features that we use for hand detection and tracking. Experimental results are discussed in Section 4. Finally, we draw the conclusions about our hand detection and tracking application.

2 AdaBoost Algorithm with LBP Features for Hand Detection

Hand detection is one of the obstacles of gesture recognition. Skin-color-based method, one of the solutions for hand detection has to face the difficult task of distinguishing the hands from other objects having skin color, such as arms and face. And it is very sensitive to the changing of light. Thus, if the background or lighting condition does not meet the needs, it would be difficult to detect the hand with skin color-based method.

In this paper, we use AdaBoost algorithm with local binary pattern (LBP) features to detect the hand. It can avoid affections of other objects with skin color, and it is also robust to lighting changes. LBP is a texture descriptor which codifies local primitives into a feature histogram. The cascade of AdaBoost architecture for objects detection is first proposed by Viola and Jones [2] to solve the problem of face detection. Their method uses four basic Haar-like features. They also use AdaBoost algorithm to select and train the classifier. AdaBoost is one of Boosting algorithms combining weak classifiers to form a strong classifier with better accuracy.

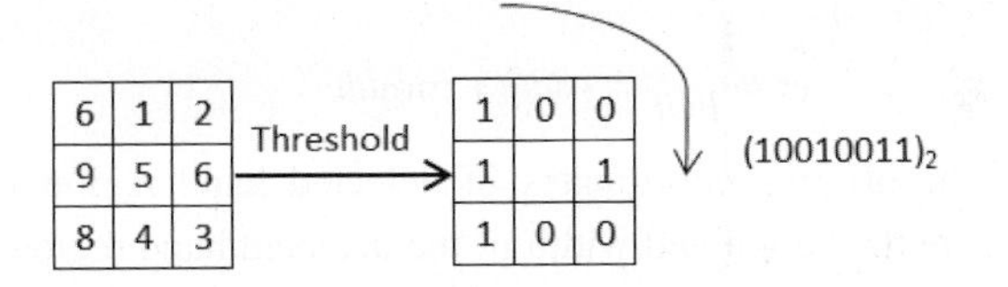

Fig. 1. The basic LBP operator

In our method, we use the LBP features rather than the Haar-like features used by Viola and Jones [2]. The LBP operator is first introduced by Ojala et al. [8] as a powerful means of texture description. The operator labels every pixel of an image by thresholding its 3x3-neighbourhood with the center value and considering the result as a binary number. Then the histogram of the labels can be used as a texture descriptor. Fig. 1 shows the LBP descriptor. When training the classifier, the hand is divided into small regions from which LBP histograms are computed as a feature.

When detecting the hand in an image, the image is scanned by a sub-window containing the LBP features. In order to detect the different size of hand, the image is sampled in different scales. In each scale, the image is smaller than the last one with a fixed factor. In our experiment, we choose the scale factor of 1/1.2. Based on the LBP features, a weak classifier ($h(x,f,p,\theta)$) is defined as:

$$h(x,f,p,\theta)=\begin{cases}1 & \text{if } pf(x)<p\theta \\ 0 & \text{otherwise}\end{cases} . \tag{1}$$

Where f is the feature, θ is the threshold, p is a polarity indicating the direction of the inequality, and x is the sub-window.

In practice, one single feature can't detect the hand with a high accuracy. The AdaBoost algorithm is used to improve the overall accuracy by combining these weak classifiers.

In order to improve detection performance and reduce the computation time Viola and Jones [2] proposed a cascade of strong classifiers. Only sub-windows that passed all the previous strong classifiers are sent to the next classifiers for further classification. Only the region passing all the classifiers is recognized as the target object.

3 Hand Tracking Based on Rotated ROIs with LBP Features

The hand tracking module is based on the hand detection algorithm in section 2.If it detects the hand successfully, we define an adaptive region of interest (ROI) based on the region of detected hand, and in the next frame we detect the hand in the ROI. In this way, it speeds up the processing time. The ROI is defined as:

$$(x_{ROI}, y_{ROI}) = (x, y) . \tag{2}$$

$$height_{ROI} = height + \alpha height . \tag{3}$$

$$width_{ROI} = width + \alpha width . \tag{4}$$

Where (x,y) is the original coordinates of detected hand region in the last frame, $height$ and $width$ are the height and width of the detected hand region in the last frame. A typical value of α is 2.

When the hand is moving, it is usually rotated by some degrees. In order to improve the system's robustness to the rotation we apply an affine transformation to the ROI. The affine transformation is expressed in (5).

$$\begin{bmatrix} X' \\ Y' \end{bmatrix} = \begin{bmatrix} \cos\theta & -\sin\theta \\ \sin\theta & \cos\theta \end{bmatrix} \begin{bmatrix} X \\ Y \end{bmatrix} . \tag{5}$$

Where (X,Y) is the coordinates of a general point. (X',Y') is the coordinates after transformation.

In our method, we set the θ as ±15°, ±30°,±45°.If detecting hand fails in the ROI, the system will detect the hand in the rotated ROIs.

Fig. 2 is the diagram of the hand tracking system.

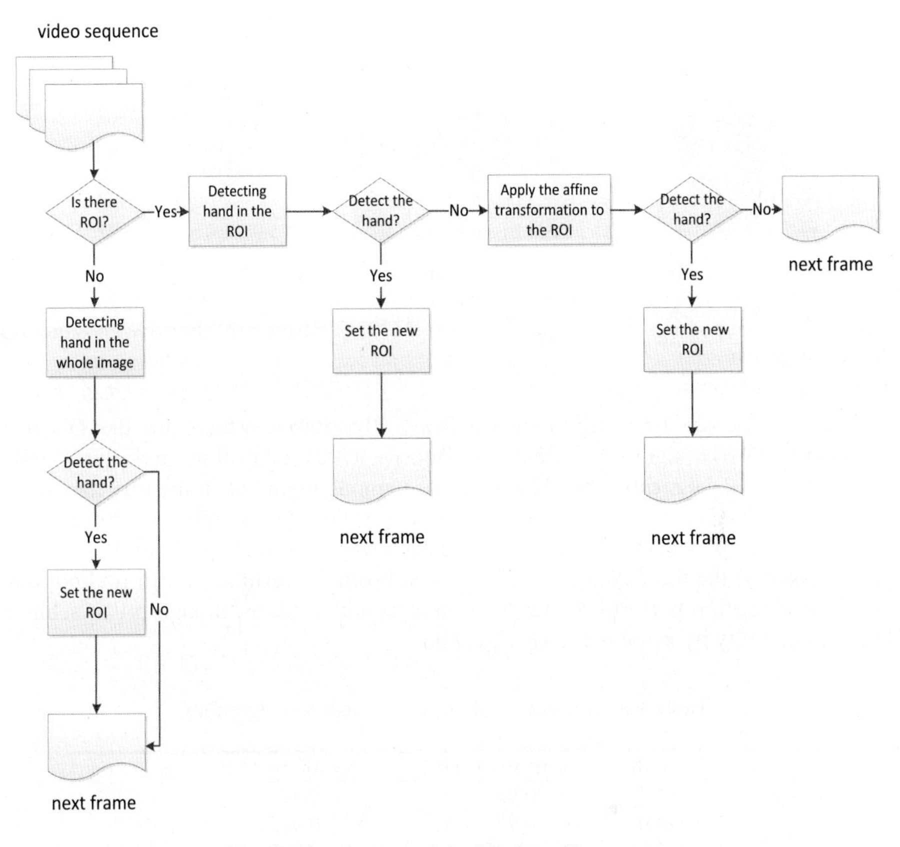

Fig. 2. The diagram of the hand tracking system

4 Experiment Results and Discussion

In our experiment, we tested three hand postures, which are the "fist" posture, the "five" posture and the "palm" posture. Fig. 3 shows these three postures. We use Logitech QuickCam Webcam as image capture device. In our experiment, we set the camera at the resolution of 640x480.

To train the classifiers, we collected 958, 724, 895 positive samples for the "fist" posture, the "five" posture and the "palm" posture. To improve the robustness of the classifiers the positive samples are captured with different scales at different places from different people and in different lighting conditions. We also collected about 2000 random images as the negative samples. The negative samples are background images which exclude the three postures above.

Once the samples are chosen, we can use them to train the classifiers. There are several parameters affecting the performance of the classifiers. In order to evaluate the performances of the classifiers with different training parameters, we collected 482 images different from the training set for testing. Each of the testing images contains at least one posture and the testing set includes 600 postures. We studied the influence of each training parameter by only varying it while leaving the other parameters unchanged.

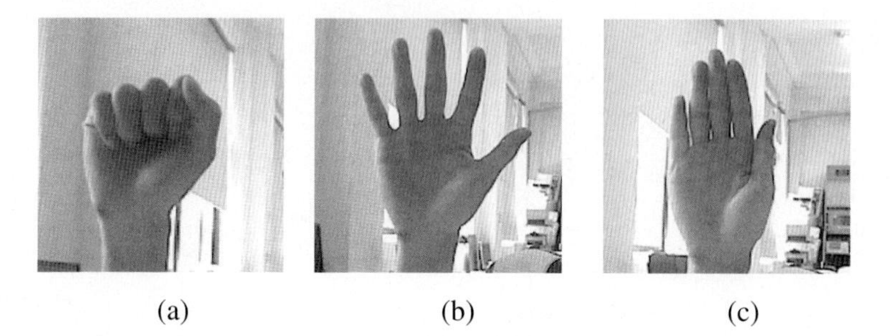

(a) (b) (c)

Fig. 3. The three postures in our experiment: (a) the "fist" posture; (b) the "five" posture; (c) the "palm" posture

First, we choose three different AdaBoost algorithms, which are the Discrete AdaBoost (DAB) algorithm, Real AdaBoost (RAB) algorithm and the Gentle AdaBoost (GAB) algorithm [9]. The difference among them lies in the way they reassign the weights in each interaction of the algorithm. The number of the cascade stages is set to 18. The max false detect rate each stage is set to 0.5. Table 1 shows the performance of the three AdaBoost algorithms. From the results, we can find out that the GAB algorithm performs better than other two algorithms, though it can achieve higher specificity by using the RAB algorithm.

Table 1. Perfomance with different AdaBoost algorithm

Posture	Algorithm	Sensitivity/TPR	Specificity/TNR	Accuracy
Fist	GAB	0.935	0.977	0.942
Fist	DAB	0.92	0.915	0.917
Fist	RAB	0.825	0.993	0.937
Five	GAB	0.942	0.949	0.947
Five	DAB	0.844	0.963	0.918
Five	RAB	0.582	1	0.843
Palm	GAB	0.829	0.92	0.893
Palm	DAB	0.16	0.925	0.86
Palm	RAB	0.703	1	0.755

The number of the stages also has an important impact on the performance of hand detection. We trained the classifiers with the stage number of 15, 18 and 20, with the algorithm of GAB, and with the false detection rate per stage of 0.5. Table 2 shows the performance of different number of stages. It can be observed that an increase of the number of the cascade stage causes a decrease of the sensitivity and an increase of the specificity.

We trained four classifiers for each posture with a false detect rate per stage of 0.45, 0.5, 0.55 and 0.6, with the algorithm of GAB and with the stages number of 18.

Fig. 4 shows the results with different false detect rate. Though an increase of false detect rate can cause an increase of the sensitivity, it also makes a rapid decrease of specificity and accuracy. Considering the results of sensitivity, specificity and accuracy, the false detect rate of 0.5 per stage can achieve the best performance.

Table 2. Performance with different number of stages

Posture	Stages	Sensitivity/TPR	Specificity/TNR	Accuracy
Fist	15	0.95	0.253	0.485
Fist	18	0.935	0.977	0.942
Fist	20	0.855	0.998	0.95
Five	15	0.978	0.736	0.827
Five	18	0.942	0.949	0.947
Five	20	0.907	0.995	0.962
Palm	15	0.954	0.8	0.845
Palm	18	0.829	0.92	0.893
Palm	20	0.657	0.979	0.885

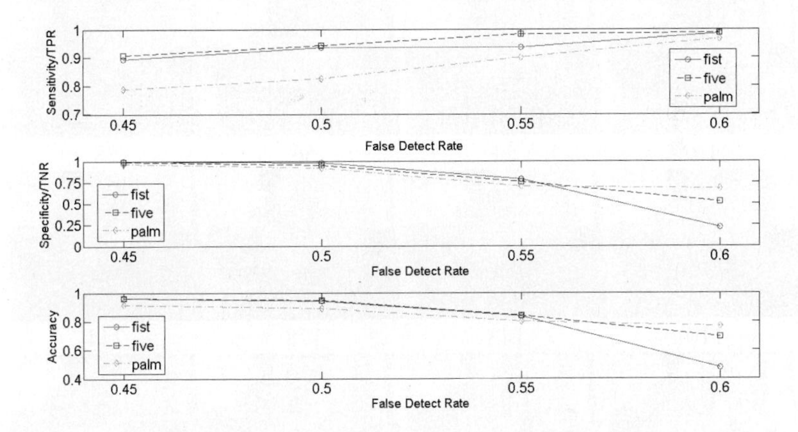

Fig. 4. Performance with different max false detect rate

After studying the influences of the parameters on the classifier performance, we chose three classifiers for hand detection and hand tracking. The parameters and performance of these three classifiers are shown in Table 3.

Table 3. Classifiers Using for Hand Detecion and Hand Tracking

Posture	Parameters		Performance	
Fist	Algorithm:	GAB	Sensitivity:	0.935
	Stages:	18	Specificity:	0.977
	False Detect Rate:	0.5	Accuracy:	0.942
Five	Algorithm:	GAB	Sensitivity:	0.907
	Stages:	20	Specificity:	0.995
	False Detect Rate:	0.5	Accuracy:	0.962
Palm	Algorithm:	GAB	Sensitivity:	0.891
	Stages:	16	Specificity:	0.889
	False Detect Rate:	0.5	Accuracy:	0.89

Fig. 5. Hand tracking result: the first two rows are the result of tracking "post" posture; the third and the forth row are the result of tracking "five" posture; the last two rows are the result of tracking "palm" posture

Having the hand detected, the method in Section 3 is used to track the hand. Some video sequences are used for testing, which are at resolution of 640x480 with 30 frames/s captured by Logitech QuickCam Webcam. The experiment runs on Core 2 Duo processor at 2.16 GHz. Fig. 4 shows some frames extracted from the video sequence. It can be observed that the tracking is successfully achieved. With 30 frames/s it satisfied the real-time requirement.

5 Conclusion

In this paper, we have presented a robust and real-time method for hand detection and tracking, which achieves high detection accuracy and satisfactory real-time tracking. The approach is focused on the posture recognition with LBP features and AdaBoost learning algorithm. A ROI is defined to reduce hand tracking time, and affine transformation is applied to improve the performance of the hand tracking. Our hand detection and tracking method can be integrated in a vision-based HCI, for example gesture-control TV and so on.

Acknowledgements. This work is supported by the Fundamental Research Funds for the Central Universities and Science and Technology Planning Project of Guangdong Province under the grand No. 2010A080402015 and No. 2009B080701060.

References

1. Wu, Y., Huang, T.S.: Hand modeling, analysis and recognition. IEEE Signal Processing Magazine 18(3), 51–60 (2001)
2. Viola, P., Jones, M.: Rapid objet detection using a boosted cascade of simple features. In: Proc. IEEE Conf. on Computer Vision and Pattern Recognition, pp. 511–518 (2001)
3. Binh, N.D., Shuichi, E., Ejima, T.: Real-Time Hand Tracking and Gesture Recognition System. In: Proc. GVIP 2005, Cairo, Egypt, pp. 19–21 (2005)
4. Manresa, C., Varona, J., Mas, R., Perales, F.: Hand Tracking and Gesture Recognition for Human-Computer Interaction. Electronic letters on computer vision and image analysis 5(3), 96–104 (2005)
5. Angelopoulou, A., García-Rodriguez, J., Psarrou, A.: Learning 2D Hand Shapes using the Topology Preserving model GNG. In: Leonardis, A., Bischof, H., Pinz, A. (eds.) ECCV 2006. LNCS, vol. 3951, pp. 313–324. Springer, Heidelberg (2006)
6. Fang, Y., Wang, K., Cheng, J., Lu, H.: A Real-Time Hand Gesture Recognition Method. In: Proc. 2007 IEEE Int. Conf. on Multimedia and Expo., pp. 995–998 (2007)
7. Chen, Q., Georganas, N.D., Petriu, E.M.: Real-time Vision-based Hand Gesture Recognition Using Haar-like Features. In: Proc. Instrumentation and Measurement Technology Conf. on IMTC 2007, Warsaw, Poland (2007)
8. Ojala, T., Pietikäinen, M., Harwood, D.: A comparative study of texture measures with classification based on feature distributions. Pattern Recognition 29, 971–987 (1996)
9. Lienhart, R., Kuranov, A., Pisarevsky, V.: Empirical Analysis of Detection Cascades of Boosted Classifiers for Rapid Object Detection. In: Michaelis, B., Krell, G. (eds.) DAGM 2003. LNCS, vol. 2781, pp. 297–304. Springer, Heidelberg (2003)

Modeling DNS Activities Based on Probabilistic Latent Semantic Analysis

Xuebiao Yuchi[1,2], Xiaodong Lee[1], Jian Jin[1], and Baoping Yan[1]

[1] China Internet Network Information Center, Computer Network Information Center, Chinese Academy of Sciences, 100190 Beijing, China
[2] Graduate University of Chinese Academy of Sciences, 100190 Beijing, China
{yuchixuebiao,lee,jinjian}@cnnic.cn, ybp@cnic.cn

Abstract. Traditional Web usage mining techniques aim at discovering usage patterns from Web data at the page level, while little work is engaged in at some upper level. In this paper, we propose a novel approach to the characterization of Internet users' preference and interests at the domain name level. By summarizing Internet user's domain name access behaviors as the co-occurrences of users and targeting domain names, an aspect model is introduced to classify users and domain names into various groups according to their co-occurrences. Meanwhile, each group is characterized by extracting the property of *characteristic* users and domain names. Experimental results on real-world data sets show that our approach is effective in which some meaningful groups are identified. Thus, our approach could be used for detecting unusual behaviors on the Internet at the domain name level, which can alleviate the work of searching the joint space of users and domain names.

Keywords: Domain Name System, Probabilistic latent semantic analysis, Co-occurrence.

1 Introduction

In order to better understand the usage of the Internet, many possible techniques have been developed, of which Web usage mining techniques [1], defined as the process of applying data mining techniques to the discovery of usage patterns from Web data, have achieved great success in various application areas such as Web personalization [2], link prediction and analysis [3] and e-commerce data analysis [4], etc.

Generally, as illustrated in Fig. 1, these techniques are aiming at discovering usage patterns at the page view level, and their applying objects are always some specific websites providing some special services such as search engine, IPTV, and B2C, etc. In other words, there is few work evolved in discovering usage patterns at the website level, or to be more generally, at the domain name level. However, this is also essential for understanding the usage of the Internet. Discovering usage patterns at the domain name level can provide us with a macroscopic view of the Internet, which is especially useful for the Internet governors. In particular, we are interested in some typical questions just as listed below:

L. Cao, J. Zhong, and Y. Feng (Eds.): ADMA 2010, Part II, LNCS 6441, pp. 290–301, 2010.

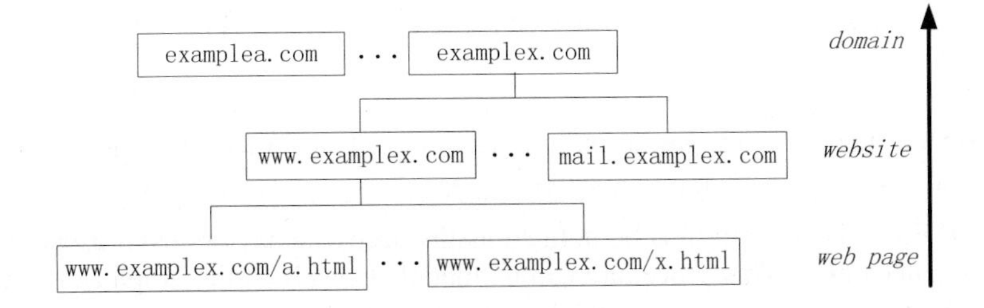

Fig. 1. Three different usage levels of the Internet from the point of view of Internet users

1. How are the users' preferences distributed among domain names? Are there similar users with respect to their preferences?
2. How are the popularities of domain names distributed among users? Which names are 'closer to' each other in terms of user preferences?
3. Do users show some underlying behavior patterns when they are surfing the Internet? How to profile them?
4. Are there some hidden correlations among users as well as between users and domain names?
5. Who are abusing the Internet or doing evil on the Internet? Which names are illegal? Can we detect them?

With these questions, we try to find an effective way to discover Internet usage patterns at the domain name level.

As a globally distributed database, the Domain Name System (DNS) [5] is used for translating worldwide unique domain names such as *www.google.com* to other identifiers. Serving as the Internet's phone book, DNS is needed by almost all Internet applications. For example, every time an Internet user visits a website, his/her computer will firstly perform a DNS query for the domain name of the website. In other words, a typical DNS query can explicitly tell who is looking for what on the Internet. Therefore, DNS activities can to some extent reflect the overall usage of the Internet applications.

However, there is little work found in analyzing this reflection so far. In this paper, based on probabilistic latent semantic analysis (PLSA) model, we propose a novel approach to the characterization of Internet users' preference and interests at the domain name level. We classify users and domain names into various groups according to their co-occurrence activities. Meanwhile, each group is characterized by extracting the property of *characteristic* users and domain names. The effectiveness of our approach is demonstrated through experiments conducted on real-world data sets.

The rest of the paper is organized as follows. In Section 2, we provide an overview of PLSA model as applied to DNS co-occurrence activities. The procedure of identifying and characterizing DNS groups are described in Section 3. We conduct our experiments in Section 4 together with explanations in Section 5. Finally, we conclude the paper in Section 6.

2 Probabilistic Latent Semantic Models of DNS Activities

The objects appeared in DNS activities can be divided into two different sets: users and domain names. Here, we define the user set as $U = \{u_1, u_2, ... u_M\}$ and the domain name set as $D = \{d_1, d_2, ... d_N\}$, where M and N are the size of these two sets respectively. DNS activities can thus be defined as the access behaviors from users in U to domain names in D. This procedure can be modeled as a bipartite graph as illustrated in Fig. 2.

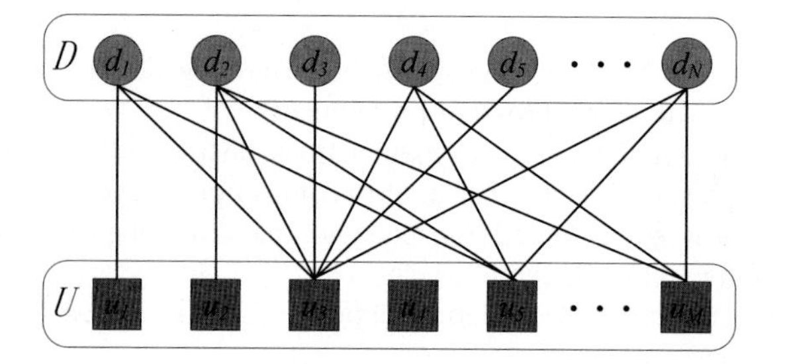

Fig. 2. The bipartite graph view of DNS activities, in which each edge represents an access from a user in U to a domain name in D

Generally, the user access interests exhibited may be reflected by the varying degree of access to different domain names during the user's session. Thus, we can represent a user session as a weighted domain name vector visited by the user during the session. The weight can be binary, representing the existence or non-existence of the domain name, or it may be a function of the occurrence of this domain name during the user's session. As a result, the overall DNS activities described above can be represented by an $M \times N$ matrix $UD = [w(u_m, d_n)]_{M \times N}$, where $w(u_m, d_n)$ represents the weight associated with the domain name d_n in the user session of u_m.

The PLSA model is originally proposed by Hofmann [6], which has been successfully used in a variety of application areas, including information retrieval [7], collaborative filtering [8] and some other related topics. The starting point for PLSA is a statistical model called the *aspect model*, which can be utilized to identify the hidden semantic relationships among general co-occurrence activities. Similarly, we can conceptually view the user sessions over domain names space as co-occurrence activities to discover the latent usage patterns in our context. The given aspect model is utilized to associate an latent (unobserved) class variable z_k ($k = 1, 2, \ldots, K$) with each access by a user to a domain name in a particular user session which is represented as an entry of the $M \times N$ co-occurrence matrix UD.

In PLSA, three fundamental schemes are implemented:

1. select a user session of u_m from U with probability $P(u_m)$,
2. pick a latent class z_k with probability $P(z_k | u_m)$,
3. generate a domain name d_n from D with probability $P(d_n | z_k)$.

As a result, the probability of an observed pair (u_m, d_n) can be obtained by a joint probability model in the following:

$$P(u_m, d_n) = P(u_m)P(d_n \mid u_m) \ , \tag{1}$$

where
$$P(d_n \mid u_m) = \sum_{k=1}^{K} P(d_n \mid z_k)P(z_k \mid u_m) \ . \tag{2}$$

By using Bayes' rule and substituting (2), (1) can be transformed into:

$$P(u_m, d_n) = \sum_{k=1}^{K} P(z_k)P(u_m \mid z_k)P(d_n \mid z_k) \ . \tag{3}$$

Following the likelihood principle, the total likelihood $L(U, D)$ of the co-occurrence data is determined as:

$$L(U, D) = \sum_{m=1}^{M}\sum_{n=1}^{N} w(u_m, d_n)\log P(u_m, d_n) \ . \tag{4}$$

In order to maximize $L(U, D)$, we use *Expectation Maximization* (EM) algorithm [9] which is a well-known approach to perform maximum likelihood estimation in latent variable models. Generally, it alternates two steps: (1) an expectation (E) step where posterior probabilities are computed for latent variables based on the current estimates of conditional probability, (2) a maximization (M) step, re-estimate the conditional probabilities to maximize the expectation of the complete data likelihood.

We describe the whole procedure in details:

1. Firstly, given the randomized initial values of $P(z_k)$, $P(u_m \mid z_k)$, $P(d_n \mid z_k)$.
2. Then, in the E-step, we can simply apply Bayes' formula to generate following variable based on co-occurrence observation:

$$P(z_k \mid u_m, d_n) = \frac{P(z_k)P(u_m \mid z_k)P(d_n \mid z_k)}{\sum_{k'=1}^{K} P(z_{k'})P(u_m \mid z_{k'})P(d_n \mid z_{k'})} \ . \tag{5}$$

3. Furthermore, in M-step, we can compute:

$$P(u_m \mid z_k) = \frac{\sum_{n=1}^{N} w(u_m, d_n)P(z_k \mid u_m, d_n)}{\sum_{m'=1}^{M}\sum_{n=1}^{N} w(u_{m'}, d_n)P(z_k \mid u_{m'}, d_n)} \ , \tag{6}$$

$$P(d_n \mid z_k) = \frac{\sum_{m=1}^{M} w(u_m, d_n)P(z_k \mid u_m, d_n)}{\sum_{m=1}^{M}\sum_{n'=1}^{N} w(u_m, d_{n'})P(z_k \mid u_m, d_{n'})} \ , \tag{7}$$

$$P(z_k) = \frac{1}{R}\sum_{m=1}^{M}\sum_{n=1}^{N} w(u_m, d_n)P(z_k \mid u_m, d_n) \ , \tag{8}$$

where
$$R = \sum_{m=1}^{M}\sum_{n=1}^{N} w(u_m, d_n) \ . \tag{9}$$

Basically, substituting (6)-(8) into (3) and (4) will result in the monotonically increasing of the total likelihood of the co-occurrence data $L(U, D)$. The executing of E-step and M-step is iterating until $L(U, D)$ is converging to a local optimal limit, which means the estimated results can represent the final probabilities of the co-occurrence data.

It is easily found that the computational complexity of this algorithm is $O(MNK)$, where M is the number of user sessions, N is the number of domain names, and K is the number of classes. Since the co-occurrence matrix is generally very sparse, the memory requirements can be dramatically reduced by using efficient sparse matrix representation of the data.

3 Identifying and Characterizing DNS Groups with PLSA

One of the main advantages of PLSA model is that it generates probabilities which can quantify relationships between users and domain names, as well as domain names and classes. From these basic probabilities, using probabilistic inference, we can derive relationships among users, among domain names, and between users and domain names. As we discussed in Section 2, we note that each latent class z_k does really represent specific aspect with co-occurrence in nature. In other words, for each class, the degrees related to the co-occurrence can be expressed by the class- conditional probability estimates. From this viewing point, we can thus utilize the class-conditional probability estimates generated by the PLSA model to partition domain names, as well as users, into various usage-based groups. For each of these groups, users will commonly reveal stronger preferences to the domain names inside, than to those of outside. Furthermore, in order to better understand users' preferences and interests in domain names within each group, we characterize each group by interpreting the property of *characteristic* users and domain names whose probabilities exceed a predefined threshold.

3.1 Identifying Groups in DNS Activities

Due to the Internet's immense size and its continuous evolution, it is impossible to track down such groups by a manual effort. Here, we describe an automatic method for identifying DNS groups based on PLSA.

As noted before, the PLSA model generates probabilities $P(z_k)$, which measures the probability of a certain class (namely, a DNS group) is chosen; $P(u_m \mid z_k)$, the probability of observing a user session given a certain class; and $P(d_n \mid z_k)$, the probability of a domain name being visited given a certain class. By applying Bayes' rule to these probabilities, we can generate the probability that a certain group is chosen given an observed user:

$$P(z_k \mid u_m) = \frac{P(u_m \mid z_k)P(z_k)}{\sum_{k'=1}^{K} P(u_m \mid z_{k'})P(z_{k'})}, \tag{10}$$

and the probability that a certain group is chosen given an observed domain name:

$$P(z_k \mid d_n) = \frac{P(d_n \mid z_k)P(z_k)}{\sum_{k'=1}^{K} P(d_n \mid z_{k'})P(z_{k'})}. \tag{11}$$

Based on these two probabilities, we can then classify each user and domain name into different groups respectively. In user case, this is achieved by firstly computing $P(z_k \mid u_m)$ for all possible values of k and then, u_m can be classified into the group whose $P(z_k \mid u_m)$ is the largest. Similarly, the group belongingness for domain name d_n can be determined as the k with the largest value of $P(z_k \mid d_n)$. In Section 4, we will present the identified groups in DNS activities from our real-world data sets.

3.2 Evaluation of Grouping Clarity

An intuitive definition of cluster in connectivity-based data, such as group, is a subset of objects whose total number of internal links is greater than the total number of external links [10]. In order to evaluate the quality of our DNS group clarity here, we compute numeric values that express the goodness of grouping according to the above definition. Let the intra-connectivity C_{ki} of a group k be the density of links between objects of that group, while the inter-connectivity C_{ko} be the density of links among objects of the k-th group and the rest of the inferred topology. That is,

$$C_{ki} = \frac{\sum_{m \in U_k, m \in D_k} w(u_m, d_n)}{|U_k||D_k|}, \tag{12}$$

$$C_{ko} = \frac{\sum_{k'=1,k'\neq k}^{K} \left(\sum_{m \in U_k, n \in D_{k'}} w(u_m, d_n) + \sum_{m \in U_{k'}, n \in D_k} w(u_m, d_n)\right)}{\sum_{k'=1,k'\neq k}^{K} \left(|U_k||D_{k'}| + |U_{k'}||D_k|\right)}, \tag{13}$$

where U_k and D_k denotes the set of users and names for group k, respectively. Hence the overall grouping clarity C_o is obtained by averaging over the clarity of all groups,

$$C_o = \frac{1}{K}\sum_{k=1}^{K} \frac{C_{ki}}{C_{ko}}. \tag{14}$$

3.3 Characterizing Groups in DNS Activities

To better understand these users' preferences and interests, we try to characterize each group in a way that is easy to interpret. One possible approach is to find the *characteristic* users and domain names that are strongly associated with a given group, yet not commonly identified as part of other groups. We call each such domain name a *characteristic* domain name for this group, denoted by d_{ch}. This definition of "*characteristic*" here has two aspects of meaning. First, given a group, a domain name which is seldom visited by users of this group cannot be a good *characteristic* domain name for this group. Secondly, if a domain name is frequently visited in this group, but is also commonly visited in other groups, the domain name is not a good *characteristic* domain name either. Note that $P(d_n \mid z_k)$ represents the conditional occurrence probability over the domain name space corresponding to a specific group, whereas $P(z_k \mid d_n)$ represents the conditional probability distribution over the group space corresponding to a specific domain name, so we define *characteristic* domain names for a group z_k as the set of all domain names, d_{ch}, which satisfy:

$$P(d_{ch} \mid z_k)P(z_k \mid d_{ch}) \geq \mu \ , \tag{15}$$

where μ is a predefined threshold.

A similar approach can be used to identify *characteristic* users for each group. We believe that a user involving only one group can be considered as the *characteristic* user for the group. So, we define *characteristic* users, u_{ch}, for a group z_k, as users which satisfy

$$P(u_{ch} \mid z_k)P(z_k \mid u_{ch}) \geq \mu \ . \tag{16}$$

By exploring and interpreting the properties of these domain names and users, we can obtain a better understanding of the nature of each group. The characterization of the identified groups from our real-world data sets will be shown next.

4 Experiments

In order to evaluate the effectiveness of the proposed method based on PLSA model and explore the discovered groups, we conduct preliminary experiments on the real-world DNS query data which is collected from a large Internet Service Provider in China. The data collection procedure has lasted for over a week from *10-June-2009 15:48:33.696* to *17-June-2009 18:05:12.119*. The total log file is over 12.6 Giga Bytes containing 147,128,488 queries. After data preprocessing, 141,331 users and 68,885 second-level domain[1] names are extracted.

We randomly sample 20,000 users as well as 50,000 domain names that are queried by at least one of these users. With these objects, the original co-occurrence matrix *UD* is given in Fig. 3. From this figure, we can see that the co-occurrence matrix is quite sparse in which the percentage of non-zero values is only 0.037% here. Based on this matrix data, we conduct our experiments to cluster these objects (namely,

[1] A second-level domain (SLD) is a domain that is directly below a top-level domain (TLD). For example, the domain name *example.com* is the second-level domain of the *.com* TLD. Specially, for those names under the country-code SLD (ccSLD) such as *.com.cn*, we convert them into third-level domains (such as *sina.com.cn*) instead of ccSLDs.

users and domain names) into different DNS groups based on our method described in Section 3.1. By applying the grouping clarity evaluation introduced in Section 3.2, we compute the overall grouping clarity C_o for each possible[2] K. As illustrated in Fig. 4, the overall grouping clarity C_o reaches its maximum when $K = 15$.

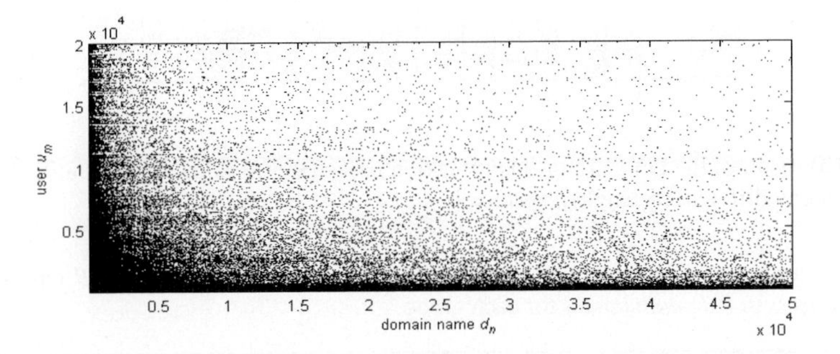

Fig. 3. The original co-occurrence matrix *UD* with a size of *20,000×50,000* where users and domain names are both ranked by their number of relevant queries

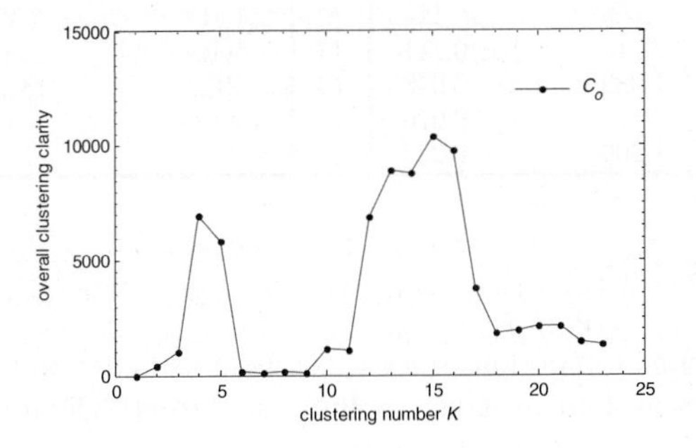

Fig. 4. The overall grouping clarity C_O for each possible grouping number K ($1 \leq K \leq 23$), where C_O reaches its best when $K = 15$

Accordingly, the grouping result of the DNS groups identification procedure in the case of $K = 15$ is depicted in Fig. 5 where the fifteen identified DNS groups are marked out by different rectangles respectively.

To be more intuitive, we list the number of users and domain names for each group in Table 1. From this table, we can see clearly that different groups can vary a lot in their number of users as well as domain names. Generally, we can divide these groups into three different categories according to their size of ratio R:

1. groups with $R<0.1$ ($k = 1, 2, 6, 7, 8, 10, 11, 14$),
2. groups with $0.1 \leq R < 1$ ($k = 3, 4, 5, 9, 12$),
3. and groups with $R \geq 1$ ($k = 13, 15$).

[2] Meaningless groups with no user or domain name will appear in our grouping results when $K > 23$.

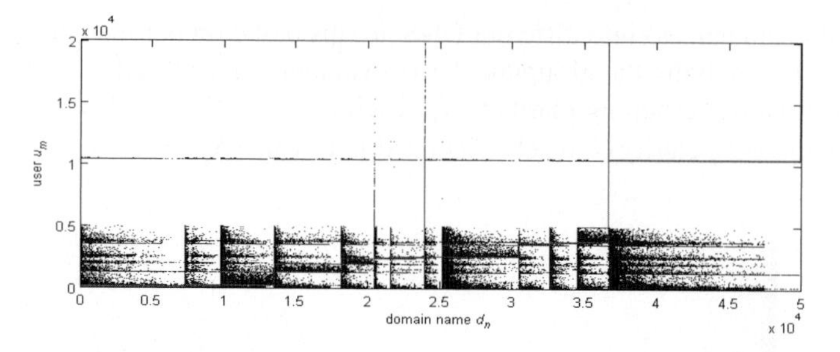

Fig. 5. The result of the DNS groups identification procedure ($K = 15$). Each identified group is marked out by different rectangle respectively.

Table 1. Number of users and domain names for each of the fifteen groups. R represents the ratio of *# users* to *# domain names* for each group.

k	*# users*	*# domain names*	*R*	*k*	*# users*	*# domain names*	*R*
1	99	7,377	0.013	*9*	969	5,295	0.183
2	119	2,419	0.049	*10*	168	2,203	0.076
3	1,024	3,770	0.272	*11*	87	1,860	0.047
4	711	4,608	0.154	*12*	1,188	2,173	0.547
5	488	2,310	0.211	*13*	5,404	8	676
6	52	1,088	0.048	*14*	102	13,262	0.008
7	63	2,417	0.026	*15*	9,493	1	9493
8	33	1,209	0.027				

5 Explanations

Group characterization is important since it can help us better understand users' preferences and interests in domain names within each group. Obviously, *characteristic* users and domain names' extraction and interpretation are effective ways to do this. By conducting the method proposed in Section 3.3, *characteristic* users and names for each of these groups are extracted where the threshold μ is predefined to be 0.01. Here, "*characteristic*" has two implications. First, if a user is classified into one group, he/she is likely to visit the *characteristic* domain names of this group. Secondly, if we find a user visits the *characteristic* domain names of a group, he/she is very likely to be classified into this group. In this section, based on the results exhibited above, we try to interpret each of these groups respectively.

$k = 1$: *IDC service.* The $k = 1$ group has only 99 users and yet as many as 7,377 domain names. As a typical group under the 1st category, the value of ratio R for this group is only 0.013 here, being the second lowest of all. There are only one *characteristic* domain name together with four *characteristic* users identified in this group. By checking this domain name *21okcdn.cn*, we find it belongs to a largest third-party independent Internet Data Center service provider in China who provides server placement for its clients. Note that the number of domain names in this group is large, thus we can consider these names to be clients of this IDC service provider.

$k = 2$: *BitTorrent.* The $k = 2$ group has four *characteristic* names, three of which are used as platform for BitTorrent seeds releasing. Therefore, we consider this group as BitTorrent interested group. Only 119 users are being classified into this group, but obviously they have stronger interests in BitTorrent related activities than the others.

$k = 3$: *living*. For the group of $k = 3$ with 1,024 users and 3,770 names, three *characteristic* users and names are identified respectively, but no obvious property is shared by them. By mapping all of these 1,024 users' IP addresses into their belonging organizations using MaxMind's *GeoIP* Organization database [11], we find that the organization belongingness of these users is extremely disperse. On the other hand, the domain names in this group also vary a lot in their uses from social network service to IT product purchasing. In other words, objects in this group are quite heterogeneous and diverse. Therefore, we consider groups of this category as "*living*" groups where both users and domain names are closer to people's living activities.

$k = 4, 5, 9, 12$: *living*. Just like the one $k = 3$, these four groups are also under the 2nd category. Again, no obvious property is found to be shared by any of these *characteristic* names within each group. The only one name which is used for online game (*w2i.com.cn*) is identified to be *characteristic* for $k = 4$. For group $k = 5$, the two *characteristic* names identified are used for B2C (*alimama.cn*) and BBS (*tianya.cn*) service respectively. For the other two groups $k = 9, 12$, applications of these *characteristic* names also vary from portal to stock exchange, from online television to anti-virus services, etc. Similar to the group $k = 3$, we label these four groups with "*living*" here. For all these five groups labeled with "*living*", we find they are similar in many ways such as distributions of users' organization, as well as names' popularities and categories[3]. On the other hand, since they are of separate groups, we believe there must be some difference among them which needs further analysis.

$k = 6$: *domain service*. The $k = 6$ group is another one under the 1st category. Only one name is identified to be *characteristic*, which is a largest domain trading agent in China providing professional domain services such as domain registration, domain parking and domain transaction, etc. Therefore, we can consider these names and users within this group to be active in this domain service.

$k = 7, 10$: *anti-spam*. The reason why we use "*anti-spam*" to generalize both of these two groups here is that they both have a *characteristic* domain name used for anti-spamming service respectively. For the $k = 7$ group, the *characteristic* name *163data.com.cn* is used by a domestic ISP *China Telecom* for spam filtering through IP address reverse resolution mechanisms. For the $k = 10$ group, the *characteristic* name *anti-spam.org.cn* is owned by *China Anti-Spam Alliance*, which is a community organization aiming at anti-spam issues. As a result, users and domain names within these groups can be deemed to be somewhat related to spam or anti-spam activities.

$k = 8$: *.CN TLD service*. The $k = 8$ group has three *characteristic* names, of which *cnnic.cn* is the biggest in its probabilistic weight. Note that *cnnic.cn* is the domain name for *CNNIC* who is the national network information center of China responsible for the operation of .CN TLD services. Therefore, we can consider users within this group have stronger interest in .CN TLD related activities such as .CN domain name's registration and *whois* searching, etc. And these domain names within this group can be exactly considered as the ones that these users are interested in.

---

[3] By using the domain name registration information database provided by *CNNIC* [12], we make classification of domain names within each group into eighteen different categories (such as *information technology*, *scientific research and education*, *business commerce*, etc).

k = 11, 14: *abuse.* The value R for the k = 14 group is the lowest of all with only 102 users and yet as many as 13,262 domain names being grouped in. By checking our query data, we find that all of these *characteristic* users have a large number of queries, as well as unique domain names that they visit. However, no *characteristic* name is identified for this group. In other words, users within this group query names randomly and arbitrarily, just like gadabouts on the Internet with no specific intension. Similar phenomenon is also found in the group k = 11 which has no *characteristic* name either. By checking these users in these groups, we find that most of their volumes of queries are high, and so are their numbers of unique names that they queried. There are at least three possible reasons that can explain for this phenomenon: web crawling, domain name scanning, or some kind of DDoS attacks [13] by querying stochastic names.

k =13, 15: *DNS forwarding service.* As groups under the 3rd category which have extremely high value of ratio R, these two groups are the most interesting ones of all. We find that most of users in these groups are from abroad. Some interesting phenomenon is also found in their *characteristic* users and names. On one hand, all of the *characteristic* users for k =15 (no *characteristic* user is identified for group k =13) belong to a same organization called *OpenDNS* [14], an organization who provides free DNS resolution service. On the other hand, the unique *characteristic* name for k =15, namely, *cstnet.cn*, is the domain of *CSTNET* [15], a major DNS resolution service provider in China. As DNS resolution service providers themselves, why could they exhibit such strong interest in other similar providers? Similar phenomenon is also found in the group k =13 where the two *characteristic* names *cstnet.net.cn* and *cnc.ac.cn* are both domains of *CSTNET* again. One possible explanation is that these users are forwarding resolution service to the others. In other words, these users are not real Internet end users at all, but some DNS resolution service providers, or simply some DNS resolution service forwarders.

From our explanations, we can see that these DNS groups identified in our data sets vary a lot between each other in many ways. Generally, they can be divided into three different categories as shown in Section 4. For the groups with ratio $R<0.1$, there are usually some *characteristic* names providing some special services which have high popularities within these groups. For the groups with $0.1\leq R<1$, they are all labeled as "*living*" which usually have a large number of users as well as domain names. And for these two groups under the last category with extremely high value of ratio R, they are both considered as groups related to DNS forwarding services.

6 Conclusions and Future Work

In this paper, we propose a novel approach to the discovery of Internet usage patterns at the domain name level. Here, the Internet user's domain name access behaviors are summarized as the co-occurrences of users and domain names. Then, an aspect model is introduced to classify users and domain names into various groups according to their co-occurrences. Meanwhile, each group is characterized by extracting the property of *characteristic* users and domain names. The effectiveness of our approach is demonstrated through experiments conducted on real-world data sets in which some meaningful usage patterns for Internet users at the domain name level are identified.

Therefore, our approach could be used for detecting unusual behaviors on the Internet at the domain name level, which can significantly alleviate the work of searching the joint space of users and domain names.

On the other hand, our work is still rough where further analysis is necessary in order to gain deeper understanding of these groups in DNS activities. In the future work of this area, we plan to conduct more research on using our proposed framework to discover in-depth usage patterns which can involve users, domain names, and some additional semantic attributes such as user's geographical location information, domain's category, etc., thus capturing users' preferences and interests at a deeper level. Furthermore, our investigation into the DNS groups identified here are still rough, as well as the DNS groups themselves. Therefore, it will be another interesting task to perform further analysis within each of these groups recursively.

Acknowledgments. This work is supported by the China Next Generation Internet Project (Project name: the Next Generation Internet's Trusted Domain Name System; No.: CNGI-09-03-04).

References

1. Srivastava, J., Cooley, R., Deshpande, M.: Web Usage Mining: Discovery and Applications of Usage Patterns from Web Data. SIGKDD Explorations Newsletter 1, 12–23 (2000)
2. Eirinaki, M., Vazirgiannis, M.: Web Mining for Web Personalization. ACM Transactions on Internet Technology 3, 1–27 (2003)
3. Getoor, L., Diehl, C.P.: Link Mining: a Survey. ACM SIGKDD Explorations Newsletter 7, 3–12 (2005)
4. Kohavi, R., Mason, L., Parekh, R., Zheng, Z.: Lessons and Challenges from Mining Retail E-Commerce Data. Machine Learning 57, 83–113 (2004)
5. Mockapetris, P.: Domain Names: Concepts and Facilities. Internet Request for Comments 1034 (1987)
6. Hofmann, T.: Probabilistic Latent Semantic Analysis. In: 15th Conference on Uncertainty in Artificial Intelligence, Stockholm (1999)
7. Hofmann, T.: Probabilistic Latent Semantic Analysis. In: 22nd Annual ACM Conference on Research and Development in Information Retrieval. ACM Press, Berkeley (1999)
8. Hofmann, T.: Latent Semantic Models for Collaborative Filtering. ACM Transactions on Information Systems 22, 89–115 (2004)
9. Dempster, A., Laird, N., Rubin, D.: Maximum Likelihood from Incomplete Data via the EM Algorithm. Journal of Royal Statistical Society B(39), 1–38 (1977)
10. Newman, M.E.J.: Detecting Community Structure in Networks. Eur. Phys. J. B. 38, 321–330 (2004)
11. MaxMind, `http://www.maxmind.com`
12. CNNIC, `http://www.cnnic.cn`
13. Mirkovic, J., Reiher, P.: A Taxonomy of DDoS Attacks and Defense Mechanisms. ACM SIGCOMM Computer Communication Review 34, 39–53 (2004)
14. OpenDNS, `http://www.opendns.com`
15. CSTNET, `http://www.cstnet.cn`

A New Statistical Approach to DNS Traffic Anomaly Detection

Xuebiao Yuchi[1,2], Xin Wang[1], Xiaodong Lee[1], and Baoping Yan[1]

[1] China Internet Network Information Center, Computer Network Information Center, Chinese Academy of Sciences, 100190 Beijing, China
[2] Graduate University of Chinese Academy of Sciences, 100190 Beijing, China
{yuchixuebiao,wangxin,lee}@cnnic.cn, ybp@cnic.cn

Abstract. In this paper, we describe a new statistical approach to detect traffic anomalies in the Domain Name System (DNS). By analyzing real-world DNS traffic data collected at some large DNS servers both authoritative and local, we find that normally the DNS traffic follows Heap's law in dual ways. Then we utilize these findings to characterize DNS traffic properties under normal network conditions. Based on these properties, we make estimations for the traffic of forthcoming. If the forthcoming traffic actually varies a lot with our estimations, then we can infer that some anomaly happens. Our approach is simple enough and can work in real-time. Experiments on both real and simulated DNS traffic anomalies show that our approach can detect most of the common anomalies in DNS traffic effectively.

Keywords: Domain Name System, Anomaly detection, Heap's law.

1 Introduction

The Domain Name System (DNS) is a fundamental part of the Internet infrastructure, which is responsible for translating domain names used by people into corresponding IP addresses needed by software and vice versa [1]. Thus, its functionality is a critical component to almost all Internet applications where two devices need to be connected remotely, such as Web, Email and P2P, etc. Due to its special significance for the Internet, DNS is always suffering various network attacks, either directly or indirectly. Therefore, DNS traffic anomaly identification and detection is important for the entire Internet as well as the DNS itself.

In this paper, based on the observations of Heap's law that we find in DNS traffic, we introduce a new statistical approach to DNS traffic anomaly detection. Our approach is simple enough and can work in real-time. Experimental results suggest that our approach is effective and have the ability to detect some common anomalies in DNS traffic.

This paper is structured as follows. In Section 2 we discuss related work. In Section 3 an overview of DNS and DNS traffic anomaly as well as Heap's law is presented. Then we give our findings of Heap's law in the DNS traffic data collected at both authoritative and local DNS servers in Section 4. On this basis, the method for DNS traffic anomaly detection is introduced in Section 5, and in Section 6 experimental

L. Cao, J. Zhong, and Y. Feng (Eds.): ADMA 2010, Part II, LNCS 6441, pp. 302–313, 2010.

results on some typical DNS anomalies, both real and simulated, are presented. Finally, we conclude our work in Section 7.

2 Related Work

DNS traffic monitoring and anomaly detection has attracted much more attention recently. In [2], a detection method based on associative feature analysis is presented to locate the anomalous name servers by analyzing the real-world DNS traffic. A context-aware clustering methodology is described in [3] where DNS traffic is clustered into three categories: canonical, overloaded and unwanted. Other methods such as Bayesian approach [4], time series analysis [5] are also referred in DNS traffic anomaly detection technique.

In general, these methods are carefully designed in order to have promising detection result. However, they can only do well in their own scenarios concentrating on detecting some particular anomalies. Moreover, their deployment issue is also a hot potato for most DNS operators without such domain-specific knowledge. Therefore, we expect a new solution that can be deployed easily enough and has the ability to detect potential DNS anomalies as quickly as possible.

3 Background

3.1 DNS and DNS Traffic Anomaly

The DNS is a distributed naming service handling domain name and IP (Internet Protocol) address resolution, and has grown into one of the largest distributed systems in the world. The DNS infrastructure consists of three different types of components: stub resolvers (usually a software library implementing the DNS protocol on end-user machines), local DNS servers (usually deployed by the resolver's organization such as a company or an ISP), and authoritative DNS servers, of which the Root, TLD (Top Level Domain) and SLD (Second Level Domain) servers are of the three different cases.

Fig. 1 shows the interactions among these three components during the process of a typical name lookup. A stub resolver first sends a query to a configured local DNS server (step 1). The local DNS server then iteratively sends queries on behalf of the resolver, following referrals given by the responses until it receives an authoritative response for this query (steps 2-4). Finally, it can respond to the querying resolver with this answer (step 5), enabling the user's program to continue the task it was performing.

Note that the DNS lookup procedure described above doesn't take caching mechanism into consideration. In fact, the caching mechanism [6] performed on local DNS servers is a key characteristic that ensure the efficiency of DNS infrastructure by reducing server load and client latency. In other words, the local DNS server can directly skip any of these intermediate steps (steps 2-4) if it has already had their corresponding responses locally cached. The size of DNS caches can be treated as unlimited due to the few bytes of a DNS record entry. However, each record has a time to live (TTL) value that specifies when it expires from the local DNS server's cache.

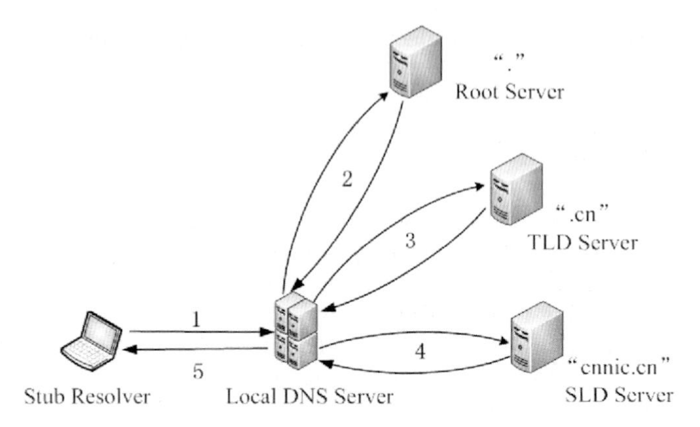

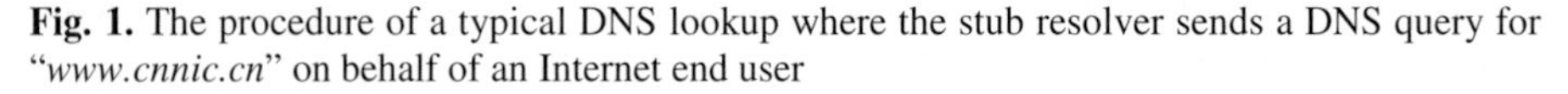

Fig. 1. The procedure of a typical DNS lookup where the stub resolver sends a DNS query for "*www.cnnic.cn*" on behalf of an Internet end user

Due to its significance to the Internet, DNS servers, both authoritative and local, have increasingly become the victims of various network attacks, such as Distributed Denial of Service (DDoS) attacks [7]. Usually these attacks try to launch tremendous bogus traffic towards targeting DNS servers for different purposes. In addition, even if their targets are not DNS servers, the DNS is also likely to be involved in trouble, which is because that DNS service is referred to by nearly all the Internet activities. For example, if the target of an attack is specified by a domain name instead of the IP address, there will be a great number of DNS queries for this name sending towards DNS servers while this attack is performing.

3.2 Heap's Law

Heap's law was first introduced in linguistics [8], which states that asymptotically, a corpus of text containing N words typically contains on the order of CN^{β} distinct words, with $0 < \beta < 1$, namely:

$$V = CN^{\beta} , \tag{1}$$

where C represents a constant and V represents the number of distinct vocabulary words present in the text. Taken logarithmic transformation on both sides, (1) can be rewritten as:

$$\log(V) = \beta \log(N) + K , \tag{2}$$

where $K = \log(C)$. As illustrated in Fig. 2, Heap's law means that as more instance text is gathered, there will be diminishing returns in terms of discovery of the full vocabulary from which the distinct terms are drawn.

Empirically, the exponent β was found to fall in the range of 0.4 and 0.6 by some authors [9], varying somewhat with the language, the type of data set, and so forth. Using the Web data set from the Web-KB (World Wide Knowledge Base) project at Carnegie Mellon University which consists of several thousand Web pages collected from Computer Science departments of various universities, researchers find a characteristic exponent equal to 0.76 [10].

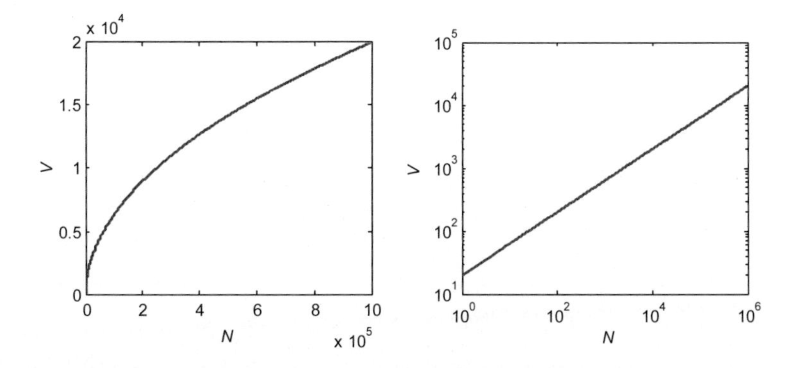

Fig. 2. A typical Heap's law plot with parameters $\beta = 0.5$ and $C = 20$. The x-axis represents the text size, and the y-axis represents the number of distinct vocabulary elements present in the text. Left: linear scale; Right: log scale.

4 Heap's Law Observed in DNS Traffic Data

A typical DNS query contains a timestamp, the sender's source IP address, the source port number, the query name to be resolved, the query class and the query type, etc. By analyzing the DNS query data collected at an authoritative DNS server of China's country code TLD *.cn* which is one of the largest TLDs in the world [11], we surprisingly find that under normal network conditions, the DNS query flow also follows Heap's law by treating source IP addresses or requested SLD names as words from vocabulary and queries as text. Fig. 3 shows our observations during a two-day's period, from which we can see that there are two Heap's law observations found in the DNS query data: the total number of DNS queries N versus the number of distinct SLD names queried V_n ; the other one is the total number of DNS queries N versus the number of distinct local DNS servers V_l .

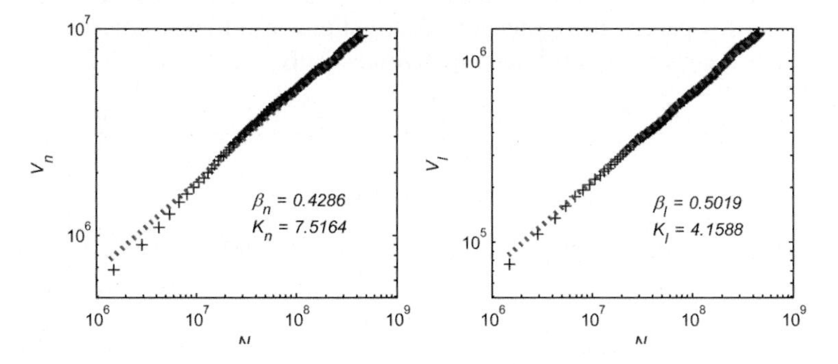

Fig. 3. Heap's law observations in the DNS query data collected on an authoritative DNS server of *.cn* TLD during a two-day's period: log-log plot of the total number of DNS queries N versus the number of distinct SLD names queried V_n (left), and the total number of DNS queries N versus the number of distinct local DNS servers presented V_l (right)

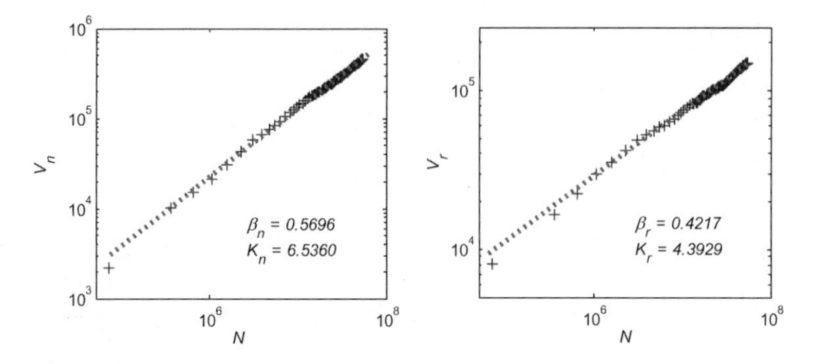

Fig. 4. Heap's law observations in the DNS query data collected on a local DNS server of *cstnet* during a three-day's period: log-log plot of the total number of DNS queries N versus the number of distinct SLD names queried V_n (left), and the total number of DNS queries N versus the number of distinct stub resolvers presented V_r (right)

Similar observations are also identified in the DNS query data collected during a three-day's period on a local DNS server of *cstnet* [12], one of the largest Internet Service Providers in China. As illustrated in Fig. 4, one exists in N and V_n with parameters $\beta_n = 0.5596$, $K_n = 6.5360$; the other one exists in N and the number of distinct stub resolvers presented V_r with parameters $\beta_r = 0.4217$, $K_r = 4.3929$.

DNS is a complex system containing millions of different objects (stub resolvers, local DNS servers and authoritative DNS servers) interacting with each other. Take the *.cn* dataset for example, there are nearly 458 million queries generated by over 1.4 million local DNS servers querying for up to 9.3 million SLD names. Meanwhile, DNS itself is also a disparate system in which access popularities among these nodes vary a lot. Our recent studies [13] [14] show that typically 0.1% of the busiest local DNS servers (we call them heavy hitters) can contribute over 50% to the overall queries received by the *.cn* authoritative DNS servers. Similarly, 0.1% of the most popular SLD names can attract nearly one half of the overall queries. For the two data sets we utilize in this paper, their popularity distributions of different objects in order of access counts are given in Fig. 5 and Fig. 6 respectively.

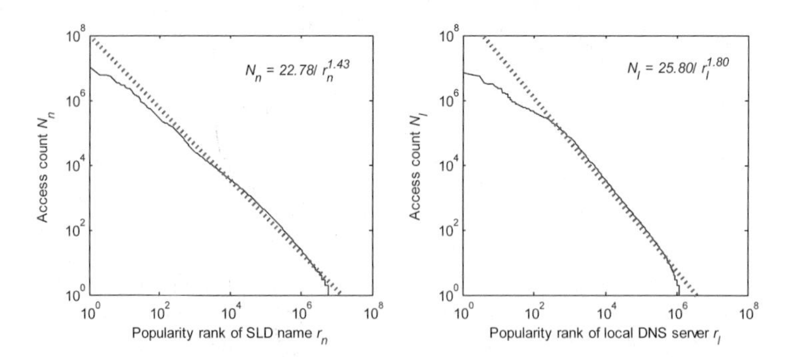

Fig. 5. Popularity distributions of SLD names (left) and local DNS servers (right) in the data set of *.cn*, both in log-log scale

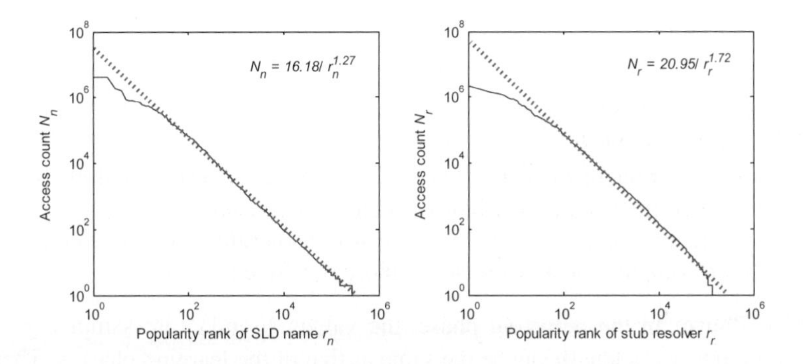

Fig. 6. Popularity distributions of SLD names (left) and stub resolvers (right) in the data set of *cstnet*, both in log-log scale

From above we can see that, similar to many other Internet systems, DNS is observed to be following Zipf's law, a well-known law describing the frequency distribution of different items in an itemset [15]. Recent studies [16] show that these two laws are related and the Heap's law can be considered as a derivative phenomenon if the system obeys the Zipf's law. Therefore, our Heap's law observations in the DNS traffic are not found occasionally. As many other complex systems [17], the two experimental laws appear together in DNS. Moreover, compared with the Zipf's law, the formula form of Heap's law makes itself easier for temporal characterization of the DNS query data.

5 Detection Method

The above section tells us that normally, the DNS query data on both authoritative and local DNS servers follow the Heap's law. In other words, the observed value $(V_n)_t$ at time t can be estimated to be $(V'_n)_t$ by:

$$\log(V'_n)_t = \beta_n \log(N_t) + K_n \ . \tag{3}$$

Similarly, the estimated value $(V'_l)_t$ for $(V_l)_t$ and $(V'_r)_t$ for $(V_r)_t$ can be obtained respectively by:

$$\log(V'_l)_t = \beta_l \log(N_t) + K_l \ , \tag{4}$$

$$\log(V'_r)_t = \beta_r \log(N_t) + K_r \ . \tag{5}$$

Under normal situations, the error between the value observed and the estimated one should be small. Therefore, once the error in any one of these two observations exceeds some threshold, we can infer that some traffic anomaly happens. On this basis, the DNS traffic anomaly detection procedure based on N and V_n can be described as follows (the detection procedure based on N and V_l is similar with no need to be detailed here).

Learning Phase: First, we periodically sample a series of value pairs $< N_t, (V_n)_t >$ (t = 1, 2, …, m) under normal network conditions, where m represents the number of value pairs we sample during the learning phase. Based on these value pairs, we calculate the parameters for this Heap's law observation (namely, β_n , K_n and the threshold Y_n ; the calculation of Y_n is detailed in the appendix). The sampling interval's length can be specified as a time window, such as one or ten minutes, being a trade-off between detective sensitivity and performance. Generally, shorter sampling intervals incur more computation and yet can be more sensitive to change.

Detecting Phase: In the detecting phase, the values N and V_n are sampled periodically, whose interval's length can be the same as that of the learning phase's. Then the estimated value V'_n for the observed V_n can be calculated by (3). If the error between V'_n and V_n exceed the threshold at some time t, namely (6) is satisfied, and then the anomaly alarm can be launched.

The overall anomaly detection procedure on DNS servers is depicted in Fig. 7. Note that the detection procedures based on the two Heap's law observations are carried out concurrently here. Therefore, the anomaly alarm would be triggered if any one of these two inequalities below is established:

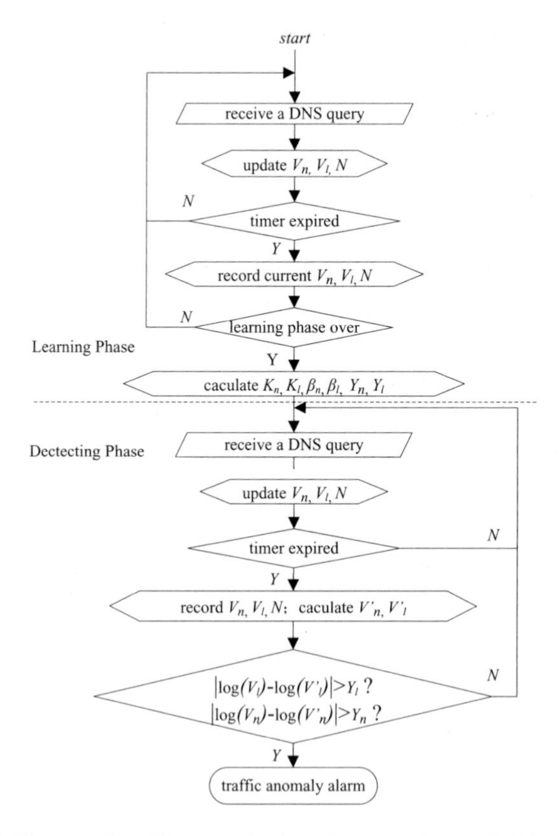

Fig. 7. The overall traffic anomaly detection procedure on DNS servers

$$|\log(V_n)_t - \log(V'_n)_t| > Y_n \ , \tag{6}$$

$$|\log(V_l)_t - \log(V'_l)_t| > Y_l \ . \tag{7}$$

6 Experiments

In this section, we present the detective results on some typical DNS anomalies, both real and simulated.

6.1 China's May 19 DNS Collapse

The real data set was collected on May 19, 2009 when a large population of Chinese Internet users suffered an Internet accessing problem late that day. This problem originally began with the collapse of a famous DNS service provider *DNSPod* due to some hacker's attack. *DNSPod* [18] is authoritative for thousands of domain names, some of which are very popular in China. The collapse further led to tremendous repeated DNS queries targeting these domain names flooding over the Internet yet with no authoritative responses, and DNS flooding happened. It is reported that there were a large number of local DNS servers collapsed during this accident. More details can be found in [19].

If this accident was detected enough early, its disastrous consequences might be well mitigated by taking some necessary measures such as traffic filtering. Note that the local DNS servers began querying for these names continually on behalf of end users after their corresponding record entries expired. These queries would be forwarded to the authoritative DNS servers as illustrated in Fig. 1. Therefore, if we could perceive these anomalous queries early enough at the DNS servers' side whether authoritative or local, it is possible for us to avoid the collapse of these local DNS servers and potential danger to the authoritative DNS servers.

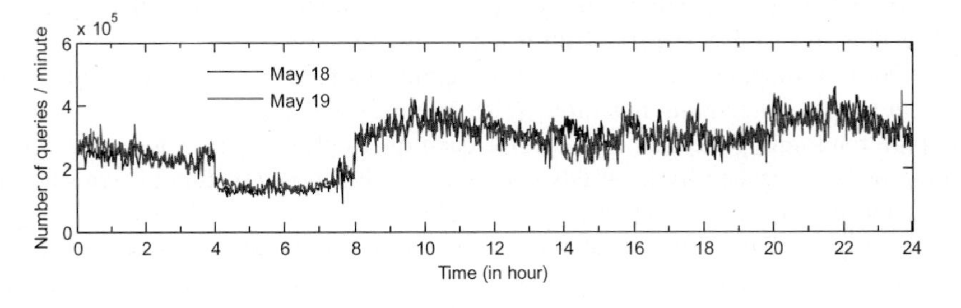

Fig. 8. DNS traffic query rate curves at an authoritative DNS server for *.cn* on May 18 and May 19, respectively

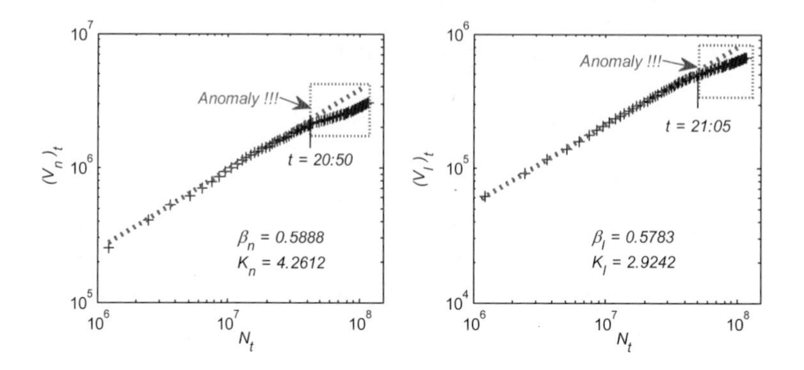

Fig. 9. Log-log plots of N_t vs. $(V_n)_t$ (left) and N_t vs. $(V_l)_t$ (right) at an authoritative DNS server for *.cn* during 18:00~24:00, May 19, 2009, where our sampling interval is five minutes' length

Firstly, in Fig. 8 we give the DNS query rate distributions of the authoritative DNS servers for *.cn* during a period of two days, namely, 00:00 18 May - 24:00 19 May, 2009. We can see that no obvious anomaly is found during this collapse day in compared with normal day. However, from Fig. 9 we can see that, from about 9:00 p.m. when DNS flooding began, there is an obvious inflexion in the left plot. Obviously, a large number of repeated queries began arriving at the authoritative DNS servers continuously, which would result in $\log(V'_n)_t - \log(V_n)_t > Y_n$ established, being a trigger of the anomaly alarm. Note that these repeated queries are mainly generated by existing local DNS servers, so anomaly can also be found in the right plot, where $\log(V'_l)_t - \log(V_l)_t | > Y_l$ establishes.

6.2 DDoS Attack by Querying Non-existent Domain Names

The purpose of the DDoS attack is to consume the target server's resource as much as possible, until the target collapses. Note that if a record entry has already been cached locally, there is no need for the local DNS server to launch additional queries to authoritative DNS servers for this entry as long as this entry is not yet expired. Therefore, the load on authoritative DNS servers can be greatly alleviated thanks to the DNS caching mechanism deployed at the local DNS servers.

In order to eliminate the effect of DNS caching, more and more attackers tend to carry out attacks by sending numerous DNS queries for non-existent domain names, which are randomly generated and haven't been queried by others before. Therefore they are unlikely cached by local DNS servers and the corresponding hit rates at the target authoritative DNS servers are greatly increased.

A simulation of this kind of attacks is set up in our intranet in which attackers launch a two hours' attack towards the authoritative DNS server utilizing a DNS performance testing tool *queryperf* [20]. During this attack, queries are launched at about 1,000 q/s with their querying names randomly generated suffixed with "*.cn*". Meanwhile, both the normal and bogus queries are captured by the logging function of DNS implementation *BIND9* [21] as depicted in Fig. 10.

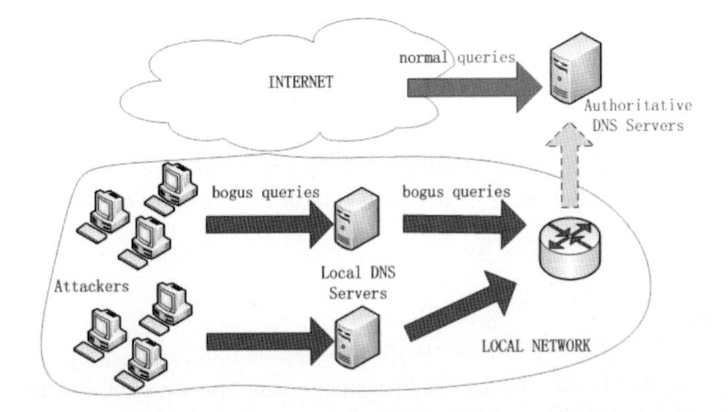

Fig. 10. Illustration for our simulation of a DDoS attack by querying non-existent domain names

The detective results for this simulation are given in Fig. 11. We can see that when the attack begins, there will be a dramatic increase in the number of distinct SLD names V_n which results in $\log(V_n)_t - \log(V'_n)_t > Y_n$ established, and the anomaly alarm can be launched. Note that anomalies of this kind can't be detected by the right plot because there is no obvious abnormal in either the total number of queries N_t or the number of distinct local DNS servers $(V_l)_t$.

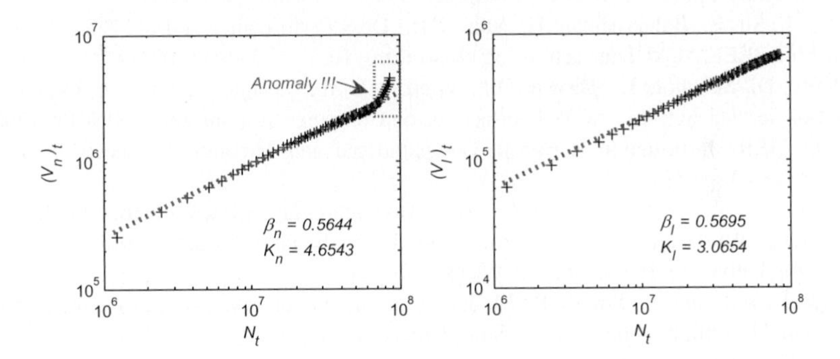

Fig. 11. Log-log plots of N_t vs. $(V_n)_t$ (left) and N_t vs. $(V_l)_t$ (right) during our simulation of DDoS attack by querying non-existence names

7 Conclusion

DNS servers, especially for larger ones, are responsible for most of the DNS resolution over the Internet and they are usually more attractive to attackers with different purposes. According to our measurements of the DNS traffic collected at some large DNS servers both authoritative and local, we find some common observations of Heap's law that exist in the DNS traffic under normal network conditions. Based on the observations, a new statistical approach to DNS traffic anomaly detection was proposed, which is simple and can be easily deployed at both authoritative and local DNS servers. Experimental results show that the approach can detect some common anomalies effectively.

Acknowledgments. This work is supported by the China Next Generation Internet Project (Project name: the Next Generation Internet's Trusted Domain Name System; No.: CNGI-09-03-04).

References

1. Mockapetris, P.: Domain Names: Implementation and Specification. Internet Request for Comments 1035 (1987)
2. Wang, Y., Hu, M., Li, B., Yan, B.: Tracking Anomalous Behaviors of Name Servers by Mining DNS Traffic. In: Min, G., Di Martino, B., Yang, L.T., Guo, M., Rünger, G. (eds.) ISPA Workshops 2006. LNCS, vol. 4331, pp. 351–357. Springer, Heidelberg (2006)
3. Plonka, D., Barford, P.: Context-aware Clustering of DNS Query Traffic. In: 8th ACM SIGCOMM Internet Measurement Conference, pp. 217–230. ACM, New York (2008)
4. Villamarín-Salomón, R., Carlos Brustoloni, J.: Bayesian Bot Detection Based on DNS Traffic Similarity. In: 2009 ACM Symposium on Applied Computing, pp. 2035–2041. ACM, New York (2009)
5. Chatzis, N., Brownlee, N.: Similarity Search over DNS Query Streams for Email Worm Detection. In: 2009 International Conference on Advanced Information Networking and Applications, pp. 588–595. IEEE Computer Society, Washington (2009)
6. Jung, J., Sit, E., Balakrishnan, H., Morris, R.: DNS Performance and the Effectiveness of Caching. IEEE/ACM Transactions on Networking 10(5), 589–603 (2002)
7. Moore, D., Shannon, C., Brown, D.J., Voelker, G.M., Savage, S.: Inferring Internet Denial-of-Service Activity. ACM Transactions on Computer Systems 24(2), 115–139 (2006)
8. Heaps, H.S.: Information Retrieval: Computational and Theoretical Aspects. Academic Press, New York (1978)
9. Araújo, M., Navarro, G., Ziviani, N.: Large Text Searching Allowing Errors. In: 4th South American Workshop on String Processing. International Informatics Series, pp. 2–20. Carleton University Press, Ottawa (1997)
10. Baldi, P., Frasconi, P., Smyth, P.: Modeling the Internet and the Web: Probabilistic Methods and Algorithms. John Wiley & Sons, Chichester (2003)
11. CNNIC, `http://www.cnnic.cn`
12. CSTNET, `http://www.cstnet.cn`
13. Yuchi, X., Wang, X., Lee, X., Yan, B.: DNS Measurements at the. CN TLD Servers. In: 6th International Conference on Fuzzy Systems and Knowledge Discovery, vol. 7, pp. 540–545. IEEE Press, Piscataway (2009)
14. Yuchi, X., Lee, X., Jin, J., Yan, B.: Measuring Internet Growth from DNS Observations. In: 2nd Future Information Technology and Management Engineering, pp. 420–423. IEEE Press, Piscataway (2009)
15. Zipf, G.: Selected Studies of the Principle of Relative Frequency in Language. Harvard University Press, Cambridge (1932)
16. Leijenhorst, D.C., Weide, T.P.: A Formal Derivation of Heaps' Law. Information Sciences 170, 263–272 (2005)
17. French, J.C.: Modeling Web Data. In: 2nd ACM/IEEE-CS Joint Conference on Digital Libraries, pp. 320–321. ACM, New York (2002)

18. DNSPod Website, https://www.dnspod.com
19. DNS-OARC Presentation, https://www.dns-oarc.net/files/workshop-200911/Ziqian_Liu.pdf
20. queryperf, http://www.freebsdsoftware.org/dns/queryperf.html
21. ISC BIND, http://www.isc.org/software/bind

Appendix

The threshold Y_n can be defined as the generalization error obtained in learning phase,

$$Y_n = \frac{1}{m}\sum_{t=1}^{m} |\log(V_n)_t - \log(V'_n)_t| \,, \tag{8}$$

where $(V_n)_t$ is the value of V_n recorded in the end of the t-th learning cycle, and $(V'_n)_t$ is the value of $(V_n)_t$ estimated by

$$\log(V'_n)_t = \beta_n \log(N_t) + K_n \,, \tag{9}$$

where N_t is the value of N recorded in the end of the t-th learning cycle.

Managing Power Conservation in Wireless Networks

Kongluan Lin[1], John Debenham[1], and Simeon Simoff[2]

[1] QCIS, FEIT, University of Technology, Sydney, Australia
[2] Computing and Mathematics, UWS, Sydney, Australia
linkongluan@gmail.com,
john.debenham@uts.edu.au,
simeon@it.uts.edu.au

Abstract. A major project is investigating methods for conserving power in wireless networks. A component of this project addresses methods for predicting whether the user demand load in each zone of a network is increasing, decreasing or approximately constant. These predictions are then fed into the power regulation system. This paper describes a real-time predictive model of network traffic load which is derived from experiments on real data. This model combines a linear regression based model and a highly reactive model that are applied to real-time data that is aggregated at two levels of granularity. The model gives excellent performance predictions when applied to network traffic load data.

Keywords: power conservation, data mining, wireless networks.

1 Introduction

Power conservation in wireless networks is attracting considerable attention. Most approaches are based either on maximising the number of stations that can be put into sleep mode, or on minimising the power for transmission when the unit is active [1]. However, the majority of approaches are not based on predictive models of network load. This paper describes work in a power conservation project conducted in collaboration between UTS and Alcatel-Lucent (Bell Labs). A component of this project addresses methods for predicting user demand load in each zone of a network. This paper describes the data mining methods used to build a predicative model that is then fed into the power regulation system.

The size and nature of communications networks suggests that forecasting methods should be distributed and react quickly. We use the term *agent* to refer to a decision making entity located in each zone of the network. Each agent has to determine: which signals to select, and how to combine the selected signals. Data mining can be an expensive business; costly solutions may be justified if they can be replicated across the network. The management of the extent to which data mining is applied is important, and is not discussed directly here.

The model described aims only to predict whether the local load is *either* generally unchanged, *or* is increasing, *or* is decreasing. This model combines:

L. Cao, J. Zhong, and Y. Feng (Eds.): ADMA 2010, Part II, LNCS 6441, pp. 314–325, 2010.

– signals derived from local observations that each agent makes *within* its zone,
– signals from agents in neighbouring zones, and
– signals derived from background information sources *external to* the network

The method for deriving predictions from local observations is particularly interesting. Here we found that a subtle combination of a linear regression based model and a highly reactive model applied to real-time data that is aggregated at two levels of granularity outperformed non-linear methods. This technique is described in detail.

Section 2 discusses the application of data mining to network management as well as distributed data mining techniques. Section 3 explains how three types of signal are combined to form our solution. Section 4 describes the experimental case study on real-time mining of local observations using real network data. The data was selected from Internet traffic at University of Technology, Sydney, where we found the data had characteristics similar to wireless networks in that it was bursty with fluctuations but bounded. Finally, Section 5 concludes.

2 Background

There is an established history of data mining in network management. We discuss work on alarm or fault related issues work about distributed data mining.

2.1 Mining Alarms and Faults in Communications Networks

Alarm Correlation. Alarms are messages produced by different components of networks. They describe some sort of abnormal situations [2]. Modern communication networks produce large numbers of alarm messages. These alarm messages traversing the network burden the network traffic, possibly lead to packet loss, latency and data retransmission, and ultimately degrade the network performance [3]. Also, due to the rapid development of hardware and software used in communication networks, the characteristics of the alarm sequences are changed as new nodes are added to the network or old ones are updated [4]. Thus, the operators may not have time to learn how to respond to each situation appropriately. In order to avoid overloading operators, alarm correlation systems are used to filter and condense the incoming alarms and diagnose the initial cause of the alarm burst [5]. Due to the dynamic nature of growing telecommunication networks, alarm correlation systems need to adapt to different topologies and extensions of network structure [5].

Neural Networks have characteristics which makes them suitable for the alarm correlation task. No expert knowledge is needed to train the neural network. Moreover, Neural Networks are resistant to noise because of their generalising capabilities [5]. The "Cascade Correlation Alarm Correlator" (CCAC) is a neural network based alarm correlator. It minimises the count of operations — no topology of the hidden layer has to be proposed during training. The system is able to treat noise with up to 25 percent of missing alarms while still achieving 99.76% correct decisions [5]. Self-Organising Maps, a type of neural network, are able to recognise input alarm patterns even when the input data is noisy, corrupted or has significant variation [6].

Bayesian Belief Networks (BBN) are well suited to automated diagnosis because of their deep representation of knowledge and their precise calculations [7]. A BBN represents cause and effect between observable symptoms and the unobserved problems. When a set of symptoms are observed, the most likely cause can be determined. However, the development of a diagnostic BBN requires a deep understanding of the cause and effect relationships in a domain [7].

Fault Prediction. The occurrence of a fault often triggers alarm signals. When two consecutive faults occur within a short time, the alarms corresponding to them may mix together. Fault identification may be a very difficult task when the operator is required to take into account the network elements up or down stream of the fault that are also issuing alarms.

The "fixed time windows" method has been used by Sasisekharan and others to predict network faults [3]. The basic idea is that a consecutive period of time is divided into two windows W_a and W_b first, then the measurements made in W_a is used to predict problems in W_b [3]. Because faults are often transient, a reasonably long period for W_b should be specified.

The "Telecommunication Alarm Sequence Analyser" (TASA) has also been used for fault prediction. By using a specialised data mining algorithm, this tool can discover recurrent patterns of alarms automatically [8]. Network specialists then use this information to construct a rule-based alarm correlation system, which can then be used to identify faults in real time. Also, TASA is capable of finding episodic rules that depend on temporal relationships between the alarms [8]. For example, it may discover the following rule: if alarms of type "link alarm" and type "link failure" occur within 5 seconds, then an alarm of type "high fault rate" occurs within 60 seconds with a probability of 0.7 [9]. Based on the rules, faults can be predicted, and counter-measures can be taken in advance.

"Timeweaver" is another tool for fault prediction. It is a genetic algorithm based machine learning system that predicts rare events by identifying predictive temporal and sequential patterns [10]. It consists of two processes. First, a Genetic Algorithm is used to extract alarm patterns and then a greedy algorithm is applied to form prediction rules. Compared with some existing methods like ANSWER, RIPPER, Timeweaver performs better at the prediction task [10].

2.2 Distributed Data Mining of Communications Networks

Traditionally data mining was based on a centralised approach. However, in distributed computing environments, such as the Internet, intranets, sensor networks, wireless networks and Peer-to-peer systems, it is often desirable to mine data that is distributed in different places [11]. In such cases, centralised data mining approach is inappropriate because of its long response time and inability to capitalise on distributed resources [11]. Distributed data mining is often indicated in distributed environments. For example, in a wireless sensor network with limited communication bandwidth and limited battery power, the central collection of data from every sensor node may create heavy traffic and consume a considerable amount of power. In contrast, distributed data mining may be more suitable because it reduces the communication load and spreads power consumption evenly across the different nodes of the sensor network.

Distributed data mining can also be used for network management and bundled service management. It offers the following advantages over centralised mining:

1. Network traffic and processing load in network management system may be both reduced by performing data processing closer to network elements.
2. Distributed methods may scale well where centralised methods will not.
3. Searches can be performed closer to the data, improving speed and efficiency.
4. Distributed network management may be inherently more robust without depending on continuous communications between the network management system and network elements [12].

Recently, agent-based distributed data mining has become a very active research area. Ogston and Vassiliadis use agents to simulate a peer-to-peer auction and a centralised auction [13] for resource allocation. They show the distributed auction exhibits price convergence behaviour similar to that of the centralised auction. Also, with a growing number of traders, the peer-to-peer system has a constant cost in the number of message rounds needed to find the market equilibrium price, while linear cost is required by central auctioneer [13]. In terms of message costs, the peer-to-peer system outperforms the central auction.

BODHI, implemented in Java, has been designed for collective DM tasks on heterogeneous data sites [14]. This framework requires low network communication within local and global data models. The mining process is distributed to the local agents and mobile agents that carry data and knowledge. A central facilitator agent is responsible for initialising and coordinating the data mining task within the agents.

Parallel Data Mining Agents (PADMA) architecture is proposed by Kargupta, Hamzaoglu and Stafford [15]. PADMA deals with the problem of distributed data mining from homogeneous data sites. At first, data cluster models are counted by agents locally at different sites. Then, the local models are collected to a central site to perform a second-level clustering to produce the global cluster model.

All the above approaches aim to integrate the knowledge which is extracted from data at different geographically distributed network sites with a minimum amount of network communication, and maximum of local computation. Concerning distributed data mining algorithms, Bandyopadhyay and others [16] have introduced a P2P K-Means algorithm for distributed clustering of data streams in a peer-to-peer sensor network environment. In the P2P K-Means algorithm, computation is performed locally, and communication of the local data models is restricted only within a limited neighbourhood. As opposed to the full synchronisation required in some other algorithms, synchronisation in P2P K-Means is restricted only within a neighbourhood. Moreover, even if some node and/or link fails, the algorithm can continue, though its performance will degrade gracefully with an increase in the number of failures.

Although distributed data mining approach could be very useful in network management, associated privacy issues are important. Roughan and Zhang proposed a distributed data mining algorithm to conduct summarisation of Internet traffic without revealing traffic volume of any ISP [17].

3 Solution Structure

Any intelligent approach to conserve power will incorporate predictive models of load. We combine three solutions to smaller problems to form the complete solution:

1. Real-time mining of local observations that predicts the load in a zone on the basis of observed variations in load on the stations in that zone.
2. Off-line data mining is applied to historic load data over a region to identify whether changes in load in one zone may, under certain conditions, signal subsequent changes in another zone. For example, the natural flow of pedestrian traffic around a building may be the underlying cause of such a relationship.
3. Background mining of data, text and news sources that are *outside* the network. For example, knowledge that a football match may be held in a certain stadium next Saturday may have significant implications for load near the stadium.

The important point is that any solution to the above problem should attempt to capitalise on these three approaches. Also, the solutions must be scalable and operate fast. Scalability suggests distributed decision making systems, and speed of operation suggests they should be based on simple models with low computational demands.

Our solution combines three classes of signals: first, signals that an agent derives from observations within its zone; second, signals that an agent chooses to import from adjacent zones, and third, signals that an agent chooses to import from the mining of background data. Traditional data mining techniques are applied to select and combine these three classes of signal — as we noted in Section 1 this is a costly process unless the solution derived can be replicated across the network.

4 Real-Time Mining of Local Observations

Our overall aim is to identify a suitable architecture for load prediction in LTE networks. To do this, we based our experiments on Ethernet load data obtained from University of Technology, Sydney (UTS), because the data is characteristic of load data in LTE networks. Our hypothesis is that the solution to the prediction of UTS Ethernet load will indicate the architecture for load prediction in LTE networks.

In each zone of UTS, an agent is used to record the network load in the zone over a period of time, and then analyses it to build a load predictive model for the zone. Based on the predictive model, each agent can then forecast its zone's network load. In this experiment, only predictive models for inbound network load are introduced, since outbound network load can be predicted similarly. Also, predictive models are built by applying linear regression and moving average[1] method is used to smooth coarse data to improve prediction accuracy. Besides, we use the goodness of fit as a measure of the accuracy of predictive models, which is defined as follows:

$$\sigma_{est} = \sqrt{\frac{\sum (Y - Y')^2}{N}} . \tag{1}$$

Where Y is an actual value and Y' is a predicted value while N is the number of values to predict.

4.1 Data Pre-Processing and Transformation

A key feature of the method described is that it operates on data aggregated at two levels of granularity. In the Internet load data used in the experiments the aggregation periods were 1 minute and 5 minutes. These granularity settings were derived as a result of visual examination of the raw data. If the method described is applied to wireless network traffic then these levels would of course be considerably shorter, and would be derived by visual examination of data as was performed in these experiments.

Table 1 shows the column headings for the raw data — some columns have the same meaning. For example, Traffic in (Volume) and Traffic in (Volume) (Raw) represent the same information, both of raws stand for the inbound network traffic volume within a minute but measured by different units, Kilobyte and Byte respectively. Since they represent the same information, only one of the two columns is considered. In our experiments, data is reconstructed with only three columns: Data Time, Traffic In, and Traffic Out.

The University of Technology, Sydney has Internet facilities deployed through a large campus. The university monitors data in a number of *zones* that vary from teaching laboratories to open access areas in the various building foyers. We inspected the data and selected six zones on the basis that they exhibited characteristics of wireless data in that it was bursty with fluctuations that were significant buy bounded. These zones are referred to in this paper as B1, B3, B4, B5 and B10 — these labels have no significance beyond identifying the data sets selected.

Initial analysis of the data applied Pearson's *sample coefficient of linear correlation* — a measure of the tendency of two variables to vary together — was applied to the five data sets. It is often denoted by r and defined as follows:

$$r = \frac{n\sum_i X_i Y_i - \sum_i X_i \times \sum_i Y_i}{\sqrt{n\sum_i X_i^2 - (\sum_i X_i)^2} \times \sqrt{n\sum_i Y_i^2 - (\sum_i Y_i)^2}} . \tag{2}$$

[1] Subset size for moving average method is arbitrarily assigned to 5, and moving average for inbound network load is calculated as: $A_{\text{in}}(T) = \frac{\text{In}(T-1)+\text{In}(T-2)+\text{In}(T-3)+\text{In}(T-4)+\text{In}(T-5)}{5}$.

Table 1. Data description

Atribute	Description
Date Time	Date time
Traffic in (Volume)	Sum of inbound network load
Traffic in (Volume) (Raw)	Sum of inbound network load
Traffic out (volume)	Sum of outbound network load
Traffic out (volume) (Raw)	Sum of outbound network load

Table 2. Correlation coefficient 1-minute time granularity

Coefficient between:	In(T)/In(T-1)	In(T)/In(T-1)	In(T)/In(T-2)
Granularity:	1 minute	5 minutes	5 minutes
B1	0.89	0.67	0.50
B3	0.88	0.66	0.48
B4	0.83	0.76	0.61
B5	0.90	0.82	0.69
B10	0.95	0.79	0.58

An r value of 1 or -1 indicates a perfect linear relationship between variable X and Y, while value of 0 means variable X is independent of Y. If $r = 1(-1)$, Y always increases as X increases. If $r = -1$, Y always decreases as X increases.

The results of this initial analysis are shown in Table 2. For 1 minute granularity the correlation coefficient between In(T) and In(T-1) is very high, ranging from 0.83 to 0.95 within the five zones. The correlation coefficient between In(T) and In(T-1) based on 1 minute granularity data is greater than that based on 5 minute granularity data. Consequently we hypothesise that current network traffic load has a strong linear relationship with preceding network traffic load. That is to say, the network load recorded in the previous minute can be used to predict the network load in the next minute. Also, data of 1 minute granularity is a better foundation for predictive models than data of 5 minute granularity.

4.2 Models Based on Moving Averages

We denote the total inbound and outbound network load within the minute T as In(T) and Out(T) respectively. An intuitive assumption is that In(T) could have strong linear relationships with its previous moving average like $A_{\text{in}}(T-1)$ and $A_{\text{in}}(T-2)$, and it may also have linear relationship with $A_{\text{out}}(T-1)$ and $A_{\text{out}}(T-2)$. Based on the above assumption, three linear predictive models were considered as follows:

Model (1): $\text{In}(T) = f(A_{\text{in}}(T-1), A_{\text{in}}(T-2))$
Model (2): $\text{In}(T) = f(A_{\text{out}}(T-1), A_{\text{out}}(T-2))$
Model (3): $\text{In}(T) = f(A_{\text{in}}(T-1), A_{\text{in}}(T-2), A_{\text{out}}(T-1), A_{\text{out}}(T-2))$

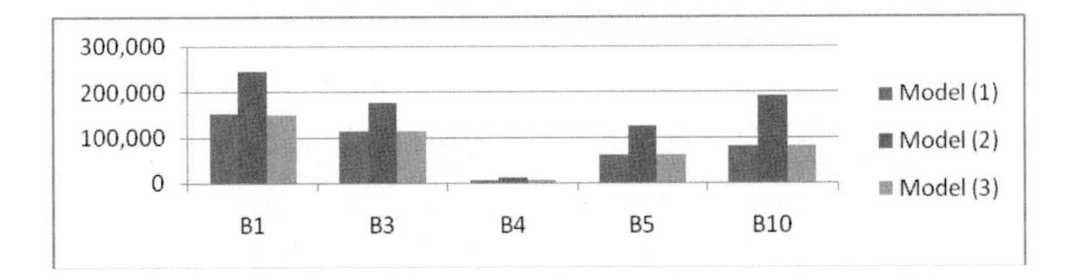

Fig. 1. Goodness of fit generated by Model (1), (2) and (3)

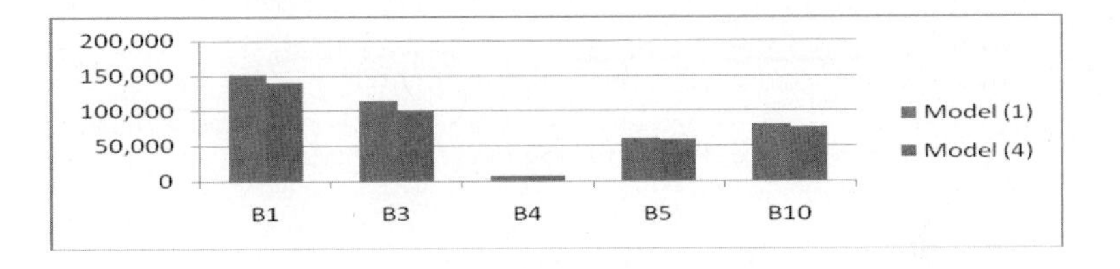

Fig. 2. Comparison of Goodness of fit between Model (1) and (4)

In Fig. 1, goodness of fit generated by Models (1), (2) and (3) are compared, with Model (3) has the smallest goodness of fit. The goodness of fit generated by Model (1) is slightly smaller than that generated by Model (1). In order to simplify our approach, we decide to build predictive models by improving Model (1). In Model (1), network traffic load is predicted based on the previous two moving averages. Then, we consider a possibility of improved performance when the model is based on the previous three moving averages. So a new model is proposed as follows:

Model (4): $\text{In}(T) = f(A_{\text{in}}(T-1), A_{\text{in}}(T-2), A_{\text{in}}(T-3))$

Goodness of fit generated by models (1) and (4) is compared in Fig. 2. It can be seen that Models (4) outperforms Model (1) because of generating smaller goodness of fit.

4.3 Models Based on Three Independent Variables

The models described in Section 4.2 have large goodness of fit values. This is due to their poor performance when there are significant oscillations in network load. To deal with these situations, we develop a new approach based on the assumption that using the previous 3 real network load values (three independent variables) for prediction will provide more accurate prediction than using the average network load data. Based on this assumption, Model (5) is proposed:

Model (5): $\text{In}(T) = f(\text{In}(T-1), \text{In}(T-2), \text{In}(T-3))$

As is displayed in Fig. 3 that Models (5) performs better than Model (4), because Model (5) has a smaller goodness of fit than Model (4).

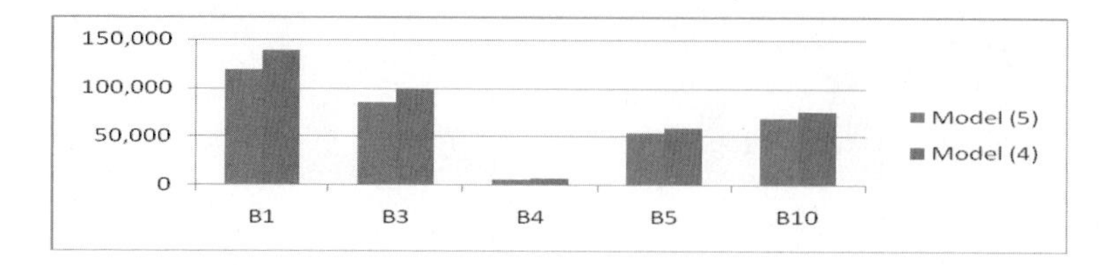

Fig. 3. Comparison of Goodness of fit between Model (4) and (5)

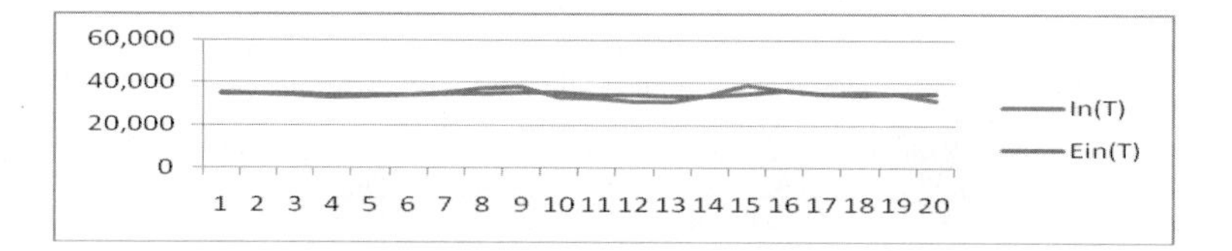

Fig. 4. Successful prediction

4.4 Model Applied on an Hour Basis

Although models (5) has the best performance within the above models, the calculated goodness of fit value is still significant. Generally, the increase of goodness of fit value is caused by sudden and dramatic fluctuations in network traffic load. It still needs to be established how well these models work on different type of data (smooth and coarse). When we consider network load data for a whole day, it is more likely that the data will be coarse. However, when we consider the same data in segments that are an hour long, most of the segments will be smooth. Therefore, Model (5) is considered an hour long segments. In this experiment, the first 40-minute data is used to build model, and then the next 20-minute network traffic load is predicted by using the model.

Numerous generated results show that Model (5) provides accurate network load forecasts for certain time periods, however for some other periods forecasts are not sufficiently accurate. Here we only display two sample examples of successful and unsuccessful forecast situations for inbound network load.

4.5 Continuous Predictive Models

In Fig. 4 and Fig. 5, $E_{\text{in}}(T)$ stands for the estimate value of the inbound network load, and $E_{\text{in}}(T) - \text{In}(T)$ is the deviation of estimated inbound network load from the real inbound network load. The two results based on different data samples (smooth and coarse data respectively) differ dramatically in predicting network load. According to the good prediction, the deviation of inbound network load is very small, with its absolute value varying from 8 KB to 2882 KB. Its maximum deviation 2882 KB is less than 9% of the average inbound load 34,000 KB. In contrast, the maximum deviation of inbound network load in the bad prediction

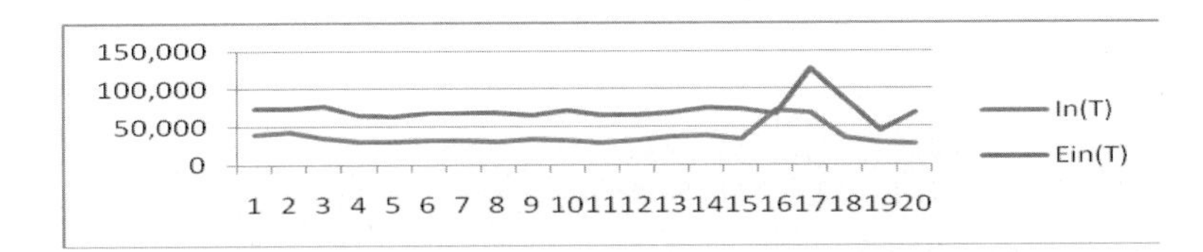

Fig. 5. Unsuccessful prediction

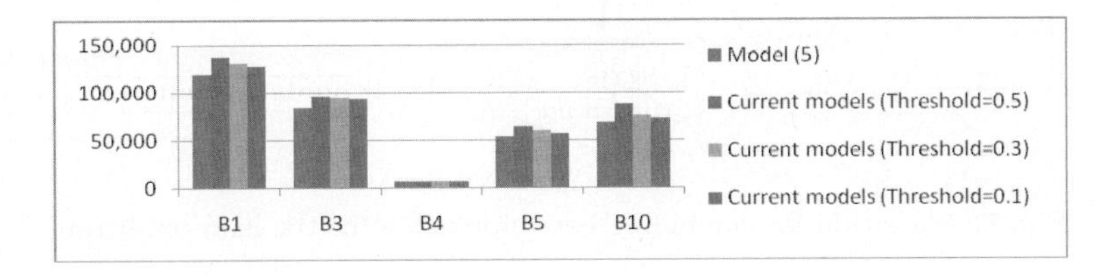

Fig. 6. Goodness of fit obtained from Model (5) and current models

incredibly reaches 57,892 KB, which even surpasses the average inbound network load 40,000 KB.

Since the models work well for the smooth data, as to how to build models based on coarse data and make it work well for prediction would be the next challenge. Besides, this model is not feasible for a continuous prediction, because the prediction of the last 20-minute network load in each hour is based on the model built with the first 40-minute network load, and the first 40-minute network load is not predicted. In this section, a predictive model is built by combining Model (5) and one instance of the same model based on constant values (Model (5) - instance A). The weight constants used in Model (5) - instance A are based on results of extensive simulations. The prediction procedure is illustrated in Fig. 7 and the combined model is explained as follows:

Model (5): $E_{\text{in}}(T) = f(\text{In}(T-1), \text{In}(T-2), \text{In}(T-3))$
Model (5) instance A: $E_{\text{in}}(T) = 0.85 \times \text{In}(T-1) + 0.1 \times \text{In}(T-2) + 0.05 \times \text{In}(T-3)$
Goodness of fit: $\text{Dr}(T) = \frac{E_{\text{in}}(T-1) - \text{In}(T-1)}{\text{In}(T-1)}$
Goodness of fit ratio threshold: Threshold: (i.e. 0.5, 0.3 and 0.1)

At first, a counter is set to 0. The counter is increased when the goodness of fit calculated for the obtained prediction is within the given threshold. When the goodness of fit is above given threshold than counter is reset to 0. Model (5) instance A is applied to predict network load whenever the counter is smaller than 5. When counter is 5, Model (5) can be built based on the previous 5 network load data and used to predict the next network load. When the counter is greater than 5 already built Model (5) is applied.

As it can be seen from Fig. 6, as threshold decreases, the combined model performs better; and when the threshold is equal to 0.1, the goodness of fit

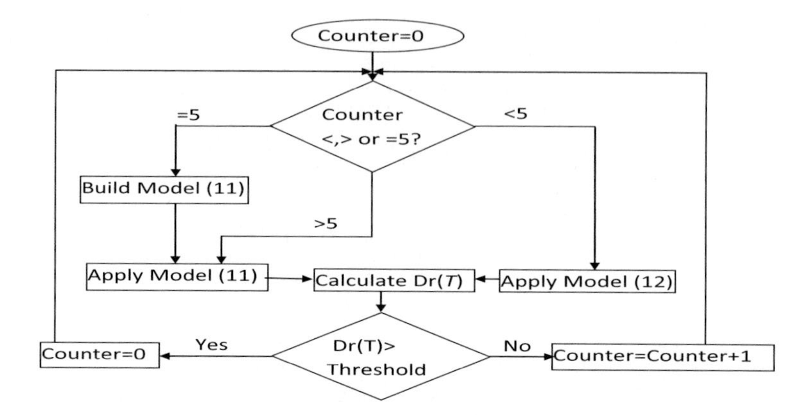

Fig. 7. Algorithm for combining two approaches for the load predicton

generated by the combined model is as small as that generated by purely using Model (5). One advantage of the combined model is that it is very reactive to sudden and dramatic changes in network traffic load. Also, it is easy to be built and implemented without requiring much storage space.

5 Discussion

We have described a predictive model that combines local signals, neighbouring signals and signals mined from background information. This paper has focussed on the predictive model for local signals as its solution is more surprising than the other components. A surprising conclusion of these experiments is that a subtle combination of a linear regression based model and a highly reactive model applied to data aggregated at two levels of granularity gives excellent performance. The use of Markov Chains and other non-linear models is not warranted. This result is to an extent a consequence of the simple goal of the local prediction task, that is just to predict whether the load is (roughly) increasing, decreasing or constant. The conclusion is significant as this simple model is easier to built, and requires less computational resources than non-linear methods. More importantly, by flipping between data granularity aggregations it is able to react quickly to sudden and dramatic changes in network traffic load.

The results described in this paper are presently being trialled in simulation experiments at the University of Technology, Sydney where distributed algorithms are being developed for regulating the power consumption of LTE networks. The solution being investigated is a multi-agent system in which an agent is located with each cluster of LTE stations in the network. A key input to support the agent's decision making is the load predications provided by the system described here. Early in 2011 we will conduct trials in the Alcatel-Lucent (Bell Labs) research laboratory located at Blackfriars on the UTS campus.

References

1. Liu, M., Liu, M.T.: A power-saving algorithm combing power management and power control for multihop ieee 802.11 ad hoc networks. International Journal of Ad Hoc and Ubiquitous Computing 4, 168–173 (2009)
2. Sterritt, R., Adamson, K., Shamcott, C., Curran, E.: Data mining telecommunications network data for fault management and development testing. In: Proceedings of Data Mining Methods and Databases for Engineering, Finance and Other Fields, Cambridge, UK, pp. 299–308. WIT Press, Southampton (2000)
3. Sasisekharan, R., Seshadri, V., Weiss, S.M.: Data mining and forecasting in large-scale telecommunication networks. IEEE Intelligent Systems 11, 37–43 (1996)
4. Hatonen, K., Klemettinen, M., Mannila, H., Ronkainen, P., Toivonen, H.: Knowledge discovery from telecommunication network alarm databases. In: Proceedings of the Twelfth International Conference, Data Engineering, pp. 115–122 (1996)
5. Hermann, W., Klaus-Dieter, T., Klaus, J., Guido, C.: Using neural networks for alarm correlation in cellular phone networks. In: Proceedings of the International Workshop on Applications of Neural Networks in Telecommunications, pp. 1–10 (1997)
6. Gardner, R., Harle, D.: Alarm correlation and network fault resolutionusing kohonen self-organising map. In: IEEE Global Telecom Conference, New York NY, USA. IEEE Computer Society Press, Los Alamitos (1997)
7. Gurer, D., Khan, I., Ogier, R.: An artificial intelligence approach to network fault management. Technical report, SRI International, California, USA (1996)
8. Weiss, G.: Data mining in telecommunications. In: The Data Mining and Knowledge Discovery Handbook, pp. 1189–1201. Springer, Heidelberg (2005)
9. Mika, K., Heikki, M., Hannu, T.: Rule discovery in telecommunication alarm data. Journal of Network and Systems Management 7, 395–422 (1999)
10. Weiss, G., Haym, H.: Learning to predict rare events in event sequences. In: Proceedings of the 4th International Conference on Knowledge Discovery and Data Mining, pp. 1–5. AAAI Press, Menlo Park (1998)
11. Park, B., Kargupta, H.: Distributed data mining. In: Ye, N. (ed.) Algorithms, Systems, and Applications in Data Mining Handbook, pp. 1–22. Lawrence Erlbaum Associates, Mahwah (2002)
12. Chen, T.: A model and evaluation of distributed network management approaches. IEEE Journal on Selected Areas in Communications 20, 850–857 (2002)
13. Ogston, E., Vassiliadis, S.: A peer-to-peer agent auction. In: Vassiliadis, S. (ed.) First International Joint Conference on Automomous Agents and Multi-Agent Systems, pp. 151–159. ACM, New York (2002)
14. Kargupta, H., Park, B., Hershberger, D., Johnson, E.: Collective data mining: A new perspective toward distributed data mining. In: Advances in Distributed and Parallel Knowledge Discovery, pp. 1–38. MIT/AAAI Press (1999)
15. Kargupta, H., Hamzaoglu, I., Stafford, B.: Scalable, distributed data mining using an agent based architecture. In: International Conference on Knowledge Discovery and Data Mining, Newport Beach, California, USA, pp. 211–214 (1997)
16. Bandyopadhyay, S., Giannella, C., Maulik, U., Kargupta, H., Liu, K., Datta, S.: Clustering distributed data streams in peer-to-peer environments. Information Sciences 176, 1952–1985 (2006)
17. Roughan, M., Zhang, Y.: Secure distributed data-mining and its application to large-scale network measurements. ACM SIGCOMM Computer Communication Review 36, 7–14 (2006)

Using PCA to Predict Customer Churn in Telecommunication Dataset

T. Sato[1], B.Q. Huang[1], Y. Huang[1], M.-T. Kechadi[1], and B. Buckley[2]

[1] School of Computer Science and Informatics, University College Dublin, Belfield, Dublin 4, Ireland
[2] Eircom Limited, 1 Heuston South Quarter, Dublin 8, Ireland
bingquan.huang@ucd.ie, takeshi.sato@ucdconnect.ie

Abstract. Failure to identify potential churners affects significantly a company revenues and services that can provide. Imbalance distribution of instances between churners and non-churners and the size of customer dataset are the concerns when building a churn prediction model. This paper presents a local PCA classifier approach to avoid these problems by comparing eigenvalues of the best principal component. The experiments were carried out on a large real-world Telecommunication dataset and assessed on a churn prediction task. The experimental results showed that local PCA classifier generally outperformed Naive Bayes, Logistic regression, SVM and Decision Tree C4.5 in terms of true churn rate.

Keywords: PCA, predict potential churners, telecommunication dataset.

1 Introduction

Retaining existing customers has become a commonplace marketing strategy to survive in ever-competitive business market nowadays [12]. As a result of recent changes in the globalisation and the liberalisation of the markets, retaining customers is becoming a challenging task. In order to overcome this problem, customer churn management has been in the spotlight to foresee customer future behaviour.

The main objective of customer churn management is to identify those customers who are intending to move to other competitive service provider. In recent years, data mining classification techniques have been very popular in building churn prediction models to support customer churn management[2,1,14]. Churn prediction model is built from a given real-world datasets but typically, the distribution of churners and non-churners are usually highly imbalanced. Imbalance distribution of class samples is an issue as it leads to poor classification performances[5]. For most services, once the number of service subscribes reaches its peak, the company places a special emphasis on the retention of valuable customers because acquiring new customer is difficult and costly. Thus, unreliable prediction performance can result in a huge financial loss. The most widely known solution to this is data sampling approach and cost-sensitive learning[7].

L. Cao, J. Zhong, and Y. Feng (Eds.): ADMA 2010, Part II, LNCS 6441, pp. 326–335, 2010.

Several data mining techniques have been applied to churn prediction problem. Most of notable classifiers in data mining were examined by many researchers and experts[14,15,6].

Wei and Chiu[14] examined an interesting model for churn prediction problem using Decision Tree C4.5. The author applied multi-classifier hybrid approach[15] to overcome the imbalance distribution problem. Hadden et al[9] employed Neural Networks, CART, and Linear Regression to build telecommunication churn prediction model. Coussement and Poe [6] investigated the effectiveness of SVM technique over logistic regression and random forests. Despite all classifiers were successfully able to build a prediction model, they are highly sensitive to the number of dimensions and the distribution of class samples. Another problem is that the solution to imbalance classification, such as data sampling, cost-sensitive learning and multi-classifier hybrid approach requires finding numerous optimal parameters selection, which could be very expensive.

The main idea of the third proposed approach is the adoption of the concept of *template matching* in pattern recognition. The approach compares the structure pattern of two datasets in terms of eigenvalues to classify future customer behaviours. This requires the help of PCA, in particular the eigenvalues of the best Principal Components. The magnitude of PC's corresponding to the eigenvalues reflects the variance/structural pattern of a dataset along the PC. The classification of unseen customers can be made based on the similarity of information pattern from training dataset.

This paper is organised as follows: the next section outlines the proposed approach on churn prediction task with imbalanced distribution. Section 3 discusses experiments and the evaluation criteria. We conclude and highlight some key remarks in Section 4.

2 Approaches

Our proposed approach uses PCA technique as classification technique. The classification is made based on the idea of template matching. Template matching finds small area of image that matches a template image. In this paper, the eigenvalues obtained from a training dataset is treated as *template* and we compare this to the eigenvalues obtained from a test instances. The objective of our approach is to avoid biased classification caused by the imbalance distribution and the size of dataset. The workflow of PCA classifier is illustrated in Figure 1.

First, we form a churn class dataset d_{ch} and non-churn class dataset d_{nch} by splitting original training dataset d_{train} according to class labels. Since d_{ch} and d_{nch} can be large we apply GA K-means clustering technique on both datasets to form K sub-datasets, respectively. Next, we apply PCA on each cluster for each class. We believe that applying PCA locally would avoid the inclusion of redundant information in principal component because of low variance within a cluster. Note that each cluster dataset is standardized and its mean,μ, and stdv,σ, are kept before PCA is applied. We extract the best PCs and their corresponding eigenvalues λ from their respective class clusters. These eigenvalues are kept in C_{eig} for churn and NC_{eig} for non-churn as templates of information pattern for comparison at prediction stage. We keep these eigenvalues because they inform and summerize information pattern contain inside the extracted PCs, which are uncorrelated variables.

The principal idea of prediction in PCA classifier is the comparison of the best λ(eigenvalues) extracted from a test instance and eigenvalues in C_{eig} and NC_{eig}. The process is described in Algorithm 1. The mean and standard deviation are represented by μ and σ, respectively. If the magnitude of information pattern obtained from a test instance is equivalent to the one of churn dataset, the test instance is related to churners, and vice versa. Note that each cluster dataset is standardized and its μ and σ are kept before PCA is applied. We explain the main steps of our approach in the following section.

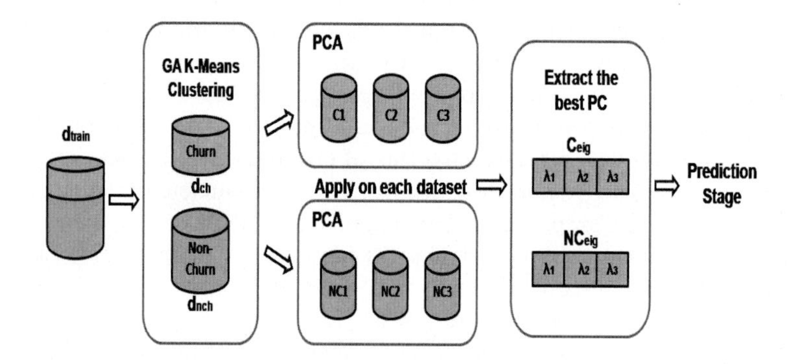

Fig. 1. The description of how the proposed approach works. In this illustration, the number of clusters is set to 3 clusters each for churn and non-churn datasets.

Algorithm 1. PCA Classifier prediction stage

1. Given a test dataset d_{test}, eigenvalue storage for churn C_{eig} and non-churn NC_{eig}.
2. For each instance i = 1 to i = $d_{test}.size$
   2.1. For each churn template j = 1 to j = $C_{eig}.size$
      2.1.1. Standardized a test instance using the μ and σ of template dataset.
      2.1.2. Apply PCA and extract the best PC and its eigenvalue λ_i^{ch}.
      2.1.3. Compute similarity measure $sim_i^{ch} = sim\left(\lambda_i^{ch}, C_{eig}^j\right)$.
   2.2. Go back and repeat Step 2.1. This time for each non-churn template NC_{eig}.
   2.3. Find $ch_i = \arg\min\left(sim_i^{ch}\right)$ and $nch_i = \arg\min\left(sim_i^{nch}\right)$.
   2.4. If $ch_i < nch_i$, instance is classified as churn else non-churn.
3. End of prediction stage.

2.1 GA K-Means Clustering Algorithm

The GA K-means algorithm is applied to d_{ch} and d_{nch} for *local* classification purpose. Traditional K-means algorithm is sensitive to the initial centroid and poor initial cluster centres would lead to poor cluster formation. Recently, Genetic Algorithm (GA)[8] is employed to avoid sensitivity problem in centre point selection. It tries to search for locally optimal clusters.

In GA K-means algorithm, a gene represents a cluster centre of n attributes and a chromosome of K genes represents a set of K cluster centres. The squared error function is employed as the GA fitness function. The GA K-means clustering algorithm works as follows:

1. **Initialization:** Randomly select K data points as cluster centres for W times from original data set as chromosomes, apply k-means with selected centroids. The initial population is composed of each of the resultant partition. Compute the fitness value for each chromosome.
2. **Selection:** The chromosomes are selected according to the roulette-wheel selection[3]. These chromosomes are subject to the survival of the fittetest by reproduction.
3. **Crossover:** The selected chromosomes are randomely paired with other parents for reproduction. The bits of chromosomes are swapped by the probability P_c.
4. **Mutation:** Compute the average fitness value F_{avg} for each chromosome. If the fitness value for one chromosome is greater than F_{avg}, the mutation operation is applied to insure diversity in the population.
5. **Elitism:** A list L_{best} is created to store the chromosome that has the best fitness value.
6. **Iteration:** Go to step 2, until the termination criterion is satisfied. The iteration process terminates when the variation of fitness value within the best chromosomes is less than a specific threshold.

2.2 Principal Component Analysis

The prediction of unseen customers in our proposed approach is made on the basis of PCA[11]. The traditional PCA has a property of searching for uncorrelated variables known as Principal Components that accounts for large part of total variance in a dataset. It also has a property to project a data *linearly* onto new orthogonal bases PC. There have been several approaches developed to extract PC[13] but the simplest form of PC extraction is employed in this work.

The simplest form of PC extraction handles eigenvalue decomposition problem of covariance matrix. Consider a dataset $X=\{x_i, i = 1, 2, \ldots, N, x_i \in \Re^N\}$ with attribute size of d and N samples. The dataset is standardized so that the standard deviation and the mean of each column are 1 and 0, respectively. PC can be extracted by solving the following Eigenvalue Decomposition Problem[11].

$$\lambda\alpha = \mathbf{C}\alpha, \text{ subject to } ||\alpha||_2 = \frac{1}{\lambda}, \tag{1}$$

where α is eigenvectors and $\mathbf{C}$ is covariance matrix, defined as:

$$\mathbf{C} = \begin{bmatrix} c_{11} & c_{12} & \cdots & c_{1d} \\ c_{21} & c_{22} & \cdots & c_{2d} \\ \cdot & & \cdot & \cdot \\ \cdot & & \cdot & \cdot \\ \cdot & & \cdot & \cdot \\ c_{d1} & c_{d2} & \cdots & c_{dd} \end{bmatrix}. \tag{2}$$

The element c_{ij} is the covariance of column i and j. Note that if a dataset is standardized, the covariance of two variables necessarily becomes a correlation measure and hence, **C** becomes correlation matrix. There are two advantages of defining PCs using correlation matrix over covariance matrix[11]:

- The results of analysis for different sets of variables are more directly comparable.
- The sensitivity of the PCs to the units used for each element in dataset X becomes moderate. Note that those variables whose variances are largest will tend to dominate the first PCs.

After solving Equation 1, one needs to sort the eigenvalues λ and its corresponding eigenvector in descending order as larger eigenvalue gives significant PC and useful information. The selected PC's eigenvalue will be significant in later stages. The transformation of X in terms of selected PCs are computed by $X_{tr} = \alpha^T X^T$, where α contains only the selected PC's eigenvectors.

PCA is applied to each clustered dataset and then make prediction. This is due to a possibility that redundant information may be included when PCA considers the *whole* data samples, hence large variance, to extract PC. From here on we refer to this approach as *local* PCA classifier, otherwise std PCA classifier if no clustering is applied.

3 Experiments

3.1 Data Description and Evaluation Criteria

139,000 customers were randomly selected from a real world database provided by Eircom for the experiments. The distribution of churner and non-churners is very imbalanced in both the training data and the testing data. These data contain respectively 6,000, resp. 2000, churners and 94,000, resp. 37000, non-churners. These data contain 122 features which describe individual customer characteristics, see [10] for more details.

In this paper, the Decision Tree C4.5 (DT), the SVM, Logistic Regression (LR) and the Naive Bayes (NB) are employed to build prediction model. The model will be compared to prediction model built by PCA classifier. The performance of prediction model is evaluated based on confusion matrix in Table 1. a_{11}, resp. a_{22} is the number of the correctly predicted churners, resp. non-churners, and a_{12}, resp. a_{21} is the number of incorrectly predicted churners, resp. non-churners. The following evaluation criteria are used in the experiments:

- the accuracy of true churn (TP) is defined as the proportion of churn cases that were classified correctly: $\mathrm{TP} = \frac{a_{11}}{a_{11}+a_{12}}$.
- the false churn rate (FP) is the proportion of non churn cases that were incorrectly classified as churn: $\mathrm{FP} = \frac{a_{21}}{a_{21}+a_{22}}$.

A good solution should have a high TP with a low FP. When no solution is dominant, the evaluation depends on the expert strategy, i.e. to favour TP or FP.

Table 1. Confusion Matrix

		Predicted	
		CHU	NONCHU
Actual	CHU	a_{11}	a_{12}
	NONCHU	a_{21}	a_{22}

We use the Receiver Operating Curve technique (ROC) and the Area under ROC curves (AUC)[4] to evaluate various learning algorithms. ROC graph shows how TP varies with FP. AUC provides single number summary for the performance of learning algorithms. We calculate the AUC threshold on FP of 0.5 as telecom companies are generally not interested in FP above 50%.

3.2 Experimental Setup

Our main objective is to see whether local PCA classifier approach can improve churn prediction results in an imbalanced dataset. We are also interested in the performance of local PCA classifier over traditional classification techniques.

We first examine the optimal cluster size of GA K-means algorithm for local PCA classification approach. We set the size of K in the range $[2 : 512]$. The classification performance is compared according to FP/TP rates.

The second experiment compares the churn prediction results of local PCA classification approach and traditional classification techniques. The cluster size K for local PCA classifier is selected from the two best results from the first experiment. The second experiment is divided into two sets. In the first set of the experiment we use the original training and testing datasets to build the model. For the second set of the experiment, we use under-sampling approach on non-churn samples to reduce imbalance distribution and then build the model. Sampling approach is one of the solution to avoid imbalance classification problem. We gradually reduce the size of non-churners from 94,000 to 500 in the training dataset and then build a prediction model.

3.3 Results and Discussion

The first part of the experiments examines the optimal cluster size K for local PCA classifier. Figure 2 illustrates the prediction performances of local PCA classifier for different K. The ROC graph in Figure 2(a) shows that most of cluster size K of local PCA classifier produced satisfactory FP/TP rates. However, although local PCA classifier performed high TP than std PCA classifier, it had inferior FP than its counterpart of 8.71%. The line in Figure 2 indicates that cluster size of $[24 : 80]$ and $[128 : 512]$ have better prediction than range $[2 : 16]$ for local PCA classifier. Since we put additional focus on True Churn, we selected two best cluster size according to TP rate. Figure 2(b) clearly shows that size 32 and 50 has the highest TP of 84.32% and 85.4%, respectively. When the cluster size is in the range of $[2 : 16]$ and $[128 : 512]$, the FP and TP both decreased.

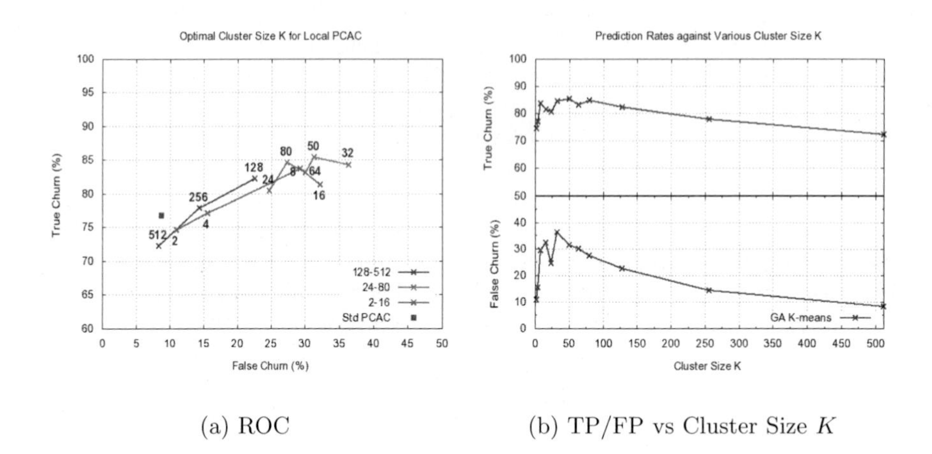

(a) ROC (b) TP/FP vs Cluster Size K

Fig. 2. ROC graph illustration on the left and variations in TP, and FP rates against size K on the right

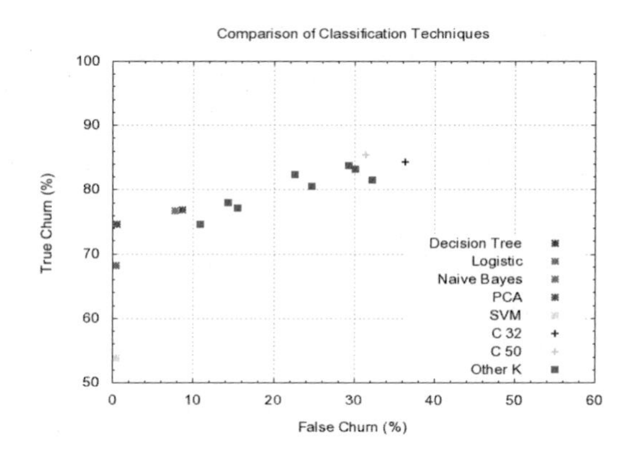

Fig. 3. ROC graph illustration of various classification techniques. The "other K" refers to local PCAC where number of clusters is not 32 or 50.

We compared the prediction performances of the selected classifiers in the first part of the second experiment. The decision tree C45, SVM, Naive Bayes and Logistic Regression were trained and made prediction based on same training/testing dataset as in PCA classifier case. Figure 3 illustrates the ROC graph of different classifiers. The local PCA classifier with optimal cluster size (i.e. C32 & C50) outperformed the rest of classifiers in terms of TP. Furthermore, all local PCA classifier prediction points (i.e. the brown squares) shows that high proportion of churners were classified correctly. However, the selected local PCA classifier generally produces inferior FP rate than the traditional classification techniques. This can be seen from top half table of Table 2. According

Table 2. The top table stores the TP/FP rates following the first part of second experiment. The table below stores the AUC of different classifiers as well as the best prediction points following under-sampling approach.

	C45	**SVM**	**Logistic**	**Naive Bayes**	**PCAC**	**PCAC-32**	**PCAC-50**
TP	74.52	53.76	68.16	76.72	76.76	84.32	85.4
FP	0.6312	0.51	0.424	7.71	8.71	36.38	31.37

AUC	0.446	0.445	0.448	0.377	0.373	0.278	0.3
TP(%)	86.464	87.176	86.576	77.744	77.3	83.85	84.84
FP(%)	12.33	13.95	14.2	9.55	9.15	34.8	31.2

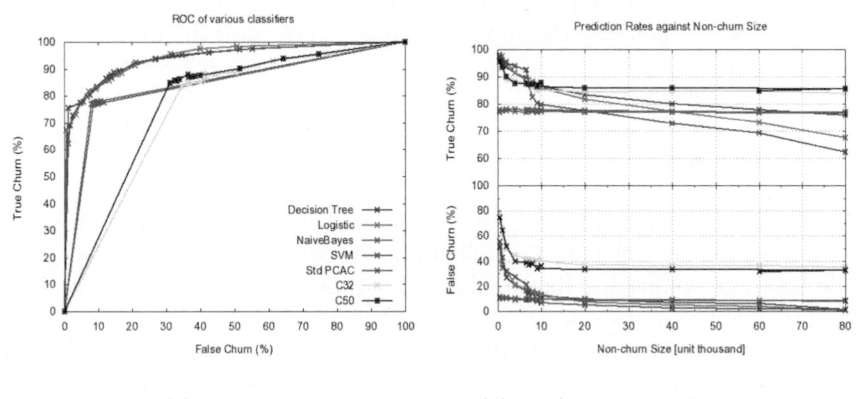

(a) ROC

(b) TP/FP vs non-churn size N

Fig. 4. A ROC graph illustration on the left and variations in TP, and FP rates against non-churn size on the right after the under-sampling approach is applied on non-churn samples

to the ROC graph, traditional classification technique and std PCA classifier are considered as better model than local PCA classifier overall. The last part of the second experiment reduces the number of non-churn size in the training dataset and then builds a prediction model based on the reduced training dataset and test dataset. The local PCA classifier's prediction results were compared to PCA classifier and traditional classifiers similar to the first part of the second experiment. Since the number of non-churn is reduced by random sampling, we generated three different reduced datasets for every reduced non-churn size N in the range of $[500 : 80000]$ and obtained the averaged results. Figure 4(a) is a ROC graph following non-churners reduction in the training dataset.

The graph shows that local PCA classifier has high TP and high FP compare to the rest of classifiers at the start and kept rising as the non-churn size N decreases. Due to this, the AUC rate of the selected local PCA classifiers was inferior than its comparative classifiers. The AUC and the pediction results of the best model of each ROC curve in Figure 4(a) are listed in the bottom half of Table

2. As the size of non-churn N reduces, most classifiers achieved improved TP rate but FP deteriorates as shown in Figure 4(a). Local PCAC in particular suffer from inferior FP as the distribution of churn and non-churn becomes even. The curve of NB classifier and standard PCAC, on the contrary, implies that under-sampling approach has less effect on the prediction results of both classifiers.

In summary, the experiments showed: 1) the best cluster size K for local PCA classifier in terms of TP was in the range of $[24 : 80]$. When the cluster size becomes too large or too small, local PCA classifier correctly classifies non-churners more but unable to classify churners correctly. 2) Local PCA classifier outperformed standard PCA classifier and traditional classifiers in TP rate but failed to classify non-churn correctly than these two techniques, 3) When under-sampling approach was applied, local PCAC produced very high FP rate than other classifiers as non-churn size N reduced and performed the worst from AUC perspective.

4 Conclusions and Future Works

In this paper, we applied PCA technique as classifier to overcome classification problem caused by imbalance distribution of class samples. The prediction is made based on comparing the similarity of eigenvalues obtained from each test instance and clustered datasets according to class labels.

The approach was tested on a telecommunication dataset on a churn prediction task. The first experiment showed that cluster size K of 32 and 50 produced the two best prediction results for local PCA classifier. The second experiment compared the classification performance of traditional classifiers to local PCA classifier. The results showed that local PCA classifier outperformed all traditional classifiers in terms of TP rate, whereas the FP rate of local/std PCA classifier is less than 50% but considerably inferior than traditional classifiers. An under-sampling approach on non-churn samples can affect prediction results. The selected Local PCA classifier suffered from high TP and FP rate than the others and hence resulted in low AUC rate.

Since we are more interested in identifying potential churners as losing a client causes greater loss for a services company, improvement in TP is a great result. Nevertheless, higher FP rate compare to the one from traditional classifier is a problem for an expert favouring low FP as the cost of high FP can be expensive for future marketing campaign. We are interested what if we have selected K that has the lowest FP and conduct similar experiments. Further experiment is required to investigat as why local PCA classifier failed predict non-churners more than traditional and std PCA classifier.

We have encountered several difficulties in this work. The difficulty of finding a locally optimal cluster size K is a problem in any type of K-means algorithm. Selecting such K requires some time spending. Moreover, the computational time of local PCA classifier for large K can take up to 5 days on Dell Vostro 410 machine. This is because PCA classifier make prediction by comparing every single eigenvalues of test instance to every eigenvalues of K clusters in both churn and non-churn class. All in all, it is necessary to 1) search a method to estimate optimal cluster size K automatically and 2) find an alternative method of local PCA classifier for larger K to avoid lengthy computational time despite the size of a test dataset and 3) Examined the prediction results of local PCA classifier for K by favoring FP over TP.

References

1. Luo, B., Shao, P., Liu, J.: Customer Churn Prediction Based on the Decision Tree in Personal Handyphone System Service. In: International Conference on Service Systems and Service Management, pp. 1–5 (2007)
2. Au, W., Chan, C.C., Yao, X.: A novel evolutionary data mining algorithm with applications to churn prediction. IEEE Transactions on Evolutionary Computation 7, 532–545 (2003)
3. Bäck, T.: Oxford Univeristy Press. ch.2. Oxford Univeristy Press, Oxford (1996)
4. Bradley, A.P.: The Use of the area under the ROC curve in the evaluation of machine learning algorithms. Pattern Recognition 30, 1145–1159 (1997)
5. Chawla, N.V., Japkowicz, N., Kotcz, A.: Editorial: special issue on learning from imbalanced data sets. SIGKDD Explor. Newsl. 6, 1–6 (2004)
6. Coussement, K., Van den Poel, D.: hurn prediction in subscription services: An application of support vector machines while comparing two parameter-selection techniques. Expert Systems with Applications 34, 313–327 (2008)
7. Domingos, P.: MetaCost: A general method for making classifiers cost sensitive. In: The 5th International Conference on Knowledge Discovery and Data Mining, pp. 155–164 (1999)
8. Goldberg, D.E.: Genetic Algorithms in Search, Optimization and Machine Learning. Kluwer Academic Publishers, Dordrecht (1989)
9. Hadden, J., Tiwari, A., Roy, R., Ruta, D.: Churn Prediction: Does Technology Matter? International Journal of Intelligent Technology 1 (2006)
10. Huang, B.Q., Kechadi, M.T., Buckley, B.: Customer Churn Prediction for Broadband Internet Services. In: Pedersen, T.B., Mohania, M.K., Tjoa, A.M. (eds.) DaWaK 2009. LNCS, vol. 5691, pp. 229–243. Springer, Heidelberg (2009)
11. Jolliffe, I.T.: Principal Components Analysis. Springer, New York (1986)
12. Kim, H.S., Yoon, C.H.: Determinants of subscriber churn and customer loyalty in Korean mobile telephony market. Telecommunications Policy 28, 751–765 (2004)
13. Oja, E.: Simplified neuron model as a principal component analyzer. Journal of Mathematical Biology 15, 267–273 (1982)
14. Wei, C., Chiu, I.: Turning telecommunications call details to churn prediction: a data mining approach. Expert Systems with Applications 23, 103–112 (2002)
15. Yen, S.J., Lee, Y.S.: Cluster-based under-sampling approaches for imbalanced data distributions. Expert Systems with Applications 36 (2009)

Hierarchical Classification with Dynamic-Threshold SVM Ensemble for Gene Function Prediction

Yiming Chen[1,2], Zhoujun Li[2,3], Xiaohua Hu[4], and Junwan Liu[2]

[1] School of Information Science and Technology, Hunan Agricultural University, Changsha, Hunan, China

[2] Computer School of National University of Defence and Technology, Changsha, Hunan, China

{nudtchenym,ljwnudt}@163.com

[3] Computer School of BeiHang University BeiJing, China

lizj@buaa.edu.cn

[4] College of Information Science and Technology, Drexel University, Philadelphia, PA, 19104, USA

thu@cis.drexel.edu

Abstract. The paper proposes a novel hierarchical classification approach with dynamic-threshold SVM ensemble. At training phrase, hierarchical structure is explored to select suit positive and negative examples as training set in order to obtain better SVM classifiers. When predicting an unseen example, it is classified for all the label classes in a top-down way in hierarchical structure. Particulary, two strategies are proposed to determine dynamic prediction threshold for different label class, with hierarchical structure being utilized again. In four genomic data sets, experiments show that the selection policies of training set outperform existing two ones and two strategies of dynamic prediction threshold achieve better performance than the fixed thresholds.

Keywords: gene function prediction, hierarchical classification, SVM ensemble, dynamic threshold.

1 Introduction

Classification is an important task in machine learning field. Give a training set $\{<x,y>|x \in F_1 \times F_2 \times ... \times F_m \subseteq X, y \in C_1 \times C_2 \times ... \times C_l \subseteq C\}$, where X and C are feature vector and label class vector space respectively, a model $H(X) = C$ can be learned from this training set and be used to predict unseen example with same structure as training example. When $l = 1$, it is traditional single-label classification with only label class C_1 including $|C_1| > 1$ classes. If $|C_1| = 2$, then the learning problem is called a binary classification problem, some algorithms, such as SVM(support vector machine) and Bayesian classifier, have been well developed. While if $|C_1| > 2$, then it is called a multi-class classification problem, which can be solved by combining two-class classifiers. When $l > 1$, the examples

L. Cao, J. Zhong, and Y. Feng (Eds.): ADMA 2010, Part II, LNCS 6441, pp. 336–347, 2010.

01 METABOLISM
01.01 amino acid metabolism
01.01.03 assimilation of ammonia, metabolism of the glutamate group
01.01.03.01 metabolism of glutamine
01.01.03.01.01 biosynthesis of glutamine
01.01.03.01.02 degradation of glutamine
02 ENERGY
02.01 glycolysis and gluconeogenesis
02.01.01 glycolysis methylglyoxal bypass
02.01.03 regulation of glycolysis and gluconeisgenes
10.01 DNA processing
10.01.01 cellular DNA uptake
10.01.01.01 bacterial competence

Fig. 1. A part of hierarchical function catalogue funcat from MIPS(Munich Information Center for Protein Sequences). The FunCat consists of 28 main functional categories (or branches) that cover general fields like cellular transport, metabolism and cellular communication/signal transduction. The main branches exhibit a hierarchical, tree like structure with up to six levels of increasing specificity. In total, the FunCat version 2.1 includes 1362 functional categories.

are associated with l labels and learning task is called multi-label classification, which was initially motivated by the tasks of document classification[6,8]. As a simple case, l labels have a flat structure, that is, there is no relationship among the labels. A more challenge case is that all the l labels construct a hierarchical structure in nature. This is called hierarchical multi-label classification and is also focus of this paper. One of the typical application domains is gene function prediction where two frequently-used functional hierarchical structure are funCat from MIPS(Munich Information Center for Protein Sequences)[12] and GO(gene ontology) compiled by GO Consortium members[1]. A little part from them are shown in fig.1 and fig. 2.

The main approach to multi-label classification are to transform it either into one or more single-label classification by enumerating all of the combinations between l label class[14]. If each label has two classes, then the number of all the combination is 2^l. Obviously, it is NP-hard and this approach becomes infeasible when number l of label is larger. Particulary, for hierarchical multi-label classification, output result is required to satisfy the hierarchy constraint, which makes classification task more difficult.

In this paper, we propose a simple and flexible hierarchical classification approach with dynamic-threshold SVM ensemble for gene function prediction. How to learn better SVM classifiers is discussed on selecting suit positive and negative examples as training set by exploring the hierarchical structure. When predicting the test examples in top-down fashion, we emphasize that it is important to

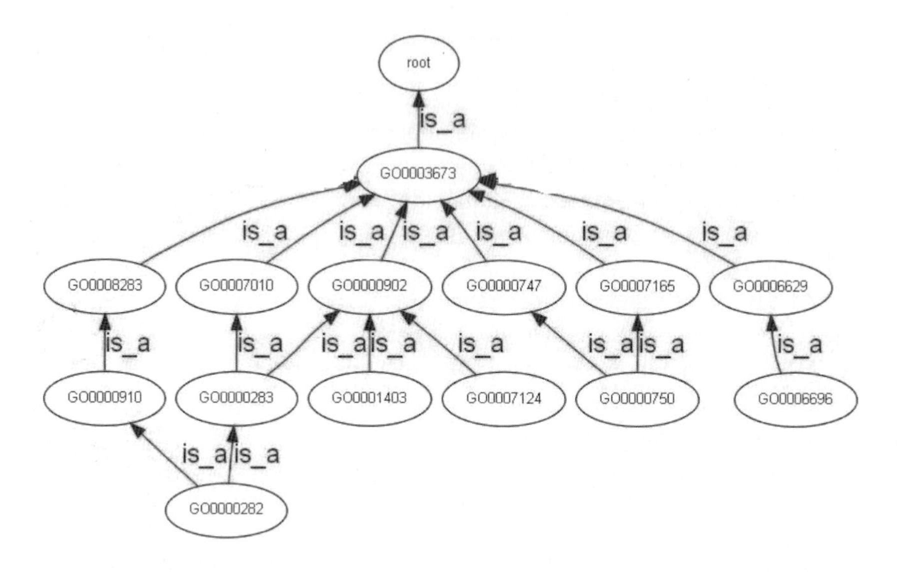

Fig. 2. A part of biological process gene ontology. The Gene Ontology project provides an ontology of defined terms representing gene product properties. The ontology covers three domains: cellular component, molecular function, and biological process. The GO ontology is structured as a directed acyclic graph, and each term has defined relationships to one or more other terms in the same domain, and sometimes to other domains.

use different threshold for different label class node in hierarchy rather than to use fixed ones. We propose two threshold selection strategies.

The rest of the paper is organized as follows. Section 2 formally describes the hierarchical multi-label classification for gene function prediction. Then, we propose hierarchical classification framework with dynamic-threshold SVM ensemble in section 3. Section 4 gives experimental analysis. Section 5 concludes the paper.

2 The Hierarchical Multi-label Classification for Gene Function Prediction

Definition 1 (Hierarchical Class Taxonomy(HCT)). *Hierarchical Class Taxonomy is a partially order set $HCT =< C, \prec>$, where C is a finite set that enumerates all label classes in the application domain, and the relation $\prec$ represents the "IS-A" relationship with following three prosperities and has only one greatest element called "root".*

$(1).\forall C_i \in C, \quad C_i \not\prec C_i.$

$(2).\forall C_i, C_j \in C, \quad if \quad C_i \prec C_j, \quad then \quad C_j \not\prec C_i.$

$(3).\forall C_i, C_j, C_k \in C, \quad if \quad C_i \prec C_j, \quad and \quad C_j \prec C_k, imply \quad C_i \prec C_k.$

For node c in HCT, some useful notation are described in Table 1. In later presentation, we denote the examples set annotated with a label class set S as

Table 1. Some notations of special set related with class node c in hierarchical class taxonomy

label set	description
Par(c)	The parents set of node c
Chld(c)	The children set of node c
Ance(c)	The ancestor set of node c
Desc(c)	The descendants set of node c
Sibl(c)	The sibling set of node c

$exs(S)$. For example, the example set annotated with label classes set $Par(c)$ can be denoted as $exs(Par(c))$.

A hierarchical class taxonomy can has two types: tree and DAG(Directed Acyclic Graph), as shown in fig.1 and fig. 2. The important difference between tree and DAG is,in contrast to $|Par(c)| = 1$ in a tree, $|Par(c)|$ can be greater than 1, which can make hierarchical classification task more challenge.

Definition 2 (Hierarchical Multi-label Classification). *Give a labeled training set* $\{< x, y > |x \in F_1 \times F_2 \times ... \times F_m \subseteq X, y \in C_1 \times C_2 \times ... \times C_l \subseteq C\}$ *where* l *label classes form a* HCT*, let* X *and* C *are feature vector and label class vector space respectively, hierarchical multi-label classification learns a hypothesis* H *so that* $H(X) = C$ *with the output* y *respects the* HCT *hierarchy.*

What the prediction output y must respect the HCT hierarchical structure means an example must be annotated with $Par(c)$ if it has predicted label class c. Hierarchical multi-label classification problem may differ in prediction depth, MLNP(mandatory leaf-node prediction) must predict down to leaf node of HCT, for example, music genre and emotion classification[5,2,16], while NMLNP(non-mandatory leaf-node prediction) may stop predicting at any non-leaf class node. They may also differ in number of prediction path of HCT, SP(single path) and MP(multiple path). Although SP prediction implies multiple label assignment for a test example, that is, for class c, if $l(c) = 1$, then $l(ance(c)) = 1$. Predicting gene function natively is NMLNP and MP because prediction process can proceed down on multiple paths and stop on any label class level.

3 The SVM Ensemble Approach with Dynamic Threshold

3.1 Hierarchical Classification Framework with SVM Ensemble

Our approach can be divided into training and test phrase. At training phrase, we train a SVM classifier τ_{c_i} for each label class node c_i in HCT. To learn high-quality SVM classifiers for each label class, selecting suit positive and negative examples as training set is very important by exploring the hierarchical structure, which will be discussed in following section. While at test phrase, for each test example e, the algorithms calculate a decision function $\tau_{c_i}(c_i|e, \theta_{c_i})$ to determine whether to accept or reject a example under c_i, where θ_{c_i} represents the decision threshold on label class node c_i. The $\tau_{c_i}(c_i|e, \theta_{c_i})$ returns 1 if $P(c_i|e) > \theta_{c_i}$ otherwise 0. In this phrase, if the label class hierarchy is not be considered, the inconsistence may occur. In other words, there may exist $c_i \in Ance(c)$ subjected to $l(c) = 1$ but $l(c_i) = 0$. Therefore, we let test process proceeds in top-down fashion to ensure prediction consistent, that is, for an example e and a label class node c_i, if $\tau_{c_i}(c_i|e, \theta_{c_i}) = 1$, then the algorithm predicts each children node class $c_j \in Chld(c_i)$, otherwise, stops predicting on all the descendant label class nodes from $Desc(c_i)$ and assigns default 0 to them. Our approach can also be called as top-down approach because of this prediction fashion and its framework is as follows.

```
 1: %Training SVM classifiers on each label class node
 2: SVM[] models
 3: for i = 1 to numLab do
 4:    selecting training set and learning SVM classifier τ_ci
 5:    models[i]=τ_ci
 6: end for
 7: %predicting each example e in top-down fashion
 8: Queue nodeQueue
 9: int[][] predRes
10: for i = 1 to numExample do
11:    nodeQueue.addlast(root)
12:    while !nodeQueue.isEmpty do
13:       c_i=nodeQueue.getfirst
14:       if τ_ci(c_i|e, θ_ci) = 1 then
15:          predRes[i][c_i]=1
16:          nodeQueue.addlast(Chld(c_i))
17:       end if
18:    end while
19: end for
```

3.2 The Selection Policies of Training Set

At training phrase, selecting a suit training set is important for a better classifier. Regretfully, to the best of our knowledge, few researches on selecting a good training set by utilizing the hierarchical structure are done although the skew property of original data set has been discovered, for example, imbalance between positive and negative examples[17,18]. Let all the original training data as Tr, Lanckriet et al. select gene examples set $exs(c)$ as positive examples, the others, that is $Tr \setminus exs(c)$, as negative examples[11], which did not consider the hierarchical structure at all. We call it "exclusive(EX)" policy. Some authors first propagate gene annotation in down-top fashion in the hierarchy, then select positive and negative example set with "exclusive" policy, because am example must be assigned to $Ance(c)$ if it is annotated with c[3,9]. According to this policy, positive example set is $exs(c) \cup exs(Desc(c))$ and negative set is $Tr \setminus (exs(c) \cup exs(Desc(c)))$, which is called as "exclusive inclusive(EI)". In addition ,we propose three policies to select suit positive and negative example set as follows.

(i) weak exclusive inclusive(WEI): positive set=$exs(c) \cup exs(Desc(c))$, negative set=$Tr \setminus exs(Ance(c)) \cup exs(c) \cup exs(Desc(c))$. We think that an example e assigned to $Ance(c)$ is not the most probable negative example because it may be annotated with c in the future and becomes a positive example of label class c;
(ii) sibling inclusive(SI): positive set=$exs(c) \cup exs(Desc(c))$, negative set=$exs(Sibl(c) \cup Desc(Sibl(c)))$. The negative set is example set assigned to sibling and descendant of sibling of label class node c , they may be most dissimilar to examples assigned c and be most probable negative examples;
(iii) weak sibling inclusive(WSI): positive set=$exs(c) \cup exs(Desc(c))$, negative set=$exs(Sibl(c))$. The negative set is example set assigned to sibling of label class node c, it does not include those examples of sibling descendant of c;

In above definition, "inclusive" means that the positive examples set includes examples assigned to $Desc(c)$, while "weak" means that negative example set includes less examples.

3.3 The Strategy of Dynamic Threshold

At test phrase, top-down prediction is done for each test example e. For label class node c, if $P(c|e) > \theta_c$, then $\tau_c(c|e, \theta_c) = 1$, e is predicted as positive and it will be predicted on $Chld(c)$ further. Otherwise, the process will stop on c. In this way, the prediction can proceed on multiple paths and stop at any level. In this process, the threshold θ_c plays a key role. If θ_c is too large, the prediction process stops quickly so that the user obtains only shallower prediction and the average R and hR(defined in section 4) become smaller. In practice, the biologist may be more interested in deeper prediction, since they provide more specific information than shallower class predictions. Furthermore, when training SVM classifiers, the deeper label classes have generally fewer annotated gene examples as positive examples and imbalance problem usually occurs, which makes the SVM classifiers have lower prediction probability for true positive examples. As a result, the threshold θ_c should be decreased gradually from top to down level in order that more test examples are predicted with deeper label classes. The existing approaches use fixed threshold 0.5 across different level[3,9]. In this paper, we propose two strategies for dynamically selecting threshold θ_c among all the label class by utilizing the hierarchical structure again. They can be used to *tree* and *DAG* structure and be calculated on valid data set.

Threshold Selection Based on Pass Rate. We define the *Passrate* to measure the rate of examples annotated with c to ones assigned to $Par(c)$.

Definition 3 (Passrate). *For a node class c in HCT, $c_i \in Par(c)$, the pass rate from c_i to c, with notation $Passrate_{c_i \to c}$, can be computed as:*

$$Passrate_{c_i \to c} = \frac{| exs(c) |}{| exs(c_i) |}. \tag{1}$$

where $| \, . \, |$ *denotes cardinality of a set.*

We calculate the θ_c with formular (2).

$$\theta_c = \frac{\sum\limits_{c_i \in Par(c)} \theta_{c_i} * Passrate_{c_i \to c}}{|Par(c)|}. \tag{2}$$

In formular (2), the θ_c is defined as average value of product of θ_{c_i} and $Passrate_{c_i \to c}$ on $Par(c)$. When the $Passrate_{c_i \to c}$ is smaller, few examples are transferred from c_i to c, the imbalance can occur easily and the SVM classifier may predict test examples with lower probability $P(c|e)$ on label class c. As a result, the θ_c should decrease quickly as $Passrate_{c_i \to c}$. This formula can be applied to tree and DAG structure. For the tree structure, the sum can be removed and the formula becomes simpler.

Threshold Selection Based on Usefulness. Since the biologist may be more interested in whether an example can be annotated with more useful label classes, the predict algorithm should gives lower threshold on them. The question that arises is how to evaluate the usefulness of a predicted class? Given that predictions at deeper levels of the hierarchy are usually more informative than the classes at shallower levels, the deeper label classes should have smaller threshold θ. In Clares work[7], the original formula for entropy was modified to take into account two aspects: multiple labels and prediction depth (usefulness) to the user. In this work we have modified the part of the entropy-based formula described in[7]and adapted the "usefulness" measure from[7] to the context of our algorithm. We adapt Clare's measure of usefulness by using a normalized usefulness value based on the position of each class level in the hierarchy. Moreover, we only use the normalized value of the Clare's equation to measure the usefulness. The usefulness can be defined as definition 4.

Definition 4 (Usefulness)

$$Usefulness(c) = 1 - \frac{log_2 treesize(c)}{max}. \tag{3}$$

where: $treesize(c) = Desc(c) + 1$ *the size of DAG with root* c*;* $max = \max_{c_i \in C} log_2 \; treesize(c_i)$ *and is used to normalize all of the usefulness into the interval* $[0, 1]$.

According to meaning of usefulness, the threshold θ_c for label class node c can be computed as follows:

$$\theta_c = \frac{\sum\limits_{c_i \in Par(c)} \theta_{c_i}}{| Par(c) |} * (1 - usefulness(c)). \tag{4}$$

As shown in definition (4), calculating the threshold θ_c can be conveniently done in tree and DAG structure too.

4 Experiments

4.1 Data Sets

To compare our approaches against existing local hierarchical classification algorithm, we have selected four bioinformatics datasets from Vens et al.[15], which use two different class hierarchy structures: tree structure (FunCat data sets) and directed acyclic graph structure (Gene Ontology data sets). The directed acyclic graph (DAG) structure represents a complex hierarchical organization, where a particular node of the hierarchy can have more than one parent, in contrast to only one parent in tree structures. Tables 2 presents details of the data sets used in our experiments. In the experiments conducted by Vens et al.[15], each data set was divided into three parts:training, validating and testing. We have used the same partitions in our experiments.

Table 2. Summary of the data sets used in our experiments

data sets	|train|	|valid|	|test|	|attributes|	|classes|
FunCat					
cellcycle	1640	952	1284	77	251
church	1632	1284	848	27	251
derisi	1612	845	1278	63	251
GO					
borat	1540	770	1155	5930	132

Before training and testing, we scale these data sets to interval $[-10, 10]$. For SVM classifiers, we use RBF(radius base function) kernel with default parameters: cost c, gamma g and the weight of negative examples $w0$. To deal with imbalance problem of training set, we search the best weights for positive examples in set $\{1000, 2000, 4000, 8000\}$. The corresponding best SVM classifiers is used to predict test examples.

4.2 Experiment Evaluation

How to evaluate hierarchical classification algorithms is a part of algorithm design and is an important issue. Most researchers used standard flat classification evaluation measures, without considering hierarchical class taxonomy. The precise rate P, the recall rate R and their combination F are frequently-used measure and defined as follows:

$$P = \frac{TP}{TP + FP}. \tag{5}$$

$$R = \frac{TP}{TP + FN}. \tag{6}$$

Table 3. The confusion matrix

	Predicted Class	
True Class	Positive	Negative
Positive	TP	FN
Negative	FP	TN

$$F = \frac{2 * P * R}{P + R}. \tag{7}$$

Where the FP, FN, TP and TN may be described as confusion matrix in table 3.

Another common evaluation measure used in binary classification problems is the ROC (Receiver Operating Characteristics) curve, which relates $sensitivity = R$ and $1 - specificity = \frac{FP}{FP+TN}$. The area under ROC curve(AUC) is usually used in the evaluation of classification algorithm: the larger the AUC, the better the classification performance[4]. In our experiments, AUC measure is not used because we may not predict an gene example on all the label classes.

The use of flat classification measures might not be enough to give us enough insight at which algorithm is really better by not taking into account the hierarchical structure of the class taxonomy. Kiritchenko et al. proposed a more relational metrics called as hierarchical precision hP, hierarchical recall hR and hierarchical f-measure hF and defined as follows[10]:

$$hP = \frac{\sum_i |P_i \cap T_i|}{\sum_i |P_i|}. \tag{8}$$

$$hR = \frac{\sum_i |P_i \cap T_i|}{\sum_i |T_i|}. \tag{9}$$

$$hF = \frac{2 * hP * hR}{hP + hR}. \tag{10}$$

where P_i is the set consisting of the most specific class(es) predicted for test example i and all its(their) ancestor classes and T_i is the set consisting of the true most specific class(es) of test example i and all its(their) ancestor classes. The summations are of course computed over all test examples. Note that these measures are extended versions of the well known metrics of precision, recall and f-measure but tailored to the hierarchical classification scenario. Furthermore, they can be effectively applied to any hierarchical classification scenario; i.e., tree-structured, DAG-structured problems.

4.3 Comparing the Selection Policies of Training Set

We compare proposed three selection policies of training set with two existing ones with fixed threshold 0.5 in top-down fashion. Table 4 lists flat and hierarchical evaluation result with average value over all the label classes.

Table 4 shows that the proposed three selection policies of training set consistently perform better than existing two ones on comprehensive performance

Table 4. Comparing the selection policies of training set by exploring the hierarchical structure

dataset	policy	flat			hierarchical		
		P	R	F	hP	hR	hF
cellcycle	EX	0.5192	0.1134	0.1385	0.7782	0.2113	0.2897
	EI	0.5336	0.1553	0.1744	0.7289	0.2375	0.3199
	WEI	0.6004	0.1975	0.2313	0.7167	0.2569	0.3368
	SI	0.5121	0.3401	0.2372	0.6510	0.3252	0.3976
	WSI	0.5011	0.1924	0.1631	0.7069	0.2721	0.3478
church	EX	0.5240	0.2398	0.2690	0.7595	0.2097	0.2936
	EI	0.4506	0.2797	0.2852	0.7595	0.2113	0.2953
	WEI	0.5506	0.2797	0.3110	0.7596	0.2113	0.2952
	SI	0.5625	0.4385	0.4711	0.6911	0.2736	0.3502
	WSI	0.5314	0.4195	0.4089	0.7116	0.2527	0.3315
derisi	EX	0.3240	0.3398	0.2717	0.7600	0.2076	0.2914
	EI	0.3505	0.4797	0.3550	0.7600	0.2091	0.2930
	WEI	0.4505	0.4797	0.4046	0.7600	0.2091	0.2930
	SI	0.4621	0.6390	0.4843	0.6912	0.2710	0.3478
	WSI	0.4312	0.5295	0.4053	0.7118	0.2500	0.3290
borat	EX	0.6123	0.1466	0.1918	0.9924	0.2241	0.3343
	EI	0.1050	0.2546	0.3444	0.9563	0.2600	0.3718
	WEI	0.1114	0.3639	0.4329	0.9437	0.2684	0.3789
	SI	0.7031	0.3745	0.5781	0.6719	0.6084	0.6047
	WSI	0.3456	0.3386	0.3835	0.6897	0.6160	0.6175

F and hF, in particular, policy SI has the highest F and hF for four data sets. As an important result, the flat evaluation has lower F value, this can be caused by stopping the top-down prediction process early with larger threshold θ. The higher hP and lower hR illustrate that the prediction process stops too early.

4.4 Dynamic vs. Fixed Threshold

We compare the proposed three dynamic threshold strategies with fixed one using the SI selection policy of training set. For the flat and hierarchical evaluation, Table 5 shows that strategies $passrate$ and $usefulness$ appropriately decrease the prediction threshold θ and achieve the better F and hF value. In most case, they not only improve R and hR significantly but also increase p and hP slightly.

5 Conclusion

The paper proposes a hierarchical prediction framework of gene function with SVM classifier ensemble. Three selection policies of training set by exploring the hierarchical structure are utilized to obtain better SVM classifier. Two dynamic threshold strategies are used to reduce prediction blocking in top-down process

Table 5. Comparing the dynamic vs fixed threshold using *SI* policy. Abbreviation: pr for passrate, use for usefulness.

dataset	policy	flat			hierarchical		
		P	R	F	hP	hR	hF
cellcycle	fix	0.5121	0.3401	0.2372	0.6510	0.3252	0.3976
	pr	0.5424	0.5185	0.4802	0.6631	0.5350	0.5322
	use	0.5849	0.5640	0.5043	0.8122	0.4699	0.5184
church	fix	0.5625	0.4385	0.4711	0.6911	0.2736	0.3502
	pr	0.6816	0.5390	0.5320	0.7366	0.3966	0.4657
	use	0.6127	0.5275	0.5070	0.8313	0.4696	0.5402
derisi	fix	0.4621	0.6390	0.4843	0.6912	0.2710	0.3478
	pr	0.4813	0.6387	0.4689	0.7366	0.3931	0.4541
	use	0.4127	0.6315	0.4392	0.8332	0.4667	0.5318
borat	fix	0.7031	0.3745	0.5781	0.6719	0.6084	0.6047
	pr	0.8003	0.4462	0.5993	0.8499	0.6700	0.6780
	use	0.7978	0.8493	0.7269	0.8359	0.6026	0.6383

to achieve useful specific prediction. In four genomic data sets, the experiments show that the proposed selection policies of training set performs consistently better than existing two ones and the dynamic threshold strategies reduce the prediction blocking and improve the prediction performance successfully.

Our approach is simple and can be applied to tree and DAG hierarchical structure. Furthermore, the basic classifier can be flexibly change from SVM to other. But it need more training and testing time. Recently, a so-called global approach has been proposed. It considers all the label classes at once and only learns a classifier which is usually obtained by modifying a basic classifier[13]. This will be our research topic in future.

Acknowledgments. This work is partially supported by NSF(national science foundation) in China with granted No. 60573057.

References

1. Ashburner, M., Ball, C., Blake, J., Botstein, D.: Gene ontology: tool for the unification of biology. Nat. Genet. 25, 25–29 (2000)
2. Barbedo, J.G.A., Lopes, A.: Automatic genre classification of musical signals. In: EURASIP Journal on Advances in Signal Processing 2007 (2007)
3. Barutcuoglu, Z., Schapire, R.E., Troyanskaya, O.G.: Hierarchical multi-label prediction of gene function. Bioinformatics 22(7), 830–836 (2006)
4. Bradley, A.P.: Use of the area under the roc curve in the evaluation of machine learning algorithms. Pattern Recognition 30, 1145–1159 (1997)
5. Burred, J.J., Lerch, A.: A hierarchical approach to automatic musical genre classification. In: Proc. Of the 6 th Int. Conf. on Digital Audio Effects, pp. 8–11 (2003)
6. Cai, L., Hofmann, T.: Hierarchical document categorization with support vector machines. In: Proceedings of the Thirteenth ACM International Conference on Information and Knowledge Management, SIGIR, pp. 78–87. ACM, Washington (2004)

7. Clare, A.: Machine learning and data mining for yeast functional genomics. Ph.D. thesis, Department of Computer Science University of Wales Aberystwyth (2003)
8. Daphne, K., Mehran, S.: Hierarchically classifying documents using very few words. In: ICML 1997: Proceedings of the Fourteenth International Conference on Machine Learning, pp. 170–178. Morgan Kaufmann Publishers Inc., San Francisco (1997)
9. Guan, Y., Myers, C.L., Hess, D.C., Barutcuoglu, Z., Caudy, A.A., Troyanskaya, O.G.: Predicting gene function in a hierarchical context with an ensemble of classifiers. Genome Biology 9, S3 (2008)
10. Kiritchenko, S., Matwin, S., Nock, R., Famili, A.F.: Learning and evaluation in the presence of class hierarchies: Application to text categorization. In: Lamontagne, L., Marchand, M. (eds.) Canadian AI 2006. LNCS (LNAI), vol. 4013, pp. 397–408. Springer, Heidelberg (2006)
11. Lanckriet, G., Deng, M., Cristianini, M., Jordan, M., Noble, W.: Kernel-based data fusion and its application to protein function prediction in yeast. In: Pac. Symp. Biocomput., pp. 300–311 (2004)
12. Mewes, H.W., Heumann, K., Kaps, A., Mayer, K., Pfeiffer, F., Stocker, S., Frishman, D.: Mips:a database for genomes and protein sequences. Nucleic Acids Res. 30(1), 31–34 (2002)
13. Rousu, J., Saunders, C., Szedmak, S., Shawe-Taylor, J.: Kernel-based learning of hierarchical multilabel classification models? Journal of Machine Learning Research 7, 1601–1626 (2006)
14. Tsoumakas, G., Katakis, I.: Multi-label classification: An overview. Int. J. Data Warehousing and Mining 2007, 1–13 (2007)
15. Vens, C., Struyf, J., Schietgat, L., Dzeroski, S., Blockeel, H.: Decision trees for hierarchical multi-label classification. Achine Learning 73(2), 85–214 (2008)
16. Xiao, Z., Dellandrea, E., Dou, W., Chen, L.: Hierarchical classification of emotional speech. Tech. rep., LIRIS UMR 5205 CNRS/INSA de Lyon/Universite Claude Bernard Lyon 1/Universit Lumiere Lyon 2/Ecole Centrale de Lyon (2007)
17. Zhao, X.M., Wang, Y., Chen, L., Aihara, K.: Gene function prediction using labeled and unlabeled data. BMC Bioinformatics 9(1), 57 (2008)
18. Zhao, X., Li, X., Chen, L., Aihara, K.: Protein classification with imbalanced data. Proteins: Structure, Function, and Bioinformatics 70(4), 1125–1132 (2008)

Personalized Tag Recommendation Based on User Preference and Content

Zhaoxin Shu, Li Yu*, and Xiaoping Yang

School of Information, Renmin University of China,
Beijing 100872, P.R. China
innocent.vivi@gmail.com, buaayuli@ruc.edu.cn, yang@ruc.edu.cn

Abstract. With the widely use of collaborative tagging system nowadays, users could tag their favorite resources with free keywords. Tag recommendation technology is developed to help users in the process of tagging. However, most of the tag recommendation methods are merely based on the content of tagged resource. In this paper, it is argued that tags depend not only on the content of resource, but also on user preference. As such, a hybrid personalized tag recommendation method based on user preference and content is proposed. The experiment results show that the proposed method has advantages over traditional content-based methods.

Keywords: Tag Recommendation, Collaborative Tagging, Web2.0.

1 Introduction

The appearance of Internet provides us a new way for information communication. Traditionally, the organization and management of information was done by the domain experts, for example, to determining the classes of book in the library. With the emerging of web2.0, thing has been changed. Including with experts, user could participate the organizing and classifying of information through collaborative tagging system. Collaborative tagging system (i.e. Delicious[1], Flickr[2], Last.fm[3]) that allows user to annotate resources with user-defined keywords is becoming a popular system for organizing and sharing online contents.

One of the important concepts in collaborative tagging is post. A post is a set containing a user, a resource and all the tags the user labelled on it. A collection of all the posts in a network refers to as a folksonomy. We can extract posts from these systems, then with some recommendation approaches, predict some tags that the user would probably use for the specifically resource. Tag recommendation not only can help user with their tagging, facilitate the tagging process, but also can show user interests and find some potential resources the user didn't know, and search for other users with the same interests to some extent.

---

* Corresponding author.

[1] delicious.com

[2] www.flickr.com

[3] www.last.fm

L. Cao, J. Zhong, and Y. Feng (Eds.): ADMA 2010, Part II, LNCS 6441, pp. 348–355, 2010.

However, most tag recommendation methods are content-based. In these methods, tag is same to keywords. But we argued that, tags are different from keywords: both of them are related to the context of tagged resources, while tags are also concerned with user preference. In other words, different people could use different tags as for the same resource. For instance, John could tag the web page "*www.sina.com*" with "*portal site*" or "*web portal*", yet Tom could use "*football news*" to label it because he is a football fan and he is just interested in the football news on *sina*. So it is important that user preference is considered while recommending tags. Hence we propose a hybrid personalized recommendation method based on user preference and resource content.

Three contributions are made in this paper:

1) Identifying that user preference should be considered when the tags are recommended for a resource.
2) Designing a hybrid tag recommendation method based on content and user preference.
3) Making experiments to show the improvement by taking user preference into consideration.

In next section, related work is surveyed in section 2. In section 3, tag recommendation in collaborative tagging system is formulated, and a basic tagging method based on content—Pop Resources (*PopRes*) is introduced. A hybrid personalized tag recommendation based on user preference and content is proposed in section 4. In section 5, the experiments are made to verify the advantage of the proposed method. Finally, the conclusions are given.

2 Related Work

Many researchers have been made on tag recommendation. The most popular techniques used currently can be roughly classified into two categories: content-based and graph-based techniques.

Most of the content-based techniques recommend tags for users through information retrieval, extracting keywords from the resource, for instance, Ref. [1] shows a simple content-based tag recommendation method. In the method, keywords from authors, abstract, title, etc. in the input bookmark are extracted, then the keywords are included in a weighted keyword-based query. This query represents an initial description of the input bookmark, then the authors search for similar bookmarks with this query, the tags that are potentially relevant for describing the input bookmarked resource are collected based on a set of similar bookmarks at last. Another content-based approach *PopRes* given in Ref. [2] is used in this paper.

One of the popular approaches in graph-based is the seminal PageRank algorithm presented by Brin and Page in Ref. [3] which reflects the idea that a web page is important if there are many pages linking to it, and if those pages are important themselves. In Ref. [4] the key idea of their FolkRank algorithm is that a resource which is tagged with important tags by important users becomes important itself.

Leandro et al. described an approach which they called relational classification based on graph-based tag recommendation in Ref. [5]. The so called Relational classification, usually considers, additionally to the typical attribute-value data of object, relational information. The idea is that, for example, a URL can be connected to another one if they were tagged by the same user. With this idea figures are drawn which the nodes are the set of user and resource while the edges are the connection between them.

Besides the above methods, there are some hybrid recommendation methods which combine content-based and graph-based methods. However, most of the methods didn't take user preference into account. So we argue that it is important to design a tag recommendation method by exploiting user preference.

3 Tag Recommendation in Collaborative Tagging System

3.1 Collaborative Tagging System

Collaborative tagging system is a collection of posts in a specific network, as shown in Fig. 1. Formally, it is a tuple $F = (U, T, R, Y)$ where U, T, R refers to users, tags, resources (in this paper means *URLs*, the same below), and Y is a ternary relation among them. For all $u \in U$ and $r \in R$, $tags(u,r) = \{t \in T | u \in U, r \in R\}$, so that a post would be described as $p = \{(u,S,r) | u \in U, r \in R, S = tags(u,r)\}$.

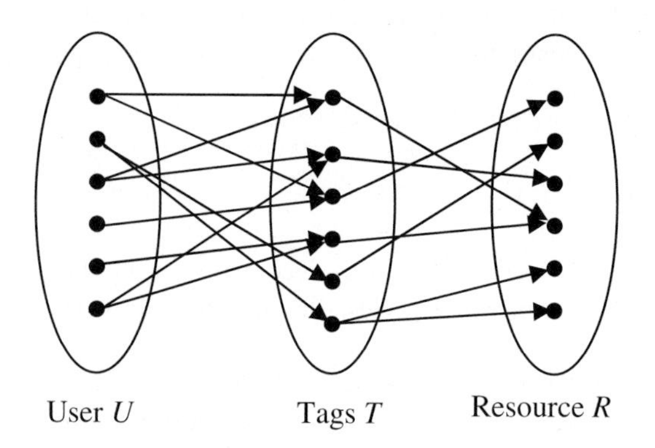

Fig. 1. Collaborative tagging system

According to the above, tag recommendation problem could be described as following. Given a set $\{u,r\}$ which means a user u is going to annotate URL r, with other user's tagging information in folksonomy, a set of tags is suggested for u.

3.2 *PopRes* Tag Recommendation

PopRes presented in Ref. [1] is one of commonly used content-based methods. A series of experiments was conducted in this paper with *PopRes* method as a baseline.

The *PopRes* is a popularity based models which relies on the frequency of a tag used. It ignores the user, and relies on the popularity of a tag within the context of a particular resource.

$$\omega(u,r,t) = \frac{f_{t,r}}{\sum_{i=1}^{i=n} f_{i,r}}. \quad (1)$$

Where $\omega(u,r,t)$ means the weight that tag t assigned to the input set$\{u,r\}$, and $f_{t,r}$ means the times that URL r is tagged by t in the database. Assuming that there are n tags, we can get a tag list $list(t_1,t_2,t_3 \dots,t_n)$ then, while $f_{i,r}$ represents the times that tag_i is attached to r. After that we rank the tags in list according to the value of ω and recommend the top k of them to u, Table 1 shows the algorithm of this process:

Table 1. *PopRes* Algorithm for Tag Recommendation

Input: An Folksonomy F, user u, Resource *URL* Output: top k tags $list(t_1,t_2,t_3 \dots,t_k)$
1. Finding the tag list $list(t_1,t_2,t_3 \dots,t_n)$ for r.
2. Ranking the $t_i, i \in [0,n]$ in *list* with $\omega(u,r,t)$.
3. Select first k tags and recommend them to u.

4 A Hybrid Personalized Tag Recommendation

4.1 Hybrid Tag Recommendation

In this section, a hybrid algorithm based on user preference and content is presented. The hybrid algorithm consists of two parts. First, *PopRes* method is used to generate candidate recommended tags. Based on it, user preference is considered to generate final recommended tags (shown in Table 2).

Table 2. Hybrid Algorithm for Tag Recommendation

Input: An Folksonomy F, user u, Resource *URL* Output: top k tags *list* $(t'_1, t'_2 \dots, t'_k)$
1. Generating candidate Tag list *CandidateTagList* and their corresponding weight according to the *PopRes* method: $\{tags, \omega'\}$= *PopRes*(F, *URL*) While: $tags = (tag_1, tag_2, \dots, tag_n)$, $\omega' = (\omega'_1, \omega'_2, \dots, \omega'_n)$
2. Searching the tags used by user u : $usedtag(u)$
3. f_j = used times of tag_j in $usedtag(u)$ by user u, (j=1,2, …, $\lvert usedtag(u) \rvert$)
4. $Max = \max\ \{ f_j, j = 1, 2, \dots, \lvert usedtag(u) \rvert\}$
5. Adjusting the weight of tags in *CandidateTagList*: $\omega''_i = \frac{f_{tag_i}}{Max}, \quad if\ tag_i \in usedtag(u)$ Otherwise $\omega''_i = 0,\ i \in [0,n]$
6. For each tag_i in *CandidatedTagList*, the two weights are combined in a certain hybrid tactics (detailed in part B).
7. Select first k tags and recommend them to u.

4.2 Hybrid Tactics

The hybrid recommendation could vary according to different hybrid tactics. In this section, three tactics were designed, respectively *Cascade Hybrid*, *Weighted Average*, and *Simple Multiplication*.

The first tactic will be used in step 6 which we called *Cascade Hybrid* is:

$$\omega_i = \omega'_i * (1 + \omega''_i), i \in [1, n] \ . \tag{2}$$

Moreover, another two hybrid methods are tested in our experiment. One of them is *Weighted Average*:

$$\omega_i = \alpha\omega'_i + (1 - \alpha)\omega''_i, i \in [1, n] \ . \tag{3}$$

The other hybrid solution is *Simple Multiplication*:

$$\omega_i = \omega'_i * \omega''_i, i \in [1, n] \ . \tag{4}$$

Although only three hybrid tactics in this paper are proposed, it is possible to design more hybrid tactics.

4.3 An Example of Hybrid Tag Recommendation

In the following part, an example is given to illustrate the recommendation process. The dataset is shown in Table 3.

Table 3. An example of Tagging dataset

User	Tag	URL
1	Menhu	www.sina.com
1	Ent.	bbs.ngacn.cc
2	Play	www.sina.com
2	Funny	www.xiaonei.com
3	Sc2	bbs.ngacn.cc
3	Play	bbs.ngacn.cc

Table 4. Statistics of BibSonomy dataset

Table name	Fields name	Statistics
tas	user, tag, content_id, content_type,	240k(181.5 k for bookmark)
bookmark	content_id, url, url_hash, description,	41k
bibtex	content_id, simhash1, title, etc.	22k

Now, it is assumed that our task is to predict which tags will be used by user 3 for the resource *www.sina.com*. First, by *PopRes* algorithm, two tags (*Menhu*, *Play*) with weight $\omega'(\frac{1}{3}, \frac{1}{3})$ will be recommended as candidate tags to user 3. According to the folksonomy, tags used by user 3 include two tags (*Sc2*, *Play*). Therefore, the adjusting weight $\omega''(0, 1)$. If *Cascade Hybrid* is exploited, the final weight combined ω' and ω'' can be achieved:

$$\omega = \omega'(1 + \omega'') = \left(\frac{1}{3} \times (1 + 0), \frac{1}{3} \times (1 + 1)\right) = \left(\frac{1}{3}, \frac{2}{3}\right) \cdot \tag{5}$$

If the number of recommended tags is one, tag *play* will be suggested to user 3.

5 Experiment

5.1 Datasets

The datasets used in this work are provided by ECML PKDD Discovery Challenge 2009[1]. It's a workshop for researchers to discuss problems concerned with collaborative tagging system. They offer *BibSonomy* database in form of SQL dump. *BibSonomy* is a web based social bookmarking system that enables users to tag web pages (bookmark) and scientific publication (bibtex), the Table 4 shows its statistics.

Discard the bibtex parts, we only focus on the bookmarks (URLs) in the dataset in our experiment. Testing data has the same tables and fields as the training data.

However there are two problems in the dataset. Firstly, for each content_id in the dataset, at most one user is associated with it, it's to say that content_id don't represent resources, so we have to join table "*tas*" and "*bookmark*" on the same content_id to get the ternary posts for our experiment. The second problem lies in the testing data. We find that nearly 90% posts in the testing data, whose resources don't exist in the training data. So we discard the testing data and extract a new one from the training data with the "*Leave One Out*" approach by ourselves. In other words, in the training data, we picked, for each user, one of his post p randomly and put it into testing data, and the remaining posts would be the new training data for this user.

In our experiment, 892 posts were extracted as new training data, and 108 posts as new testing data.

5.2 Evaluation Methodology

We will use the F1-Measure common in information retrieval to evaluate the recommendations. F1-Measure is defined as follow:

$$f1 = \frac{2*precision*recall}{precision+recall} . \qquad (6)$$

And as for Precision:

$$precision = \frac{1}{|U|}\sum_{u \in U}\frac{|T_{u,o} \cap T_{u,r}|}{|T_{u,r}|} . \qquad (7)$$

Where $T_{u,o}$ means the tags the user u has originally assigned to this post. Likewise, $T_{u,r}$ means the recommended tags for u. The Recall is defined as:

$$recall = \frac{1}{|U|}\sum_{u \in U}\frac{|T_{u,o} \cap T_{u,r}|}{|T_{u,o}|} . \qquad (8)$$

Apparently the Precision and Recall would be different with different number of recommended tags. We recommend 5 tags at most in our experiment.

5.3 Experiment Results

The performances of *PopRes* and *Cascade Hybrid* approaches are shown in Fig. 2, 3 and 4. If the number of recommended tags is 5, average F1-Measure value is 0.2382 in

---

[1] http://www.kde.cs.uni-kassel.de/ws/dc09/

PopRes approach, while average F1 is 0.2536 for personalized approach. In order to understand the effect of the number of recommended tags on experiment result, more experiments are made with varied numbers of recommended tag. As shown in Fig.6, no matter how many tags are recommended, hybrid personalized approach has a better performance than *PopRes*. With a decreasing number of recommended tags, F1-Measure value increased slightly with personalized method while it decreased with the *PopRes* method.

In general, personalized approaches are better than simple content-based approach. We obtain an improvement by 6.47% (0.2536 compares to 0.2382) of F1-Measure value if we recommend 5 tags. Furthermore, the improvement is more notable if we reduce the number of recommended tags. In other words, personalized approach becomes more reliable with fewer tags.

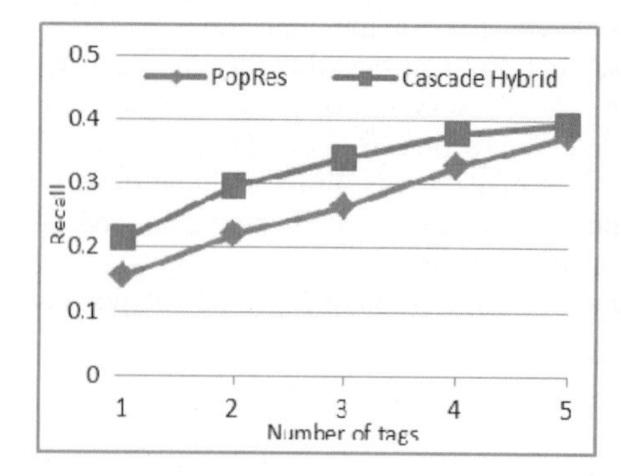

Fig. 2. Recall comparison of *PopRes* and *Cascade Hybrid*

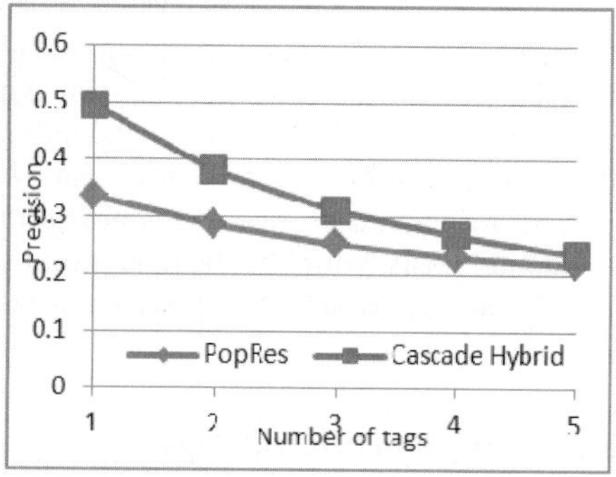

Fig. 3. Precision comparison of *PopRes* and *Cascade Hybrid*

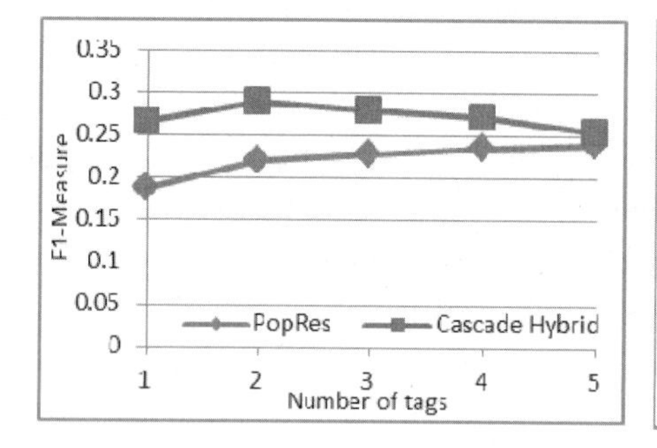

Fig. 4. F1-Measure comparison of *PopRes* and *Cascade Hybrid*

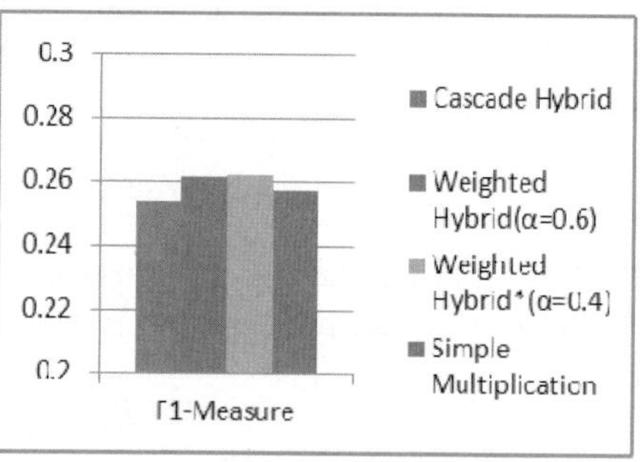

Fig. 5. F1-Measure Comparison between different hybrid methods

The Recall, Precision, F1-Measure of the three hybrid methods with 5 tags recommended were shown in Fig.5. The *weighted hybrid* method with á = 0.4 performs best in our experiment (F1-Measure= 0.2619). Apparently, every hybrid method is better than *PopRes*.

6 Conclusions

In this paper, a hybrid personalized tag recommendation based on user preference and content is proposed. By taking the tagging information and habit of the specific user into consideration, we are trying to emphasize the importance of user characteristics and personalities in tag recommendation and collaborative tagging. The experiment results show that our proposed method has advantages over the method only based on resource content. It is rational and necessary to consider user's characteristics in tag recommendation.

However, the main drawback of our approach is the cold-start problem: not only a big enough training data is necessary, but also our algorithm will stop working if the input resource doesn't exist in training data. We are going to modify the process with keywords retrieval methods or graph-based methods in the future.

Acknowledgments. This research was partly supported by National Science Foundation of China under grants No.70871115, RUC Planning on Publishing in International Journal under grants No.10XNK090. We also like to thank anonymous reviewers for their valuable comments.

References

1. Iván, C., David, V., Joemon, M.J.: Measuring Vertex Centrality in Co-occurrence Graphs for Online Social Tag Recommendation. In: Proceedings of ECML PKDD Discovery Challenge 2009, Bled, Slovenia (2009)
2. Jonathan, G., Maryam, R., Thomas, S., Laura, C., Bamshad, M.: A Fast Effective Multi-Channeled Tag Recommender. In: Proceedings of ECML PKDD (The European Conference on Machine Learning and Principles and Practice of Knowledge Discovery in Databases) Discovery Challenge 2009, Bled, Slovenia (2009)
3. Sergey, B., Lawrence, P.: The Anatomy of a Large-Scale Hypertextual Web Search Engine. Computer Networks and ISDN Systems 30(1-7), 107–117 (1998)
4. Robert, J., Leandro, M., Andreas, H., Lars, S.T., Gerd, S.: Tag Recommendations in Folksonomies. In: Workshop Proceedings of Lernen Wissensentdeckung Adaptivität LWA 2007, pp. 13–20. Martin-Luther-Universität Halle-Wittenberg (2007)
5. Leandro, B.M., Christine, P., Lars, S.T.: Relational Classification for Personalized Tag Recommendation. In: Proceedings of ECML PKDD Discovery Challenge 2009, Bled, Slovenia (2009)

Predicting Defect Priority Based on Neural Networks

Lian Yu[1], Wei-Tek Tsai[2,3], Wei Zhao[4], and Fang Wu[1]

[1] School of Software and Microelectronics, Peking University, Beijing, 102600, PRC
[2] Department of Computer Science and Engineering, Arizona State University, USA
[3] Department of Computer Science and Technology, Tsinghua University, Beijing, 100193 PRC
[4] IBM China Research Lab, Beijing, 100193, PRC

Abstract. Existing defect management tools provide little information on how important/urgent for developers to fix defects reported. Manually prioritizing defects is time-consuming and inconsistent among different people. To improve the efficiency of troubleshooting, the paper proposes to employ neural network techniques to predict the priorities of defects, adopt evolutionary training process to solve error problems associated with new features, and reuse data sets from similar software systems to speed up the convergence of training. A framework is built up for the model evaluation, and a series of experiments on five different software products of an international healthcare company to demonstrate the feasibility and effectiveness.

Keywords: Defect priority, evolutionary training, artificial neural network, attribute dependency, convergence of training.

1 Introduction

Software development and evolution constantly face resource constraints, and numerous faults or defects might be identified during the process. It is necessary for software engineers to determine fix priorities for those defects or faults. Defect fix priority (hereafter, simply called as defect priority) can also play an important role in regression testing 1 where high priority test cases can be selected for testing before low priority test cases. Traditionally, defect priority is determined manually by testers through examining system requirements. Recently researchers explore the approaches to automatically setting the defect priorities by examining previous test cases and determining test case dependency 2. This paper proposes a process to perform the prediction: first manually classifying defects, followed by automated learning by artificial neural networks (ANN, www.wikipedia.com), and predicting defect priorities with the learned ANN, as shown in Figure 1:

1. *Data Pre-process*: Analyze each bug report and extract the values of the pre-defined attributes, which are defined as *milestone*, *category*, *module*, *main workflow*, *function*, *integration*, *frequency*, *severity*, and *tester*. Every bug record corresponds to a structured data set, called defect data.
2. *Priority Prediction*: Get the defect data and loads ANN model parameters. If the bug data are training data, calculate the parameters of ANN model, then calculate the model error and adjusts the model parameters; if the bug data are new data,

L. Cao, J. Zhong, and Y. Feng (Eds.): ADMA 2010, Part II, LNCS 6441, pp. 356–367, 2010.

calculate the priority with the current model parameters, and deliver the predicted priority to the next procedure as prediction results.

3. *Exporting Results*: Sort the prediction results in decreasing order, and then export the prioritized records to managers/developers.

The paper is organized as follows. Section 2 describes the priority prediction model based on testing domain knowledge and neural network techniques. Section 3 demonstrates the experimental results. Section 4 surveys the related work. Finally, Section 5 concludes this paper.

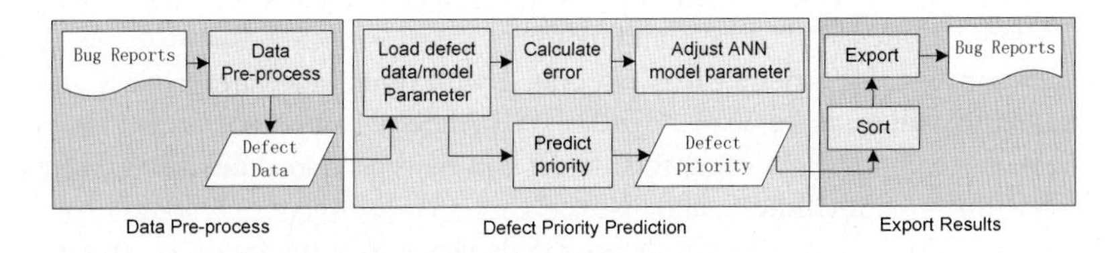

Fig. 1. Process of Defect Priority Prediction

2 Predicting Defect Priority

ANN can recognize complex relationship between inputs and output with high accuracy. This paper creates an ANN model to predict defect priorities where an extra layer is added to resolve the dependency among the attributes of *tester* and *severity*; and adopts interactive training instead of one-way training to handle adding new feature problems.

2.1 Building an ANN Model for Prediction

In this paper, the inputs of the ANN model are based on dynamic testing results, such as testing milestones and defect categories, while the output is the defect priority prediction. The following four steps illustrate the process to build the ANN model.

2.1.1 Defining Levels of Defect Priority

Based on the study of a bug management system for a software product of a CT medical device, we define four levels of defect priorities, P1, P2, P3, and P4. P1 means developers should drop everything else and fix the reported defect as soon as possible. The defects of P2 level should be fixed before the next build to be tested. The defects of P3 level should be fixed before the final release. The defects of P4 level are essentially improvement suggestions, and if there is a plenty time, developers will fix them.

2.1.2 Extracting Attributes from Defect Reports

This paper extracts 9 attributes from the dynamic defect reports as inputs of ANN model, which are *milestone*, *category*, *module*, *main workflow*, *function*, *integration*, *frequency*, *severity*, and *tester*. Take RIS (Radiology Information System) software as an example:

- Milestone: Indicate which milestone this project has passed, and its value can be M1, M2, M3, and M4.
- Category: Show which testing category this defect belongs to, and its value can be functional, performance, usability, and more.
- Module: Specify which module this defect belongs to, and its value can be *Register* module, *Report* module, and more.
- Main Workflow: Designate whether this defect was found in the main workflow, and its value is "Yes" or "No".
- Function: Signify which function point this defect is located in, and its value can be "Submit", "Reset", "Close", and more.
- Integration: Denote whether it is integrated with other software product and what it is, and its value can be "No", "EA21", "CPACS20", and more.
- Frequency: Suggest how often this defect can be reproduced, and its value can be "Every time", "Sometimes", and "Never".
- Severity: Connote how badly it incurs the damages to software users, and its value can be "Critical", "Major", "Average", and "Minor".
- Tester: Specify who is the defect submitter, and its value is the submitter's name.

Consider these input attributes as enumerate types. Let n_p be the number of values of the pth attribute. Define the pth attribute as vector $X_p=(a_{p,1},\ldots,a_{p,np})$, where variable $a_{p,q}$ ($p=1,\ldots, 9$, $q=1,..,n_p$) represents whether p^{th} attribute equals q^{th} value; when $a_{pq}=1$, it means the pth attribute equals to the qth value, and when $a_{pq}=0$, it means not equal. For example, if the value of 1st attribute "Milestone" equals M3, then vector $X_1=(0, 0, 1, 0)$, thus values $a_{11}=0$, $a_{12}=0$, $a_{13}=1$ and $a_{14}=0$ are part of the inputs.

2.1.3 Creating a Three-Layer Model of ANN

The ANN used in this paper has three layers: input layer, middle layer and output layer. L1, L2, and L3 represent the numbers of neurons in the input layer, middle layer and output layer, specifically in this case, $L1=\sum n_j$ ($j=1,\ldots,9$), $L2=9$, and $L3=1$. Let W1 be the weight matrix of middle layer. The entry element is $w1_{ij}$ ($i\in[1, L1]$, $j\in[1, L2]$), representing the weight between the ith neuron of the input layer and j^{th} neuron of the middle layer. Likewise, let W2 be the weight matrix of output layer. The entry element is $w2_{jk}$ ($j\in[1, L2]$, $k\in[1, L3]$), representing the weight between the jth neuron of the middle layer and the kth neuron of the output layer.

The *transfer* function of the middle layer can be linear, sigmoid or Gauss functions. In this paper, the *transfer* functions both in the middle and the output layers are using linear function because it is simple and meets the characteristics of the application domain. Therefore, the *transfer* function of the middle layer with 9 neurons is defined as follows:

$$b_j=\sum_{p=1}^{9}\sum_{q=1}^{n_p} w1_{ij}a_{pq},\quad j=1,2,\ldots 9, i=\sum_{a=1}^{p-1}n_a+q \tag{1}$$

The transfer function of the output layer with only 1 neuron is defined as follows:

$$c=\sum_{j=1}^{9} w2_j b_j \tag{2}$$

The threshold function of the output layer is using Gauss function. In our case, the defect priority level has the same range as the error does, i.e., the range [-0.5, 0.5]. Therefore, we define the activation function as follows:

$$\text{output } G(c) = \frac{1}{\sigma\sqrt{2\pi}} e^{-(c-\mu)^2/(2\sigma^2)} = [c+0.5] \tag{3}$$

[c+0.5] signifies that the defect priority is between 0 and 1.

2.1.4 Error Function and Backward Propagation

Once we have the output *c*, we calculate the error *E*, and use *c* and E to adjust weights in the output and middle layers using backward propagation (BP) approach.

1) Error Function *E*: Let c_{exp} be the expect result of the output, then error function *E* is defined as follows:

$$E = 1/2(c_{exp} - c)^2 \tag{4}$$

2) Back propagation (BP):

a. Adjust the output weight vector W2:

$$\Delta w2_j = -\eta \frac{\partial E}{\partial w2_j} = -\eta \frac{\partial E}{\partial c} \frac{\partial c}{\partial w2_j} = \eta \delta_j b_j, \text{ where } \delta_j = c_{exp} - c \tag{5}$$

b. Adjust the middle weight matrix W1:

$$\Delta w1_{ij} = -\eta \frac{\partial E}{\partial w1_{ij}} = -\eta \frac{\partial E}{\partial c} \frac{\partial c}{\partial b_j} \frac{\partial b_j}{\partial w1_{ij}} = \eta \delta_{ij} a_{pq}, \text{ where } \delta_{jk} = \delta_j w2_j \tag{6}$$

c. Setting learning speed η: To find out the most suitable learning speed, we performed several experiments and selected η equal to 0.1, 0.2, 0.3, 0.4, 0.5, 0.6, 0.7, 0.8, 0.9 and 1.0 respectively. We picked up 4 representative results to show the influence of different η as shown in Figure 2.

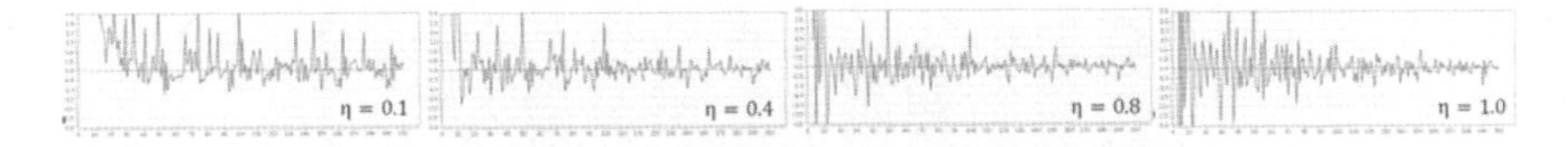

Fig. 2. Learning speed comparisons

When η equals to 0.1, the learning is very slow, and takes a long time to converge. When η equals to 0.4, it shows an obvious convergence trend. On the other hand, when η equals to 1.0, the sway of the error curve is larger than that when η equals to 0.8. In our experiments, we set η equal to 0.8.

2.2 Enhancing the Prediction Model

The classic BP neural network built in Section 2.1.3 is not entirely suitable for the collected data. We carry out several experiments to improve it. Initially we set the elements of weight matrixes W1 and W2 to 1, and performed the model training on 4 data sets separately. We utilized the mean square deviation of sample error to evaluate the convergence of the model training. The convergence of the error is not obvious from sample size 50 to sample size 170 as shown in Figure 3.

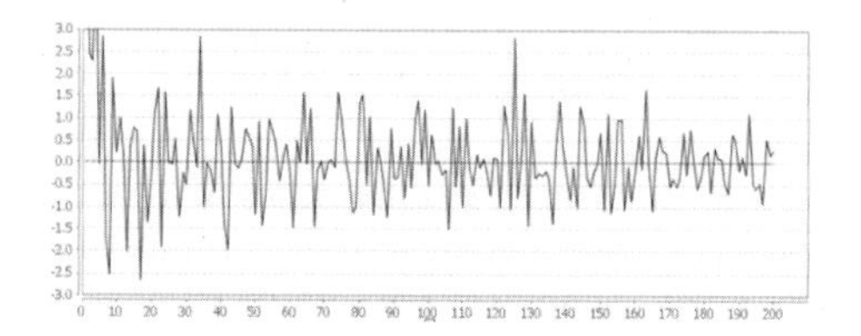

Fig. 3. Error statistic results

By examining the data characteristics, we found that there is a correlation between the "*tester*" input and the "*severity*" input. Each tester has its own prejudice over severity assessment due to different knowledge and experience. We selected a total of 60 defects from a medical project with 15 defects from each of the 4 severity levels. The severities of these defects were assessed by three experienced testers beforehand. We invited 10 testers to assess the severity. Tester1 and Tester2 have more than one year of testing experience, and the other eight testers are new. The evaluation results are shown in Figure 4. It is observed that the ratings of Tester1 and Tester2 are similar to the ratings by the experienced testers (that is, about 15 defects in each severity level), while the rest of testers have significant fluctuations among themselves and also deviate from what the experienced testers rated. This rating difference is considered as *subjective impact* because different testers have their own assessment preferences.

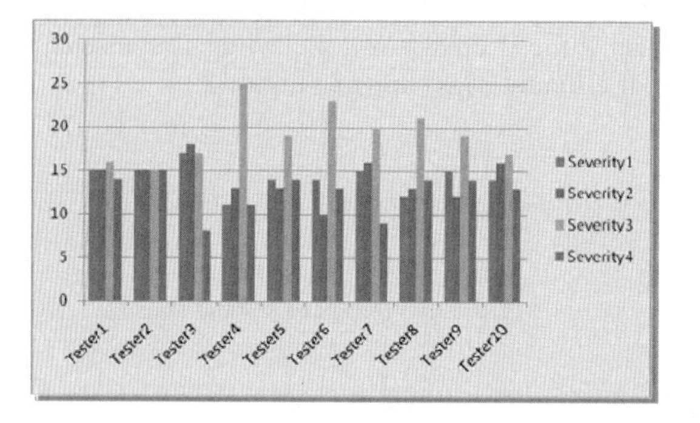

Fig. 4. Severity & Tester relations

In the formal ANN model, we can adjust the tester's lower or higher tendency in general, but the assessed severity level is more complicated, it is not always lower or higher than the actual severity level. For example, Tester4 has the tendency to assess the 1-level and 2-level of defects to lower, but 3-level of defects to much higher.

We noticed that one character of BP neural networks requires that there is no interaction between the neurons in a same layer. To fix the subjective impact problem between "*tester*" neuron and "*severity*" neuron, and maintain the net structure, we propose to add an extra layer, and move the tester neurons to a new layer. As a result, the tester neurons and the severity neurons are changed from parallel to serial as shown in Figure 5:

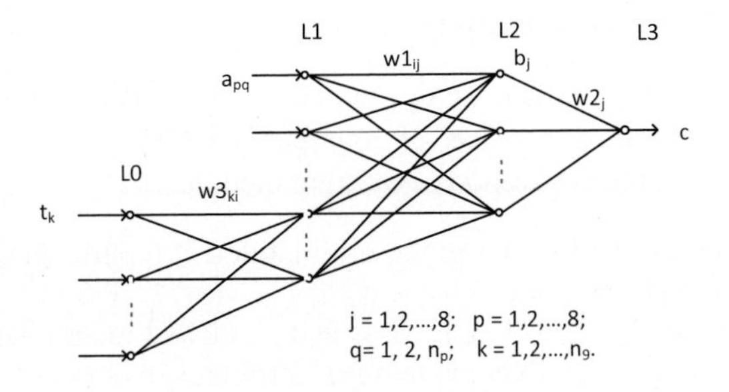

Fig. 5. Dynamic Prediction Model

Let W3 be the weight matrix on the extra layer. Element $w3_{ki}$ ($i \in [1, 4]$, $k \in [1, n_9]$) of W3 represents the weight between the ith severity neuron on L1 layer and the kth neuron of the extra layer L0 where n_9 is the number of testers. We define the input of severity attributes as transfer function of severity neurons.

Our expectation is that the subjective impact does not exist, hence initially we set $w3_{ki} = 1$. The difference between the sample severity s and the predicted one s_p is a value between -3 to 3. So the adjusting weight function on the extra layer is:

$$w3_{ki} = 1 - 1/3(s - s_p) \tag{7}$$

With the enhanced prediction model, we use the same samples as in Figure 3 to verify the enhancement efficiency as shown in Figure 6. It is observed that the error of samples from 50th to 170th is smaller than Figure 3, and the error curve looks more smoothly.

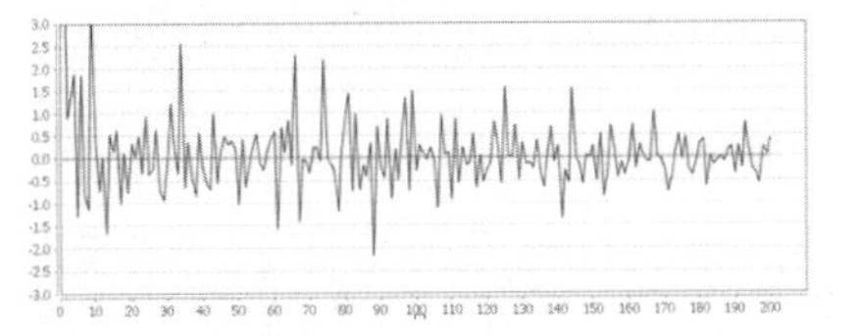

Fig. 6. Error Statistic graph

3 Experiments

To evaluate the proposed approaches, we developed an experiment platform, which provides GUI to interact with the underlined system. We did three experiments to demonstrate the ways that we applied neural networks to achieve the goals: 1) In Section 3.2, by reusing the training data from similar systems as initial values of the model, we can accelerate the convergence. 2) In Section 3.3, we adopted an iterative approach and accommodated feedback into training to reduce the errors in the prediction. 3) In Section 3.4, we compared the results using neural network with those using Bayes.

3.1 Building an Experiment Platform

To evaluate the prediction model, we developed an experiment platform. The overall architecture of the platform consists of four parts: GUI, data pre-process, defect priority prediction module, and reports exporting module.

- GUI: It is a portal for loading in the original defect reports and displaying error curve after prediction.
- Data pre-process: It parses bug records using a defect record parser and extracts the corresponding data (ANN model input attributes in our case) for prediction.
- Defect priority prediction: It has the training module and prediction module with the enhanced ANNs.
- Reports exporting: It sorts bug records in descending order according to predicted priority and then exports it as a new file of defect reports.

3.2 Reusing Training Data from Similar Systems

This experiment addresses the effort of initial value setting on the training convergence. To get a faster convergence, we did several experiments to investigate whether reusing the training model of other software product as the initial model of a new software product will raise the efficiency in ANN model training.

By default, initial values are set up as follows: at the output layer, $w2_8 = 2$ (corresponding to "severity" and "tester"), and $w2_i = 1$ ($i=1,\dots,7$); and elements of weight matrixes W1 and W3 are set to all 1.

We collected four data sets DS1, DS2, DS3 and DS4 from different software products, including HIS (Hospital Information System), software development tools, ERP system and word processing software from an international medical device maker. The attributes and the number of values are listed in Table 1.

Table 1. Data Set complexity description

	DS1	DS2	DS3	DS4
Milestone*	4	4	4	4
Severity*	4	4	4	4
Frequency*	3	3	3	3
Main Workflow*	2	2	2	2
Category	4	4	4	4
Integration	4	1	0	2
Module	4	3	4	3
Function	6	5	5	4
Tester	4	4	5	4

The attributes marked with "*" symbol signify that these attributes values are in common between the earlier software product and the new software product. So the training results of these attributes can be applied to a new software prediction. To evaluate the effectiveness, we performed the training for new software RIS (Radiology Information System) with the 5th data set using default initial values, and performed the trainings again using the training results of the 4 data sets as initial values. The error curves of prophase training are shown in Figure 7.

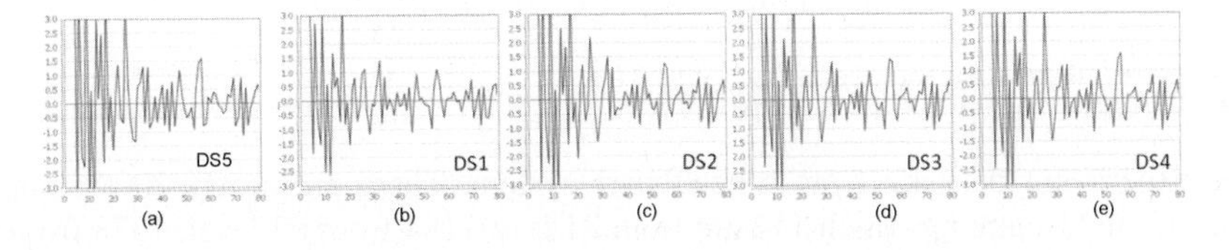

Fig. 7. Training beginning comparison

Figure 7(a) shows the training with default initial values. The other four figures show the trainings using the DS1, DS2, DS3 and DS4 date sets as initial values. Compared with the default one, the other four results all show smaller errors in certain sample points, and the DS1 is more effective than other data sets. Because DS1 (from Hospital Radiology Information System) and DS5 (from Radiology Information System) belong to same software type, medical management system, from the same company, thus they have similar characteristics and the result is as expected.

3.3 Handling New Feature Problems

Most large-scale software development adopts an iterative development model, which brings the new feature problem. At the beginning of the experiment, although faults occurred in the 4th build, no adjustment to the ANN model was performed, thus we got faults again on the new feature in the 5th build. This issue will continuously reduce the accuracy of the model. To fix this problem, and to adjust the model with the migration on which the software development focused, we decided to adopt an *iterative* training. After every build of prediction, we use the feedback to train the model again iteratively, and in the next build, the defect related to these new features will be predicted more precisely, as shown in Figure 8.

The left chart in Figure 8 is the predicted result of one-way training， and the right one is the predicted result of iterative training. Before the 300th sample defect is the training period where defect data are from the 1st to 3rd builds; between 300th defect and 375th is the prediction period 1, where defect data are from the 4th build; between 375th defect and 450th is the prediction period 2, which defect data are from the 5th build.

The training data associated with the 1st to 3rd builds cover 5 values of the function attribute and 5 values of the module attribute. In prediction periods associated with 4th and 5th builds, there are 6 values of the function attribute and 6 values of the module attribute, i.e., in the prediction period, there is a new feature added. It is observed that the prediction results for both 4th and 5th builds shown in the left chart of Figure 8 have large errors. Taking 4th build as training data iteratively, the prediction of 5th build has no large errors as shown on the right chart of Figure 8.

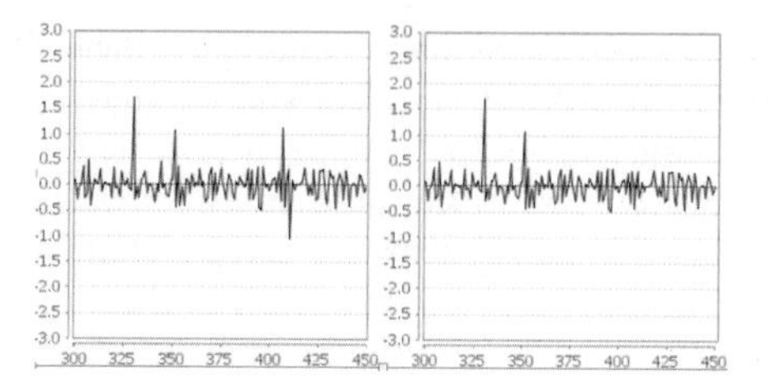

Fig. 8. Error curve of New Feature problem

3.4 Comparing with Bayes Approach

The experiment data are from a healthcare system. There are nearly 2000 samples from 13 builds, among which 112 are from P1 level, 184 from P2 level, 1378 from P3 level, and 763 from P4 level. We selected 600 samples as the testing data set, which are 30 from P1 level, 100 from P2 level, 300 from P3 level and 170 from P4 level. There are three ways to prepare the data sets:

- Closed test: All the sample data for training, taking part as a test set;
- Opening test: All samples will be divided into two parts, one part as the training set, and the other part as the test set;
- N-fold cross test: Divide all samples into N divisions, and pick N-1 divisions for training sets, the other one as a testing set with N times closed or opening tests. Calculate the average of the N times results as the evaluation result.

We use 3-fold cross test for both the closed test (CT) and opening tests (OT) to predict the defect priorities. Three metrics, Recall (R), Precision (P) and F-measure (F) are used to compare the results between ANN model with iterative training method and Bayes method. Let A be the amount of defects that are predicted with correct results, B be the total amount of defects that are predicted as priority level C, and T be the total amount of defects that are with priority level C in the experiment.

$$\begin{aligned} P &= A / B \\ R &= A/T \\ F1 &= (2 \times P \times R) / (P + R) \end{aligned} \tag{8}$$

Table 2 **Error! Reference source not found.**shows the prediction results by enhanced ANN and Bayes on data sets using CT and OT. Regarding to the enhanced ANN, CT approach has better results over that of OT in terms of precision and recall. It is observed that P3 and R1 of Bayes reach the highest value 1.00. P3=1.00 means the sample defects which are classified into priority level 3 are all correct. R1=1.00 means the sample defects which belong to priority level 3 were all classified into level 3. These two criterions only demonstrate part of the model performance. We should consider both of them to evaluate the model efficiency. Although P3 and R1 of Bayes equal to 1.00, R3 are only 0.71 (CT) and 0.63 (OT), which means 29% and 37% defects of level 3 are not classified correctly, and P1 are 0.21 and 0.19, which means 79% and 81% of the defects which had been classified into this level are wrong. So the F1 criterion can reflect the comprehensive ability of models. From the statistic of F1, ANNs show better performance than Bayes model in the classification of every priority level.

Table 2. Dynamic Prediction experiment results

	Enhanced ANN		Bayes	
	CT	OT	CT	OT
P_1	0.88	0.76	0.21	0.19
R_1	0.88	0.62	1.00	1.00
P_2	0.85	0.84	0.30	0.31
R_2	0.87	0.82	0.89	0.92
P_3	0.95	0.96	1.00	1.00
R_3	0.96	0.91	0.71	0.63
P_4	0.89	0.80	0.33	0.22
R_4	0.83	0.82	0.98	0.97
$F1_1$	0.880	0.683	0.347	0.319
$F1_2$	0.861	0.830	0.449	0.464
$F1_3$	0.955	0.934	0.830	0.773
$F1_4$	0.859	0.810	0.494	0.359

The test results of opening test reflect the generalization ability of models, and insufficient training will cause the low generalization ability. In Table 2, $F1_1$ of OT is much lower than $F1_2$, $F1_3$, $F1_4$, and it is caused by insufficient training. We added 30 samples into the testing set, and we got a better performance, the $F1_1$ of OT was raised to 0.742.

The test results of opening test reflect the generalization ability of classification model. Over-training will raise the efficiency of opening test result, but the generalization ability will be reduced. So the F1 can reflect the comprehensive ability of classification model. Based on the above experiment results, the efficiency of ANN model used in the defect priority prediction platform is better than the Bayes model.

4 Related Work

Because of its high accuracy and the ability to recognize the complex relationship between attributes, neural network has been widely used in many fields including software testing. In this paper, we present the ways to predict the priority of fixing defects with the neural network algorithms. We can find the similar work in some others papers, but all these works focus on prediction of the amounts of bugs or finding bugs in code.

Rocha et al 3 present two hybrid *EC/ANN* algorithms in datasets experiments. In the paper, the *Multilayer Perceptron*, a popular of *ANN* architecture, is the core of their research, which can merge the evolutionary computation (EC). The combination, called *Evolutionary Neural Networks (ENN)*, is a suitable candidate for topology design and can create more accurate models comparing other data mining algorithms. The authors present two *ENNs*, including *TENN* (*Topology-optimization Evolutionary Neural Networks*) and *SENN* (*Simultaneous Evolutionary Neural Network*). The major tasks of this work were the following: 1) compare with other kind data mining algorithms, such as regression; 2) propose the evaluation of several methods to build

ANN. The writer does not give the exact usages of all the algorithms mentioned in the paper to predict the bugs, which will be explored in our future work.

Bug patterns detectors 45, which are implemented by using BCEL 6, can only detect the bugs in one program. The paper gives a way to find bugs, rather than a way to determine the priority of bugs. Using the method to the experiment, the detectors can find the faults of code correctly.

The statistics method in many papers is always used to predict the number of bugs, such as the z-Ranking 7 and negative binomial regression 8. Ostrand presents the negative binomial regression to locate the faults and make testers concentrate on the testing and developers on developing.

Khoshgoftaar et al 9 use discriminate analysis to develop two classification models that predict whether or not a module will be fault-prone, rather than attempting to predict the actual number of faults.

Graves et al. 10 present a study using the fault history for the modules of a large telecommunication system. They focus on identifying the most relevant module characteristics or groups of characteristics and comparing the apparent effectiveness of the proposed models, rather than predicting faults and using them to guide testing.

Mark et al 11 give the induced data mining models of tested software which can be utilized for recovering missing and incomplete specifications, designing a minimal set of regression tests, and evaluating the correctness of software outputs when testing new, potentially flawed releases of the system.

In Engler's paper 2, they present a new rule to finding program faults. The paper demonstrates techniques that automatically extract such checking information from some source code, rather than the programmer, thereby avoiding the need of a priori knowledge of system rules.

Kremenek et al. 12 suggest that effective error report ranking provides a complementary method to increasing the precision of the analysis results of a checking tool. The tools PREfix and MC employ various rule-based ranking heuristics to identify the most reliable reports 13. There are some dynamic and static techniques to check the defects, such as the system DIDUCE 1415.

5 Conclusions

This paper proposes to use ANN techniques to predict defect priorities. The contributions of this paper include building a neural network model of the prediction, and proposing several strategies to increase the accuracy of the model based on the testing domain knowledge. This paper proposes three strategies to handle different aspects in applying ANN techniques: (1) add an extra layer to resolve the dependency between attributes *severity* and *tester*; (2) adopt evolutionary training to accommodate new features of software by adding feedback into next round training; and (3) reuse training data from similar systems as initial values to speed up the training for new data set. The 3-fold cross-test in both the closed test and opening test ways are executed. Compared with Bayes algorithm, the ANN model shows better qualification in terms of recall, precision and F-measure. The comparison is carried on the RIS2.0 software project with sample size about 2000 bug reports.

Acknowledgments. This work is partially supported by the National Science Foundation of China (No.60973001), IBM China Research Lab (No.20090101), and U.S. Department of Education FIPSE project. The authors would thank Yang Cao, Jingtao Zhao and Jun Ying for working on data collection and the empirical study described in this paper.

References

1. Rothermel, G., Untch, R.H., Chu, C., Harrold, M.J.: Prioritizing Test Cases for Regression Testing. IEEE Transactions on Software Engineering (2001)
2. Dawson, E., David, Y.C., Seth, H.: Bugs as Deviant Behavior: A General Approach to Inferring Errors in Systems Code (2003)
3. Miguel, R., Paulo, C., José, N.: Evolution of neural networks for classification and regression. Neurocomputing (2007)
4. David, H., William, P.: Finding Bugs is Easy. ACM SIGPLAN Notices (2004)
5. Michael, D.: Static and dynamic analysis: synergy and duality. IEEE Computer Society, Los Alamitos (2005)
6. The Byte Code Engineering Library, `http://jakarta.apache.org/bcel/2003`
7. Ted, K., Dawson, E.: Using Statistical Analysis to Counter the Impact of Static Analysis Approximations. In: 10th International Symposium (2003)
8. Thomas, J., Ostrand, E.J., Weyuker, R.M.: Predicting the Location and Number of Faults in Large Software Systems. IEEE Transactions on Software Engineering (2005)
9. Khoshgoftaar, T.M., Allen, E.B., Kalaichelva, K.S.: Early Quality Prediction: A Case Study in Telecommunications, IEEE Software, pp. 65–71 (1996)
10. Graves, T.L., Karr, A.F., Marron, J.S.: Predicting Fault Incidence Using Software Change History. IEEE Tran., Software Reliability Eng. 26, 653–661 (2000)
11. Mark, L., Menahem, F., Abraham, K.: The Data Mining Approach to Automated Software Testing. ACM, New York (2003)
12. Kremenek, T., Ashcraft, K., Yang, J.: Correlation Exploitation in Error Ranking. In: Proceedings of the 12th ACM SIGSOFT Twelfth International (2004)
13. Bush, W., Pincus, J., Sielaff, D.: A static analyzer for finding dynamic programming errors. In: Software: Practice and Experience, pp. 775–802 (2000)
14. Hallem, S., Chelf, B., Xie, Y.: A system and language for building system-specific, static analysis. In: PLDI (2002)
15. Hallem, S., Lam, M.S.: Tracking down software bugs using automatic anomaly detection. In: International Conference on Software Engineering (2002)

Personalized Context-Aware QoS Prediction for Web Services Based on Collaborative Filtering

Qi Xie, Kaigui Wu, Jie Xu, Pan He, and Min Chen

College of Computer Science,
Chongqing University,
Chongqing 400044, China
{xieqi,kaiguiwu,hepan,chenmin}@cqu.edu.cn,
scsjx@leeds.ac.uk

Abstract. The emergence of abundant Web Services has enforced rapid evolvement of the Service Oriented Architecture (SOA). To help user selecting and recommending the services appropriate to their needs, both functional and nonfunctional quality of service (QoS) attributes should be taken into account. Before selecting, user should predict the quality of Web Services. A Collaborative Filtering (CF)-based recommendation system is introduced to attack this problem. However, existing CF approaches generally do not consider context, which is an important factor in both recommender system and QoS prediction. Motivated by this, the paper proposes a personalized context-aware QoS prediction method for Web Services recommendations based on the SLOPE ONE approach. Experimental results demonstrate that the suggested approach provides better QoS prediction.

Keywords: Context, QoS Prediction, Collaborative Filtering, Web Service.

1 Introduction

The Service Oriented Architecture (SOA) is "an architecture that represents software functionality as discoverable services on the network" [1]. Web Services are the dominant implementation platform for SOA, it uses a set of standards, SOAP, UDDI, WSDL, which enable a flexible way for applications to interact with each other over networks [2]. The increasing presence and adoption of Web Services call for effective approaches for Web Service selection and recommendation, which is a key issue in the field of service computing [3]. Not only functional but also nonfunctional quality of services (QoS) should be taken into account to help users selecting and recommending the services. The QoS includes service reliability, availability, performance metrics (e.g., response time), scalability, (transactional) integrity, security, trust and execution cost, etc.

Web Service QoS prediction is an important step in selecting services [4]. Since changes of QoS happen due to various reasons, such as number of concurrent users, performance of back end systems, network latency, invocation failure-rate, etc. In addition, there is a problem that QoS changes are observed at run time. So it is critical

L. Cao, J. Zhong, and Y. Feng (Eds.): ADMA 2010, Part II, LNCS 6441, pp. 368–375, 2010.

to predict the QoS before using it. Currently, many researches focus on this prediction problem according to users' QoS date acquired by user. Though this method can obtain accurate results, there are some shortcomings. For example, it is impossible for user to use all candidate services because it requires extra cost on service invocation and provision. Furthermore, most of service users are not experts on the Web Service evaluation, and the common time-to-market constraints limit an in-depth evaluation of the target Web Services [5].

To solve these problems, collaborative filtering (CF) method is introduced to predict QoS for unused services [5]. This method is based on the assumption that the service users, who have similar historical QoS experience on the same set of Web Services, would have similar experience on other services. But these differences on the quality of the same Web Services are caused by many context factors, such as location and time. Users/Web Services in near location may have similar experience on Round-Trip Time (RTT), failure-rate than others. However, existing approaches generally do not consider context in real QoS date prediction with CF. To improve the accuracy of prediction of the QoS, in this paper, we present our work on personalized context-aware QoS prediction based on SLOPE ONE.

The major contributions of this work include:

* we propose a personalized context-aware QoS prediction framework.
* we employ an effective and simple collaborative filtering algorithm (SLOPE ONE) to improve the QoS prediction accuracy.
* A personalized context-aware QoS prediction based on SLOPE ONE is designed for Web Service prediction, which significantly improves the QoS prediction accuracy and remarkably reduces the computing complexity.

The remainder of this paper is organized as follows. Section 2 overviews some concepts e.g., the CF algorithms and context of Web Services. Section 3 presents the approach for personalized context-awareness QoS prediction. The implementation, experiments, and the results of our method are discussed in Section 4. Conclusion of this paper is described in Section 5.

2 Background

CF is a widely used personalized recommendation method in recommender systems. Basically, there are two kinds of CF algorithms, user based and item based approaches [6, 7]. User-based CF is the most successful recommending technique to date, and is extensively used in many commercial recommender systems [6]. It predicts a test user's interest in a test item based on rating information from similar user profiles [7]. Multiple mechanisms such as Pearson Correlation and Cosine based similarity are widely used. Item-based CF methods are similar to the user-based methods. Item-based CF employs the similarity between items and then to select the most similar items for prediction.

The SLOPE ONE is a typical item-based CF. It works on comparing the intuitive principle of a popular differential between items rather than similarity between items [8]. Daniel Lemire defined the SLOPE ONE the average deviation (1) and prediction (2) as the following:

$$dev_{i,j} = \sum_{u \in U(i) \cap U(j)} \frac{r_{u,i} - r_{u,j}}{|U(i) \cap U(j)|}, \quad (1)$$

$$p_{u,i} = \overline{r_u} + \frac{\sum_{j \in R_u} dev_{i,j}}{|R_u|}. \quad (2)$$

The $dev_{i,j}$ is computed by the average difference between item arrays of i and j (1). $\overline{r_u}$ is the vector of average of all known ratings rated by user u. R_u are the all items' ratings rated by u except i. $|R_u|$ is the number of R_u. $|U(i) \cap U(j)|$ is the number of users who rate both item i and item j.

Context is any information that can be used to characterize the situation of an entity [9]. Context -aware computing refers to the ability of a software application to detect and respond to changes in its environment [10]. The idea of the context-aware CF is based that similar users in similar context like similar items [11]. Context is of three types in our method: user (U-context), Web Service (WS-context), and service provider (SP-context).

3 Personalized Context-Aware QoS Prediction

3.1 Deployment

Fig. 1 illustrates the framework that is proposed to predict the QoS for Web Services based on Context-aware CF. The Personalized Context-aware QoS Prediction based on CF (PCQP) can automatically predict the QoS performance of a Web Service for a test user by using historical QoS information from other similar service users on the same set of commonly-invoked Web Services.

The framework consists of the following procedures: 1) Query. A service user submits a Web Service request. 2) Select. The candidates that meet the functional service quality attributes proposed by the user are retrieved by employing Web Service discovery mechanism. 3) Context. Personal U-context and WS-context are considered. 4) Find similar users and services from the training data. 5) Predict. The values of QoS are generated for test users using PCQP. 6) Recommender. It provides test user the optimal Web Services according to the predicted QoS values.

The core parts in this framework are how to find the most similar users and Web Services, and how to predict the missing QoS values more efficiently and effectively. Due to space restrictions, this paper only researches the similar Web Service item and context of service, the others will be studied in future work.

3.2 Identification of Context

Context-awareness allows software applications to use information beyond those directly provided as input by users [12]. In Web Services scenario, the information includes such as location, date, time related to users and services.

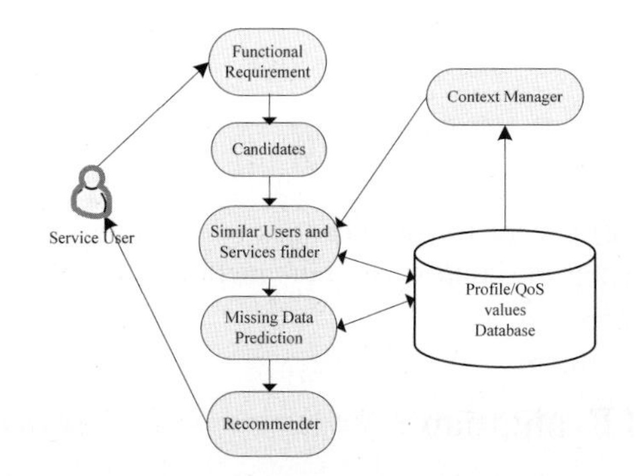

Fig. 1. Personalized context-aware QoS prediction framework overview

We adopt recommendation space (RS) [13] to present PCQP. In RS, the key dimensions could be used as QoS values. For example, if the RS includes user, service item and service location, the QoS value q(u,i,l) lies in a three-dimensional space U×I×L. Every entry in RS presents the QoS values (e.g., RTT), which is observed by the service user u on the Web Service item i that belongs to l. The q(u,i,l)=0 if user u did not use the Web Service item i in location l before.

Existing approaches for QoS prediction based on CF tend to compute the similarity in the whole user-service matrix. Our work is very different from these methods since we employ the information of context related to users and services. According to the reduction RS of similarity computation, our method minimizes the computational cost while satisfying the QoS value prediction accuracy of Web Services.

3.3 Deviation Computation

In this work, the SLOPE ONE based on context for the deviation computation of different service items is defined as:

$$dev_{i,j^*} = \sum_{u \in U(i) \bigcap U(j^*)} \frac{r_{u,i} - r_{u,j^*}}{| U(i) \bigcap U(j^*) |} \tag{3}$$

The difference between (1) and (3) is the RS of deviation computation. The RS of SLOPE ONE based context is the three-dimensional space U×I×L while the one without context is two dimensional space U×I. For example, to predict the $P_{u,i}$, we should compute the deviation between i and other service item j. In our context method the space of j becomes the item j* that belongs to the location in which item i is provided.

3.4 Missing Value Prediction

The SLOPE ONE methods use differential service items to predict the missing value (QoS) for test users by using the following equation:

$$p_{u,i} = \overline{r_{u*}} + \frac{\sum_{j^* \in R_{u^*}} dev_{i,j^*}}{|R_{u^*}|} \quad (4)$$

This is the same as (2) except for the changes of variables space(e.g., $dev_{i,j}$, r_u, R_u). Owing to the applying of context in our method, the RS is limited to three-dimensional space U×I×L.

4 Experimental Evaluation

4.1 Data Set and Evaluation Metric

In the experiment we use the dataset from the WSRec including a user-contribution mechanism for Web Service QoS information collection, which is implemented by java language and deployed to the real-world environment [5, 14]. This dataset includes 100 Web Services and 150 computer nodes (users) which are distributed in 22 countries. We obtain a 150×100 user-service item matrix from this database, where each entry in the matrix represents the QoS value (RTT).

The error metric used most often in the CF literature is the Mean Absolute Error (MAE) , which is described as:

$$MAE = \frac{\sum_{i,j} |r_{i,j} - \hat{r}_{i,j}|}{N}, \quad (5)$$

where $r_{i,j}$ is the involved QoS value of service item j experienced by user i, $\hat{r}_{i,j}$ presents the predicted QoS value for missing value. In addition, N is the number of predicted values.

Because the different QoS properties of Web Services have different value ranges [5], we employed the Normalized Mean Absolute Error (NMAE) to measure the deviation between predictions and QoS values (RTT). To compare the NMAE employed in other methods more effectively, we adopt the NMAE proposed by ZiBin Zheng [5] as follows:

$$NMAE = \frac{MAE}{\sum_{i,j} r_{i,j} / N}. \quad (6)$$

The lower the NMAE is, the higher the prediction quality is.

4.2 Performance Comparison

To research the predicted results of QoS values, we first compare the SLOPE ONE with other prediction approaches used widely in CF, such as: UPCC, IPCC. Then SLOPE ONE and personalized context-aware prediction method based on the SLOPE ONE (PCSP) will be compared in section 4.3. The total item of QoS values are divided into two parts randomly, one as the training set and the other as the test set. A parameter λ is employed to present the percentage of the matrix entries which are selected to be the training set.

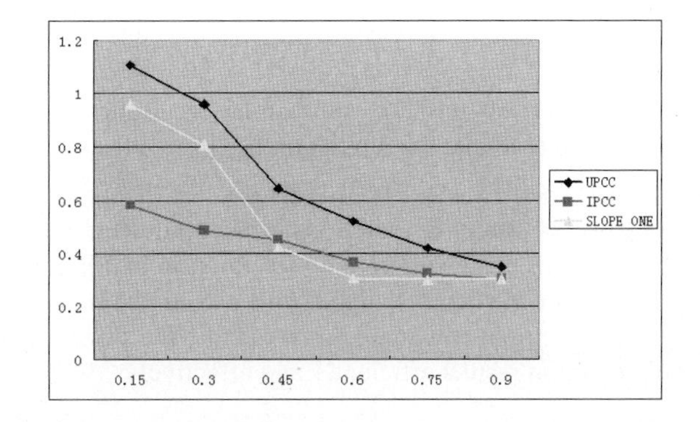

Fig. 2. Normalized mean absolute error rank for three prediction methods

In the experiment, we set λ from 0.15 to 0.9, with 0.15 as increment. The x-axis represents different value of λ, and the y-axis shows the NMAE values. As shown in Fig. 2, the prediction accuracy of UPCC is worse than other methods. The NMAE of SLOPE ONE is higher than IPCC at first two steps and consistently lower than IPCC in the following three steps. The NMAE of SLOPE ONE reaches the optimum value (0.297) when the λ is 0.75. The reason is that when the number of training matrix is small, the number of differential service items of SLOPE ONE is very limited and with the λ increasing more differential service items will appear which will provide more service items with low deviation and make the accuracy improvement more significant. After the step of 0.75 both the SLOPE ONE and IPCC perform worse. This is because too many training data appear diversity influencing the prediction performance.

4.3 Impact of the Context of Service Items

Context plays an important role in our approach, which is quite different from other methods. To research the performance of the context, we employ a simple experimental scenario as Fig. 3.

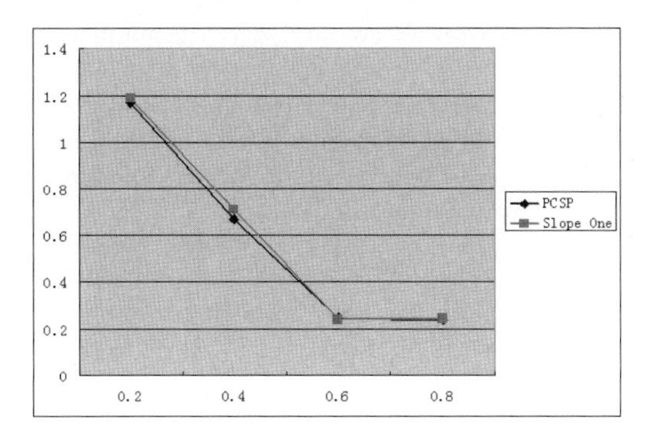

Fig. 3. Impact of the Context of Service Items

In Fig. 3, the Web Service items only belonging to China and Spain will be chose to constitute a new dataset with Web Service item size of 80-150. The new dataset is also divided into training set and test set. The value of λ is set from 0.2 to 0.8 with 0.2 as increment. From Fig. 3 we note that the prediction accuracy consistently become better with the increase of λ value. The performances of PCSP get very close to the SLOPE ONE. The PCSP obtains the minimal NMAE value (0.2356) when λ is set to 0.8.

Owing to the calculations of deviation and prediction in PCSP are implemented in the three-dimensional space U×I×L, the PCSP will efficiently decrease the computational complexity. These experiments data represent that the PCSP algorithm is helpful to improve the prediction results efficiently and effectively.

5 Conclusion

With regard to the context, an effective personalized Context-aware QoS Prediction framework for Web Services based on Collaborative Filtering is proposed. The experiments analysis presents the efficiency and effectiveness of our approach.

However, there is no similarity weight computation in our work which may enhance the accuracy with missing value. One reason is that our method only focuses on the impact of the context. Using too many factors to impact the predictions may compromise the experimental results. In future work, different QoS properties and multi-context will be integrated with collaborative filtering method to predict QoS values.

Acknowledgments. This work is supported by the Major Research Project of the National Natural Science Foundation of China under Grant No.90818028 and the Third Stage Building of "211 Project" (Grant No. S-10218).

References

1. Klusch, M., Kapahnke, P.: Semantic Web Service Selection with SAWSDL-MX. In: 2nd Workshop on Service Matchmaking and Resource Retrieval in the Semantic Web (SMR2), pp. 3–18 (2008)
2. Sathya, M., Swarnamugi, M., Dhavachelvan, P., Sureshkumar, G.: Evaluation of QoS based Web Service Selection Techniques for Service Composition. J. International Journal of Software Engineering
3. Zhang, L.J., Zhang, J., Cai, H.: Services computing. Springer and Tsinghua University Press, Beijing (2007)
4. Shao, L.S., Zhang, J., Wei, Y., Zhao, J., Xie, B., Mei, H.: Personalized QoS prediction for web services via collaborative filtering. In: ICWS, pp. 439–446 (2007)
5. Zheng, Z.B., Ma, H., Michael, R.: L., King, I.: WSRec: A Collaborative Filtering based Web Service Recommender System. In: 7th IEEE International Conference on Web Services (ICWS 2009), Los Angeles, CA, USA, pp. 6–10 (2009)
6. Resnick, P., Iacovou, N., Suchak, M., Bergstrom, P., Riedl, J.: GroupLens: An open architecture for collaborative filtering of Netnews. In: Conference on Computer Supported Cooperative Work, pp. 175–186. Chapel Hill, NC (1994)
7. Breese, J.S., Heckerman, D.: Kadie. C.: Empirical analysis of predictive algorithms for collaborative filtering. In: UAI, pp. 43–52 (1998)

8. Lemire, D., Maclachlan, A.: Slope One Predictors for Online Rating-Based Collaborative Filtering. Society for Industrial Mathematics (2005)
9. Dey, A.K.: Understanding and using context. J. Personal and Ubiquitous Computing 5, 4–7 (2001)
10. Roman, G.C., Julien, C., Murphy, A.L.: A Declarative Approach to Agent-Centered Context-Aware Computing in Ad Hoc Wireless Environments. In: 2nd Inter. Work. Software Engineering for Large-Scale Multi-Agent Systems (SELMAS 2002), Orlando, Florida, USA (2002)
11. Gao, M., Wu, Z.H.: Personalized Context-Aware Collaborative Filtering Based on Neural Network and Slope One. In: Luo, Y. (ed.) Cooperative Design, Visualization, and Engineering. LNCS, vol. 5738, pp. 109–116. Springer, Heidelberg (2009)
12. Dey, A.K., Abowd, G.D., Salber, D.: A Conceptual Framework and a Toolkit for Supporting the Rapid Prototyping of Context-aware Applications. J. Human Computer Interaction Journal 16, 97–166 (2001)
13. Weng, S.S., Lin, B.S., Chen, W.T.: Using Contextual Information and Multidimensional Approach for Recommendation. J. Expert Systems with Applications, 126–127 (2009)
14. Zhang, Y.L., Zheng, Z.B., Michael, R.: L.: WSExpress: A QoS-aware Search Engine for Web Services. In: 8th International Conference on Web Services, USA, pp. 5–10 (2010)

Hybrid Semantic Analysis System – ATIS Data Evaluation

Ivan Habernal and Miloslav Konopík

University of West Bohemia, Department of Computer Sciences,
Univerzitní 22, CZ - 306 14 Plzeň, Czech Republic
{habernal,konopik}@kiv.zcu.cz

Abstract. In this article we show a novel method of semantic parsing. The method deals with two main issues. First, it is developed to be reliable and easy to use. It uses a simple tree-based semantic annotation and it learns from data. Second, it is designed to be used in practical applications by incorporating a method for data formalization into the system. The system uses a novel parser that extends a general probabilistic context-free parser by using context for better probability estimation. The semantic parser was originally developed for Czech data and for written questions. In this article we show an evaluation of the method on a very different domain – ATIS corpus. The achieved results are very encouraging considering the difficulties connected with the ATIS corpus.

Keywords: semantic analysis, semantic parsing, spoken language understanding, ATIS.

1 Introduction

Recent achievements in the area of automatic speech recognition started the development of speech-enabled applications. Currently it starts to be insufficient to merely recognize an utterance. The applications demand to understand the meaning. Semantic analysis is a process whereby the computer representation of the sentence meaning is automatically assigned to an analyzed sentence.

Our approach to semantic analysis is based upon a combination of expert methods and stochastic methods (that is why we call our approach a hybrid semantic analysis). We show that a robust system for semantic analysis can be created in this way. During the development of the system an original algorithm for semantic parsing was created. The developed algorithm extends the chart parsing method and context-free grammars. Our approach is based upon the ideas from the Chronus system [1] and the HVS model [2].

At first the hybrid semantic analysis method is described in this article. Then we test how the method can be adapted to a domain (ATIS corpus, English data, spoken transcriptions) which is very different from the original data (LINGVOSemantics corpus, Czech data, written questions). The last part of the article shows the results achieved on both domains and it compares our results with a state-of-the-art semantic analysis system [15].

L. Cao, J. Zhong, and Y. Feng (Eds.): ADMA 2010, Part II, LNCS 6441, pp. 376–386, 2010.

2 Related Work

A significant system for stochastic semantic analysis is based on HVS model (hidden vector-state model) [2]. The system was tested on the ATIS and DARPA corpora, recently the system was also used for semantic extraction from bioinformatics corpus Genia [13]. The first model traning was based on MLE (maximum likelihood estimation), however, the discriminative training has also been proposed. According to our knowledge, the system presented in [13] achieved the state-of-the-art performance on the ATIS corpus.

An extension of the basic HVS Parser has been developed in the work of [16]. The improvement is achieved by extending the lexical model and by allowing left-branching. The system was tested on Czech human-human train timetable corpus and it is public available.

SCISSOR (Semantic Composition that Integrates Syntax and Semantics to get Optimal Representations) is another system which uses the syntactic parser enriched with semantic tags, generating a semantically augmented parse tree. Since it uses the state-of-art syntactic parser for English, the Collin's parser, we suppose, that it can not be easily adapted to other languages.

In [18], the generative HMM/CFG composite model is used to reduce the SLU slot error rate on ATIS data. Also a simple approach to encoding the long-distance dependency is proposed. The core of the system is based conditional random fields (CRF) and the *previous slot context* is used to capture non-local dependency. This is an effective and simple heuristic but the system requires a set of rules to determine whether the previous slot word is a filler or a preamble. Thus, it is difficult to port the system to other domain.

3 Semantic Representation

There are several ways how to represent semantic information contained in a sentence. In our work we use tree structures (see Figure 1) with the so-called *concepts* and *lexical classes*. The theme of the sentence is placed on the top of the tree. The inner nodes are called concepts. The concepts describe some portion of the semantic information contained in the sentence. They can contain other sub-concepts that specify the semantic information more precisely or they can contain the so-called lexical classes. Lexical classes are the leaves of the tree. A lexical class covers certain phrases that contain the same type of information. For example a lexical class "date" covers phrases "tomorrow", "Monday", "next week" or "25th December" etc.

The described semantic representation formalism uses the same principle as it was originally described in [2]. The formalism is very advantageous since it does not require annotation of all words of a sentence. It makes it suitable for practical applications where the provision of large scale annotation training data is always complicated.

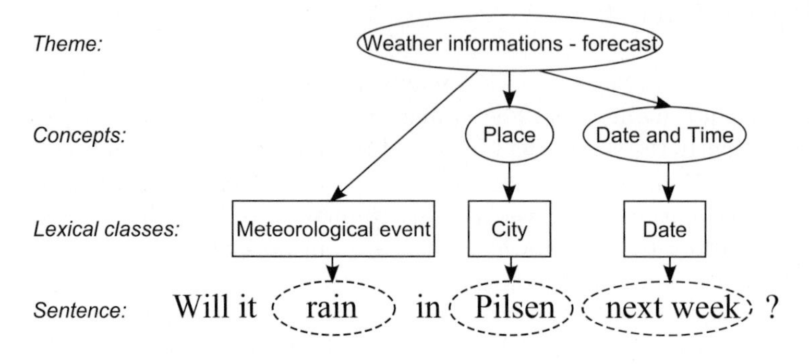

Fig. 1. An example of a semantic annotation tree

4 Data

This section describes two corpora used for training and testing. The Czech corpus (LINGVOSemantics corpus) was used at the beginning for the development of the method. The English corpus (ATIS) was used to find out whether the designed method is universal and can be successfully used for a different corpus. The second reason for using the ATIS corpus is to compare our method with the state-of-the-art system that has been also tested on the ATIS corpus.

4.1 LINGVOSemantics Corpus

The data used during the development are questions to an intelligent Internet search engine. The questions are in the form of whole sentences because the engine can operate on whole sentences rather than just on keywords as usual. The questions were obtained during a system simulation. We asked users to put some questions into a system that looked like a real system. In this way we obtained 20 292 unique sentences. The sentences were annotated with the aforementioned semantic representation (Section 3). More information about the data can be found in [11].

An example of the data follows (*How warm will it be the day after tomorrow?*):
WEATHER(Jaká EVENT(teplota) vzduchu bude DATETIME(DATE (pozít?í))?)

4.2 ATIS Corpus

One of the commonly used corpora for testing of semantic analysis systems in English is the ATIS corpus. It was used for evaluation in e.g. [2], [3], [4] and [5]. The original ATIS corpus is divided into several parts: ATIS2 train, ATIS3 train, two test sets etc. [6].

The two testing sets NOV93 (448 sentences) and DEC94 (445 sentences) contain the annotation in the semantic frame format. This representation has the same semantic expressive ability as the aforementioned semantic tree representation (Section 3). Each sentence is labeled with a goal name and slot names with associated content.

The corpus does not contain any fixed training set. Originally in [2], 4978 utterances were selected from the context independent training data in the ATIS2 and ATIS3 corpora and abstract semantics for each training utterance were derived semi-automatically from the SQL queries provided in ATIS3.

At this point we have to thank Y. He for sharing their data. It allowed us to test our system on the same testing data that uses their state-of-the-art system for semantic analysis. However, deep exploration revealed that the training data are specially tailored for the HVS model. The data were in the form of HVS model stacks and the conversion from stacks to proper trees was ambiguous and difficult. However, we still were able to use the test data (the test data are stored in the semantic frame format) and the plain sentences from the training data.

Instead of trying to convert the training data from HVS stacks or obtaining the original SQL queries and converting them we decided to annotate a part of the ATIS corpus using the abstract semantic annotation (see section 3). Using the methodology described in [13] we have initially created an annotation scheme[1] from the test data. In the first step, 100 sentences from ATIS2 train set were manually annotated. Thereafter the system was trained using this data. Another set of sentences was automatically annotated and then hand-corrected (this incremental methodology of annotation is called *bootstrapping*). In total, we annotated 1400 random sentences from both ATIS2 and ATIS3 training set.

5 System Description

The system consists of three main blocks (see Figure 2). The *preprocessing* phase prepares the system for semantic analysis. It involves sentence normalization, tokenization and morphological processing. The *lexical class analysis* is explained in Section 5.1 and the *probabilistic parsing* is explained in Section 5.2.

5.1 Lexical Class Identification

The lexical class identification is the first phase of the semantic analysis. During this phase the lexical classes (see Section 3) are being found in the input sentence.

The lexical class identification consists of two stages. First, several dedicated parsers are run in parallel. During this stage a set of lexical classes are found. In the second stage the lexical classes are stored in a lattice. Then the lattice is converted into possible sequences of lexical classes. Only the sequences that contain no overlapping classes are created.

During the first step the lexical classes are being found as individual units. We found in our data two groups of lexical classes:

[1] An annotation scheme is a hierarchical structure (a tree) that defines a dominance relationship among concepts, themes and lexical classes. It says which concepts can be associated with which super-concepts, which lexical classes belong to which concepts and so on. More in [13].

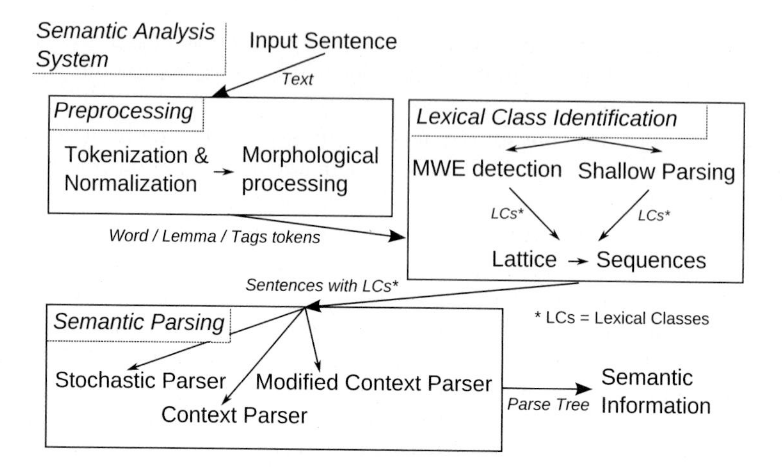

Fig. 2. The structure of our semantic analysis system

1. Proper names, multi-word expressions, enumerations.
2. Structures (date, time, postal addresses, ...).

To analyze the first group we created a database of proper names and enumerations (cities, names of stations, types of means of transport etc). Since a lexical class can consist of more than one word it is necessary to look for multiple word expressions (MWEs). To solve the search problem effectively a specialized searching algorithm was developed.

The main ideas of the searching algorithm are organizing the lexical classes in the trie structure [12] and using parallel searching. The algorithm ensures that all lexical classes (possibly consisting of more words) are found during one pass with a linear complexity $O(n)$ (where n is the number of letters of the sentence). The algorithm is explained in [11] in details.

For the analysis of complicated phrases, such as dates, numbers, time, etc., the LINGVOParser [9] was developed. It is an implementation of a context-free grammar parser. The parser has some features suitable for semantic analysis. It uses the so-called *active tags*. Active tags contain processing instructions for extracting semantic information (for more information see [9]).

Another feature of the LINGVOParser is the one we call *partial parsing*. Turning the partial parsing on causes the parser to scan the input sentence and build the partial parse trees wherever possible (a standard parser usually requires to parse the whole input from the start to the end). Partial trees do not need to cover whole sentences and they are used to localize the lexical classes described by a grammar.

During the second stage the lexical classes found in the first stage are put into a lattice. Then the lattice is walked through and the sequences of lexical classes that do not overlap are created. The result of the algorithm is the sequences of lexical classes. The algorithm uses the dynamic programming to build the sequences effectively.

The active tags used in the LINGVOParser allow the system to formalize the content of lexical classes. By formalizing we mean for example the transformation of date time information into one unified format. We can also transform spoken number into their written forms etc. This step is crucial for practical applications where it is required to express the same type of information in the same way [11].

5.2 Semantic Parsing

In the previous section the process of finding lexical classes was described. In this section we will presume that the lexical classes are known and the semantic tree is being built. The structure of the tree is shown in Figure 1. The task of the parsers described here is to create the same trees as in the training data.

Stochastic Parser. The parser works in two modes: training and analysis. The training phase requires annotated training data (see Section 4). During the training the annotation trees are transformed to context free grammar rules in the following way. Every node is transformed to one rule. The node name makes the left side of the rule and the children of the node make the right side of the rule (for example see node "Place" in Figure 1, this node is transformed into the rule `Place -> City`). In this way all the nodes of the annotation tree are processed and transformed into grammar rules. Naturally, identical rules are created during this process. The rules are counted and conditional probabilities of rule transcriptions are estimated:

$$P(N \to \alpha|N) = \frac{\text{Count}(N \to \alpha)}{\sum_{\gamma} \text{Count}(N \to \gamma)} , \tag{1}$$

where N is a nonterminal, α and γ are strings of terminal and nonterminal symbols.

The analysis phase is in no way different from standard stochastic context-free parsing. The sentence is passed to the parsing algorithm. The stochastic variant of the active chart parsing algorithm (see e.g. [7]) is used. The lexical classes identified in the sentence are treated as terminal symbols. The words that are not members of any lexical class are ignored. The result of parsing – the parse tree – is directly in the form of the result tree we need.

During parsing the probability of the so far created tree $P(T)$ is computed by:

$$P(T) = P(N \to A_1 A_2 ... A_k|N) \prod_i P(T_i) , \tag{2}$$

where N is the top nonterminal of the subtree T, A_i are the terminals or nonterminals to which the N is being transcribed and T_i is the subtree having the A_i nonterminal on the top.

When the parsing is finished a probability is assigned to all resulting parse trees. The probability is then weighted by the prior probability of the theme and the maximum probability is chosen as the result:

$$\hat{T} = \arg\max_i P(S_i)P(T_i) , \tag{3}$$

where $\hat{T}$ is the most likely parse tree and $P(S_i)$ is the probability of the starting symbol of the parse tree T_i.

Context Parser. The context parser looks at other words of the sentence rather than looking at lexical classes only. For this purpose it was necessary to extend both the training algorithm and the analysis algorithm.

The training phase shares the same steps with the training of the previous parser in Section 5.2. The node is thus transformed into the grammar rule and the frequency of the rule occurrence is counted. However, instead of going to the next node, the context of the node is examined. Every node that is not a leaf has a subtree beneath. The subtree spans across some terminals. The context of the node is defined as the words before and after the span of the subtree. During training the frequency of the context and a nonterminal (Count($word, nonterminal$)) are counted. The probability of a context given a nonterminal is computed via MLE as follows:

$$P(w|N) = \frac{\text{Count}(w, N) + \lambda}{\sum_i \text{Count}(w_i, N) + \lambda V} , \tag{4}$$

where λ is the smoothing constant, V is the estimate of the vocabulary size, w is the actual context word and w_i are all the context words of nonterminal N.

Additionally, to improve the estimate of the prior probability of the theme (the root node of the annotation) we add words to the estimate as well:

$$P(w|S) = \frac{\text{Count}(w, S) + \kappa}{\sum_i \text{Count}(w_i, S) + \kappa V} , \tag{5}$$

where κ is the smoothing constant, w_i are the words of the sentence and S is the theme of the sentence (the theme constitutes the starting symbol after annotation tree transformation).

The analysis algorithm is the same as in the previous parser but the probability from formula 2 is reformulated to consider the context:

$$P(T) = \sum_i P(w_i|N)P(N \to A_1 A_2 ... A_k|N) \prod_j P(T_j) . \tag{6}$$

Then the best parse is selected using context sensitive prior probability:

$$P(\hat{T}) = \arg\max_i P(S_i) \prod_j (P(w_j|S)P(T_i) , \tag{7}$$

where S_i is the starting symbol of the parse tree T_i and w_j are the words of the analyzed sentence.

Modifications for Morphologically Rich Languages. We tried to further improve the performance of parsing algorithms by incorporating features that consider the specific properties of the Czech language. The Czech language is a morphologically rich language [8] and it has also a more flexible word order than for example English or German. To deal with specific properties of the Czech language *lemmatization* and *ignoring word order* features were incorporated to the Context Parser.

6 Performance Tests

6.1 Results on the LINGVOSemantics Corpus

Figure 3 shows the results for the LINGVOSemantics corpus (Section 4.1). The figure shows performance of the base line parser (Stochastic Parser – section 5.2), the novel parser (Context Parser – section 5.2) and the modifications of the Context Parser (section 5.2).

The results are measured in the *accuracy* and *f-measure* metrics. Both the metrics use the slot-value pairs for computation. The slot is the name of a slot and its path in the slot hierarchy and the value is the value of the slot. Accuracy and F-measure are standard metrics for multiple outputs (one semantic tree or semantic frame consists of more slot-value pairs) and the formulas are defined in [2]. We use the same metrics as in [2] for sake of mutual comparison of our results and the results in [2].

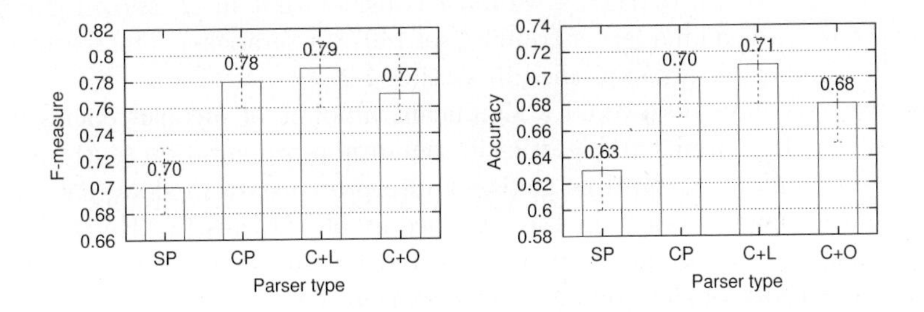

Fig. 3. Results of Semantic Parsers. *SP* = Stochastic Parser, *CP* = Context Parser, *C+L* = Context + Lemma, *C+O* = Context + word Order. Confidence intervals are computed at confidence level: $\alpha = 95\%$.

6.2 Results on the ATIS Corpus

The adaptation of the system to the ATIS corpus consisted of two steps. First, an appropriate English context-free grammar covering the date, time and numbers was created for the LINGVOParser (see section 5.1). Aditionally, the multiword expression identificator was re-trained using the data from ATIS `*.tab` files that contain cities, airports, etc. Second, the semantic parser was trained using the training set described in section 4.2.

Figure 4 shows the results on the ATIS corpus depending on training data size. The training sentences were chosen randomly, ten times from each set. The

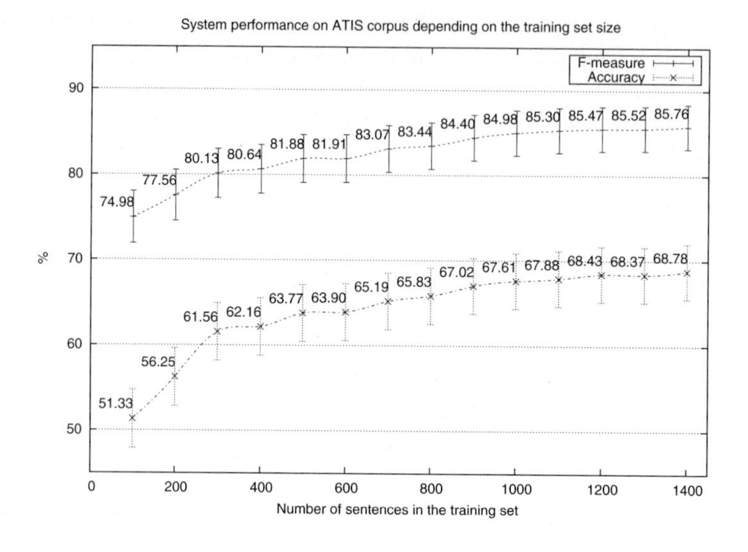

Fig. 4. System performance on the ATIS corpus with various amount of training data. Confidence intervals are computed at confidence level: $\alpha = 95\%$.

best result achieved on 1400 training sentences was 85.76%. When compared to [2] (89.28%) or [15] (91.11%), we must consider that in [2] and in [15] their systems were trained on a larger training set (4978 utterances). The reasons that we used a smaller set are explained in section 4.2.

However, we have discovered a significant amount of inconsistencies in the ATIS test set. It contains ambiguities in semantic representation (e.g. two same slots for one sentence), multiple goals, or interpreted data in slots (e.g. airport names which do not appear in the sentence at all).[2] Thus, we think that the performance testing on this corpus is affected by the testing set inconsistency and the objectivity of the evaluation is compromised.

7 Conclusions

This article described the hybrid semantic parsing approach. The tests performed on ATIS data show that we almost reached the performance of the state-of-the-art system on a reduced training data set. It is probable that by fine-tuning the system and by annotating the full training set, the system could be capable of reaching the state-of-the-art performance. We, however, consider the results sufficient to prove that the hybrid approach with the context parser is worth of further development.

To compare our system with a very similar system (described in [15]) it can be concluded that a significant progress was made in two areas. Firstly, the

[2] The examples are shown at http://liks.fav.zcu.cz/mediawiki/index.php/Interspeech_2010_Paper_Attachment

annotation methodology was improved. It is is now faster and more fault-proof. Secondly, our system is prepared to be used in practical applications by using data formalization. In [15] the lexical classes are automatically learned without the possibility of data formalization. We however use a hybrid approach where the data formalization is used. In the near future we are going to publish papers on the results achieved under real conditions.

Acknowledgements. This work was supported by grant no. 2C06009 Cot-Sewing. We would like to thank Mrs. Yulan He for sharing the ATIS semantic corpus.

References

1. Pieraccini, R., Tzoukermann, E., Gorelov, Z., Levin, E., Lee, C.-H., Gauvain, J.-L.: Progress Report on the Chronus System: ATIS Benchmark Results. In: Proc. of the workshop on Speech and Natural Language, pp. 67–71. Association for Computational Linguistics, New York (1992)
2. He, Y., Young, S.: Semantic processing using the Hidden Vector State model. Computer Speech and Language 19(1), 85–106 (2005)
3. Iosif, E., Potamianos, A.: A soft-clustering algorithm for automatic induction of semantic classes. In: Interspeech 2007, Antwerp, Belgium, pp. 1609–1612 (2007)
4. Jeong, M., Lee, G.: Practical use of non-local features for statistical spoken language understanding. Computer Speech and Language 22(2), 148–170 (2008)
5. Raymond, C., Riccardi, G.: Generative and discriminative algorithms for spoken language understanding. In: Interspeech 2007, Antwerp, Belgium, pp. 1605–1608 (2007)
6. Dahl, D.A., et al.: ATIS3 Test Data, Linguistic Data Consortium, Philadelphia (1995)
7. Allen, J.: Natural Language Understanding. Benjamin/Cummings Publ. Comp. Inc., Redwood City, California (1995)
8. Grepl, M., Karlík, P.: Skladba ?e?tiny, Olomouc, Czech Republic (1998)
9. Habernal, I., Konopík, M.: Active Tags for Semantic Analysis. In: Sojka, P., Horák, A., Kopeček, I., Pala, K. (eds.) TSD 2008. LNCS (LNAI), vol. 5246, pp. 69–76. Springer, Heidelberg (2008)
10. Jurafsky, D., Martin, J.: Speech and Language Processing. Prentice-Hall, Englewood Cliffs (2000)
11. Konopík, M.: Hybrid Semantic Analysis, PhD thesis, University of West Bohemia (2009)
12. Knuth, D.: The Art of Computer Programming, 2nd edn. Sorting and Searching, vol. 3. Addison-Wesley, Reading (1997)
13. Habernal, I., Konopík, M.: Semantic Annotation for the LingvoSemantics Project. In: Matoušek, V., Mautner, P. (eds.) TSD 2009. LNCS (LNAI), vol. 5729, pp. 299–306. Springer, Heidelberg (2009)
14. Konopík, M., Habernal, I.: Hybrid Semantic Analysis. In: Matoušek, V., Mautner, P. (eds.) TSD 2009. LNCS (LNAI), vol. 5729, pp. 307–314. Springer, Heidelberg (2009)
15. He, Y., Zhou, D.: Discriminative training of the hidden vector state model for semantic parsing. IEEE Trans. on Knowl. and Data Eng. 21(1), 66–77 (2009)

16. Jurčíček, F.: Statistical approach to the semantic analysis of spoken dialogues. PhD thesis, University of West Bohemia, Faculty of Applied Sciences (2007)
17. Kate, J., Mooney, R.J.: Using string-kernels for learning semantic parsers. In: ACL- 44: Proceedings of the 21st International Conference on Computational Linguistics and the 44th Annual Meeting of the Association for Computational Linguistics, Morristown, pp. 913–920 (2006)
18. Wang, Y., Deng, L., Acero, A.: An introduction to statistical spoken language understanding. IEEE Signal Processing Magazine 22(5), 16–31 (2005)

Click Prediction for Product Search on C2C Web Sites

Xiangzhi Wang, Chunyang Liu, Guirong Xue, and Yong Yu

Department of Computer Science and Engineering, Shanghai Jiao Tong University, Shanghai 200240, China
{xzwang,cyliu,grxue,yyu}@apex.sjtu.edu.cn

Abstract. Millions of dollars turnover is generated every day on popular ecommerce web sites. In China, more than 30 billion dollars transactions were generated from online C2C market in 2009. With the booming of this market, predicting click probability for search results is crucial for user experience, as well as conversion probability. The objective of this paper is to propose a click prediction framework for product search on C2C web sites. Click prediction is deeply researched for sponsored search, however, few studies were reported referred to the domain of online product search. We validate the performance of state-of-the-art techniques used in sponsored search for predicting click probability on C2C web sites. Besides, significant features are developed based on the characteristics of product search and a combined model is trained. Plenty of experiments are performed and the results demonstrate that the combined model improves both precision and recall significantly.

Keywords: Click Prediction, Logistic Regression, Ecommerce, C2C.

1 Introduction

The past decade is an exponentially increasing period for online ecommerce business. China, as one of the fastest growing markets in the world, generated 30 billion dollars trade volume from online C2C web sites in 2009. More than 200 million registered users attract more than 200,000 sellers set up online-store on popular C2C web sites every month. With the booming of products in the database, each search might result in thousands of items matching user's query. While more and more products can be selected in front of the users, it is increasingly hard for products search engine to rank the item list and predict correct item for buyer to click and go deeper after that action. We summarize that the problem of predicting the click probability on search results is crucial for product search based on three fold motivations:

- Click probability can be used as an important score to rank the list of search results. As the cascade model illustrated in [1], the higher position the item with more click probability is ranked with, the less energy will be cost for user to make click decision, which in turn improves user experience.

L. Cao, J. Zhong, and Y. Feng (Eds.): ADMA 2010, Part II, LNCS 6441, pp. 387–398, 2010.

– Simply, the conversion probability can be defined as:

$$P(purchase|item) = P(purchase|click) \cdot P(click|item) \ . \qquad (1)$$

So if item with more click probability can be selected in to the result page, a bargain will be concluded more possibly, which in turn increase the revenue of the company. Of course this point might be debatable as C2C web sites should also consider the fairness of presenting items on search result page.
– The click prediction information can be leveraged in both browsing and searching advertisements.

Therefore, in this paper, we address the problem of predicting the click probability on items for online C2C web sites. One of the most straightforward methods is to predict click probability from historical CTR(click through rate), however, this historical information is unavailable for new items published. Moreover, historical information is sometimes biased while product popularity changing frequently with time passed.

Click prediction has been deeply researched in domain of sponsored search recently. [2,3,4,5] proposed state-of-the-art methodologies which can be referred as possible solutions. We build a model with features extracted from the log of a C2C web site as proposed in these papers and validate the performance of the model on search results of C2C web site. The evaluation results prove that the techniques used in sponsored search also work in C2C product search. However, we conclude the characteristics of online C2C web site differentiate from web search engine site as follows:

– Unlike users of web search engine with definite goals to populate queries, more than half of online buyers are surfing on the C2C web sites without any definite searching objective. Buyers can even make a search without any query terms by clicking links while browsing the web pages.
– With complicated search patterns provided by C2C web sites, online buyers have more control on the result page. As shown in Fig. 1, most C2C web sites allow buyers to select *sort type* for result list, including price ascending, price descending, credit ascending, etc. Also, buyers can add various *filters* for the retrieval process, such as price range, item location and so on.
– As shown in Fig. 1, the result page of C2C web site is more delicate than web search result page. With item picture, price, rank of seller credit, shipping rate and other item related information being presented on the search result page, buyers can compare items comprehensively before a click action delivered.

Based on above points, a novel model more suitable for product search can be built. We explore a significant set of features based on the characteristics of C2C web sites as described above and combine them to the original model. A large data corpus is collected from a real online C2C web site, and plenty of experiments are performed. The results demonstrate that significant improvements are obtained after the C2C site-based features is combined to the model. To analyze

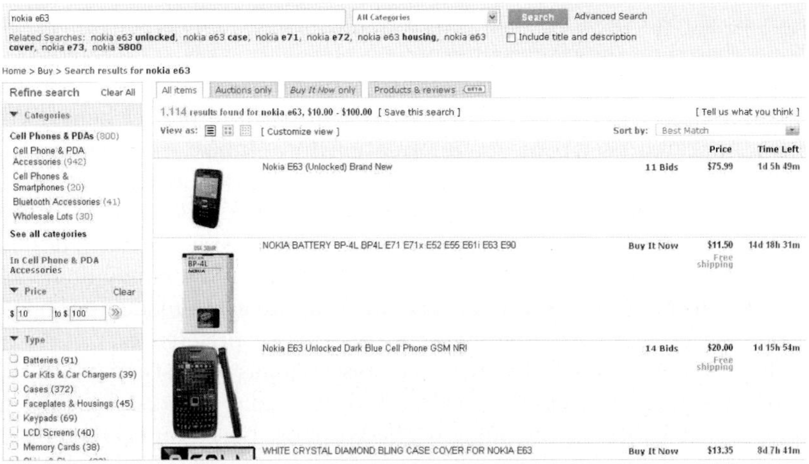

Fig. 1. Search result page in C2C web site

the features more deeply, we group them into four different dimensions: *search, buyer, seller and item.* Models trained with features of each group are tested to clarify the difference of contribution of each feature group. We also test models trained with each single feature and present the top 5 models to demonstrate the most important features. Finally, an interesting discussion is presented on the problem of unbalanced class distribution encountered in our research. The rest of this paper is organized as follow. Section 2 provides a brief description of related work. The detail of our methodology is presented in Section 3. We address the data set, evaluation metrics and experiments in Section 4. Finally, conclusion and future work is discussed in Section 5.

2 Related Work

Predicting the probability that a user click on an advertisement is crucial for sponsored search because the probability can be used to improve ranking, filtering, placement, and pricing of ads. Thus click prediction problem in domain of sponsored search is deeply researched recently years. Moira *et al.* proved different terms have different potential of receiving a sponsored click and estimated click probability based on term level CTR [2]. CTR of clusters of related terms were also used for less frequent or novel terms. Besides term level information, Matthew *et al.* explored various features related to ads, including the page the ad points to, and statistics of related ads and built a model to predict the click probability for new ads [3]. Different from user-independent models in previous work, Cheng *et al.* developed user-specific and demographic-based features and built a user-dependent model to predict click probability [4]. The topic of predicting click probability for advertisements is very similar to the problem addressed in this paper, however, as we discussed in Section 1, the characteristics of product search make this task more complicated. So valuable features can be developed based on these specialties.

As people search the web, certain of their actions are logged by a search engine. Click logs are representative resources to extract patterns of user behavior. Xue *et al.* incorporated user behavior information to optimize web search ranking [6,7]. In [1], Nick *et al.* discussed the click bias caused by document position in the results page and modeled how probability of click depends on position. We adopted his proposal and adjust the bias for CTR related features. [8] developed explicit hypothesis on user browsing behavior and derived a family of models to explain search engine click data. [9] generalized approach to model user behavior beyond click-through for predicting web search result preferences.

To the best of our knowledge, not much research has been studied focusing on domain of product search in online C2C web sites. And this task becomes more and more crucial with the booming of the business. Based on characteristics of C2C web sites, Wu *et al.* developed similar features as our work to predict conversion probability [10]. Though both this paper and [10] focus on online ecommerce business, the problems to be solved are completely different. Moreover, the models are different because of the different formulations of problems. Wu *al et.* formulated their problem as $P(purchase|item)$ which is independent of query, search and buyer, while our formulation is $P(click|query, search, buyer, item)$ that is related to query, search and buyer contrarily.

3 Click Prediction

The framework of our click prediction system is straightforward, though the process for user to make click decision is complicated. As shown in Fig. 2, a set of items is selected by product search engine according to the query user submitted. The complicated process of item retrieval is out of the scope of this paper. Here we assume the input of our system is a set of query-item pairs with features extracted from log data. Click probability is estimated by the model trained during offline phase. Finally, this score can be used in ranking, filtering or advertisement matching components.

3.1 Logistic Regression Model

We formulate click prediction as a supervised learning problem. To classify a query-item pair into *CLICK/NON-CLICK* class, logistic regression model is

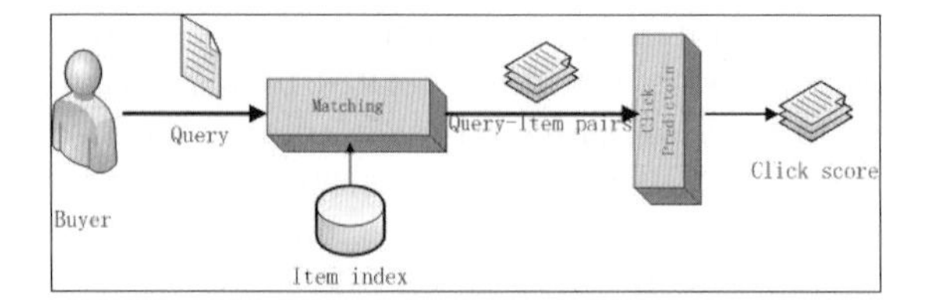

Fig. 2. System Framework

utilized for this task. It is a generalized linear model used for binomial regression, which makes it perfectly suitable for this problem. In statistics, logistic regression is used for predicting the probability of occurrence of an event by fitting data to a *logit* function logistic curve [11]. The simple logistic model has the form:

$$logit(Y) = \log \frac{p}{1-p} = \beta_0 + \sum_{i=0}^{n} \beta_i \cdot X_i \ . \tag{2}$$

where p is the probability of occurrence of an event Y and $X_1, X_2, ..., X_n$ are the independent variables(predictors); β_0 is called the *intercept* and β_i is called *regression coefficients* of x_i. Maximum likelihood method is used to learn the intercept and regression coefficients. Applied the antilog on both sides, we can transform the predictor values into probability:

$$p = P(Y|X = X_1, X_2, ..., X_n) = \frac{e^{\beta_0 + \sum_{i=1}^{n} \beta_i \cdot X_i}}{1 + e^{\beta_0 + \sum_{i=1}^{n} \beta_i \cdot X_i}} \ . \tag{3}$$

where each of the regression coefficients β_i represent the contribution of corresponding predictor value X_i. So a positive regression coefficient β_i means larger(or smaller) X_i is associated with larger(or smaller) logit of Y, which in turn increases(or decreases) the probability of the outcome.

With the regression model, we can formulate our click prediction model as:

$$P(click|query, item) = \frac{\exp(\boldsymbol{w} \cdot \boldsymbol{f})}{1 + \exp(\boldsymbol{w} \cdot \boldsymbol{f})} \ . \tag{4}$$

where $\boldsymbol{f}$ represents the feature vector extracted from the query-item pair, while $\boldsymbol{w}$ is the corresponding regression coefficient vector.

3.2 Features

Previous work proposed a large scope of features used for building click prediction model in sponsored search. Besides of that, we analyze the characteristics of product search as well as user behavior and develop a significant set of online C2C web site-based features. To make it more clear and systematic, we group the features into four dimensions as: *search, buyer, item and seller*, which are corresponding different roles of product search on C2C web sites. In Section 4, we compare the different models built with features of each dimension. To make a clear distinction for C2C site-specified features from features that are widely used in sponsored search, we mark the C2C site-specified features with underline in the rest of section.

3.2.1 Search Features

Unlike web search engine, besides text query submitted to search engine, users are allowed to define complicate search patterns. We group this kind of features into search features to represent user behavior information.

Sort Type. To compare items more intuitively, most C2C web sites allow users to select sort pattern for result list. We believe the sort pattern user selected can reflect the status of the user when submitting the query. Our statistic data indicates different CTR is obtained according to different sort type selected. Nominal values including *price ascending, price descending, time left and seller credit* are extracted for each query-item pair.

Search Filter. Abundant types of filters are provided in C2C web sites.

- *Category*, user can indicate of which category items should be returned.
- *Price interval*, user can define the price range to filter items.
- *Sale type*, when an item is published to online store, the seller needs to decide that the item is sold with auction style or fixed price style.

View Type. Users can choose presentation form for the result list: *list* means put the items into a list style; *grid* makes the items presented in a matrix style with enlarged picture on each position.

Query. The complication degree of query reflects the clarity of purchasing intention for buyers. We extract query term count, unique term count separately to represent this info. Besides of that, we also extract historical CTR on each term to represent the click potential of specific term buyer submitted :

$$CTR_t = \frac{\sum_{q,t\in q} c(q)}{\sum_{q,t\in q} pv(q)} . \tag{5}$$

where q is query that contains term t, while $c(q)$ and $pv(q)$ represent the click count and page view count received from the query respectively.

3.2.2 Item Features

While search features determine to a certain degree of click potential of the search, item features impact a certain extent that which item is more possibly for user to click.

Price. One of the most significant factors impacts a user to perform a click is the price of item. Under the same condition of other aspects, user apparently will choose the cheapest one among the items list. We calculate mean price of item list for the query. And for each item, the deviation from mean price is calculated. Certainly the shipping rate of item is an important factor, especially for those relatively cheap items. So same method is applied for shipping rate.

Category CTR. For C2C web sites, one of the most important differences from web search engine is that all the items or documents are labeled with product category by sellers. CTR on category reflects click potential of items that belong to this category. As category can be very sparse, according to EBay, 30,000 leaf categories are maintained on US site, we extract CTR of root category for each item:

$$CTR(rc) = \frac{\sum_{j\in rc} c(j)}{\sum_{j\in rc} pv(j)} . \tag{6}$$

where rc is root category, and each j is item whose root category is rc.

Explicit Matching. Features representing the degree of apparent matching of each query-item pair are also extracted. This kind of features represent presentational relevance of query-item pair on the result page.

- *Text similarity*, cosine similarity is calculated between item title and query to represent lexical match
- *Location match*, though not all of users apply location filter for their queries, items in the same location as the user are preferred.

3.2.3 Buyer Features

As illustrated in [4], personalized model performs effectively in sponsored search for predicting click probability. We refer this proposal and extract both demographic and user-specific features for each buyer.

Demographic. Gender, age and location of the buyer are extracted as nominal values.

Buyer CTR. Click-through rate of the buyer is calculated for a period of one week.

Purchase Count. Times of the buyer purchased product on the site successfully.

Buyer Credit Rank. Some C2C web sites provide mechanism to rank credit or reputation for buyers and sellers through their transaction feedback logs. The rank is determined by a score which is calculated by both count of positive feedbacks and negative feedbacks from historical transaction records.

3.2.4 Seller Features

We develop seller related features for mainly two fold reasons: firstly, the information of seller reflects the devotion of the seller to the online store, which can implicitly impact the attractiveness of his items on the result list; secondly, buyers are concerned for the reputation or credit of sellers.

Demographic. Gender, age and location of the seller are extracted as nominal values

Seller CTR. Click-through rate of the seller is calculated for a period of a week.

Product Count. Reflecting the business scale of the seller. Professional online sellers always possess of large number of items.

Seller Credit Rank. As described in Buyer features.

Attractiveness Degree. Popular C2C web sites provide the functionality of 'Save the store to favorite' or 'Watch the item'. The number of being stored up reflects the attention of buyers paid to this seller.

3.3 Emendation on Position Biased CTR

Various CTRs are extracted in previous section. However, as proved in [1], position bias also exists in buyers' click actions. The probability of click is influenced by the position of the item presented in the result list. Buyers prefer to click items ranked on the top of the result list. We use a position-normalized statistic known as clicks over expected clicks (COEC) to take account for this position bias for all features CTR related[4].

$$COEC = \frac{\sum_{pos=1}^{R} c(pos)}{\sum_{pos=1}^{R} pv(pos) \cdot CTR_{avg}(pos)} . \quad (7)$$

where the numerator is the number of clicks performed on each position pos; the denominator is expected clicks that would be received averagely after being presented $pv(pos)$ times at pos, and $CTR_{avg}(pos)$ is the average click-through rate for pos in the result page.

4 Experiments

We implemented the click prediction system as described in previous section. A large data corpus was collected from a real popular C2C web site. Plenty of experiments were performed to evaluate our methodology. In this section, we first give a brief description on our data set. Then the evaluation metrics are introduced. Finally, we present evaluation results of experiments performed on different aspects.

4.1 Data Set

The training and test data used in our experiments were sampled from a real popular C2C web site within a period of a week. Each sample is a query-item pair labeled with *click* or *non-click* which was extracted from the click log of this web site. We removed the records of users who searched or clicked more than 300 times a day to clear robot's data or spam data. In the other hand, records of users with less than 5 actions a day were also removed as this kind of data contains little statistical information and can be treated as noise. After that, 235,007 page views were randomly collected, with 130,865 clicked items and 9,504,443 non-clicked items.

As most other C2C web sites, more items are presented on each result page than web search engine, which makes the problem of unbalanced class distribution more critical. Less than 1.5% items were labeled as clicked in this corpus. Unbalanced data is a common problem in machine learning. If we build the model in the usual way, the model would get more than 98% accuracy by predicting all the items as non-click under the aim to minimize error rate. Our approach to adjust the modeling is to down sample the non-clicked samples to even up the classes. While tuning the proportion of clicked samples, the important thing to consider is the cost of mis-classification, which is the cost of incorrectly classifying a click sample as non-click, vice versa. We leave it to marketing strategy as the business consideration is out of scope of this paper. An experiment was performed on evaluating the performance changing with the growth of positive examples proportion in the training data set and a brief discussion is presented in Section 4.5.

4.2 Evaluation Metrics

As a binary classification problem, we defined clicked and non-click item as positive example and negative example respectively. So *TP rate*, *FP rate*, *Precision*, *Recall* and *F-Measure* score were referred as evaluation metrics. The *TP(True Positive)* rate is the proportion of examples which are classified as class x, among all examples which truly have class x. The *FP(False Positive)* is the proportion of examples which are classified as x, but belong to a different class, among all examples which are not of class x. *F-Measure* score is defined as:

$$F - Measure = \frac{2 \cdot Precision \cdot Recall}{Precision + Recall} . \tag{8}$$

Effectiveness on both positive and negative class are interested and corresponding results are provided in the rest of this section. All evaluation results presented below were obtained through 10-fold cross validation.

4.3 Performance Evaluation

Features proposed and used for sponsored search described in Section 3.2(without underline marked) were extracted to build a model as baseline which we referred as *Web* in the rest of this paper. The C2C site-based features(marked with underline) were combined to the *Web* model, which is referred as *C2C* model. With 50% positive examples sampled in the training data set, 10-fold cross validation was performed on both of the two models. The performance evaluation are presented in Table 1.

Table 1. Evaluation

	TP rate		FP rate		Precision		Recall		F-Measure	
	C2C	*Web*	*C2C*	*Web*	*C2C*	*Web*	*C2C*	*Web*	*C2C*	*Web*
Click	0.546	0.533	0.329	0.401	0.624	0.571	0.546	0.533	0.582	0.551
Non-Click	0.671	0.599	0.454	0.467	0.597	0.562	0.671	0.599	0.632	0.580

From Table 1, it turns out that state-of-the-art techniques used in sponsored search are also suitable for predicting the click probability for product search on C2C web site. Precision for *Click* and *Non-Click* is promising while the recall of *Click* class is a little poor. After the combination of C2C site-based features, the performance got improved significantly on all of the evaluation metrics. Precision and recall for *Click* is improved by 9.3% and 2.4% respectively; for *Non-Click*, 6.2% and 12.0% improvements are obtained for precision and recall separately. The results definitely prove the effectiveness of C2C site-specific features in click prediction problem.

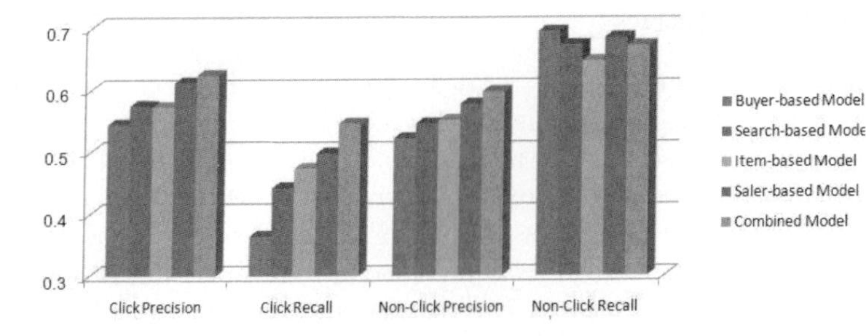

Fig. 3. Feature group comparison

4.4 Feature Analysis

According to different roles of product search, we grouped features into four different dimensions: *search, item, buyer, seller*. To analyze the contribution of each dimension, we trained different models based on each group features. The comparison is presented in Fig. 3.

From the figure, we can conclude that features related to items and sellers contribute most to the combined model. For commercial search, user-based model did not work as effective as in sponsored search. However, it is not out of our expectation. For sponsored search, the goal of user to search on web search engine is to find relevant information related to the query rather than browsing advertisements. The habits, age, gender and other user-specific features certainly will impact the click through rate for advertisements presented [4]. Contrarily, for product search, users are focusing on comparing items and sellers. Though with different demographic info or personal characteristics, the criterion of evaluating an item is relative common: better price, trustworthy seller, attractive description and so on. So the contribution of user-specific features got weakened while the influence of item-related features got enhanced. Also, from the figure, we validated our claim that search styles reflecting status of the user impact click actions.

Besides analysis on feature groups, we also evaluated models trained with each feature separately to rank the importance of each single feature. The top 5 important features are *Seller CTR, Sort Type, Item Left Time, Root Category CTR, Seller Credit*. Out of our expectation, features related to price and explicit matching did not show its effectiveness as we supposed.

4.5 Unbalance Data Re-sample

As we discussed in Section 4.1, the proportion of positive examples impacts performance of the prediction for both positive and negative classification. Though we do not expect to make decision of what is the best cut-off point, we still analyzed the change of performance according to the re-sample emendation process. In Fig. 4, with increasing of positive example ratio, the prediction on positive instances becomes more accurate while the performance of negative classification

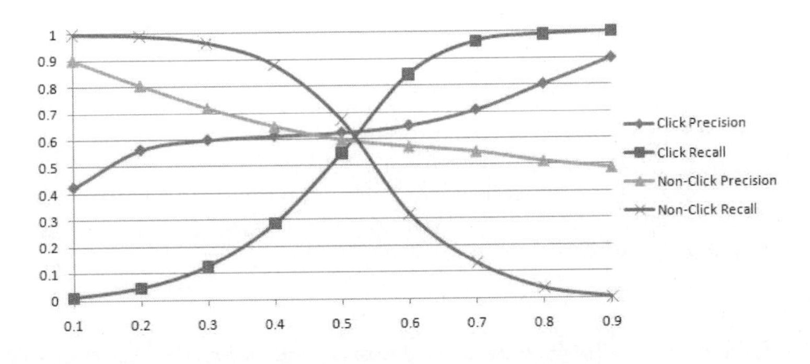

Fig. 4. Unbalanced data re-sample. Horizontal axis represents the proportion of positive examples in the training data set.

decreases as expected. We believe the scope of 45% to 55% for the proportion of *Click* examples is a reasonable range for an online prediction system.

5 Conclusion and Future Work

In this paper, we proposed a click prediction modeling solution for product search based on the characteristics of online C2C web sites. Both C2C-based features and state-of-the-art features used in sponsored search are developed and regression model is utilized with the feature set. To summarize, we conclude the contribution of this paper as follows:

- We present a novel problem in domain of product search. To the best of our knowledge, few studies were reported on this domain.
- We validate the feasibility of transforming techniques used in sponsored search to the domain of product search.
- Significant features based on characteristics of C2C web sites are developed and experiments prove promising improvements are obtained from the combined model.
- Our methodology for predicting click probability is general and extensible for applying to other C2C web sites

However, as a started up problem in this domain, more potential data can be mined in this task in the future. For example, buyers are more easily attracted by distinctive item with delicate picture or description containing plenty of adjective words. There is a trend that C2C web sites import B2C stores in their business which might impact the search style and user behavior. And an interesting problem is how to adjust ratio of positive examples under the consideration of business factors.

References

1. Nick, C., Onno, Z., Michael, T., Bill, R.: An Experimental Comparison of Click Position-bias Models. In: 1st ACM International Conference on Web Search and Data Mining, California (2008)

2. Regelson, M., Fain, D.: Predicting Click-through Rate Using Keyword Clusters. In: 2nd Workshop on Sponsored Search Auctions (2006)
3. Matthew, R., Ewa, D., Robert, R.: Predicting Clicks: Estimating the Click-through Rate for New Ads. In: 16th International World Wide Web Conference, Banff (2007)
4. Haibin, C., Erick, C.: Personalized Click Prediction in Sponsored Search. In: 3rd ACM International Conference on Web Search and Data Mining, Barcelona (2010)
5. Massimiliano, C., Vanessa, M., Vassilis, P.: Online Learning from Click Data for Sponsored Search. In: 17th International World Wide Web Conference, Beijing (2008)
6. Gui-Rong, X., Hua-Jun, Z., Zheng, C., Yong, Y., Wei-Ying, M., WenSi, X., WeiGuo, F.: Optimizing Web Search Using Web Click-through Data. In: 13rd ACM International Conference on Information and Knowledge Management, Washington (2004)
7. Eugene, A., Eric, B., Susan, D.: Improving Web Search Ranking by Incorporating User Behavior Information. In: 29th Annual International ACM SIGIR Conference on Research and Development in Information Retrieval, Seattle (2006)
8. Georges, D., Benjamin, P.: A User Browsing Model to Predict Search Engine Click Data from Past Observations. In: 31st Annual International ACM SIGIR Conference on Research and Development in Information Retrieval, Singapore (2008)
9. Eugene, A., Eric, B., Susan, T.D., Robert, R.: Learning User Interaction Models for Predicting Web Search Result Preferences. In: 29th Annual International ACM SIGIR Conference on Research and Development in Information Retrieval, Seattle (2006)
10. Xiaoyuan, W., Alvaro, B.: Predicting the Conversion Probability for Items on C2C Ecommerce Sites. In: 18th ACM Conference on Information and Knowledge Management, Hong Kong (2009)
11. David, W.H., Stanley, L.: Applied Logistic Regression. Wiley Series in Probability and Statistics. Wiley-Interscience Publication, Hoboken (2000)

Finding Potential Research Collaborators in Four Degrees of Separation

Paweena Chaiwanarom[1,2], Ryutaro Ichise[2], and Chidchanok Lursinsap[1]

[1] Advanced Virtual and Intelligent Computing (AVIC) Center
Department of Mathematics, Chulalongkorn University, Bangkok 10330, Thailand
{cc.paweena,lchidcha}@gmail.com

[2] Principles of Informatics Research Division, National Institute of Informatics
2-1-2 Hitotsubashi, Chiyoda-ku, Tokyo 101-8430, Japan
ichise@nii.ac.jp

Abstract. This paper proposed a methodology for finding the potential research collaborators based on structural approach underlying co-authorship network and semantic approach extends from author-topic model. We proposed the valuable features for identifying the closeness between researchers in co-authorship network. We also proved that using the combination between structural approach and semantic approach is work well. Our methodology able to suggest the researchers who appear within the four degrees of separation from the specific researcher who have never collaborated together in the past periods. The experimental results are discussed in the various aspects, for instance, top-n retrieved researchers and researcher's community. The results show that our proposed idea is the applicable method used for collaborator suggestion task.

Keywords: co-authorship network, author-topic model, graph mining, digital library, research collaboration, social network analysis, information retrieval.

1 Introduction

When a researcher would like to start a work in a new research topic, a problem usually encountered by the researcher is who I should collaborate with. Searching the potential collaborators is arduous because most academic search engines obtain the outcomes only in forms of document search rather than people search. The effective measures should be designed for ranking the potential collaborators for a given researcher.

In our definition, the potential collaborators for the specific researcher are not necessary the most well-known or familiar researchers, but the most research similarity and reachability.

From a collection of published papers, we can visualize the relationship among researchers by using the collaboration network called co-authorship network

L. Cao, J. Zhong, and Y. Feng (Eds.): ADMA 2010, Part II, LNCS 6441, pp. 399–410, 2010.

where nodes represent authors and edges represent collaboration between authors. If we get node X as a given researcher, the rest of nodes in graph are the candidates who can be selected as the potential collaborators for node X.

However, the graph structure is often very sparse and visiting every nodes in a large graph must be exhaustive. To reduce the time complexity, it is reasonable to consider a smaller graph by limiting the number of hops from the given node. Interestingly, Liu et al. [1] and Sharma and Urs [2] found the average distance of co-author in digital library conferences (DL) is 3.6 and 3.5 respectively. Thus, the candidates researchers in our experiment will be limited to four degrees of separation with respect to the given researcher.

In this paper, we proposed an effective measure used for ranking the most potential collaborators from the set of candidates based on structure of co-authorship graph called *structure approach*. Furthermore, the classic author-topic model [3] based on the contents of papers will be considered for calculating the knowledge similarity between a pair of researchers called *semantic approach*. Finally, both approaches will be combined to a *hybrid approach*.

This paper is organized as follows. Section 2 briefly mentions the related works. In section 3, our methodology is being discussed. Section 4 describes the experiments and results. Finally, section 5 conclude the paper.

2 Background and Related Work

Nowadays, almost researches in academic collaboration domain focus on browsing and searching based on the analyzed statistical information from network structure. In digital libraries, many approaches have been proposed to provide useful tools to researchers, e.g., citation recommendation [4], paper recommendation [5]. A few researches concentrate on mining and prediction tasks [6], [7], especially collaborator suggestion.

The similar views to ours is reported in [7], [8] proposed the link predictor based on both structure approach and semantic approach. Wohlfarth and Ichise [7] transformed the problem to finding collaborators in a link prediction problem. They combined the *structural attributes* from the co-authorship network with *non-structural attributes* from research titles and meeting place of researchers for predicting the new collaborations. This methods do not make any significant use of node properties and rely on link information. Then, Sachan and Ichise [8] utilized node properties such as abstract information, community alignment and network density to improve the predictor of [7]. By counting the number of words in common between all the abstracts of the previous research papers, they can introduce a new feature based on the semantics of the research besides the network structure.

Our more related research was proposed in *DBconnect* system [9]. The list of recommended collaborators could be selected based on the relation from topics without co-authorship. On the other hand, the list of related researchers is obtained from relationships derived from co-authorships and conferences. This paper used the number of publications for representing the weights of edges in their designed graph. There is some overlap between the list of related researchers and the list of recommended collaborators calculated from the different features. Unfortunately, the time evaluation was ignored and the efficiency of method was not evaluated in this paper.

Steyvers et al. [3] introduced a probabilistic algorithm, called author-topic model, to automatically extract information of authors, topics, and documents. This model was utilized by Ichise et al. [10] for research area mapping system which could interactively explore a research area to obtain knowledge for research trends. Moreover, Hassan and Ichise [11] extended idea of the classic author-topic model to classify more authors in a given research domain by introducing a distance matrix.

The objective of our work is to find the list of potential collaborators similar to the work in [9]. The co-authorships and the relation from topics will be combined together in ours. Not only the number of publications but also other features will be proposed for calculating the weight of edge in co-authorship graph. Moreover, author-topic model will be utilized for calculating the research knowledge similarity between authors.

3 Proposed Methodology

3.1 Approach

In order to find the potential collaborators in four degrees of separation from a given researcher, we propose a new methodology based on structure and semantic approaches. Our methodology mainly consist of four parts. Firstly, for a given researcher, we retrieve his neighbor researchers within four degrees of separation. Next, a new model which is a kind of *structure approach* based on co-authorship network for measuring the closeness between the given researcher and his neighbors is introduced. We obtain the probability distributions over topics of each author from the classic author-topic model for calculating the research knowledge similarity between the given researcher and his neighbors. This part is called *semantic approach*. Finally, the *structure approach* and *semantic approach* are combined as a *hybrid approach*. Our hypothesis is that the results from *hybrid approach* should be better than the results from either *structure approach* or *semantic approach*. The overview of our proposed methodology is shown in Fig.1 and four methods are considered as follows.

3.2 Neighbors in Four Degrees of Separation

From a collection of published papers, we construct the co-authorship graph where nodes represent authors and edges represent collaboration between authors. Starting from a given author (node), we use depth-first search (DFS) algorithm for traversing the graph and retrieving the neighbor nodes within four degrees of separation.

A potential collaborator in the future may be someone used to collaborate in the past period or may be someone has never collaborated before. Then, the

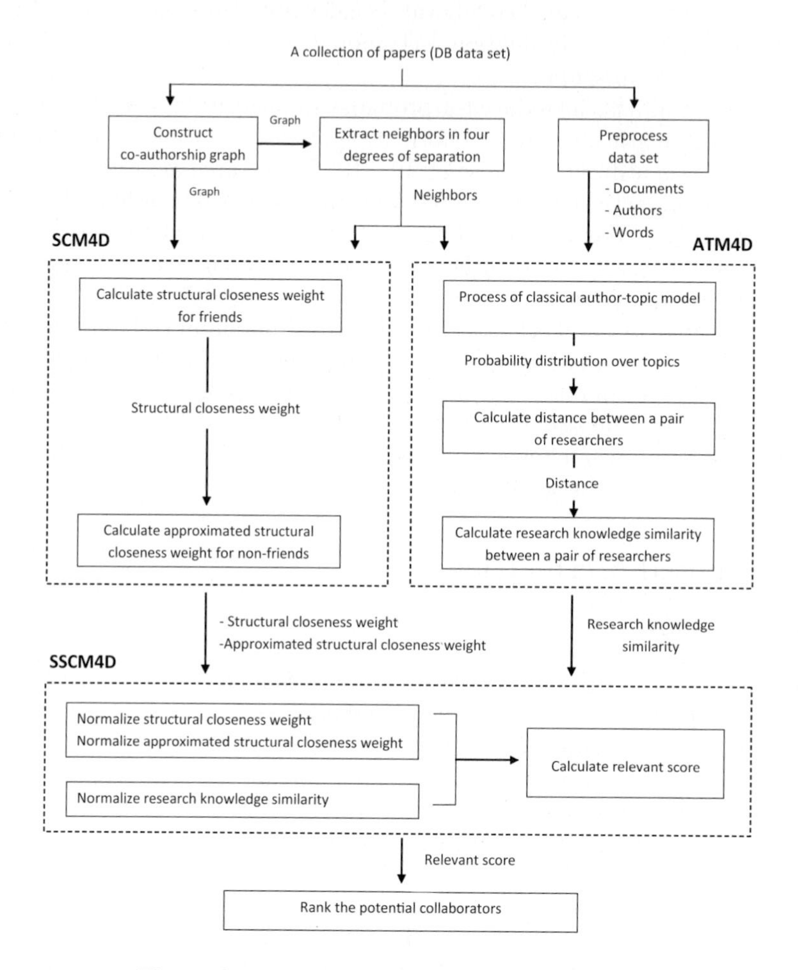

Fig. 1. Overview of the proposed methodology

retrieved neighbors can be divided in two communities, i.e., *friend community* and *non-friend community*. The *friend community* refer to the set of adjacency nodes which located in one degree of separation from the given node. In other word, the researchers in this set are co-authors who have written at least one paper with the given researchers. Inversely, the *non-friend community* refer to the set of nodes which located between two and four degrees of separation from the given node. The researchers in this set have never collaborated with the given researcher.

Let $\mathbf{V} = \{v_1, v_2, \ldots, v_n\}$ be a set of authors and v_i is a given researcher. The definition of *neighbors*, *friend community*, and *non-friend community* can be defined as follows.

Definition 1: Neighbors. Neighbors denoted as $R(v_i)$ is a set of r authors who appear within four degrees of separation from v_i for $r \leq n$, and $R(v_i) \subset V$.

Definition 2: Friend Community. Friend community denoted as $Q(v_i)$ is a set of q authors who have collaborated with author v_i for $q \leq r$, and $Q(v_i) \subseteq R(v_i)$.

Definition 3: Non-friend Community. Non-friend community denoted as $\overline{Q(v_i)}$ is a set of p authors who have never collaborated with author v_i for $p = r - q$, and $\overline{Q(v_i)} \subset R(v_i)$.

3.3 Structural Closeness Model in Four Degrees of Separation(SCM4D)

From the existing co-authorship network and the set of neighbors, e.g. v_i has a set of neighbors $R(v_i)$, the main question is *how to determine the closeness weight between* v_i *and his neighbors in* $R(v_i)$. This weight will be represented as edge label between them. We split this model into two measures based on community of the neighbors, i.e., *structural closeness weight* for friend community and *approximated structural closeness weight* for non-friend community.

Structural Closeness Weight. We propose the measure for identifying the weight between a pair of a given researcher v_i and his friend in $Q(v_i)$ by using three features:

1. *Frequency of co-authorship*: authors that frequently co-author should have a higher closeness between them.
2. *Total number of co-authors in paper*: the closeness between them should be weighted more if a paper has small authors.
3. *Year of publication*: authors that recently co-author should have a higher closeness.

All features will be comprehensively analyzed for determining the closeness between a pair of adjacent authors. That is the first two features will be dynamically considered based on the evolution over time in the last features.

Let $\mathbf{P} = \{p_1, p_2, \ldots, p_m\}$ be a set of published papers. Suppose authors v_i and v_j co-authored h papers in set $\mathbf{C}_{v_i,v_j} = \{p_{k_1}, .., p_{k_h}\}$, where $\mathbf{C}_{v_i,v_j} \subseteq P$. The *structural closeness weight* between v_i and v_j, denoted by $\mathcal{S}_{v_i,v_j}$, based on h papers can be defined in the following formula (1).

$$\mathcal{S}_{v_i,v_j} = \sum_{p_k \in C_{v_i,v_j}} w_{v_i,v_j,p_k} \ . \tag{1}$$

$$w_{v_i,v_j,p_k} = (z_{v_i,v_j,p_k}) \times (t_{v_i,v_j,p_k}) \ . \tag{2}$$

Term z_{v_i,v_j,p_k} is calculated from the total number of co-authors on papers. If a paper has few authors, the closeness between co-authors should be weighted more. Assume that authors v_i and v_j are co-authors in paper p_k, and $f(p_k)$ is the number of co-authors of paper p_k. The z_{v_i,v_j,p_k} between v_i and v_j on paper p_k is calculated by using formula (3). The maximum value of z_{v_i,v_j,p_k} is one when the paper has two authors. Note that, our experiment filters out all papers having only one author.

$$z_{v_i,v_j,p_k} = \frac{1}{f(p_k) - 1} . \tag{3}$$

Term t_{v_i,v_j,p_k} is calculated based on the year of publication. Authors that recently co-author should have a higher closeness. Suppose authors v_i and v_j co-author in paper p_k published in year α and they expect to collaborate together in year β, the weight between v_i and v_j on paper p_k is calculated by using formula (4), where $\varphi = \beta - \alpha - 1$. The maximum value of t_{v_i,v_j,p_k} is one when they publish a paper in the last year.

$$t_{v_i,v_j,p_k} = \frac{1}{2^{\varphi}} . \tag{4}$$

Approximated Structural Closeness Weight. Due to the neighbors in *non-friend community* not having collaborated with the given researcher, the *structural closeness weight* between them have to be approximated by using the transitive property which can be indirectly computed through *structural closeness weight* in formula (1).

There could be several interaction paths between the given researcher and the specific non-friend researcher not directly collaborate but join work through a number of the other researchers in the network. We propose the measure for identifying *approximated structural closeness weight* based on the following concept.

Let v_i is a given researcher, v_λ is a non-friend researcher of v_i where $v_\lambda \in \overline{Q(v_i)}$, v_j is a researcher in path l, $v_i \neq v_j$, and Δ_l be the number of edges along path l between v_i and v_λ, $1 \leq \Delta_l \leq 4$. Each path consists of a set of edges. The first edge connects v_i with a node on path l but the last edge connects a node on path l with v_λ. For any path l, let δ be the degree of separation from v_i to v_j, $1 \leq \delta \leq 4$ and $\delta \leq \Delta_l$. Prior to finding the value of the *approximated structural closeness weight*, the average *structural closeness weight* of each path l, denoted by s_l must be computed by using formula (5). Suppose the *structural closeness weight* S_{v_i,v_j} is 0.3 and S_{v_j,v_λ} is 0.8 shown in Fig. 2, the *approximated structural closeness weight* between v_i and v_λ (dash line in Fig. 2) can be calculated as $2/\left(\frac{1}{0.3\left(\frac{1}{2^1}\right)} + \frac{1}{0.8\left(\frac{1}{2^2}\right)}\right)$.

$$s_l = \frac{\Delta_l}{\sum_{\delta=1}^{\Delta_l} \frac{1}{S_{v_i,v_j}\left(\frac{1}{2^{\delta}}\right)}} . \tag{5}$$

The factor $\frac{1}{2^{\delta}}$ attenuates the *structural closeness weight* of a further node v_j. This implies that the further distance v_j is from v_i, the less *structural closeness*

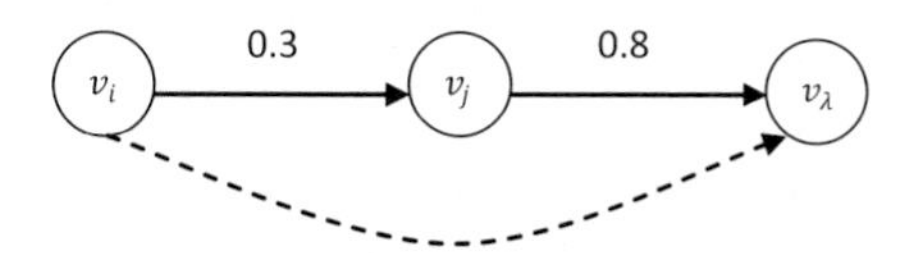

Fig. 2. Example of the relationship between author v_i, v_j, and v_λ

weight of v_j is assigned. Suppose there are η possible paths from v_i to v_λ. Thus, the *approximated structural closeness weight*, denoted by $\tilde{S}_{v_i,v_\lambda}$ is

$$\tilde{S}_{v_i,v_\lambda} = \max(s_1, \ldots, s_\eta) . \tag{6}$$

3.4 Author-Topic Model in Four Degrees of Separation (ATM4D)

The classical author-topic model will assign the authors to a likely probability for only one topic. Although the likely topics between two researchers are different, it is quite possible that the nature of research might be similar. Fortunately, each author in $\mathbf{V}$ can be represented by a vector of probability distribution of μ topics. We obtain the relationship among author and topic from classical author-topic model as the input for our ATM4D. Thus, we utilize the probability distribution of all topics rather than only likely topic to calculate the research knowledge similarity between author v_i and his r neighbors in $R(v_i)$.

Firstly, calculate the distance between author v_i and his all neighbors in $R(v_i)$. For each neighbor v_j of author v_i where $v_j \in R(v_i)$, we compute the distance between author v_i and v_j by using Euclidean distance in formula (7). The term $v_{i\kappa}$ and $v_{j\kappa}$ are the probability distribution of topic κ for v_i and v_j, respectively where $\kappa \in \mu$. If every elements of probability distribution vector for v_i and v_j is the same, the distance between them is equal to zero and the similarity is equal to one.

$$d(v_i, v_j) = \sqrt{\sum_{\kappa=1}^{\mu} (v_{i\kappa} - v_{j\kappa})^2} . \tag{7}$$

Since a short distance represents more similarity, we use the distance from formula (7) for calculating the similarity between author v_i and his all neighbors in $R(v_i)$. Thus, the *research knowledge similarity* between a given researcher and his neighbors are calculated by using formula (8).

$$\mathcal{A}_{v_i,v_j} = \frac{1}{d(v_i, v_j) + 1} . \tag{8}$$

3.5 Semantic Structural Closeness Model in Four Degrees of Separation(SSCM4D)

The SSCM4D is a hybrid model based on SCM4D and ATM4D for calculating the *relevant score*. Using either SCM4D or ATM4D may be misled for selecting the collaborators since both models focus on the different viewpoints. The SCM4D concentrates on the graph structure but overlooks the research knowledge of the researchers. In contrast, ATM4D pays attention to find the most relevant research knowledge who may be difficult to reach. It is often the case that the researchers separated by very few links might still not collaborate due to the differences of research topic they do, even though the experience of research work might be the same. Not only researchers who have written papers together, but also researchers far apart in the graph might collaborate in the future if they have the similar research experience. Thus, it is reasonable to add ATM4D for making precise decision. The combination of SCM4D and ATM4D should be designed for finding the suitable collaborators rather than using only one of them.

There are two measures of SSCM4D depending on community of the neighbors, i.e., formula (9) for *friend community* and formula (10) for *non- friend community*. The first factor in both formulas are normalized weights from SCM4D and the second are normalized weights from ATM4D. For (10), we adjust the both factors since we would like to pay more attention to the semantic approach than structural approach. Thus, the normalized ATM4D will be multiplied by 1.75 for increasing semantic weight, whereas weight from normalized SCM4D is decreased. Finally, the *relevant score* between v_i and his friend can be obtained from formula (9) and the *relevant score* between v_i and his non-friend can be obtained from formula (10).

$$\mathcal{H}_{v_i,v_j} = \frac{\mathcal{S}_{v_i,v_j}}{\sum\limits_{v_j \in Q(v_i)} \mathcal{S}_{v_i,v_j}} \times \frac{\mathcal{A}_{v_i,v_j}}{\sum\limits_{v_j \in Q(v_i)} \mathcal{A}_{v_i,v_j}} . \quad (9)$$

$$\mathcal{H}_{v_i,v_\lambda} = 0.25 \left(\frac{\tilde{\mathcal{S}}_{v_i,v_\lambda}}{\sum\limits_{v_\lambda \in \overline{Q(v_i)}} \tilde{\mathcal{S}}_{v_i,v_\lambda}} \right) \times 1.75 \left(\frac{\mathcal{A}_{v_i,v_\lambda}}{\sum\limits_{v_\lambda \in \overline{Q(v_i)}} \mathcal{A}_{v_i,v_\lambda}} \right) . \quad (10)$$

4 Experiment

4.1 Experimental Data

To implement our ideas, we conducted experiments using real bibliography data. We obtained freely available data set in excel format from Scopus [12] which is the largest abstract and citation database. We prepared data for our experiment by the following methods. Firstly, we defined 14 research domains (topics) in computer science which observed from the several computer science ranking websites. Next, we selected the authors who published at least ten papers for the particular domain. Thus, we obtained about 100 authors per a domain. These authors are called *core authors*. Next, we selected the papers which were published between year 2000 and 2009 written by core authors in each domain. We split these papers into two partitions, depending on the year of publishing.

The first part consist of published papers during 2000 – 2007 and the second part are published papers during 2008 – 2009. Next, we extracted the core authors who appeared during both time periods, called *active core authors*. The papers without any active core authors were filtered out from the experiment. Finally, the number of data used in our experiments are shown in Table 1. The set of unique authors contains the active core authors and their co-authors for selected papers.

4.2 Experimental Models

From Table 1, we used papers in *DB data set* for build our designed models described in Section 3 and papers in *test data set* for testing the models.

Firstly, we construct the *structural closeness model in four degrees of separation(SCM4D)*. Each paper in *DB data set* contains a list of co-authors and year of published papers for creating a co-authorship graph. We set each *active core author* as a given author and retrieved his neighbors in four degrees of separation along with calculated structural closeness weight, $\mathcal{S}_{v_i,v_j}$, for *friend community* and approximated structural closeness weight, $\tilde{\mathcal{S}}_{v_i,v_\lambda}$, for *non-friend community* described in Section 3.2 – 3.3. This process was repeated 1,168 iterations for discovering all neighbors of 1,168 *active core authors*. As results, the built graph contains 14,002 nodes of unique authors and 23,501 edges labeled by $\mathcal{S}_{v_i,v_j}$.

Next, we obtain a contents of titles, abstract, author keywords, and index keywords in *DB data set* for creating *author-topic model in four degrees of separation (ATM4D)*. For better semantic analysis, we extracted the set of nouns from this contents by using SharpNLP library [13]. Moreover, Porter Stemming algorithm [14] was applied for removing the commoner morphological and inflectional endings from nouns. The numbers of keywords were 966,822 whereas the unique keywords were 17,450. These keywords along with set of authors were entered to the input of classical author-topic model. The output in this step are the probability distribution over 14 topics for 14,002 unique authors. We utilized these outputs for calculating the *research knowledge similarity*, $\mathcal{A}_{v_i,v_j}$, between 1,168 *active core authors* and their neighbors by the steps described in Section 3.4.

Finally, the SCM4D and ATM4D were combined to SSCM4D by using the method described in Section 3.5. The output of SSCM4D called *relevant score* for *friend community*, $\mathcal{H}_{v_i,v_j}$, and for *non-friend community*, $\mathcal{H}_{v_i,v_\lambda}$ were calculated as the results.

Table 1. Statistics of experimental data

	DB data set (2000-2007)	Test data set (2008-2009)
#Papers	11,661	4,124
#Authors	37,798	14,227
#Unique authors	14,002	7,514
#Active core authors	1,168	1,168
#Unique author pairs	23,501	11,051

4.3 Evaluation Methods

For a given researcher, we used the *relevant score* based on SSCM4D for selecting the potential collaborators from his neighbors. The neighbors were ranked in descending order, so the most potential collaborator will be ranked in the first order. We applied several thresholds (TH) for retrieving the number of neighbors, i.e., $10, 20, 30, 50, 100$. In order to evaluate our methodology, we classified the $11,051$ author pairs in *test data*, accumulated by the collaboration between *active core author* and his actual collaborators, into three groups of researcher. The investigated output shows $3,179$ collaborators being friend researchers, $2,943$ collaborators being non-friend researchers, and $4,929$ collaborators being unknown researchers. Obviously, about 44% of the new collaboration arise with the new researcher who have not published any papers in *DB data set*. Thus, the $6,122$ author pairs will be used for evaluation since the unknown researchers will be ignored. The *research knowledge similarity* output from classical author-topic model without four degrees of separation limitation (ATM all degree in Table 2) and our three proposed models, i.e., ATM4D, SCM4D, SSCM4D, limited in four degrees of separation were compared by precision/recall measurement in the next section.

4.4 Experimental Results

Table 2 shows the average of precision/recall from four models where threshold denoted by TH and true positive denoted by TP. We split the table into three groups, i.e., friend community, non-friend community, and Neighbors. For each group, the column ATM all degree and column ATM4D of precision and recall was calculated based on author-topic model. The difference is the latter column retrieved only the researchers in four degrees of separation, whereas the former column has no limitation (retrieved every degree of separation). Obviously, for neighbors, the precision and recall obtained from ATM4D are higher than the precision and recall obtained from ATM all degree. Thus, retrieving neighbors within the scope of four degrees of separation is work well.

Interestingly, if we retrieve all of neighbors in four degrees of separation (no threshold), the recall for *friend community* is always 100% whereas *non-friend community* is 77.78% since approximate 22% of actual collaborators in this community appear beyond the scope of four degrees. The total retrieved neighbors are $3,365,016$, the average neighbors per *active core author* is $3,365,016/1,168 \approx 2,881$, especially in *non-friend community* is $3,341,516/1,168 \approx 2,860$. It is exhausted task to select $2,943$ actual collaborators from $3,341,516$ of *non-friend community*. This is the reason why we obtained small precision/recall from *non-friend community*.

If we use only structural approach based on SCM4D, the precision and recall from all groups is better than ATM4D. In other word, the researchers select their collaborators based on structural relation more than based on semantic. The results of SSCM4D model which the combination of ATM4D and SCM4D was shown in the last column. The SSCM4D in table neighbors shows the most

Table 2. Average precision/recall from four models grouped by neighbor's community

Friend Community								
	Precision				Recall			
TH	ATM all degree	ATM4D	SCM4D	SSCM4D	ATM all degree	ATM4D	SCM4D	SSCM4D
100	19.39	16.54	17.76	15.01	39.32	58.04	93.24	98.43
50	21.21	17.77	19.13	16.01	30.54	47.81	89.27	96.63
30	22.51	18.43	20.34	17.24	25.13	40.04	85.18	93.83
20	23.11	18.90	21.80	18.76	21.08	33.88	81.66	89.53
10	24.71	20.97	25.01	23.04	15.54	25.51	70.12	74.80
NO	-	13.53	13.53	13.53	-	100.00	100.00	100.00
Non-friend Community								
	Precision				Recall			
TH	ATM all degree	ATM4D	SCM4D	SSCM4D	ATM all degree	ATM4D	SCM4D	SSCM4D
100	0.23	0.34	0.50	0.57	8.56	10.94	15.05	16.21
50	0.28	0.46	0.77	0.85	5.10	7.03	10.30	10.02
30	0.30	0.52	1.00	1.17	3.16	4.62	6.69	6.25
20	0.32	0.55	1.33	1.68	2.24	3.13	4.69	4.08
10	0.41	0.56	1.87	2.82	1.36	1.43	1.60	1.05
NO	-	0.07	0.07	0.07	-	77.78	77.78	77.78
Neighbors								
	Precision				Recall			
TH	ATM all degree	ATM4D	SCM4D	SSCM4D	ATM all degree	ATM4D	SCM4D	SSCM4D
100	1.29	2.06	3.23	3.42	24.53	35.40	55.65	58.90
50	1.92	3.20	5.82	6.23	18.31	28.21	51.31	55.00
30	2.55	4.27	8.80	9.60	14.57	23.02	47.45	51.73
20	3.16	5.24	12.25	13.29	12.02	19.10	44.66	48.45
10	4.58	7.47	19.93	21.09	8.72	13.93	37.18	39.35
NO	-	0.16	0.16	0.16	-	89.32	89.32	89.32

precision and most recall compared with the first three models. Hence, our proposed methodology can prove that using hybrid approach is better than using either structure or semantic approach. The recall of *friend community* from SSCM4D in top-100 and top-10 are 98.43% and 74.80%, although there are $(3,179 \times 100)/23,501 = 13.53\%$ of author pairs in *DB data set* re-collaborate in the *test data*. This statistic express our metholody give the high efficiency to select the collaborators in *friend community*. In summary, the recall of SSCM4D top-100 is 58.90% compared with the actual collaboration in *test data set* and $(58.90 \times 100)/89.32 = 65.94\%$ campared with the actual collaboration in four degrees of sepatation.

5 Conclusions

In this paper, we propose a methodology for finding the potential research collaborators with respect to a given researcher based on structural approach and semantic approach. For structural approach, we utilize co-authorship network for calculating the structural closeness weight between researchers. We found that frequency of co-authorship, total number of co-authors in a paper, and year of publication play the important role for this approach. For semantic approach, we obtain the probability distributions over topic of each authors from the classic author-topic model for calculating the research knowledge similarity between researchers. Finally, we combined the both approach to hybrid approach. From our studied, 89.32% of actual collaborators can be discovered in four degrees of separation from a given researcher and the recall from our hybrid approach is 65.94% of this set. The results comparision shows that using our proposed hybrid approach give more efficiency than using either structural approach or semantic approach.

References

1. Liu, X., Bollen, J., Nelson, M.L., Sompel, H.V.: Co-Authorship Networks in the Digital Library Research Community. Information Processing and Management-An International Journal 41(6), 1462–1480 (2005)
2. Sharma, M., Urs, S.R.: Small World Phenomenon and Author Collaboration: How Small and Connected is the Digital Library World? In: Goh, D.H.-L., Cao, T.H., Sølvberg, I.T., Rasmussen, E. (eds.) ICADL 2007. LNCS, vol. 4822, pp. 510–511. Springer, Heidelberg (2007)
3. Steyvers, M., Smyth, P., Rosen-Zvi, M., Griffiths, T.: Probabilistic Author-Topic Models for Information Discovery. In: 10th ACM SIGKDD, pp. 306–305 (2004)
4. McNee, S.M., Albert, I., Cosley, D., Gopalkrishnan, P., Lam, S.K., Rashid, A.M., Konstan, J.A., Riedl, J.: On the Recommending of Citations for Research Papers. In: CSCW 2002, pp. 116–125 (2002)
5. Torres, R., McNee, S.M., Abel, M., Konstan, J.A., Riedl, J.: Enhancing Digital Libraries with TechLens+. In: JCDL 2004, pp. 228–236 (2004)
6. Huang, J., Zhuang, Z., Li, J., Giles, C.L.: Collaboration Over Time: Characterizing and Modeling Network Evolution. In: WSDM 2008, pp. 107–116 (2008)
7. Wohlfarth, T., Ichise, R.: Semantic and Event-Based Approach for Link Prediction. In: Yamaguchi, T. (ed.) PAKM 2008. LNCS (LNAI), vol. 5345, pp. 50–61. Springer, Heidelberg (2008)
8. Sachan, M., Ichise, R.: Using Abstract Information and Community Alignment Information for Link Prediction. In: ICMLC 2010, pp. 61–65 (2010)
9. Zaiane, O.R., Chen, J., Goebel, R.: Mining Research Communities in Bibliographical Data. In: Zhang, H., Spiliopoulou, M., Mobasher, B., Giles, C.L., McCallum, A., Nasraoui, O., Srivastava, J., Yen, J. (eds.) WebKDD 2007. LNCS, vol. 5439, pp. 59–76. Springer, Heidelberg (2009)
10. Ichise, R., Fujita, S., Muraki, T., Takeda, H.: Research Mining Using the Relationships among Authos, Topics and Papers. In: IV 2007, pp. 425–430 (2007)
11. Saeed-Ul-Hassan, Ichise, R.: Discovering Research Domains Using Distance Matrix and Coauthorship Network. In: LACTS 2009, pp. 1252–1257 (2009)
12. The scopus databased (2010), http://www.scopus.com/
13. SharpNLP - open source natural language processing tools (2006), http://sharpnlp.codeplex.com/
14. Porter stemming algorithm (2010), http://tartarus.org/~martin/PorterStemmer/

Predicting Product Duration for Adaptive Advertisement

Zhongqi Guo, Yongqiang Wang, Gui-rong Xue, and Yong Yu

Computer Science Department,
Shanghai Jiao Tong University,
Dongchuan Road 800, Minhang District, Shanghai 200240, China
{guozq,wangyq,grxue,yyu} @apex.sjtu.edu.cn

Abstract. Whether or not the C2C customers would click the advertisement heavily relies on advertisement content relevance and customers' searching progress. For example, when starting a purchasing task, customers are more likely to click the advertisements of their target products; while approaching the end, advertisements on accessories of the target products may interest the customers more. Therefore, the understanding of search progress on target products is very important in improving adaptive advertisement strategies. Search progress can be estimated by the time spent on the target product and the total time will be spent on this product. For the purpose of providing important information for product progress estimation, we propose a product duration prediction problem. Due to the similarities between the product duration prediction problem and user preference prediction problem (e.g. Large number of users, a history of past behaviors and ratings), the present work relies on the collaborative filtering method to estimate the searching duration of performing a purchasing task. Comparing neighbor-based, singular vector decomposition(SVD) and biased SVD method, we find biased SVD is superior to the others.

Keywords: Collaborative filtering, user interest, personalize advertisement.

1 Introduction

Measuring time consuming of purchase behavior in consumer to consumer (C2C) business plays a significant role in industry and research area. First, understanding the time consuming is essential to advertisement releasing. Generally, the queries of C2C customers at the beginning of the purchase behavior, are always different with those at the end of the purchase behavior. For instance, the customers would always search for the products highly related to their target when they start the purchase behavior, however, they would always search for the accessories for the target product or the product they have purchased. Therefore, it is necessary to release advertisement on products to meet the customers' different interests. However, the conventional strategies on advertisement releasing emphasize the customers' current queries only, but did not take into account the

L. Cao, J. Zhong, and Y. Feng (Eds.): ADMA 2010, Part II, LNCS 6441, pp. 411–418, 2010.

various interests in different purchasing procedure. Lacking of pertinence invalidates the advertisement, since the information may not be the customers' interest. Second, understanding the time consuming is important in understanding the customer interests, since the length of time consuming of purchase behavior is found to be the most important criterion to gauge the customers' interest [1]. In a view of the research gap noticed and the significance of the time consuming, the present work aims at predicting the time consuming of the customer's purchase behavior, and further providing references to the advertisement releasing. A better fit between the advertisement and the demand of the customers brings more enjoyment to the customers and higher profits to the websites.

The framework of this paper will be as follows. We will introduce related works in Section. 2. Then, in Section. 3 we describe the problem and how we formulate this problem. Models are introduced in the Section. 4. In Section. 5 and Section 6, we provide a description of our experiment set and experimental results. And conclusions and future work introduction will be listed Section. 7.

2 Related Works

In the previous works, some factors were found to influence the releasing of personalized advertisements. The query histories significantly influence the personalized advertisements [2]. Xiao et. al provided a model to address this issue based on user history queries, user history views of advertisements, user history clicks of advertisements. In the previous work, product duration has never been considered in exploring the personalized advertising. However, time is an important criterion predicting the efficiency of advertisements, as it significantly represents the user interest. In order to fill in the research gap, the present work therefore tests the impact of product duration on the advertising.

Collaborative filtering is a technique using the observed preferences of a certain group of users to make predictions for the unobserved preferences. Collaborative filtering tasks face many challenges, such as data sparsity, scalability, etc. [3], which are highly related to our problem. Memory-based collaborative filtering uses the user rating data to calculate the similarity between users or items in order to make predictions [4]. Nowadays, Koren et. al, improve approaches by introducing latent factor adjustments which is very helpful [5] in movie recommending system. Therefore, we are encouraged to introduce collaborative filtering in solving our problem.

3 Product Duration Time Calculation

In modern product search engines, customers search their target by inputting the queries which are chronically categorized by the product taxonomy. Therefore, it is reasonable to measure the costumer interest in categories. Since the customers' target product is constant during a certain period of time (viz. one searching session), their queries are assembled by category in one searching session. With this character, we can calculate a user u's time consumption t on one category c within one session s in this way:

$$t(u,c,s) = \textit{last search time}(u,c,s) - \textit{first search time}(u,c,s) + \sigma \ . \quad (1)$$

which *last search time*(u,c,s) is last time of perform search action in session at category c. Correspondingly, *first search time*(u,c,s) is that of first time in this session. σ is a constant for the user will keep his focus on a category for a while after a search action on product search engine. In our work, we assign $\sigma = 5$, which means when user leave a search, a extra 5 seconds' attention is paid on that query. So we can calculate a user's time consuming on a category, by cumulate the time for each sessions,

$$t(u,c) = \sum_{s} t(u,c,s) \ . \quad (2)$$

Then we introduce implicit interest score with natural logarithm of time:

$$s(u,c) = \ln t(u,c) \ . \quad (3)$$

There are two reasons to define a logarithm-based measure: First and the most important, the same difference value of consuming time does not show the interest difference when the comparison base is not same. The difference should be outstanding when the base time is small and vice vera. Second, with this transformation, the value interval of the data become more concentrated which is more fit for our rating problem.

4 Model

In this section, we will describe our proposed duration predict problem. Then, three collaborative filtering estimator, neighbor based, SVD based and biased-SVD are build and tested.

4.1 Problem Description

At first, we reserve some special term: we have a set of subject users $\{u \in U\}$ and a set of product categories $\{c \in C\}$. With a group of observed implicit interest scores $\{s(u,c)|(u,c) \in O\}$, we try to predict a group of unobserved implicit interest scores $\{s(u,c)|(u,c) \in U\}$.

4.2 Basis Estimator

The elements in this model are users and categories. In the statistical sense, it is common for some users spend more time on each category to achieve a careful selection; it is also common for some categories attracts more attention time than the others. Considering these effects, we can give a basis estimation of implicit interest score,

$$b(u,c) = \mu + o_u + o_c \ . \quad (4)$$

Where μ is a constant value, the average of all product duration; o_u denotes the user effect offset, and o_c denotes the category effect offset. We use this as our basis estimator. Learning the estimation of product duration can be performed by solving the following formula:

$$\min_{o*} \sum_{(u,c)\in T} (s(u,c) - \mu - o_u - o_c)^2 + \lambda(\sum_{u\in U} o_u^2 + \sum_{c\in C} o_c^2) . \tag{5}$$

where the first part of the formula is the square error on test; the second part is the regular term of avoiding overfitting. We call λ as the anti-overfitting coefficient,

$$\begin{aligned} o_u &\leftarrow o_u + \delta(r_{u,c} - \lambda o_u) , \\ o_c &\leftarrow o_c + \delta(r_{u,c} - \lambda o_c) . \end{aligned} \tag{6}$$

We do not direct use this as a estimator because it ignores the factor of individual (user, category) pair. Estimating offset against average isn't take the user preference into account.

4.3 Neighbor-Based Method

Neighborhood-based collaborative filtering is based on the similarity of users or items. Item-based neighborhood method is on the hypothesis that users give similar rating for similar items. User-based neighborhood method takes the advantage of similar user shows similar interest on the same item. In collaborative filtering field, item-based method has been proved to be more effective for its robustness on reducing impact of abnormal users [3]. We performed K-neighbors item-based method by introducing a adjustment factor on the basis estimator,

$$\widehat{s}(u,c) = b(u,c) + \frac{\Sigma_{c\in\kappa(u,c)} Sim(c,c') \cdot (s(u,c') - b(u,c'))}{\Sigma_{c\in\kappa(u,c)} Sim(c,c')} . \tag{7}$$

For a target category, we use weighted average of difference value between K-neighbors' predictive value and true value as the adjustment of predictive value. Where $Sim(c,c')$ is the similarity between the target category and it's neighbors. We involve the Pearson's correlation coefficient as the similarity weight, which is best fit to measure user's time consuming.

4.4 SVD Based Method

Singular value decomposition is a mathematical transition of matrices which minimize approximation error. In simple word, SVD can be used to build latent factors for both users and categories [7]. For a better understanding, latent preference factor are stored in dimensions in a user vector; correspondingly latent properties are stored in the category vector. Then the cumulate of matching degree on each dimension construct the offset term of the estimator. Therefore, the estimation of stay time s can be formulated:

$$\widehat{s}(u,c) = \mu + p_u * q_c . \tag{8}$$

where p_u and q_c is the user vector and the category vector respectively. The product of these vector describes the user's preference on the category, which is the adjustment for the average. Rather than build the vector directly, a gradient descent learning can be performed loop all the training cases by solving following formula:

$$\min_{p,q*} \sum_{(u,c)\in T} (i(u,c) - \mu - p_u \cdot q_c)^2 + \lambda_2(\sum_{u\in U} ||p_u||^2 + \sum_{c\in C} ||o_c||^2) . \tag{9}$$

The parameters are updated with following assignment:

$$\begin{aligned} p_u &\leftarrow p_u + \delta_2(r_{u,c} \cdot q_c - \lambda_2 p_u) , \\ q_c &\leftarrow q_c + \delta_2(r_{u,c} \cdot p_u - \lambda_2 q_c) . \end{aligned} \tag{10}$$

For each loop of all training cases, one dimension of user vector and corresponding dimension of query vector will be updated.

4.5 Biased SVD Method

The weakness of SVD method is the offset against average is directly measure by the matching degree. In Koren et. al's work[5], they introduce a statistical term(basis estimator) into the SVD,

$$\widehat{s}(u,c) = b(u,c) + p_u * q_c . \tag{11}$$

As well, we can learn the parameters by solving the following formula with a gradient descent according (13),

$$\min_{o,p,q*} \sum_{(u,c)\in T} (i(u,c) - \mu - o_u - o_c - p_u \cdot q_c)^2 + \lambda(\sum_{u\in U} o_u^2 + \sum_{c\in C} o_c^2) + \lambda_2(\sum_{u\in U} ||p_u||^2 + \sum_{c\in C} ||o_c||^2) . \tag{12}$$

$$\begin{aligned} o_u &\leftarrow o_u + \delta_3(r_{u,c} - \lambda_3 o_u) , \\ o_c &\leftarrow o_c + \delta_3(r_{u,c} - \lambda_3 o_c) , \\ p_u &\leftarrow p_u + \delta_4(r_{u,c} \cdot q_c - \lambda_4 p_u) , \\ q_c &\leftarrow q_c + \delta_4(r_{u,c} \cdot p_u - \lambda_4 q_c) . \end{aligned} \tag{13}$$

5 Experiment

In this section, we provide a description of data preprocessing experiment design. In order to build a reliable prediction system, we extracted 3 month span of log data from a C2C web site(from March 1, 2010 to May 31, 2010). Due to the large amount of customers, this data set is difficult to process. We therefore use sampled subjects, in order to select the customers who have regular purchasing habits, we set a threshold on visited product number limiting costumers. Finally, a total number of 123, 000 users are chosen through a random sampler.

5.1 Data Preprocessing

Behavior Grouping and Session Segmentation. Without loss of generality, queries are committed to describe a target product. Therefore it is reasonable for us to group users' queries into groups. In our method, we tagged each query with the category label which appeared most times in clicked item list after the query. The common method of session segmentation is to put a boundary between two sessions by setting a timeout threshold between two successive activities [6]. In our work, we adopt this method of segment sessions with a threshold of 30 minutes.

Data Source. We filter with the threshold of users viewing at least 30 of categories. With this log data, we build a rating data base of 123,000 users within 4,000 categories. We collected overall 4, 000, 000 ratings into our user preference database, average less than 35 categories for each user.

5.2 Evaluation Metric

We use Root Mean Square Error(RMSE) as our metric of evaluation predict precision. RMSE [5] is a commonly-used measure of the differences between values predicted and the values actually observed in user preference predicting systems. The RMSE metric can be given as follows:

$$RMSE = \sqrt{\frac{\sum_{(u,c)\in TestSet}(i_{(u,c)} - \widehat{i}_{(u,c)})^2}{|TestSet|}} \,. \tag{14}$$

6 Results

In this section, we will provide the results of experiments and insight of the methods performance. We found biased-SVD shows the most powerful of modeling our problem. However, parameters of biased-SVD greatly influence modeling.

6.1 Methods Comparison

We evaluate three different methods of duration estimating according to the formula, namely Neighbor based, SVD based and biased-SVD. We note them as neigb, svd, bsvd respectively. For each method, we have set different set of learning rate, anti-overfitting coefficient. We list the best result of each method in the Fig. 1.

In Fig.1, the x axis is the iteration time of the gradient descent, and the y axis displays the RMSE after corresponding x time iteration. This figure shows that item-based method error does not descent much after 1000 iterations, but it is stable when the iteration continues. In both SVD based method or biased SVD method, the error has experience a great descent time in the first 1500 iterations. But after 1500, SVD based method shows up it's weakness, the parameters of the model start to overfit the training data. We tried to punish more over fitting

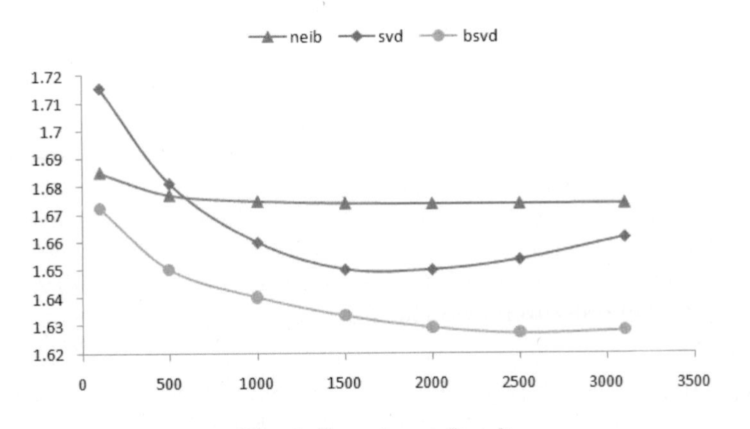

Fig. 1. Experiment Result

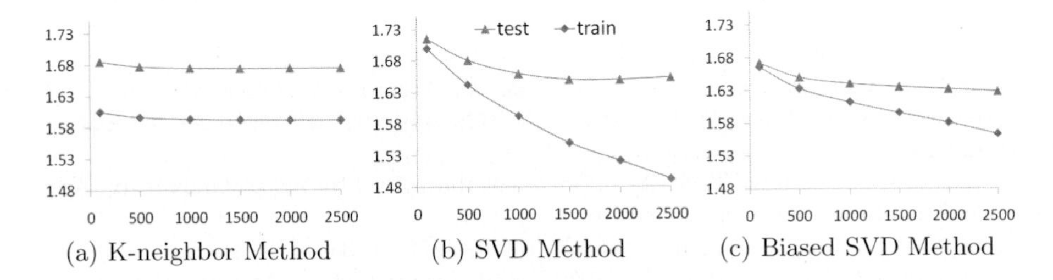

(a) K-neighbor Method (b) SVD Method (c) Biased SVD Method

Fig. 2. RMSE on different methods

by enlarge anti-overfitting coefficients λ_2, but it does not work beyond 0.015. However, biased SVD showed it's greatness of modeling product duration. After 1500 iterations, the error still keep going down slowly. It achieves a much better result than the other two methods.

In Fig.2, we listed the data of different methods RMSE on both training and test data. Neighbor-based method is performs badly on both training and test data. Both SVD based method and biased SVD is performs much better on the training data. That is to say, it is much greater power of SVD based method and biased SVD for fitting the training data. But SVD faces much horrible overfitting problem, which biased SVD is not much troubled. The success of biased SVD can be concluded with it combined both basis estaminet and latent matching degree. The former decrease the overall error from user and category aspect, the later adjust the answer with a more elaborate user, category matching degree.

7 Conclusion and Future Work

In our work, we propose a product duration predicting problem. As mentioned in Section.2, we find many similarities between user preference predicting problem and product duration predicting problem. Based on these similarities, we introduce collaborative filtering methods to solve our problem. Evaluations of these methods are performed on three month data of C2C sites. The results show biased SVD works much better in our problem than the other methods.

Future work can be directed in two aspects. First, we focus on three month's data and establish a basic model in the present work. There are some variables which can influence the customers queries, such as season, location, gender, etc. Future research can involve these variables to make a more accurate prediction on time consuming. Second, we measure the consuming time based on each category in our work. Future research can specify the current categories into some more detailed groups in order to provide further suggestions on advertising.

References

1. Al halabi, W.S., Kubat, M., Tapia, M.: Time spent on a web page is sufficient to infer a user's interest. In: IMSA 2007: IASTED European Conference on Proceedings of the IASTED European Conference, pp. 41–46. ACTA Press, Anaheim (2007)
2. Xiao, G., Gong, Z.: Personalized Delivery of On—Line Search Advertisement Based on User Interests. In: Li, Q., Feng, L., Pei, J., Wang, S.X., Zhou, X., Zhu, Q.-M. (eds.) APWeb/WAIM 2009. LNCS, vol. 5446, pp. 198–210. Springer, Heidelberg (2009)
3. Su, X., Khoshgoftaar, T.M.: A survey of collaborative filtering techniques. In: Adv. in Artif. Intell. 2009, vol. 2 (2009)
4. Sarwar, B., Karypis, G., Konstan, J., Reidl, J.: Item-based collaborative filtering recommendation algorithms. In: WWW 2001: Proceedings of the 10th International Conference on World Wide Web, pp. 285–295. ACM, New York (2001)
5. Koren, Y.: Factorization meets the neighborhood: a multifaceted collaborative filtering model. In: KDD 2008: Proceeding of the 14th ACM SIGKDD International Conference on Knowledge Discovery and Data Mining, pp. 426–434. ACM, New York (2008)
6. Jones, R., Klinkner, K.L.: Beyond the session timeout: automatic hierarchical segmentation of search topics in query logs. In: CIKM 2008: Proceeding of the 17th ACM Conference on Information and Knowledge Management, pp. 699–708. ACM, New York (2008)
7. Funk, S.: Netflix update: Try this at home, `http://sifter.org/~simon/journal/20061211.html`

An Algorithm for Available Bandwidth Estimation of IPv6 Network

Quanjie Qiu[1,*], Zhiguo Li[2], and Zhongfu Wu[3]

[1] Information and Network Management Center, Chongqing University, Chongqing 400030, China
[2] Shanghai Baosight Software Corporation, Chongqing, 400039, China
[3] College of Computer Science, Chongqing University, Chongqing 400030, China
{qqj,lizhiguo,wzf}@cqu.edu.cn

Abstract. Based on the analysis of the measurement principle of IPv4 network bandwidth and with the combination of the next-generation network protocol IPv6, we put forward a one-way and different-length packet pair subtraction algorithm for available bandwidth estimation of IPv6 network. An IPv6 network available bandwidth estimation prototype system was designed and programmed by using the flow label field of IPv6 messages to control the sequence path of tested messages. The test results show that the algorithm is feasible for IPv6 network with the estimation error less than 0.1M. The estimation results are more stable and better to reflect the real-time correlation of network available bandwidth and time delay, providing a useful means of network monitoring and performance estimation while effectively reducing the rate of network congestion.

Keywords: IPv6, flow label, bandwidth estimation, algorithm.

1 Introduction

The rapid growth of Internet population leads to the shortage of IP addresses. The next generation IP protocol, IPv6, was proposed around mid-ninety to accommodate the problem [1]. Network bandwidth refers to the maximum bit number of data messages that the network link bandwidth can send in the unit time. It is generally classified into bottleneck bandwidth and available bandwidth. Available bandwidth refers to the maximum data transmission rate that a path can provide for a new single link without affecting the transmission rate of existing links [2]. In recent years, a great number of bandwidth test algorithms and estimation systems have been designed [3]-[7].

Based on the combination of the next-generation network protocol IPv6, we put forward a one-way and different-length packet pair subtraction algorithm for available bandwidth estimation of IPv6 network. by using the flow label field of IPv6 messages to control the sequence path of tested messages, An IPv6 network available bandwidth

* This work is supported by key fund project of science and technology research supported by Chongqing Education Commission (KJ08A07) and the Third Stage Building of "211 Project" (Grant No. S-10218).

L. Cao, J. Zhong, and Y. Feng (Eds.): ADMA 2010, Part II, LNCS 6441, pp. 419–426, 2010.

estimation prototype system was designed and programmed. The test results show that the algorithm improves the precision of IPv6 network available bandwidth estimation.

2 Principles of Network Bandwidth Estimation

Definition: l was the hop count of the link; d_i the physical distance between router i and router $i+1$; v_i the transmission speed of the electrical signals (bit stream) in the transmission medium; b_i the bandwidth of link i; s^k the length of packet k; q_i^k the time from packet k to link i. A packet passed（l-1）links to reach router l, then the time delay of the packet was:

$$t_l^k = t_0^k + \sum_{i=0}^{l-1} \left(\frac{s^k}{b_i} + \frac{d_i}{v_i} + q_i^k \right). \tag{1}$$

Here, $\frac{s^k}{b_i}$ was the transmission time delay of the packet, $\frac{d_i}{v_i}$ the propagation time delay, q_i^k the queuing and waiting time delay of the packet in the router, and t_0^k the time from packet k to link 0. Here we only calculated the arrival time to the router. For a single packet, if the round trip time (RTT) would be calculated, we needed to add the back trip time delay or the transmission time delay of ICMP timeout packet. According to the formula and in the case of other variables being known, we could determine the value bandwidth (b_i) of the link. It was just the fundamental principle of network link bandwidth estimation [2], [3], [4].

3 IPv6 Network Bandwidth Estimation Algorithm

3.1 Basic Idea of the Algorithm

A same hypothesis has been made by all the current IPv4 protocol bandwidth estimation algorithms: in the estimated measurement process, the path is a fixed and unique one, that is, there is no routing change and also no multi-link forwarding. In fact, IP is often described as a connectionless protocol, just like any of the packet-switched networks. IP is designed to allow separate routing for each packet to reach its destination, and each packet is treated separately. As a result, two packets sent from the same data source to the same destination can take different routes to cross the entire network. Therefore, the premise hypothesis is difficult to be tenable for the current estimated measurement of IP network bandwidth. Based on the deep research on the next-generation network IPv6 protocol, we solved the problem that the path and method of IPv4 bandwidth estimation were variable in this paper. The basic ideas for realization were as follows:

(1)The flow label regarded the measuring data packets as a part of the same traffic flow of a series of source and destination addresses. All data packets in a same traffic flow had the same flow label, which enabled the measuring data packets to have the fixed and unique paths;

(2)It was to firstly determine one path from source host *h1* to destination host *h2*, then note down IPv6 addresses of the routers along the path and make the statistics of hop counts (with the common tool tracert6);

(3)Based on IPv6 addresses of the routers obtained and with the use of IPv6 message header flow label entry, a fixed transmission path of the measuring data packets was established in the path from source host *h*1 to destination host *h*2 to ensure that two data messages (message packet) sent to each router along the way could transmit along the same path;

(4)According to the hop count preset, that was, the hop count value written into the message headers of IPv6 protocol, the message packets sending different hop counts (2 messages of different length), by using ICMPv6 messages, returned to IPv6 addresses and timestamps recorded and then deducted the delay of the measured messages to reach two routers, and the result produced thereby was not the absolute time but the relative time, namely the difference of the arrival time between two data packets of source node and destination node;

(5)The bandwidth formula could be deduced by the algorithm, and then the bandwidths of all links 1……*i* in the path could be calculated too. The bottleneck bandwidth was the minimum one of all values, min(*i*), while the available bandwidth was the maximum one, max(*i*).

3.2 Algorithm

According to formula (1), taking into account various link layer protocols and the error caused by clock skew, Especially under the case that there were many tunnels like IPv6 over IPv4 in IPv6 network ,another formula for time delay calculation could be drawn, the more common formula was(here d^* is the length of the link layer protocol header; l' is the length of the tunnel IPv4 message header):

$$t_{i+1}^{n-1} = t_i^{n-1} + \frac{d_i}{v_i} + \frac{s_i^{n-1} + d^* + l'}{b_{i,t1}} + q_i^{n-1} + c_i^{n-1}, \tag{2}$$

$$t_{i+1}^{n} = t_i^{n} + \frac{d_i}{v_i} + \frac{s_i^{n} + d^* + l'}{b_{i,t2}} + q_i^{n} + c_i^{n}. \tag{3}$$

Supposing two data messages sent $s^{n-1} < s^n$, that was formula (3) deducted formula (2), we could get the following formula,

$$\begin{aligned} t_{i+1}^{n} - t_{i+1}^{n-1} &= t_i^{n} + \frac{d_i}{v_i} + \frac{s_i^{n} + d^* + l'}{b_{i,t2}} + q_i^{n} + c_i^{n} \\ &- (t_i^{n-1} + \frac{d_i}{v_i} + \frac{s_i^{n-1} + d^* + l'}{b_{i,t1}} + q_i^{n-1} + c_i^{n-1}) . \end{aligned} \tag{4}$$

At this time, we supposed $\Delta t_{i+1}^{n,n-1}$ as the time difference of two adjacent data packets n and n-1, that is $\Delta t_{i+1}^{n,n-1} = t_{i+1}^{n} - t_{i+1}^{n-1}$; and supposed $\Delta c_i^{n,n-1} = c_i^n - c_i^{n-1}$ and $\Delta q_i^{n,n-1} = q_i^n - q_i^{n-1}$ at the same time, thus:

$$\Delta t_{i+1}^{n,n-1} = \Delta t_i^{n,n-1} + \frac{s_i^n + d^* + l'}{b_{i,t2}} - \frac{s_i^{n-1} + d^* + l'}{b_{i,t1}} + \Delta c_i^{n,n-1} + \Delta q_i^{n,n-1}. \quad (5)$$

Now let's discussed formula (5), in the right of the formula, the denominator of the second sub-formula, $b_{i,t1}$, represented the real-time bandwidth that data packet *n*-1 passed through link *i* at *t*1, and the denominator of the third sub-formula, $b_{i,t2}$, represented the real-time bandwidth that data packet *n* passed through link *i* at *t*2. If the interval between *t1* and *t2* was small enough, yet $b_{i,t1} \approx b_{i,t2}$, where there were other data packets between two data packets. If only assuming that the transmission speed of link *i* between *t1* and *t2* was unchanged, that was to say the network bandwidth was not variable or there was an average transmission bandwidth, this assumption was reasonable, namely $\overline{b}_i^{t1,t2} = b_{i,t1} \approx b_{i,t2}$, in which, $\overline{b}_i^{t1,t2}$ was the average transmission bandwidth of link *i* from *t*1 to *t*2. From formula (5), what could be drawn was:

$$\Delta t_{i+1}^{n,n-1} - \Delta t_i^{n,n-1} = \frac{1}{\overline{b}_i^{t1,t2}} (s_i^n - s_i^{n-1}) + \Delta c_i^{n,n-1} + \Delta q_i^{n,n-1}. \quad (6)$$

In the existing bandwidth estimation and measurement algorithm, the queuing time delay of the data packet was neglected, and the influence of system time deviation and link layer protocols on bandwidth estimation was also not considered. In fact, however, these factors would affect the actual measurement. If the processing time delay of data packets was taken into account, the value was smaller when the processing time delay values of two data packets deducted each other, thus it could be ignored. The error thereby would be much smaller than that one due to neglecting the processing time delay of data packets itself. Then, $\Delta q_i^{n,n-1} \approx 0$. Although the system time of various routers or hosts might be wrong in the same path and it was available to correct the time by using software or GPS, it was still impossible to ensure the absolutely correct and consistent time. Therefore, after subtraction, the time finally involved in calculation was the relative time of the same host or router, and it was very small and could be neglected completely. Thus, $\Delta c_i^{n,n-1} \approx 0$. As for the impact of the link layer protocol, there was no impact on measuring bandwidth when d^* was subtracted. Therefore, the above formula became:

$$\Delta t_{i+1}^{n,n-1} - \Delta t_i^{n,n-1} = \frac{1}{\overline{b}_i^{t1,t2}} (s_i^n - s_i^{n-1}). \quad (7)$$

That was,

$$\overline{b}_i^{t1,t2} = \frac{s_i^n - s_i^{n-1}}{\Delta t_{i+1}^{n,n-1} - \Delta t_i^{n,n-1}}, \quad (8)$$

$$b_{i,t} = \frac{s_i^n - s_i^{n-1}}{\Delta t_{i+1}^{n,n-1} - \Delta t_i^{n,n-1}} \,. \tag{9}$$

The final formula (9) showed that the relative time Δt participated in the calculation rather than the absolute time t, and it was the difference of arrival time or timestamp between two adjacent data packets on the source node and the next node. Besides, from this formula, the bandwidth of link i could be calculated easily because of only knowing the size of the data packets sent to link i and their time difference to arrive at routers i and $i+1$ in the link. It was more important that it only needed to make use of ICMP messages to return the timestamp time of two measured data packets concerned to reach routers i and $i+1$ in the link, other than their own arrival time of ICMP messages. Therefore, the one-way algorithm obviously reduced the problem of path nonconformity between return messages and measured messages.

3.3 Algorithm Analysis

In the common mechanism, the source host sent probe packets to routers till the destination node, which could find out IP addresses of all routers on the path from H1 to H2 through some tools just like the command "tracert", and then wrote into IP addresses of these routers on probe packets, returning the time information needed through the specific ICMP datagram. After that, the bandwidth of each link on this path could be calculated. Here, we supposed:

$\Delta t_{t,t1}^{n,n\ 1}$ stood for the time difference between data packet n and data packet n-1 to reach router i in the time period of *t1*. According to formula (9), the bandwidth of the first link was:

$$b_{0,t1} = \frac{s_0^2 - s_0^1}{\Delta t_{1,t1}^{2,1} - \Delta t_{0,t1}^{2,1}} \,. \tag{10}$$

It showed that ICMP message recorded the time that the data packets s_0^1 and s_0^2 reached router 1 after being sent from source node *A*, and returned the time to calculate the bandwidth of link 0. At the time period of *t2*, data packets s_1^1 and s_1^2 were sent to router 2, similarly, two timestamps of router 2 were returned, and reply2 returned to *A*. From what mentioned before, we could calculate the bandwidth of link 1 according to the obtained two pairs of time.

$$b_{1,t1} = \frac{s_1^2 - s_1^1}{\Delta t_{2,t2}^{2,1} - \Delta t_{1,t2}^{2,1}} \,. \tag{11}$$

$\Delta t_{2,t2}^{2,1}$ and $\Delta t_{0,t2}^{2,1}$ could be obtained directly from measurement, and according to formula (11), we could have:

$$b_{0,t2} = \frac{s_0^2 - s_0^1}{\Delta t_{1,t2}^{2,1} - \Delta t_{0,t2}^{2,1}} \,. \tag{12}$$

Then,

$$\Delta t_{1,t2}^{2,1} = \frac{s_0^2 - s_0^1}{b_{0,t2}} + \Delta t_{0,t2}^{2,1}. \tag{13}$$

Taking formula (13) into the denominator of formula (12), there was:

$$b_{1,t2} = \frac{s_1^2 - s_1^1}{\Delta t_{2,t2}^{2,1} - \Delta t_{1,t2}^{2,1}} = \frac{s_1^2 - s_1^1}{\Delta t_{2,t2}^{2,1} - \Delta t_{0,t2}^{2,1} - \dfrac{s_0^2 - s_0^1}{b_{0,t2}}}. \tag{14}$$

Hereby, the value of $b_{1,t2}$ could be calculated through simple substitution and using the previous results, and then the bandwidth of link 2:

$$b_{2,t3} = \frac{s_2^2 - s_2^1}{\Delta t_{3,t3}^{2,1} - \Delta t_{2,t3}^{2,1}}. \tag{15}$$

$\Delta t_{3,t3}^{2,1}$ and $\Delta t_{0,t3}^{2,1}$ could be obtained directly from measurement, and according to formulas (15) and (15), the calculation formula of $b_{2,t3}$ could be deduced:

$$b_{2,t3} = \frac{s_2^2 - s_2^1}{(\Delta t_{3,t3}^{2,1} - \Delta t_{0,t3}^{2,1}) - (s_2^2 - s_2^1)(\dfrac{1}{b_{1,t3}} + \dfrac{1}{b_{0,t3}})}. \tag{16}$$

The numerical value of network bandwidth should remain stable in the course of measurement for the bandwidth measurement being finished within a short time. In other words, the bandwidth of various links remained unchanged basically in the course of measurement. This, there was:

$$b_{1,t3} \approx b_{1,t2} \approx b_{1,t1} \approx b_{1,t} \qquad b_{0,t3} \approx b_{0,t2} \approx b_{0,t1} \approx b_{0,t}$$

Here, t stood for a time period for finishing once measurement, ant by extending formula (16) to the bandwidth calculation of each link on each path, there was:

$$\begin{aligned} b_{i,t} &= \frac{s_i^n - s_i^{n-1}}{(\Delta t_{i,t}^{n,n-1} - \Delta t_{0,t}^{n,n-1}) - (s_i^n - s_i^{n-1})(\dfrac{1}{b_{i-1,t}} + \dfrac{1}{b_{i-2,t}} + \ldots + \dfrac{1}{b_{0,t}})} \\ &= \frac{s_i^n - s_i^{n-1}}{(\Delta t_{i,t}^{n,n-1} - \Delta t_{0,t}^{n,n-1}) - (s_i^n - s_i^{n-1})\sum_{i=0}^{i-1}\dfrac{1}{b_{i,t}}}. \end{aligned} \tag{17}$$

4 Realization of IPv6 Network Bandwidth Estimation Algorithm and Analysis on the Results

The two routers supporting IPv6 protocol were set up with PC and accessed to cernet2 network, and then the prototype system for estimation of available bandwidth on both ends of IPv6 was established by programming with the bandwidth estimation and measurement algorithm. The algorithm-based prototype system, was used to test the actual IPv6 network 3 twisted pair and 3 fiber optic links of the available bandwidth one by one, test interval to 20 minutes, The results are as follows the figure 1. The figure showed that for IPv6 network making use of IPv6 flow label to control the path, as the result of bandwidth estimation on one fixed path obtained by using the one-way different-length packet pair subtraction bandwidth estimation algorithm, the error could be less then 0.1M. The test data showed that the result of the measurement was true and effective.

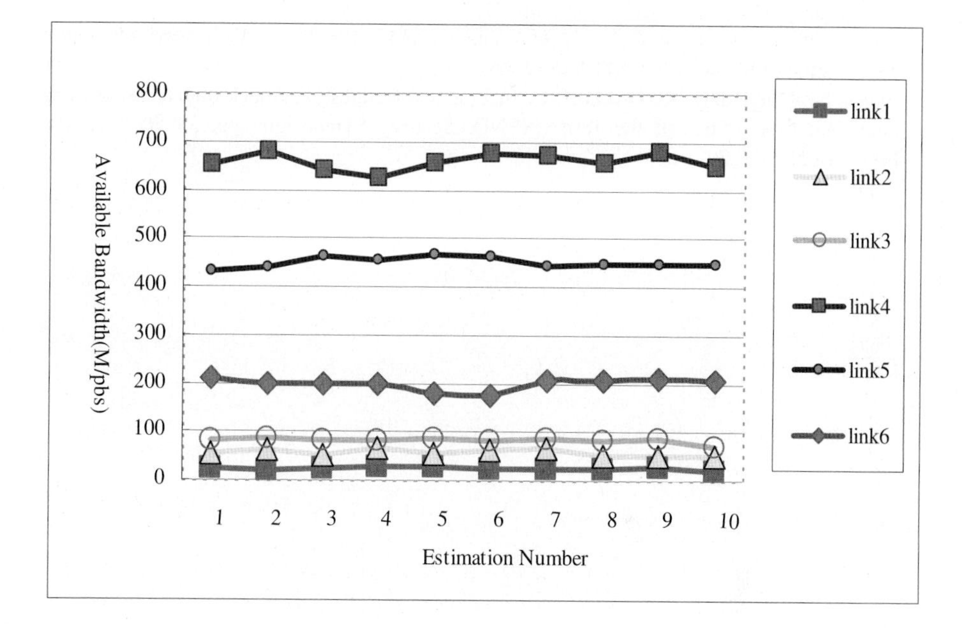

Fig. 1. The result of bandwidth estimation of 3 twisted pair and 3 fiber optic link (The horizontal axis is the estimation number, vertical axis is the available bandwidth)

5 Conclusions

In this paper, we put forward a IPv6 Protocol-based one-way different-length pocket pair subtraction algorithm, applied the flow label of IPv6 message headers to realize the path control of the measured message sequence, an IPv6 network available bandwidth estimation algorithm was designed and its prototype system is established by programming. The test results show that the algorithm is feasible for IPv6 network with the estimation error less than 0.1M.

References

1. Che, H., Chung, C.: Parallel Table Lookup for Next Generation Internet. In: Proceeding of 32nd Annual IEEE International Computer Software and Applications Conference, pp. 52–59. IEEE Press, New York (2008)

2. Min, L., Cheng, L., Jing, S.: Priority-based Available Bandwidth Measurement Method for IPv6 Network. Journal of Computer Research and Development 41, 1361–1367 (2004)
3. Strauss, J., Katabi, D., Kaashoek, F.: A Measurement Study of Available Bandwidth Estimation Tools. In: Proc. of ACM Internet Measurement Conference (IMC), pp. 39–44. MIT Press, Florida (2003)
4. Ribeiro, V., Riedi, R., Navratil, J.: PathChirp: A Light-Weight Available Bandwidth Estimation Tool for Network-Aware Applications. In: Proceedings of 2003 LACSI Symposium, pp. 1–12. LACSI Press, New Mexico (2003)
5. Hu, N., Steenkiste, P.: Evaluation and Characterization of Available Bandwidth Probing Techniques. IEEE J. Sel. Areas Commun. 21, 879–974 (2003)
6. Min, L., Cheng, L., Bing, G.: An End-to-end Available Bandwidth Estimation Methodology. Journal of Software 17, 108–116 (2006)
7. Evans, N., Dingledine, R., Grothoff, C.: A Practical Congestion Attack on Tor using Long Paths. In: Proceedings of the 18th USENIX Security Symposium, pp. 33–50. USENIX Press, Montreal (2009)

A Structure-Based XML Storage Method in YAFFS File System

Ji Liu, Shuyu Chen, and Haozhang Liu

College of Computer Science, Chongqing University,
Chongqing 400044, China
rabbitrobbin@163.com

Abstract. The design of Flash File system YAFFS aims at dealing with problems at start time, memory consumption and wear balance etc. It cannot manage great amount of complex, structural and semi-structural data. The key reason for these points exists in the fact that YAFFS file system treats data as bit stream without semantics. And YAFFS file system uses special applications to operate the inner structures and contents of files, which finally put the control of file to a bad grain level. We propose a structure-based XML storage method in YAFFS file system. The experiment on embedded Linux proved that, with our method, the XML structure-based information can be stored and managed effectively.

Keywords: YAFFS file system, XML, embedded Linux, structure-based XML storage.

1 Introduction

XML is self-descriptive and portable [1]. XML is stored in flat text in forms of serialized data. It is strong in managing XML data with XML document, but lacks efficiency in accessing and updating. This is because that in file system, file is the unit of managing and organizing information, including user data and metadata. As a complement to file data, metadata is not user data themselves. In file system, information is denoted and associated in forms of metadata. Useful information can be extracted from different files. As a result, the ability of describing information mostly comes from its support for metadata [2].

Flash memory has several characteristics such as non-volatility, fast access speed, kinetic shock resistance and low power. These characteristics explain the popularity of flash memory in portable devices and embedded devices. YAFFS (Yet Another Flash File System) is a dedicated flash file system [3]. It can run on Linux, uClinux and Wince. YAFFS has its own flash driver. With no operating system, it provides embedded system with API that can access file system directly and perform operation on files. This design targets at start time, memory consumption and wear balance etc. But it lacks efficiency in organizing and managing XML structural information.

The focus of our work is proposing an XML data storage method based on semantic structure in YAFFS file system. While storing XML data and structure information effectively, it can manage and control through XML data, and enable YAFFS file

L. Cao, J. Zhong, and Y. Feng (Eds.): ADMA 2010, Part II, LNCS 6441, pp. 427–434, 2010.

system with the ability of scanning and updating. In Section 2, we give a storage method supporting XML data management by improving YAFFS file system logic layout and semantic structure of storage file. It is also realized by providing the ability of information retrieval and scanning. In Section 3, we provide experimental statistics on embedded Linux platform to show the new method's advantage in supporting direct XML data access, scanning and searching.

2 Semantic XML Data Storage Method on YAFFS File System

XML flat file storage method uses text file to store serialized data, which is easy to realize without database and storage management [4]. Other alternatives are being stored in tables of relational database [5] [6] and in OO/OR database [7]. How to realize the convenience of XML storage and the flexibility of Native XML? We referred to SFS (Semantic File System) [8], and put forward a data storage method based on semantic structure on YAFFS file system. This section talks about the XML data storage model and mapping and related algorithms on YAFFS file system.

2.1 XML Storage Model

In XML storage model, we usually talks about element, attribute and CDATA.

Definition 1. XML data model based on semantic structure: this paper uses directed map G to describe the semantic model of XML data: $C =< V \; E >$, in which $V = \{oid \mid oid \in SV\}$, *SV* is the set of nodes:
$E = \{< x, y > \mid P(x, y) \wedge (x, y) \in SV\}$. Predicate $P(x, y)$ means the arc between node x and node y. It represents father-child relation. Here, node set *SV* only uses four kinds of node: root node, Element Node, Text Node and Attribute Node. Root nodes represent the whole XML document. Each node has three child nodes: element, attribute and text. Attribute nodes and text nodes are leaf nodes. In this model, the number of element nodes, attribute nodes and text nodes are decided by their position in the document. It is shown as Figure 1.

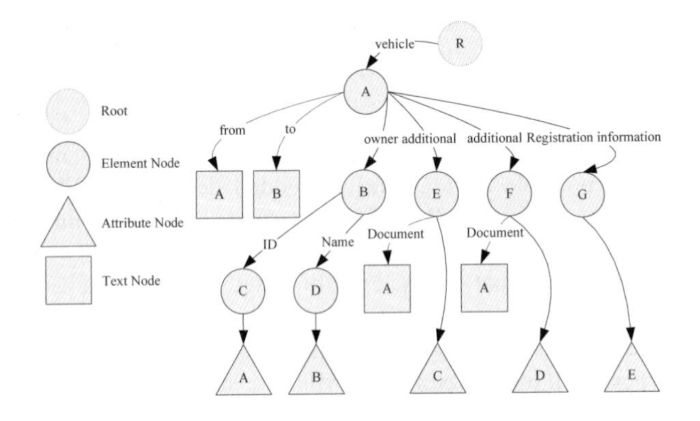

Fig. 1. Semantic structural model for XML data

XML data model is the organizing structure of the elements, attributes and CDATA of XML. But XML document structure only contains part of element nodes, attributes and CDATA nodes. It does not contain root elements. Instead, it uses the first element to represent root. The selection and modification of nodes conforms to the grain level of information that operating system needs. The purpose of data model is to support data operations, while the operation of document structure is supporting the inner structural control of document. From this, we suppose the following constrain: Apart from the element of the bottom level, no element has CDATA data.

In order to store this semantic structural information in XML document directly, there will be some key questions: (1) How to get the largest structural information possible in XML file? (2) How to manage these structural objects? (3) How to arrange these relatively small objects logically? (4) How to store that information on flash memory in order to reach the most efficient access control and the least space cost?

2.2 XML Storage's Mapping on YAFFS

To solve the questions of the precious section, we improve YAFFS file system. Firstly, we introduce a XML model. By adding XML Parser, Indexer and Object Manager at the YAFFS file system management level, XML data can be parsed so that we can get every semantic structural object. This is shown in figure 2.

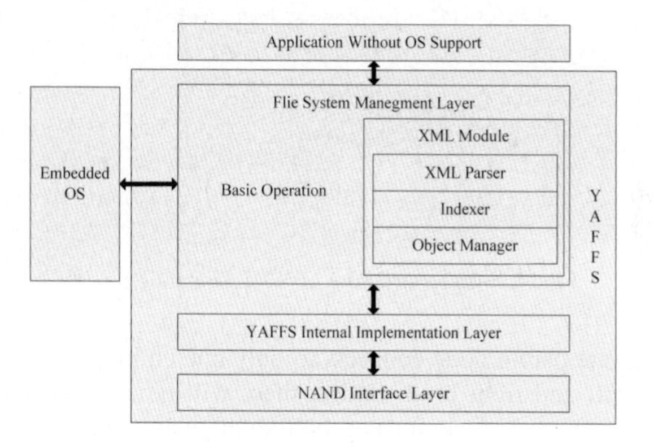

Fig. 2. Improved YAFFS structure

Next, we arrange these structures into a tree and manage them together. Besides, file management in YAFFS file system consists of metadata management and text data management. The main logical layout is divided into boot block, super block, inode table and data section. We approach is improving YAFFS file system's direct locating method. We introduce a new element/attribute node table field, and add a $XML_Struct attribute field in order to locate semantic structural root nodes parsed by XML data. We also introduce a data attribute $DA to locate char string. The layout after improvement is as the following:

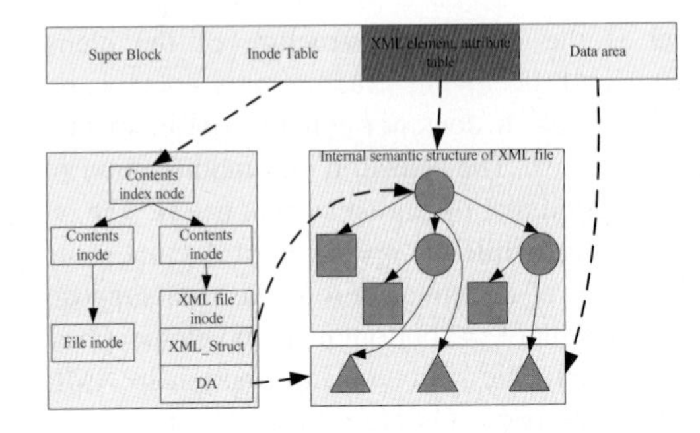

Fig. 3. Improved YAFFS logical layout

If element/attribute nodes are stored directly in the way shown in figure 3, there will be serious scalability problem because of the uncertainty of number of child nodes. So we refer to the idea of transformation of binary tree and transform the data model in figure 3.

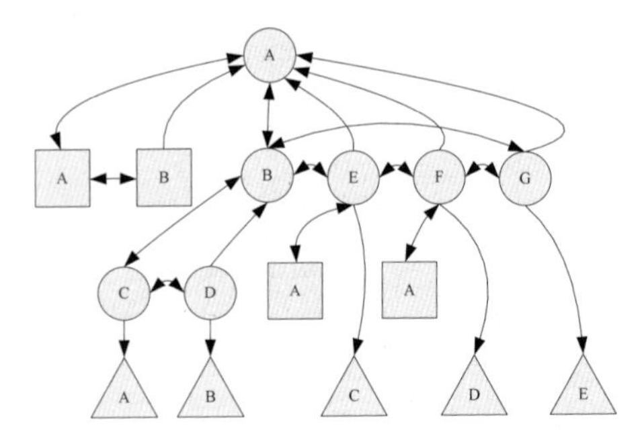

Fig. 4. Storage structure of XML semantic model

In this model, one parent node connects to only one child node. The other children connect to their left and right brothers. So, there will be at most one attribute child node and one element child node for non-bottom element nodes. And there will be at most one attribute leaf and one CDATA leaf. For an element node, it has five pointers that point to other nodes. They are parent node, element child node, attribute child node, left brother node and right brother node. For an attribute node, it has three pointers that point to other nodes. They are parent node, left brother node and right brother node. Brother nodes use two-way circulation linked list to locate a certain node quickly.

In this paper, we will use long record to store the information about every element or attribute. We use indirect addressing in CDATA node text, which means that the location information of CDATA is stored in $DA attribute in document. If we want to get CDATA, we would first read the text and find the description about data. After it, we will seek to locate data according to the location.

2.3 Realization

For element and attribute nodes, we use the following data structure to define:

```
program Inflation (Output)
{
 uint16   Eid ;//Serial number and label
 uint16   Fnode ;//Father node
 uint16   SubEnode ;//Child element node
 uint16   SubAnode ;//Child attribute node
 uint16   Lnode ;//Left brother
 uint16   Rnode ;//Right brother
 uint64   Dpos ;//Document's location information
 uint64   Dlen ;//Length of document
 uint8    Enlen ;//Length of element's name
 uint8    Name ;//Name of element
 uint16   Value ;//value of element
}//256 byte in total
struct Attr_Str//Attribute node
{
 uint16   Aid ;// Serial number and label
 uint16   Fnode ;//Father node
 uint16   Lnode ;//Left brother
 uint16   Rnode ;//Right brother
 uint8    Anlen ;//Length of attribute name
 uint8    Avlen ;//Length of attribute value
 uint16   Name ;//Name of attribute
 uint32   Value ;// Value of attribute
}//128 byte in total
```

We use the following algorithm for creating XML structure record:

Step 1. Call XML parsing function to parse XML document and build an object tree.
Step 2. We use preorder to traverse the object tree.
Step 3. Relocate to the beginning of data structure and write information about maintenance tree according to each record's length and value.
Step 4. Write $XML_Struct

We add API about element and attribute in file system management layer on YAFFS file system. Its algorithm is as follows:

Algorithm of Searching Element Nodes

Step 1. Read the current node's content by the node's index, locate to $XML_Struct;
Step 2. Read $XML_Struct, we use preorder to traverse element nodes and find the corresponding element node;
Step 3. If the control information of element node can not be read, return unsuccessful; else, write data into buffer according to the request for processing control information.

Algorithm of Searching Attribute Node

Step 1. Read the current node's content by the node's index, locate to $XML_Struct;
Step 2. We use preorder to traverse element nodes and find the corresponding element node;
Step 3. We use element control information to determine the possibility of reading the attribute of the element. If it is not possible, return false.
Step 4. Read the attribute child node and search the attribute node with the corresponding name.
Step 5. If it exists, write data into buffer, or return false.

Algorithm of Updating Element Node

Step 1. Read the current node's content by the node's index, locate to $XML_Struct;
Step 2. We use preorder to traverse element nodes and find the corresponding element node;
Step 3. If the control information of element node cannot be read, return unsuccessful;
Step 4. Write data into file after data is processed according to the request for processing control information is met.
Step 5. Re-solve the text specified by the element. Update corresponding $XML_Struct record;

3 Experimental Results and Analysis

3.1 Experimental Environment

Platform configuration of experiment in this paper:

CPU	S3C6410	533MHz
Memory	256MB	DDR
Flash	8GB	SD card
Operating System	Linux2.6.8	
Database	Berkeley DB4.3.27	

The document we use in this experiment is based on a DTD defining car owner's information description. It contains 6 elements and 2 attributes. We create documents of different sizes based on this DTD. These documents have better structure. It is also easy for database to transform them.

3.2 Experimental Results and Analysis

We first compare the efficiency between three storage methods.

From table 1, we can see in relational database storage method, it is effective when we do not count the time of changing modes. In structure-based storage method, when XML document objects are smaller and have more semantic structures, their inner semantic information grows as the document becomes larger.

Table 1. Comparison of first writing time of XML document(Unit μS)

XML length	Text	Relational database	Structure
2 KB	136	169	109
32 KB	1836	2563	2668
128 KB	5308	9857	12842

Table 2. Comparison of reading XML document time(Unit ìS)

XML length	Text	Relational database	Structure
2 KB	146	167	159
32 KB	694	794	668
128 KB	1356	1287	994

Table 2 shows the comparison of reading times

Because NAND Flash can perform random read and write on each block, accessibility will not be affected as the times of reading flash block increases.

Table 3. Comparison of reading XML document time(Unit ìS)

XML length	Text	Relational database	Structure
2 KB	175	198	108
32 KB	2798	2468	1799
128 KB	13656	10698	7165

Table 4. Comparison of updating XML document time(Unit ìS)

XML length	Text	Relational database	Structure
2 KB	569	467	259
32 KB	5945	5794	2268
128 KB	24396	23687	12894

Table 3 and table 4 show that when it comes to searching and updating, structure-based storage is obviously better than text files and relational databases. This is because those text files need to call DOM or SAX resolving module to parse XML document, thus making the efficiency relatively low. And relational databases need to transform data and thus increase times cost.

4 Conclusion

A contribution in this paper is that we present a new semantic XML data storage method based on semantic structure in YAFFS file system. With this method, we can store XML data and structure information effectively. Also, we can manage and control through XML data, and the method enables YAFFS file system with the ability of scanning and updating. We give a storage method supporting XML data management by improving YAFFS file system logic layout and semantic structure of storage file.

The comparative experiment is done based on ARM 11 embedded Linux platform with text file and embedded relational database Berkeley DB 4.3.27. The results show that in searching and updating, semantic storage method is obviously better than text file and relational database.

In the future, we intend to explore improvements on storage model. This will lead to more complex storage algorithm, but may enhance storage efficiency and reduce maintenance cost, which has prominent significance for flash file system YAFFS.

Acknowledgement. This work is partially supported by Natural Science Foundation Project of CQ CSTC, 2008BB2307 and the Third Stage Building of "211 Project" (Grant No. S-10218).

References

1. Harold, E.R., Scott Means, W.: XML in a Nutshell. O'Reilly, Sebastopol (2001)
2. Gopal, B., Manber, U.: Integrating content-based access mechanisms with hierarchical file systems. In: Symposium on Operating System Design and Implementation (OSDI), New Orleans, pp. 265–278 (1999)
3. Chung, T.-S., Park, D.-J., Park, S.-w., Lee, D.-H., Lee, S.-W., Song, H.-J.: A survey of Flash Translation Layer. Journal of Systems Architecture 55, 332–343 (2009)
4. Abiteboul, S., Clued, S.: Querying and Updating the File. In: Proc. of VLDB, Morgan Kaufmann Publishers Inc.73248, San Francisco (1993)
5. Floreseu, D., Kossman, D.: Storing and Querying XML Data Using an RDBMS. In: Proc. of ACM SIGMOD 2002. ACM Press 2042215, New York (2002)
6. Shanmugasundaram, J., Tufte, K., Zhang, C.: Relational Databases for Querying XML Documents: Limitations and Opportunities. In: Proc. of VLDB. Morgan Kaufmann Publishers Inc.3022314, San Francisco (1999)
7. Kanne, C., Moerkott, G.: Efficient Storage of XML Data. In: Proc. of ICDE 2000, IEEE Computer Society, Washington (2000)
8. Gifford, D.K., Jouvelot, P.: Semantic File System. In: ACM Symposium on Operating System Principles, pp. 16–25 (1991)

A Multi-dimensional Trustworthy Behavior Monitoring Method Based on Discriminant Locality Preserving Projections

Guanghui Chang[1], Shuyu Chen[2], Huawei Lu[1], and Xiaoqin Zhang[1]

[1] College of Computer Science, Chongqing University, Chongqing 400030, China
[2] School of Software Engineering, Chongqing University, Chongqing 400030, China
cquteam@yahoo.com.cn

Abstract. Trustworthy decision is a key step in trustworthy computing, and the system behavior monitoring is the base of the trustworthy decision. Traditional anomaly monitoring methods describe a system by using single behavior feature, so it's impossible to acquire the overall status of a system. A new method, called discriminant locality preserving projections (DLPP), is proposed to monitor multi-dimensional trustworthy behaviors in this paper. DLPP combines the idea of Fisher discriminant analysis (FDA) with that of locality preserving projections (LPP). This method is testified by events injection, and the experimental results show that DLPP is correct and effective.

Keywords: multi-dimensional trustworthy behavior monitoring, anomalies discrimination, discriminant locality preserving projections.

1 Introduction

The Internet has become one of the most important infrastructures in the modern society. Meanwhile, the system abnormal behaviors brought out by the internal faults and external security risks of various entities in the network have become more and more serious, and these problems also lead to the distrust of the network services [1].

The research on the trust management shows that the dynamic decision is a key step in trustworthy computing, and it is mainly based on the monitoring and processing of dynamic behavior data [2,3]. Effective handling of the multi-dimensional dynamic behavior data is essential to reliable monitoring, while it has not been well designed till now. System behavior monitoring often need to deal with a wide range of anomalies behavior data which could be expressed as multi-dimensional vector. In order to monitor a running system status, useful information could be refined from these data by dimension reduction [4]. Principal Component Analysis (PCA) and Fisher Discriminant analysis (FDA) are the most popular dimension reduction methods, which are widely implemented in industrial production, image recognition, and system diagnosis etc. PCA [5,6,7] is a method for the sake of optimal reconstruction of sample data, and the FDA method is not only considered the global structure of

L. Cao, J. Zhong, and Y. Feng (Eds.): ADMA 2010, Part II, LNCS 6441, pp. 435–442, 2010.

data but also the label information to determine its inherent structure [8,9,10]. Recently, a number of manifold learning algorithms have been proposed to discover the nonlinear structure of the manifold by investigating the local geometry of samples, such as LPP [11,12]. LPP, computing a low-dimensional embedding of high-dimensional data to preserve the local neighborhood structure of data manifold, has achieved successful applications. However, LPP does not make use of the class label information, so it cannot perform well in classification. In this paper, discriminant locality preserving projections (DLPP), is proposed to monitor multi-dimensional trustworthy behaviors. DLPP combines the idea of FDA and LPP. DLPP considers not only the local structure of data but also the class label information. Thus, it makes the monitoring more accurate.

2 Formalized Presentation of Dynamic Behavior Monitoring

A trustworthy monitoring system contains several modules such as a data collector, a feature extractor, an analysis processor and a trustworthy strategy lib. It can be presented as follow in Fig 1:

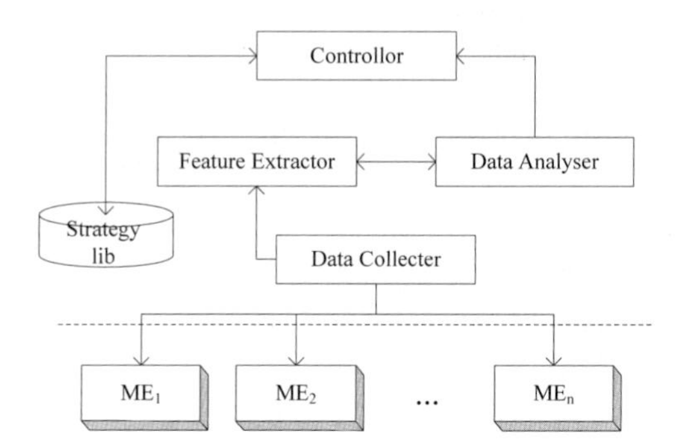

Fig. 1. Trustworthy monitoring system

ME is a Monitor Entity in a network system. In a network composed by MEs, whenever a ME is attacked or crashes itself, it will show some kinds of anomalies. So we can detect a ME's behaviors to induce its status. In this paper we mainly focus on the data processing, feature extraction and anomalies discrimination after data collection.

2.1 Formalized Description of Trustworthy System Status

The system status could be presented by some features which could be measured, and a system status could be denoted as a feature vector. Then all these feature vectors at different running stage form a multi-dimensional random vector space, which could be represented as X , and $X \subset R^{d \times n}$. Here, d and n means the number of dimensions and data samples, respectively. The value of d is decided by the chosen system features. Traditional methods dealing with multi-dimensional data are time-costly, and this is unbearable to a real-time trustworthy monitoring task, so it is necessary to reduce the dimension of original data and then recognize them.

Assume $X = [x_1, x_2, ..., x_n] \subset R^{d \times n}$, x_i could be presented as a d dimension column vector, and n is the total number of samples. Then a sample matrix of original monitoring data can be defined as following:

$$X = \begin{bmatrix} x_{11} & x_{12} & \cdots & x_{1n} \\ x_{21} & x_{22} & \cdots & x_{2n} \\ \vdots & \vdots & \ddots & \vdots \\ x_{d1} & x_{d2} & \cdots & x_{dn} \end{bmatrix}$$

It includes all the information of the monitoring samples. And the aim of dimension reduction is to seek an optimal transformation matrix V and project Y onto it as: $Y = V^T X$, here $Y \subset R^{c \times n}$ and $c \ll d$.

3 Locality Preserving Projections in Dynamic Behaviors Discrimination

3.1 Fisher Discriminant Analysis and Locality Preserving Projections

Given a training sample set $X = [x_1, x_2 ... x_n] \subset R^{d \times n}$, which belongs to C_1 , C_2 , $\cdots C_l$ classes. The main idea of FDA is to best take advantage of the labels of the known samples so that the aggregation of samples in the same class can be denser while on the other hand the data in different classes could be further away. Obviously the effective usage of label information is significant to the task of discrimination. Based on FDA, it is necessary to find a transformation matrix W , which maximizes the between-class scatter while minimizes the within-class scatter.

Let the between-class scatter matrix S_b and the within-class scatter matrix S_w be defined as:

$S_b = \sum_{i=1}^{l} n_i (\mu_i - \mu)(\mu_i - \mu)^T$, $S_w = \sum_{i=1}^{l} \sum_{x_k \in X_i} (x_k - \mu_i)(x_k - \mu_i)^T$ .Here, n_i is the total number of samples belonging to X_i, and μ_i is the mean vector of X_i, while μ is the mean vector of all samples. Depending on the requirement of FDA, we can get the following objective function:

$$\max \left| W^T S_b W - W^T S_w W \right| = \max \left| W^T (S_b - S_w) W \right| . \tag{1}$$

FDA takes advantage of the label information in the original samples to preserve the entire structure and discrimination information of the data, but some locality structure is disturbed after projection. So in this paper a thought of local structure preserving is inducted to overcome this defect.

The idea of locality structure preservation is to preserve the local neighborhood structure of data manifold. So Y after transformation should satisfy:

$$\min \sum_{i,j} \left\| y_i - y_j \right\|^2 S_{ij} \tag{2}$$

In formula (2), y_i is the projected vector from x_i, and S_{ij} is a similarity matrix which describes the relationship between samples.

$$S_{ij} = \begin{cases} e^{-\left(\left\| x_i - x_j \right\|^2 / t \right)}, & \text{if } x_i \text{ is among } k \text{ nearest neighbors of } x_j \\ 0 & \text{otherwise} \end{cases}$$

Here, we use K-nearest neighbor (KNN) to determine the value of S_{ij} and t is an experience parameter. Assume W is a transformation operator of X, and then we have $Y = W^T X$. After algebraic transformation we could get:

$$\begin{aligned} & \frac{1}{2} \sum_{i,j} \left\| y_i - y_j \right\|^2 S_{ij} \\ & = \frac{1}{2} \sum_{i,j} \left(W^T x_i - W^T x_j \right)^2 S_{ij} \\ & = \sum_{i,j} W^T x_i S_{ij} x_i^T W - \sum_{i,j} W^T x_i S_{ij} x_j^T W \\ & = W^T X D X^T W - W^T X S X^T W \\ & = W^T X (D - S) X^T W \\ & = W^T X L X^T W \end{aligned}$$

Here we have $X = [x_1, x_2 \cdots x_n]$, D is a diagonal matrix, $D_{ii} = \sum_j S_{ij}$, and $L = D - S$ is a Laplacian matrix. By using similarity matrix the locality relationship in original samples would be preserved after transformation, and the objective function could be presented as follow:

$$\min_{W} W^T X L X^T W \tag{3}$$

3.2 DLPP for Trustworthy Indicator Space

In order to take advantage of the label information, we could conduct a new optimized method DLPP (Discriminant Locality Preserving Projections) by combining FDA with LPP. Obviously this is a multi-objective optimization problem, which can be transformed to single-objective optimization by using evaluation function technique. Here we get the optimized model for DLPP by using multiply and divide:

$$W_{opt} = \arg\max_{W} \frac{W^T (S_b - S_w) W}{W^T XLX^T W}. \tag{4}$$

This optimization problem could be solved by using Lagrangian operator. First we construct a Lagrangian function: $\psi(W) = W^T (S_b - S_w) W - \lambda(W^T XLX^T W - 1)$.

By differentiating the function with respect to W, it could be transformed to a generalized eigenvalue problem, as follow:

$$\frac{\delta\psi(W)}{\delta W} = (S_b - S_w)W - \lambda XLX^T W \tag{5}$$

Setting formula (5) to zero, we can get:

$$(S_b - S_w)W = \lambda XLX^T W \tag{6}$$

If we have $S_L = XLX^T$ and $S_{FDA} = S_b - S_w$, then:

$$S_L^{-1} S_{FDA} W = \lambda W \tag{7}$$

Obtain the eigenvalues of $S_L^{-1} S_{FDA}$, then select k biggest eigenvalues $\lambda_1 > \lambda_2 > ... > \lambda_k$, and these k eigenvectors form a projection matrix: $W = [w_1, w_2 \cdots w_k]$.

4 Anomalies Discrimination

The statistical presentation of anomalies discrimination can be stated as follow: in a set Z, there are several classes $X_1, X_2 \cdots X_l$, based on some given test samples ξ, to discriminate a specific class X_j by using some discrimination function ϕ. In this paper, we discriminate the class which the sample should be in by the criterion of nearest distance to center. Specify sets $X_1, X_2 \cdots X_l$ as classes, and assume that the corresponding mean values are $\mu_1, \mu_2 \cdots \mu_l$. Choose a test sample ξ, and the distance from ξ to X_i is computed as $d(\xi, \mu_i) = \|\xi - \mu_i\|^2$. If $d(\xi, \mu_j) = \min(d(\xi, \mu_1), d(\xi, \mu_2), ..., d(\xi, \mu_n))$, so we could discriminate that ξ belongs to class X_j.

Mahalanobis distance is used in the trustworthy monitoring, and it's defined as: $d^2(x,y)=(x-y)^T\Sigma^{-1}(x-y)$. Where x and y are samples from Z, and Σ is the covariance matrix of Z. In practical applications, Σ is always unknown, and it needs to be estimated by sample covariance. It should be noticed that the inverse covariance matrix Σ used in the Mahalanobis distance, and it is necessary that the estimation of covariance must be invertible. As the number of sample data is far larger than the number of dimension in trustworthy indicator space, it does not have the problem that the covariance matrix isn't invertible. Anderson statistic is used to estimate the covariance matrix in discrimination and analysis.

Assume that x_i is a sample from class X_j, and n_j is the total number of samples in the class, and then we have:

$$\overline{x}=\frac{1}{n_j}\sum_{i=1}^{n_j}x_i \quad , \quad A_j=\sum_{i=1}^{n_j}(x_i-\overline{x})(x_i-\overline{x})^T \quad .$$

And the Z covariance estimation is: $\Sigma=\frac{1}{\sum n_j - j}\sum A_j$.

If there are l classes of training sample sets, after a random test sample ξ is selected, then the anomalies discrimination function can be defined as:

$$\phi(\xi,j)=\begin{cases}1 & if\ d(\xi,\mu_j)=\min(d_1,d_2\cdots d_l);\\ -1 & if\ \exists d(\xi,\mu_k)=d(\xi,\mu_j)=\min(d_1,d_2\cdots d_l)\ while\ k\neq j;\\ 0 & otherwise;\end{cases}$$

In the function above, d_i is short for $d(\xi,\mu_i)$, ξ is a test sample, j is the label of a class. If $\phi(\xi,j)$ equals 1, then it could be discriminated that ξ belongs to a class labeled by j. If $\phi(\xi,j)$ equals 0, the afterwards classification process would continue. When the minimum value is not stable, then the classification is not decided, and at this moment, $\phi(\xi,j)$ equals -1.

5 Experiments and Analysis

Taking different number of samples as training sets, we could get the different DLPP veracities under separated circumstances in which there're different number of training samples. The result comparison is showed in Fig 2.

From Fig 2, we could easily find out that the discrimination accuracy is promoted by the augment of the training sample amount. And when the number of training samples is 500, the discrimination accuracy reaches 84.8%, and this could satisfy practical usage in anomaly monitoring applications.

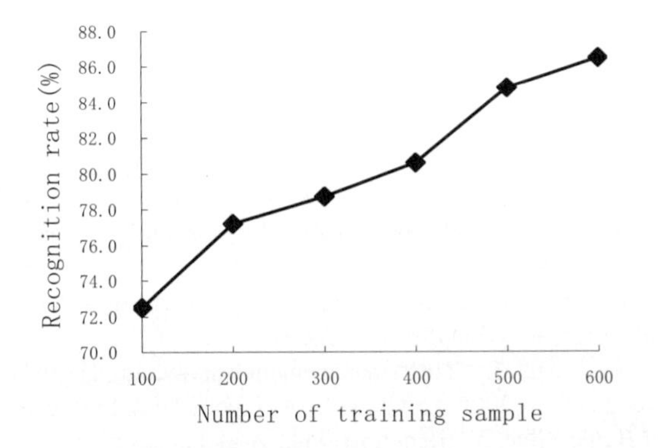

Fig. 2. Recognition rate (different number of training samples)

And took the same training sets and test sets, the loss rates are showed in fig 3:

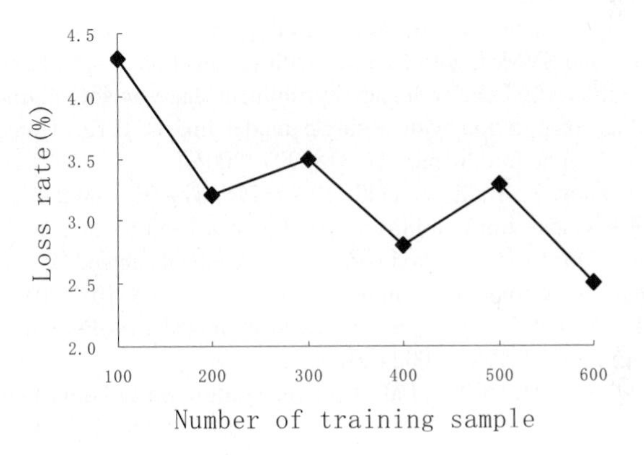

Fig. 3. Loss rate (different number of training samples)

In Fig 3, it shows that the loss rate of DLPP is decreased by increasing the training sample amount. Though the rate is fluctuating, the result is acceptable to practical applications.

6 Conclusion

A new multi-dimensional trustworthy monitoring method, called discriminant locality preserving projections (DLPP), is proposed in this paper. DLPP combines the idea of FDA and LPP. It considers not only the local structure of data but also the class label information. The experimental results show that this method could effectively deal with the monitoring and discriminating problems which contain a large number of multi-dimensional complicated data.

Acknowledgement. This work is partially supported by Natural Science Foundation Project of CQ CSTC (Grant No. 2008BB2307).

References

1. Lin, C., Peng, X.H.: Research on Trustworthy Networks. Chinese Journal of Computers 28, 751–758 (2005)
2. Ji, M., Orgun, M.: Trust management and trust theory revision. IEEE Transactions on Systems, Man and Cyberbetics 36, 451–460 (2006)
3. Ruohoma, S., Kutvonen, L.: Trust management survey. In: Herrmann, P., Issarny, V., Shiu, S.C.K. (eds.) iTrust 2005. LNCS, vol. 3477, pp. 77–92. Springer, Heidelberg (2005)
4. Tenenbaum, J.B., de Silva, V., Langford, J.C.: A global geometric framework for nonlinear dimensionality reduction. Science 290, 2319–2323 (2000)
5. Zhang, R., Wang, W.J., Ma, Y.C.: Approximations of the standard principal components analysis and kernel PCA. Expert Systems with Applications 37, 6531–6537 (2010)
6. Martinez, A.M., Kak, A.C.: PCA versus LDA. IEEE Transactions on Pattern Analysis and Machine Intelligence 23, 228–233 (2001)
7. Gumus, E., Kilic, N., Sertbas, A., et al.: Evaluation of face recognition techniques using PCA, wavelets and SVM. Expert Systems with Applications 37, 6404–6408 (2010)
8. Kim, T.K., Kittler, J.: Locally linear discriminant analysis for multimodally distributed classes for face recognition with a single model image. IEEE Transactions on Pattern Analysis and Machine Intelligence 27, 318–327 (2005)
9. Celenk, M., Conley, T., Willis, J., et al.: Predictive Network Anomaly Detection and Visualization. IEEE Transactions on Information Forensics and Security 5, 288–299 (2010)
10. Sugiyama, M.: Dimensionality reduction of multimodal labeled data by local fisher discriminant analysis. Journal of Machine Learning Research 8, 1027–1061 (2007)
11. Zhang, L.M., Qiao, L.S., Chen, S.C.: Graph-optimized Locality Preserving Projections. Pattern Recognition 43, 1993–2002 (2010)
12. He, X.F., Yan, S.C., Hu, Y.X., et al.: Face recognition using Laplacianfaces. IEEE Transactions on Pattern Analysis and Machine Intelligence 27, 328–340 (2005)

NN-SA Based Dynamic Failure Detector for Services Composition in Distributed Environment

Changze Wu, Kaigui Wu, Li Feng, and Dong Tian

College of Computer Science, ChongQing University, ChongQing, China
{wuchangze,kaiguiwu}@cqu.edu.cn,
joycelee1027@163.com, tiandong@cqu.edu.cn

Abstract. Neural network(NN) and simulation annealing algorithm(SA), combined with adaptive heartbeat mechanism, are integrated to implement an adaptive failure detector for services composition in distributed environment. The simulation annealing algorithm has the strong overall situation optimization ability, therefore in this article a NN-SA model, which connect simulation annealing algorithm and the BP neural network algorithm, is proposed, to predict heartbeat arrival time dynamically. It overcome the flaw running into the partial minimum of the BP neural network. Experimental results show the availability and validity of the failure detector in detail.

Keywords: failure predict, service composition, neural network, simulation annealing.

1 Introduction

Service Composition[1][2] involves the development of customized services often by discovering, integrating, and executing existing services. This can be done in such a way that already existing services are orchestrated into one or more new services that fit better to your composite application. One service can be invoked by several service composition, while each service composition usually invokes multiple services. This brings challenges to assure reliability, because the relationships between service composition and the supporting services are complex and dynamic.

Failure detector is an essential component for building reliable service composition in distributed environment, and many ground-breaking advances have been made on failure detectors [3]. Moreover, with the emerging of large-scale, dynamic, asynchronous composition application in distributed environment, adaptive failure detectors which can adapt to changing network conditions have drawn much attention of literature, such as [4-6].The idea of adaptive failure detection is that a monitored process p periodically sends a heartbeat message ("*I'm alive*!"). A process q begins to suspect p if it fails to receive a heartbeat from p after some timeout. Adaptive failure detection protocols change the value of the timeout dynamically, according to the network conditions measured in the recent past [7]. Doing so, they are able to cope adequately with changing networking conditions, and maintain a good compromise between how

L. Cao, J. Zhong, and Y. Feng (Eds.): ADMA 2010, Part II, LNCS 6441, pp. 443–450, 2010.

fast they detect actual failures, and how well they avoid wrong suspicions. Nevertheless, existing adaptive failure detectors (e.g.[4-6]) almost employ statistical methods to predict heartbeat arrival time dynamically, which require the sample data present in normal distribution, making them unsuitable for highly dynamic service composition in distributed environment.

In this paper, we present a novel implementation of adaptive failure detector. It follows the adaptive heartbeat strategy, but employs a quite different method, that is, combined Neural Network (NN) algorithm with a simulated annealing (SA) controller to modify network weight. By use of an improved NN-SA prediction method, we can predict the next heartbeat arrival time promptly through the sample data, and do no assumption on the distribution of sample data. Furthermore, experimental results demonstrate the validity and availability of our method.

2 Architecture Overview

2.1 System Model

A Simplified service composition

The service composition in distributed environment is complex. It can be have several branch. It hard to assure the relation of two service in different branch. To deal with this problem, we break up a complex service composition into several "flat" service composition, where each "flat" service composition represents a path through the complex workflow. Thus, to guarantee the availability of a complex service composition, we only need to guarantee the same availability for all its "flat" service composition.

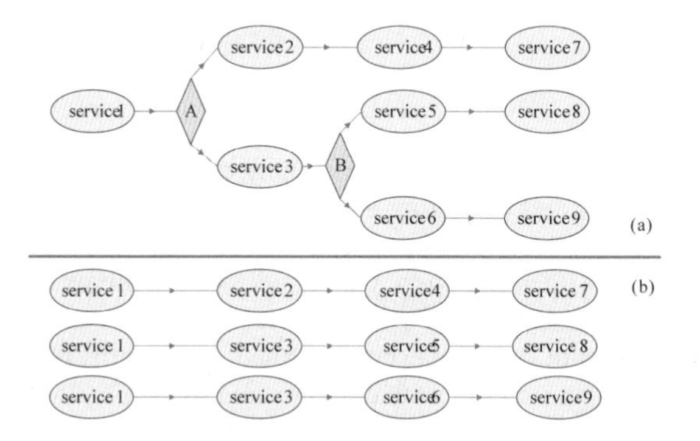

Fig. 1. Service composition simplified, the complex service composition is transformed into three separate "flat" service composition. The things begin to simple, we treat each "flat" service composition as a complete service composition.

B Failure detector model

For the "flat" service composition, similar to the model of Chen *et al.* [4], we consider a simple asynchronous model consisting of two processes p and q. if a process, say p, sends an infinite number of messages to process q and q is correct, then q eventually receives an infinite number of message from p.

In addition, processes have access to some local physical clock giving them the ability to measure time. We assume nothing regarding the synchronization of these clocks.

In the remainder of the paper, we consider the situation where process q monitors process p.

2.2 Architecture of Failure Detector

As is shown in Fig. 2, the implementation of failure detector on the receiving side can be decomposed into two basic parts as follows.

1) *Prediction layer.* An adaptive neural network predictor is constructed (see in section 3). Upon receiving a new heartbeat message, the adaptive predictor calculates the next heartbeat message arrival time, generates both predict value and error amount. According to the prediction, the simulator annealing modify the network weights of NN.
2) *Execution layer.* The final prediction value is the interval of the heartbeat message. Then, actions are executed as a response to triggered failures based on the execution layer whether or not receiving the next heartbeat within prediction time. This is normally done within applications.

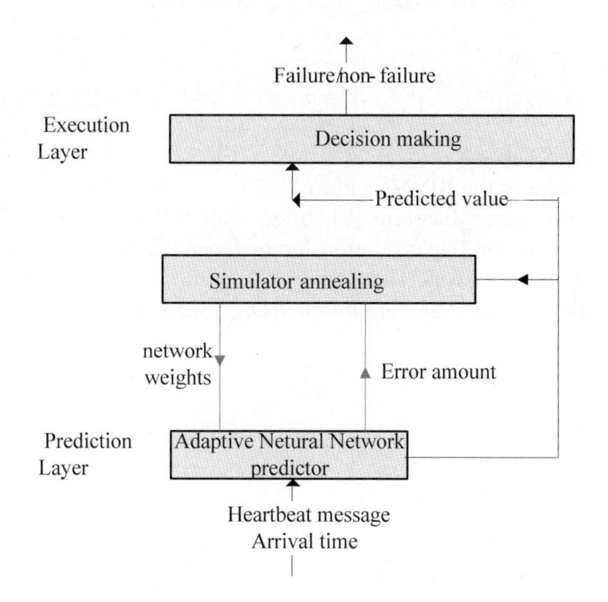

Fig. 2. The architecture of failure detector

3 Prediction Layer

Existing adaptive failure detectors (e.g.[7-8]) almost employ statistical methods, which need a large volume of sample data and require the sample data present in normal distribution. If the samples are distributed at random, the prediction values may become unstable and make the system predict inaccurately.

To avoid the above limitations, we explore NN-SA model to predict heartbeat arrival time dynamically. The algorithm are depicted as:

Step1: Get current time sequence
Collecting the small number of the just passed heartbeat arrival time and message loss rate as prediction samples to form current time sequence, which denote by:

$$t^{(0)} = (t^{(0)}(1), t^{(0)}(2), t^{(0)}(3), \cdots, t^{(0)}(n), l_M(t)) , \quad (1)$$

where n: the number of samples. $l_M(t)$:message loss rate

$$l_M(t) = \frac{M - N}{M} \times 100\% , \quad (2)$$

where M is the heartbeat sending number by q, N is the heartbeat receiving number by p.
Step 2: Normalized of $t^{(0)}$
Defined $t^{(1)}$ as:

$$t^{(1)} = (t^{(1)}(1), t^{(1)}(2), t^{(1)}(3), \cdots, t^{(1)}(n), l_M(t)) , \quad (3)$$

where

$$t^{(1)}(k) = \{t^{(0)}(k) - 0.5(t^{(0)}{}_{\max} + t^{(0)}{}_{\min})\} / 0.5(t^{(0)}{}_{\max} - t^{(0)}{}_{\min}) \quad k = 1, 2, 3, \cdots, n . \quad (4)$$

Step 3: Form NN-SA model
(a) Network Structure of the BP Algorithm
According to Kolmogorov theorem [9], three-layer *BP* with sigmoid function can represent any linear or non-linear relationship between the input and output. So in our research, a three-layer network and sigmoid is used. It consists of an *input layer (IO)*,a *hidden layer (HL)*, and an *output layer (OL)* (See Fig.3).

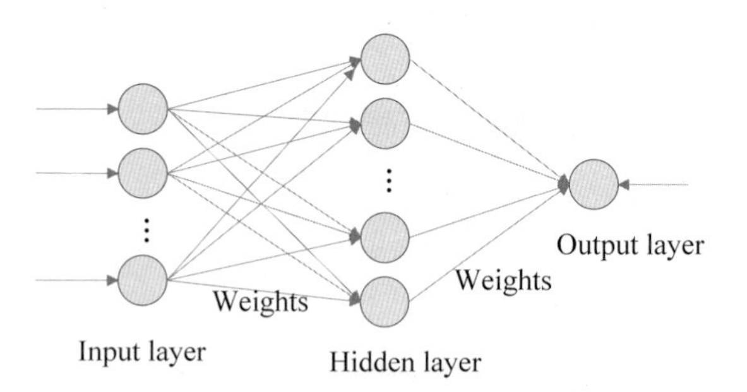

Fig. 3. the Network Structure of the BP Algorithm

- The number of neural nodes n in the *IO* is the number of heartbeat time.
- The number of the *HL* is $2n+1$ according to Kolmogorov theorem.
- There is only one node in the *OL*. The output value of this node represented the last prediction for heartbeat time.

(b) Active Function
In the *NN*, the activation function is used to express the relationship process between the input and output. There are many links to a neural node. We regard the weighted sum of this links as the input of the node. We use the most commonly used sigmoid function [9]

$$f(x)=1/(1+e^{-x}), \tag{5}$$

as the active function for the weight training. Here *x* is the input of the node, and *f(x)* is the output.

(c) Weight Adjustment
During training, the adjustment of the weights uses simulation annealing algorithm.
1)Initialize network weights w_{ij} and high temperature T_0， the temperature drop function is

$$T_{k+1}=\lambda T_k, \tag{6}$$

whereëis the temperature drop ratio.

2) For each sample *t(k)*,compute the output *y(k)* in *OL* and *HL*;
3)compute the error *E*
If error is sufficiently small
then save the results.
Else

$$w_{ij}^*=w_{ij}+\Delta w_{ij}, \tag{7}$$

whereÄw is a small random perturbation;

4) For each sample *t(k)*,compute the output *y(k)* in *OL* and *HL*;
5) compute the error E*;

If error is sufficiently small
then save the results.
Else If ($E^*<E$) **then** $w_{ij}=w_{ij}^*$
Else $w_{ij}=w_{ij}^*$ as probability $P=e^{-\Delta E/T_i}$ where, $\Delta E=E^*-E$, T_i is the current temperature

6) This process of 1)-5) is repeated until the error for all data is sufficiently small;
7) The process of 1)-6) is repeated until the $T_i=0$ or $T_i=T_e$,where target of low temperature
8) At last, we get the weights.

Step 4: Obtain the next heartbeat arrival time
According to the output y(k) of NN-SA, we can get

$$t = 0.5(t_{\max} - t_{\min})y(k) + 0.5(t_{\max} + t_{\min}), \tag{8}$$

where t is the next heartbeat arrival time.

Step 5: Form new prediction model.
Upon receiving the ($n+1$)th heartbeat, the monitoring process p reads the process clock and stores the heartbeat rank and arrival time into a sliding window (thus discarding the oldest heartbeat), and form new prediction model as follows.

$$t_{new}^{(0)} = \{t^{(0)}(2), t^{(0)}(3), \cdots, t^{(0)}(n), t^{(0)}(n+1)\}. \tag{9}$$

Then, repeat steps 2- 4 to predict the (n+2)th heartbeat arrival time, and so on.

4 Experimental Results

Analog to [3,6,7], our experiments involved two computers, with one locates in ChongQing University (CERNET), and the other locates outside of ChongQing University (Internet). All messages are transmitted with UDP protocol. Neither machine failed during the experiment.

The sending host locates outside of ChongQing University, the IP address is 221.5.183.108. It is equipped with a Celeron IV processor at (1.7GHz) and the memory is 256 MB, the operating system is Red Hat Linux 9 (kernel 2.4.20).

While the receiving host is locates in ChongQing University. It is equipped with a Pentium IV processor at 2.4 GHz and the memory is 512MB, the operating system is also Red Hat Linux 9 (kernel 2.4.20).

- **Experimental Results**

Phase 1: Recording heartbeat arrivals
The experiment lasts for three weekdays, during which heartbeat message is generated every 400ms.

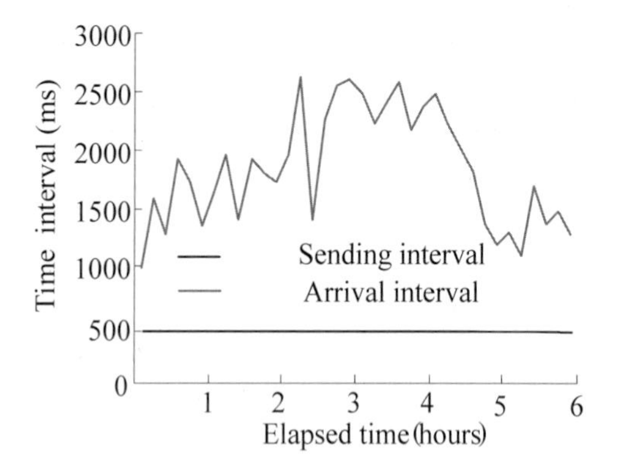

Fig. 4. Depicts the curves of the mean heartbeat sending interval and arrival interval from 10am to 4pm in the three days. We can know that message delay is a bit high, the highest message delay is 2428ms, and the lowest, 989ms. We think this is result from the transmit speed between CERNET and Internet.

Fig. 5. It shows the mean message loss rate of the first two days, in bursting hours, the message loss rate is almost nearly 20% owing to the low transmit speed between CERNET and Internet

Phase 2: Simulating failure detectors
Experiment 1: *determine the size of prediction sample space.*
We set the size of sample space from 20 samples to 500 samples, and measured the accuracy obtained by the failure detector running during 10am and 4pm.

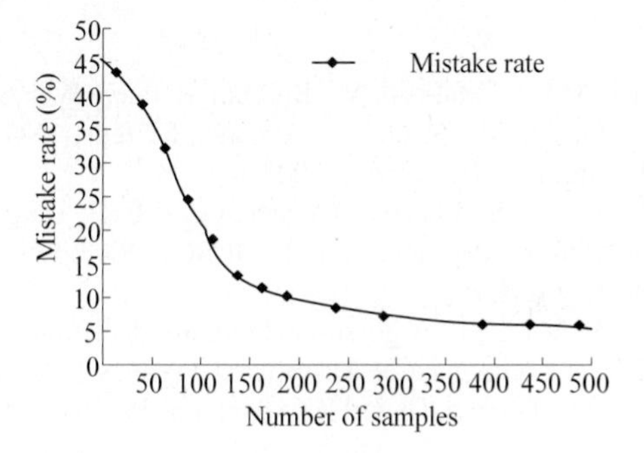

Fig. 6. The mistake rate of our failure detector improves as the sample size increases. The curve seems to flatten slightly when the size is more than 200.

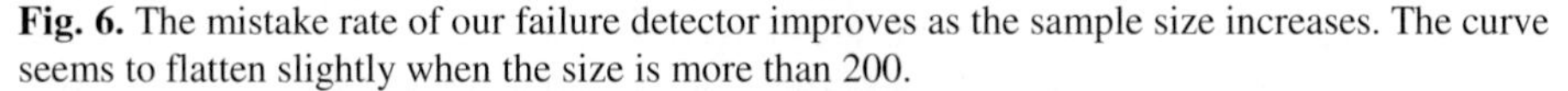

Experiment 2: *comparison with Chen's failure detector*[2]

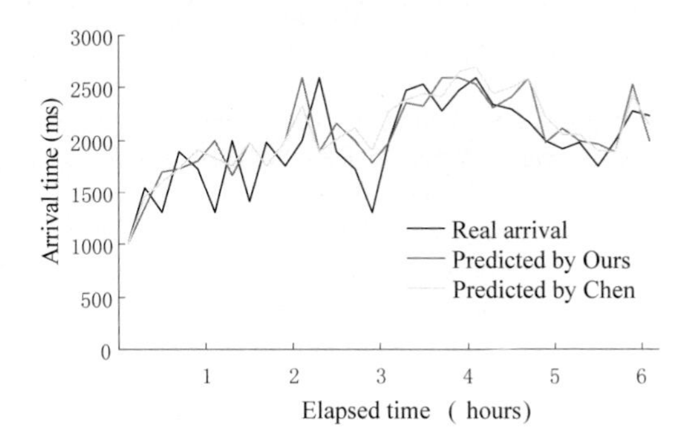

Fig. 7. For the two failure detectors, we set the size of sample space as 300, the prediction value of the two methods is almost same, which means that our method is valid and available

5 Conclusion and Future Work

Failure detection is a fundamental building block for ensuring fault tolerance of services composition in distributed environment. In this paper, combining BP neural network and simulation annealing algorithm, we present a novel implementation of adaptive failure detector. That is, using simulation annealing algorithm to modify the network weights of neural network. In doing so, we do no assumption on the distribution of sample data. Moreover, it overcome the flaw running into the partial minimum of the BP neural network, we can acquire the confidence of failure/non-failure of services composition in distributed environment. At last, experimental results demonstrate the validity and availability of our method in detail. In the near future, we will implement a failure detection middleware based on the algorithms present in this paper.

Acknowledgment. This work is supported by the major program of the Chinese National Science Foundation, under contact No.90818028, the Third Stage Building of "211 Project of china" (Grant No. S-10218).

References

1. Pires, P.F., Benevides, M.R.F., Mattoso, M.: Building Reliable Web Services Compositions. In: Chaudhri, A.B., Jeckle, M., Rahm, E., Unland, R. (eds.) NODe-WS 2002. LNCS, vol. 2593, pp. 59–72. Springer, Heidelberg (2003)
2. May, K.S., Judith, B., Johan, S., Luciano, B., Sam, G.: A Fault Taxonomy for Web Service Composition. In: Di Nitto, E., Ripeanu, M. (eds.) ICSOC 2007. LNCS, vol. 4907, pp. 363–375. Springer, Heidelberg (2009)
3. Robbert, R., Yaron, M., Mark, H.: A gossip-style failure detection service. In: Proceedings of the IFIP International Conference on Distributed Systems Platforms and Open Distributed Processing, The Lake District, United Kingdom, pp. 55–70 (2009)
4. Chen, W., Toueg, S., Aguilera, M.K.: On the quality of service of failure detectors. IEEE Transactions on Computers 51(2), 13–32 (2002)
5. Bertier, M., Marin, O., Sens, P.: Implementation and performance evaluation of an adaptable failure detector. In: Proc. IEEE Intl. Conf. On Dependable Systems and Networks (DSN 2002), pp. 354–363 (2002)
6. Hayashibara, N., Défago, X., Yared, R., Katayama, T.: The φ accrual failure detector. In: Proc. 23nd IEEE Intl. Symp. On Reliable Distributed Systems (SRDS 2004), pp. 66–78 (2004)
7. Hayashibara, N., Défago, X., Katayama, T.: Implementation and performance analysis of the φ-failure detector. Research Report IS-RR-2003-013, Japan Advanced Institute of Science and Technology, Ishikawa, Japan (2003)
8. Hayashibara, N., Défago, X., Katayama, T.: Two-ways adaptive failure detection with the φ-failure detector. In: Proc. Workshop on Adaptive Distributed Systems (WADiS 2003), Sorrento, Italy, pp. 22–27 (2003)
9. Ilunga, M., Stephenson, D.: Infilling streamflow data using feed-forward backpropagation (BP) artificial neural networks: Application of standard BP and pseudo Mac Laurin power series BP techniques. Water SA 31, 171–176 (2005)

Two-Fold Spatiotemporal Regression Modeling in Wireless Sensor Networks

Hadi Shakibian and Nasrollah Moghadam Charkari

Faculty of Electrical Engineering and Computer Science
Tarbiat Modares University, Tehran, Iran
{h.shakibian,moghadam}@modares.ac.ir

Abstract. Distributed data and restricted limitations of sensor nodes make doing regression difficult in a wireless sensor network. In conventional methods, gradient descent and Nelder Mead simplex optimization techniques are basically employed to find the model incrementally over a Hamiltonian path among the nodes. Although Nelder Mead simplex based approaches work better than gradient ones, compared to Central approach, their accuracy should be improved even further. Also they all suffer from high latency as all the network nodes should be traversed node by node. In this paper, we propose a two-fold distributed cluster-based approach for spatiotemporal regression over sensor networks. First, the regressor of each cluster is obtained where spatial and temporal parts of the cluster's regressor are learned separately. Within a cluster, the cluster nodes collaborate to compute the temporal part of the cluster's regressor and the cluster head then uses particle swarm optimization to learn the spatial part. Secondly, the cluster heads collaborate to apply weighted combination rule distributively to learn the global model. The evaluation and experimental results show the proposed approach brings lower latency and more energy efficiency compared to its counterparts while its prediction accuracy is considerably acceptable in comparison with the Central approach.

Keywords: Wireless sensor network, spatiotemporal regression, particle swarm optimization.

1 Introduction

In many scenarios of wireless sensor networks (WSNs), a large amount of data is generated in each node. For example, in weather conditions monitoring for a geographical area [1], sensor nodes capture phenomenon of interest on defined time intervals. As the collected data is increased during sensing process, some methods will be needed to extract useful information from the raw data.

Regression analysis due to its importance and usefulness is addressed in this paper. Having regressor of the network, a compressed description of the data is available and predictions can also be conducted. There are some well-known regression techniques in machine learning literature [2]. But all of them basically work in a centralized environment where both data and processing are centrally available. In a WSN, data are distributed among the nodes as well as processing resources. In addition, the limited

L. Cao, J. Zhong, and Y. Feng (Eds.): ADMA 2010, Part II, LNCS 6441, pp. 451–462, 2010.

power supply and bandwidth capacity of nodes are accomplished the difficulty of doing regression in WSNs.

In a simple solution, Central approach henceforth, one may force the sensor nodes to send their own raw data to a fusion center. Afterwards, a common regression technique could be applied. Although, the Central approach can achieve a high accuracy, but its heavy data transmissions makes it almost inapplicable particularly when the network grows in size. For this reason, distributed approaches are needed to be developed.

Regression is considered as an optimization problem in [3]. Accordingly some distributed approaches has been proposed based on gradient descent [4-6] and Nelder-Mead simplex [7, 8] optimization techniques. They all generally work over a pre-established Hamiltonian path among the nodes. Incremental gradient (IG) [4] applies gradient descent method distributively and incrementally in such a way that every iteration of the gradient method is mapped into a network cycle (travelling from the first node to the last node). In this way, many network cycles are needed to obtain an acceptable accuracy which leads to the significant decreasing in the network lifetime. In [5] an improvement is made by clustering the network and employing the incremental gradient method within the clusters. By setting up Hamiltonian path among cluster heads, convergence rate is increased [6]. An incremental optimization approach based on Nelder-Mead Simplex (IS) is proposed in [7] and [8] which is benefited by boosting and re-sampling techniques, respectively. They reported better accuracy and convergence rate than gradient based approaches.

Usually in a WSN, the sensor nodes collect their measurements across time as well as space in which data are collected at regular time intervals. Therefore, data analysis has to take account both spatial and temporal dependencies between the data points. Accordingly the focus of this paper is spatiotemporal regression modeling in sensor networks. The idea of this paper is to cluster the network and learn the clusters' regressors first. For this purpose, learning the temporal part is separated from learning the spatial part for each cluster's regressor. In the former step, member nodes are required to learn their local temporal models which are then combined by the cluster head. In this way, an accurate cluster's temporal model is obtained as it has been emphasized in many papers [3] that there are close similarities with the sensor readings of a cluster. Afterwards, the cluster head tries to learn the spatial part by using particle swarm optimization (PSO). Finally, the global regression model is constructed through collaborations between the cluster heads in which weighted averaging combination rule, from multiple classifier systems concept, is applied distributively. We have compared the performance of the proposed approach with the Central and IG approaches. While clustering the network brings substantial decreasing in latency, the final prediction accuracy, thanks to good accuracy of the clusters' regressors, is considerably acceptable compared to the Central model. Also the communication requirement for the proposed approach is quite better than two other approaches.

Section 2 provides an overview of regression as a supervised learning problem and particle swarm optimization as well. Problem statement is formulated in section 3. Proposed approach is described in details in section 4. Analysis and experimental results are discussed in section 5 and the last section is conclusion of the work.

2 Preliminaries

2.1 Regression Analysis

Given a data set DS = $\{(x_1, y_1),\ldots,(x_N, y_N)\}$ as a training set, the aim of the supervised learning is to extract a mapping function from the input space ($X=\{x_i\}$, $i=1,\ldots,N$) into the output ($Y=\{y_i\}$, $i=1,\ldots,N$). X and Y are also called features and labels, respectively. The learning program is called regression if the label takes its value from a continuous space and it is called classification when a discrete value is given to the label [9]. Let $g(x|R)$ be a model defined up over X and R be its parameters. When the form of the $g(.)$ is known, regression is parametric, otherwise non-parametric. While in non-parametric type, predictions can be conducted based on data similarities, the parametric regression has to optimize the coefficients of g, namely R, such that g can predict Y with the least possible error:

$$R^* = \arg\ \min_R \{\sqrt{\frac{1}{N}\sum_{i=1}^{N}[g(x_i \mid R) - y_i]^2}\} \tag{1}$$

In an environmental monitoring system, the sensor nodes usually collect their measurements according to both time and space. In more words, before the sensing process starts, regular time intervals are pre-defined. At every time slot, the sensor node i captures phenomenon of interest (temperature in our study), and stores it as a quadruple $< x_i, y_i, t_{ij}, T_{ij} >$ in which (x_i, y_i) is the sensor's location and T_{ij} is the sensed temperature at time slot t_{ij}. In this way, applying the parametric regression is straightforward: x,y and t are as features while T is the label. Now, it is interested to build $g(x,y,t|R)$ such that given a location in the network field and a desired time slot, $g(.)$ can predict temperature with the most possible accuracy.

2.2 Particle Swarm Optimization

Particle swarm optimization (PSO), is one the efficient stochastic search/optimization algorithms which is successfully employed in many engineering optimization problems as well as in many problems from WSN context [10-12]. It has a good ability to explore real valued search spaces such as regression analysis. In this regard, it has recently been applied for linear regression analysis [13, 14]. They have obtained quite better accuracy in comparison with the standard LSE method.

PSO is a population based optimization technique in which every candidate solution is called particle. A swarm is composed of many particles and its size depends on the problem at hand. The particles move (fly) within (over) the search space based on two components, *gbest* and *pbest*. The former one stands for the best particle encountered in the swarm while the other component is the memory of each particle which reminds the best previously visited position by the particle. The particles use the relations (2) and (3) to compute their own velocity vectors and update their new positions, respectively:

$$v_{ij}(t+1) = w.v_{ij}(t) + c_1 r_1(t)[pbest_{ij}(t) - x_{ij}(t)] + c_2 r_2(t)[gbest_j(t) - x_{ij}(t)] \tag{2}$$

$$x_{ij}(t+1) = x_{ij}(t) + v_{ij}(t+1) \tag{3}$$

Here, j refers to dimension, w is inertia weight, c_1 and c_2 are acceleration coefficients and $r_1(t)$, $r_2(t)$ are two uniform random numbers. When determined stopping condition is satisfied, the particles stop their flying over the search space.

3 Assumptions and Problem Statement

We consider a sensor network with n nodes in which every node captures m temperature measurements temporally and spatially, based on time slots and node's location, as mentioned in section 2.1. The network is partitioned into k clusters designating one cluster head for each one CH_c $(c=1,\ldots,k)$. In different clustering algorithms, different desired properties should be found for generated clusters. We consider the case where every cluster head can negotiate with its members directly. This brings both spatial and temporal similarities between the cluster measurements. Of course, since clustering is not the subject of this paper, we assume the network is clustered via a well known clustering algorithm. A full review on clustering algorithms over WSNs can be found in [15]. Also we supposed the network nodes can localize themselves through using a well-known node localization algorithm [22]. Figure 1 depicts the communication topology within the clusters and between the cluster heads as well.

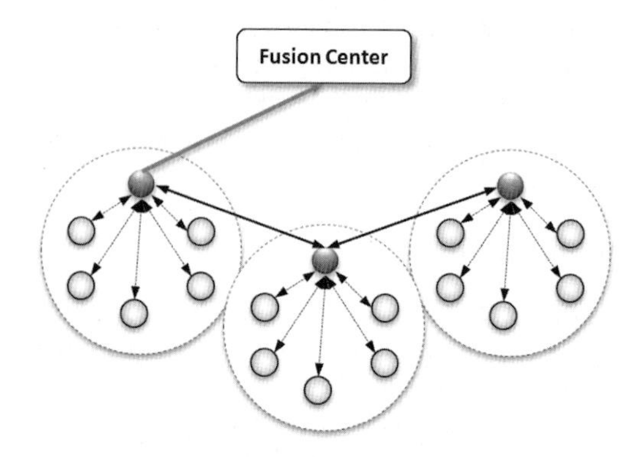

Fig. 1. The communication topology within the network. One of the cluster heads is allowed to directly communicate with the fusion center in order to deliver the final network model. Other cluster heads only negotiate with their immediate neighbors.

The objective is to fit a model on the network data. We follow [16] to choose a spatiotemporal model. They have suggested some polynomial models and we have chosen a linear space and quadratic time model as it has the lowest RMS error among the other models:

$$g(x, y, t \mid R) = r_1.x + r_2.y + r_3.t^2 + r_4.t + r_5 \tag{4}$$

where $R=< r_1, r_2, r_3, r_4, r_5 >$ is the unknown coefficients vector. Within the cluster c, member nodes are required to learn R_c such that the RMS error of the cluster's regressor, g_c, (5) becomes the least possible:

$$RMSE(g_c) = \frac{1}{n_c.m}\sum_{i=1}^{n_c}\sum_{j=1}^{m}[g_c(x_i, y_i, t_{i,j} \mid R_c) - T_{i,j}]^2 \tag{5}$$

where n_c denotes to the size of the cluster c. Minimizing (5) needs all the cluster data which is distributed among the member nodes. Although, one may force each member node to transmit its own raw data to other cluster nodes, but that brings considerable data communications and consequently energy consumption. In the next section we describe how we can obtain g_c $(c=1,\ldots,k)$ as well as G_{net} in an efficient distributed manner.

4 Proposed Approach

Global model is constructed in two folds. First, the regressor of each cluster should be achieved. We divide each cluster's regressor into two temporal and spatial parts. Within a cluster, as member nodes compute their own local temporal models, they are combined by the cluster head and the cluster's temporal model is obtained. Then the cluster head starts its attempt to build the spatial part by employing PSO algorithm. Secondly, when all the clusters finish their in-cluster learning processes and achieve their regressors, the cluster heads collaborate over a sequential path starting from CH_1 through CH_k and vice versa, to construct the final regressor. For the sake of simplicity, we focus on learning the regressor of one cluster (name it c). The other clusters work similarly.

4.1 Learning the Regressors of Clusters

4.1.1 Learning Temporal Part

Within the cluster c, each member node $s_{c,i}$ $(i=1,\ldots,n_c)$ is required to compute its local temporal model:

$$\tau_{c,i} = a_{c,i}.t^2 + b_{c,i}.t + c_{c,i} \tag{6}$$

based on its local data set. The pattern of temperature readings for a randomly selected sensor from Berkeley Intel Lab data set [17] is depicted in Figure 2. The inference of quadratic behavior of the temperature variations over 1500 consecutive time slots can be drawn. So, each member node has to do simply quadratic regression on its local data set $LD_{c,i}$ [18]. Of course, there is no any limitation on the form of the defined model and it only depends on the phenomenon under study. Afterwards, every member node sends its local temporal model to the cluster head, CH_c.

According to our assumptions about clustering the network, we considered the case where every cluster head could be able to do direct transmissions with its members. This normally causes cluster nodes to capture their measurements close to each other.

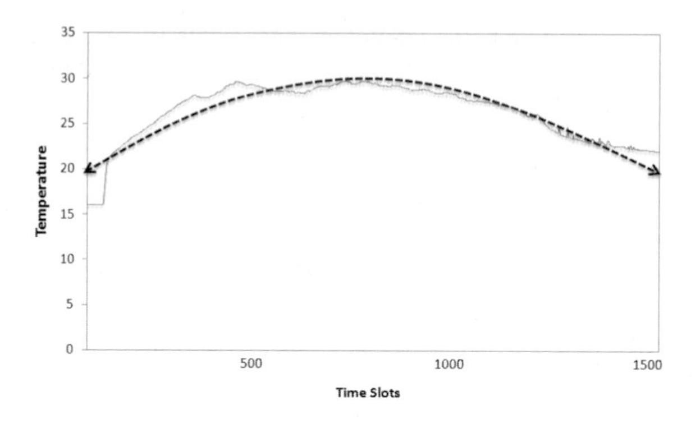

Fig. 2. Temperature pattern for a randomly selected sensor node from Berkeley network over 1500 sensor readings and an approximated quadratic model

Respect to this fact, the received local temporal models from the member nodes could efficiently be combined by the cluster head to compose the temporal part of the cluster's regressor as applying simple voting combination rule [9]:

$$g_c^{temporal} = < r_{c,3}, r_{c,4}, r_{c,5} > = \frac{1}{n_c} \sum_{i=1}^{n_c} \tau_{c,i} \tag{7}$$

To complete R_c two remaining coefficients $<r_{c,1}, r_{c,2}>$, which stands for the spatial part of the g_c, should be achieved which is expressed in the following.

4.1.2 Learning Spatial Part

Obtaining local temporal models, the member nodes send their models to the cluster head. To obtain the spatial part, they also send their own locations as well. Recalling the relation (5), the regressor of cluster should be able to predict cluster data as accurate as possible. Instead of gathering all the cluster data in the cluster head, we use re-sampling technique [8] and let the cluster head CH_c to regenerate a data view of the cluster. In other words, at the end of the temporal learning phase, CH_c receives the following pairs as < *location, temporal model* > from its members:

Node location	**Local temporal model**
$(x_{c,i}, y_{c,i})$	$(\tau_{c,i})$
...	...
(x_{c,n_c}, y_{c,n_c})	(τ_{c,n_c})

At this point, CH_c uses the pairs one by one and regenerates new data points from data space of each relevant member node. Creating this in-cluster data view helps the cluster head CH_c to employ the relation (5) to optimize the spatial part of the cluster's regressor, g_c. To this end, PSO is employed as an efficient optimization method for real valued search spaces. Every particle in the initial swarm represents a candidate solution for the coefficients vector R_c. Note that every particle is a quintuple vector which its last three components are fixed as the temporal part is recently achieved.

The Equations (2) and (3) are used by every particle until the stopping criterion is satisfied.

4.2 Learning Global Model

The second step involves constructing the global network model. One may use simple voting, like as used within the clusters to combine the temporal models, for combining the regressors of clusters. But due to diversity of sensor measurements from one cluster to another one, it is not a good option. Instead, if a top down point of view is taken, in fact we are confronted with a multiple classifiers system (MCS). The ultimate goal in a MCS is to obtain a global and more accurate learner through combination of several base learners. In [19] it is discussed why and how multiple classifiers are appeared in a system. In our problem, many regressors (real valued classifiers) are constructed due to the presence of several disjoint data sets, one for each cluster. On the other hand, some combination rules have been proposed to combine multiple classifiers. As the performance of some popular rules has been compared in a number of papers (e.g. [20, 21]), it is validated that the weighted averaging combination rule is a well qualified option. Of course the other combining rules might bring better accuracy compared to the weighted averaging [20], but in this paper averaging rule is employed due to its simpler distributed implementation.

Accordingly we apply the weighted averaging combination rule to combine the clusters' regressors. The main challenge in this way is to assign appropriate weights to each cluster's regressor. We use RMS error of each regressor to assign the weights. For this purpose, first the RMS error of each regressor is required to compute. Within the cluster c, after g_c is achieved, CH_c returns its regressor to all its members. The member node $s_{c,i}$ tests the cluster's regressor on its local data set and computes a sum of squared error for it $SE_{c,i}$ (i=1,...,n_c). Receiving all the local squared errors, the cluster head will be able to computer the actual RMS error of the cluster's regressor:

$$error_c = RMSE(g_c) = \sqrt{\frac{1}{n_c.m}\sum_{i=1}^{n_c} SE_{c,i}}$$

Now, we let every cluster's regressor takes its initial weight as the inverse of its RMS error,

$$\omega_c = 1/error_c,\ (c=1,\ \ldots,\ k)$$

These initial weights should be normalized before applying on the regressors. After computing the normalized weight, the global model G_{net} can be constructed as the linear combination of the weighted regressors:

$$G_{net} = \sum_{c=1}^{k} \hat{w}_c.g_c \tag{8}$$

where $\hat{w}_c$ denotes to the normalized weight of g_c. Since every cluster head just knows its own ω_c, the relation (8) should be computed distributively. It can be simply done through two traversing the cluster heads (Figure 3). During the first traversing, starting from CH_1, the last cluster head CH_k could be known σ and consequently its own $\hat{w}_k$. At this point the second travelling is started in the reverse direction. This time,

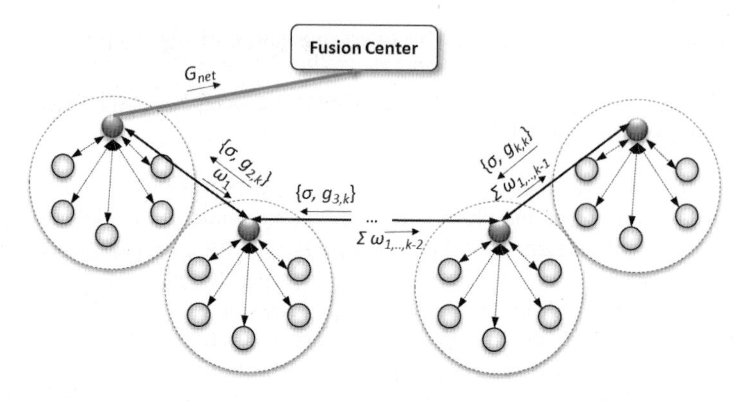

Fig. 3. The cluster heads collaborate to construct the global network model over a sequenial path. The σ is the sum of the initial regressors' errors $\{\omega_i\}$, $i=1,...,k$.

starting from CH_k, CH_j receives σ as well as the regressor achieved from CH_k through CH_{j+1}, denoted as $g_{k,(j+1)}$, computes its $\hat{w}_j$, obtains $g_{k\,j}$ by adding up $\hat{w}_j\times\, g_j$ to $g_{k,(j+1)}$, and sends the result to the next cluster head on path CH_{j-1}. So, the first cluster head CH_1 could be able to deliver the final global model G_{net} to the fusion center.

5 Evaluation and Experimental Results

To see the performance of the proposed approach, we used Berkeley Intel Lab data set [17]. It has 54 sensor nodes with two corrupted ones which are distributed in a bi-dimensional area. Two portions from the data set have been chosen. In the first one, DS1, there are 100 data points for each node (m = 100 and measuring for 50 minutes) and in the second one, DS2, there are 2880 data points for each node (m = 2880 and measuring for one day). So, there are 52×100 and 52×2880 data points for DS1 and DS2 in total, respectively. Also the network has been partitioned into 5 clusters. We compared our approach with the Central and IG approaches regarding to prediction accuracy, latency and communication cost. In order to see the efficiency of the PSO algorithm for spatial learning phase, the behavior of the PSO algorithm for one cluster is shown in Figure 4. The convergence rate is high and the swarm converges to an accurate cluster's model before the iteration number of 100 is elapsed. Also, the improvement of the global model is depicted in Figure 5. As mentioned in section 4.2, the last cluster head, CH_5, starts constructing the global model in an incremental manner. As shown in Figure 5, the quality of the global model is improved as more clusters are visited during this path.

5.1 Prediction Accuracy

The accuracy of three approaches is shown in Table 1. With the assumption of the possibility of long distance transmissions, the Central approach is performed in a single node (fusion center) in which a high accuracy can be achieved respect to the availability of all data points. On the other hand the IG has to pass many network cycles to obtain an average prediction accuracy. As it is obvious in Table 1, the proposed approach achieves acceptable accuracy compared to the Central case.

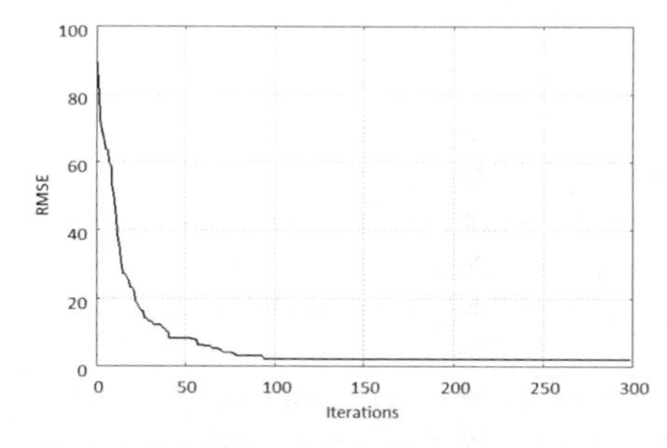

Fig. 4. The convergence behavior of the PSO algorithm during the spatial learning phase for one randomly selected cluster (using DS2)

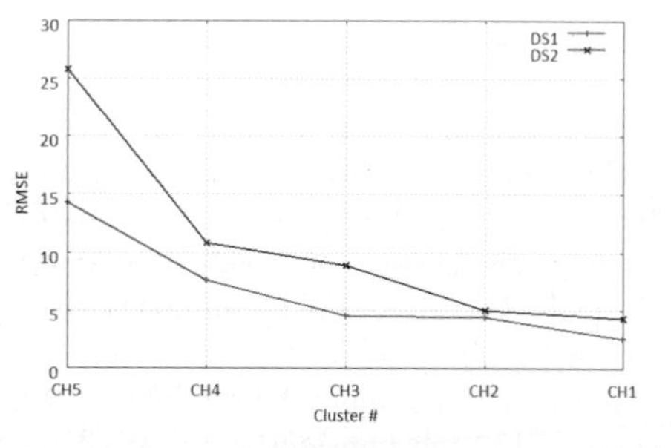

Fig. 5. Improvement of the global model through second traversing the cluster heads

Table 1. Comparison of prediction accuracy based on RMS error between Central, Proposed Approach, and IG

Approach	Data Set1	Data Set2
Central	0.835	2.536
Proposed Approach	2.468	4.319
IG	17.480	21.549

5.2 Latency

The constructed model will be valuable for some periods of time (e.g. some hours). So, it is essential to rebuild the model when the network data is refreshed. In this regard, it is important how much time is needed to construct the global model. Therefore, latency, which is defined as the time needed to visit all the network data for the first time, becomes an important metric. In the Central approach all the network data are visited during one step [5]. So, its latency is $O(1)$. On the other hand, in the IG, n steps are needed to traverse all the network nodes one by one. Thus, its latency is in the order of $O(n)$. In the proposed approach, all the clusters work in parallel and cluster nodes perform their operations simultaneously. In other words, during the temporal learning phase, all the network data are seen for the first time. Consequently the latency of the proposed approach is $O(1)$ which is equivalent to the Central approach.

5.3 Communication Requirements

We follow [4] to compare the communication order of the three approaches which depends on the average distance as well as the length of the parameter transmitted in every transmission. Following [4] we consider the case where the network is deployed in a unit square area. In this way, the average distance of transmissions for the Central approach is $O(1)$ as every sensor node makes direct transmissions to the fusion center. On the other hand, every data point which is transmitted from the sensor nodes to the fusion center is a record by the size of 4. Since every sensor node is required to send all its m measurements, the communication requirement for the Central approach is equal to $O(n \times m \times 4 \times 1)$.

Data transmissions for the IG approach are happened between two consecutive nodes on the Hamiltonian path in which the average distance is equal to $(log^2 n / n)^{1/2}$. In practice, the IG needs to meet many cycles through the network. During every cycle, each sensor node sends a parameter vector by the size of L to the next one on the path. Therefore, the communication order of the IG is equal to $O(cycles \times n \times L \times (log^2 n / n)^{1/2})$.

In order to compute the communication cost of the proposed approach, we act the same as [5] and consider the case where the n network nodes are clustered into $n^{1/2}$ clusters and $n^{1/2}$ nodes per cluster. There are two types of transmissions during the learning process: inter-cluster transmission and intra-cluster transmission. The first type is happened during the first step (sec 4.1) and the second type is for learning the global model (sec. 4.2). The average distance of transmissions within the clusters is $(1/n^{1/2})$ while the average distance between two consecutive cluster heads equals to $((log^2 n^{1/2})/n^{1/2})^{1/2}$. Besides that, one additional transmission is required to deliver the network model to the fusion center over an average distance of $O(1)$. Accordingly, we compute the communication order of our approach based on the type of data transmissions as follows:

- $O(Sec.$ 4.1) for one cluster:
  1. Each member node transmits a vector which is composed of its temporal model by the size[1] of v as well as its location over a distance of $(1/n^{1/2}) = O(n^{1/2} \times L \times (1/n^{1/2}))$.
  2. The cluster head returns the obtained cluster's regressor to the members for computing its RMS error = $O(1 \times L \times (1/n^{1/2}))$.
  3. Members send their partial computed squared errors to the cluster head = $O(n^{1/2} \times 1 \times (1/n^{1/2}))$.
- $O(Sec.$ 4.2):
  1. During the first traversing the cluster heads, from CH_1 through CH_k, one real number is sent = $O((n^{1/2}-1) \times 1 \times ((log^2 n^{1/2})/n^{1/2})^{1/2}))$.
  2. During the second traversing the cluster head, from CH_k through CH_1, a combined regressor by the size of L and one real number (σ) are sent = $O((n^{1/2}-1) \times (L+1) \times ((log^2 n^{1/2})/n^{1/2})^{1/2}))$.
  3. The first cluster head, CH_1, sends the final global model by the size of L to the fusion center = $O(1 \times L \times 1)$.

So, the total communication order of our approach will be equal to $O(Sec.$ 4.1)$\times(n^{1/2})+O(Sec.$ 4.2). Table 2 compares the total communication requirements for the three approaches. As usually $m >> cycles$, the IG is more energy efficient than the Central approach. On the other hand, since *cycles* practically tends to a large value, it can be found that the proposed approach is obviously more energy efficient that the IG and consequently prolongs the network lifetime.

[1] Note that the upper bound of v, L, was taken into account.

Table 2. Communication order for Central, IG, and proposed approach

Approach	Avg. distance of transmissions	Communication order
Central	1	$O(n\times m\times 4\times 1)$
IG	$\sqrt{\log^2 n/n}$	$O(cycles\times n\times L\times\sqrt{\log^2 n/n})$
Proposed Approach	Members to cluster head = $1/\sqrt{n}$ Between cluster heads = $\sqrt{\log^2\sqrt{n}/\sqrt{n}}$ CH_l to the fusion center = 1	$2L+\sqrt{n}(L+1)+$ $(\sqrt{n}-1)(L+2)(\log\sqrt{n}/n^{1/4})$

6 Conclusion

In this paper, we proposed a distributed approach for regression analysis over sensor networks. At first, the clusters' regressors are learned in parallel and are then combined through collaborations of the cluster heads. In order to learn the regressor of each cluster, in one hand measurements similarities are taken into account for learning the regressor's temporal part and on the other hand PSO algorithm is employed for learning the spatial part. Utilizing from cluster level parallelism leads to the significant decreasing in the latency while by benefitting from weighted averaging combination rule, an acceptable accurate global model is achieved. Also, regarding to the communication cost, our approach works better than its counterparts. However, compared to the Central approach, the estimation accuracy should be improved even further. For combining the clusters' regressors, we used weighted averaging combination rule. It is essential to employ other popular combination rules such as multiplying classifiers or ensemble learning methods. The use of PSO in the whole learning process is also placed in our future plan.

Acknowledgments. The authors would like to thank Iran Telecommunication Research Center (ITRC) for partial funding of this research (Grant No. 17581/500).

References

1. Pierce, F.J., Elliott, T.V.: Regional and On-farm Wireless Sensor Networks for Agricultural Systems in Eastern Washington. Elsevier, J. Computer and Electronics in Agriculture, 32–43 (2008)
2. Kononenko, I., Kukar, M.: Machine Learning and Data Mining: Introduction to principles and algorithms. Horwood publishing, Chichester (2007)
3. Predd, J.B., Kulkarni, S.R., Poor, H.V.: Distributed Learning in Wireless Sensor Networks. IEEE Signal Processing Magazine, 56–69 (July 2006)
4. Rabbat, M., Nowak, R.: Distributed Optimization in Sensor Networks. In: International Symposium on Information Processing in Sensor Networks, Berkley, California, USA, ACM Press, New York (2004)

5. Son, S.H., Chiang, M., Kulkarni, S.R., Schwartz, S.C.: The Value of Clustering in Distributed Estimation for Sensor Networks. In: Proceedings of International Conference on Wireless Networks, Communications and Mobile Computing, Maui, Hawaii, vol. 2, pp. 969–974 (2005)
6. Charkari, N.M., Marandi, P.J.: Distributed Regression Based on Gradient Optimization in Wireless Sensor Networks. In: Proceedings of First Iranian Data Mining Conference, Tehran, Iran, IDMC (2007)
7. Marandi, P.J., Charkari, N.M.: Boosted Incremental Nelder-Mead Simplex Algorithm: Distributed Regression in Wireless Sensor Networks. In: Proc. of International Conference on Mobile and Wireless Communications Networks, Toulouse, France, Springer, Heidelberg (2008)
8. Marandi, P.J., Mansourizadeh, M., Charkari, N.M.: The Effect of Resampling on Incremental Nelder-Mead Simplex Algorithm: Distributed Regression over Wireless Sensor Network. In: Li, Y., Huynh, D.T., Das, S.K., Du, D.-Z. (eds.) WASA 2008. LNCS, vol. 5258, pp. 420–431. Springer, Heidelberg (2008)
9. Alpaydm, E.: Introduction to machine learning. MIT Press, Cambridge (2004)
10. AlRashidi, M.R., El-Hawary, M.E.: A Survey of Particle Swarm Optimization Applications in Electric Power Systems. IEEE Trans. On Evolutionary Computation 13(4), 913–918 (2009)
11. Wimalajeewa, T., Jayaweera, S.K.: Optimal Power Scheduling for Correlated Data Fusion in Wireless Sensor Networks via Constrained PSO. IEEE Trans. On Wireless Communications 7(9), 3608–3618 (2008)
12. Wang, B., He, Z.: Distributed Optimization over Wireless Sensor Networks using Swarm Intelligence. In: Proc. Of International Symposium on Circuits and Systems, pp. 2502–2505 (2007)
13. Satapathy, S.C., Murthy, J.V.R., Prasad Reddy, P.V.G.D., Misra, B.B., Dash, P.K., Panda, G.: Particle swarm optimized multiple regression linear model for data classification. Elsevier, J. Applied Soft Computing, 470–476 (2009)
14. Behnamian, J., Fatemi Ghomi, S.M.T.: Development of a PSO–SA hybrid metaheuristic for a new comprehensive eregression model to time-series forecasting. Elsevier, J. Expert Systems with Applications, 974–984 (2010)
15. Abbasi, A.A., Younis, M.: A survey on clustering algorithms for wireless sensor networks. Elsevier, J. Computer Communications, 2826–2841 (2007)
16. Guestrin, C., Bodi, P., Thibau, R., Paskin, M., Madde, S.: Distributed Regression: An Efficient Framework for Modeling Sensor Network data. In: Proceedings of Third International Symposium on Information Processing in Sensor Networks, Berkeley, California, USA, pp. 1–10 (2004)
17. `http://berkeley.intel-research.net/labdata/`
18. Draper, N.R., Smith, H.: Applied Regression Analysis, 3rd edn. Wiley, Chichester (1998)
19. Duin, R.P.: The combining classifier: to Train or Not to Train? In: Proceeding of 16th International Conference on Pattern Recognition, vol. 2, pp. 765–770 (2002)
20. Tax, D.M.J., van Breukelen, M., Duin, R.P.W., Kittler, J.: Combining multiple classifiers by averaging or by multiplying? J. Pattern Recognition, 1475–1485 (2000)
21. Kittler, J., Alkoot, F.M.: Sum versus Vote Fusion in Multiple Classifier Systems. IEEE Trans. on Pattern Analysis and Machine Learning 25(1) (January 2003)
22. Stoleru, R., He, T., Stankovic, J.A., Luebke, D.: A high-accuracy low-cost localization system for wireless sensor networks. In: Proceedings of ACM Conference on Embedded Networked Sensor Systems, (SenSys) (2005)

Generating Tags for Service Reviews

Suke Li[1,2], Jinmei Hao[3], and Zhong Chen[1,2]

[1] School of Electronics Engineering and Computer Science, Peking University, China
[2] Key Laboratory of High Confidence Software Technologies (Peking University), Ministry of Education
{lisuke,chen}@infosec.pku.edu.cn
[3] Beijing Union University, China
haomei99@yahoo.com.cn

Abstract. This paper proposes an approach to generating tags for service reviews. We extract candidate service aspects from reviews, score candidate opinion words and weight extracted candidate service aspects. Tags are automatically generated for reviews by combining aspect weights, aspect ratings and aspect opinion words. Experimental results show our approach is effective to extract, rank, and rate service aspects.

Keywords: service review tagging, service aspect weighting, opinion mining.

1 Introduction

Web users have published a lot of reviews about products and services on the Web attributed to dramatic development of Web 2.0 techniques. Potential consumers like to read relevant Web reviews before making service purchasing decisions. However, a service provider may have hundreds of relevant reviews even in the same website for example *Tripadvisor*[1]. Web users suffer from spending much time in reading reviews. The challenge is whether it is possible for consumers to know the major service aspects that are mentioned in service reviews and to only read a small number of sentences with important service aspects to grasp the essential points of service quality instead of browsing all the contents of relevant reviews. To address this issue, this paper proposes a tagging algorithm that can highlight important service aspects for users who browse online service reviews.

Recent years social bookmarking systems have emerged and are becoming more and more popular, for example *Delicious*[2] which lets users annotate objects with tags of their own choosing. A fundamental premise of tagging systems is that regular users can organize large collections for browsing and other tasks

[1] http://www.tripadvisor.com
[2] http://delicious.com/

L. Cao, J. Zhong, and Y. Feng (Eds.): ADMA 2010, Part II, LNCS 6441, pp. 463–474, 2010.

using uncontrolled vocabularies [5]. Because different users have different understanding of the tagged objects (e.g. Web reviews), it is hard to judge whether these keywords are consistent with target objects or valid to be labels for them. Since many online service reviews have not been tagged, this paper tries to find a way to tag Web reviews automatically. This work focuses on extracting, ranking and rating service aspects. Specifically, we propose an algorithm of tagging service related Web reviews. Our approach has three subtasks which are

- extracting candidate service aspects in the second-order context window around each candidate opinion word in reviews;
- scoring candidate opinion words and weighting extracted candidate service aspects in terms of weights of their associated candidate opinion words;
- and generating tags for service reviews. In the second step, we propose mutual information-based method, probability-based method and information gain-based method to obtain these weights.

2 Related Work

To the best of our knowledge, little research work focuses on the automatic Web review tagging. However, there are some related publications in the areas of opinion mining, keyphrase extraction and web page tagging. Hu and Liu [7] extracted product features according to their frequencies using association mining method. Gamon et al. [3] mined topics and sentiment orientation jointly from free text customer feedback. This work was based on a TF-IDF related approach. Unlike opinion summarization systems that focus on product information summarization, for example Opinion Observer [8], our method focuses on tagging a single review according to knowledge mined from the global data set. Lu et al.'s work [9] studied the problem of generating a "rated aspect summary" of short comments comes with an overall rating. They used PLSA [6] based method to decompose the overall ratings for the major aspects. However, our work is different to [9].

Graph-based keyphrase extraction is related to our work. For instance, recent research work [11] employed graph-based method on small number of nearest neighbor documents to provide more knowledge to improve single document keyphrase extraction. Apart from [11], most of keyphrases methods don't focus on service aspect extraction and weighting.

There are some publications that give methods to automatically generate tags for web pages [2] or documents [10]. Chirita et al. [2] proposed a method to automatically tag web pages in a personalized way, however the method is based on the hypothesis that there are very rich documents of personal information which can and should be exploited for generating personalized tags. In our work, we are not going to grasp personal information to generate tags, because it is hard to know whether a Web user has plentiful of documents related to products or services. Song et al. [10] is another essential work related to our work, and it is different from our work in that our work will generate complex tags with rating, opinion list for each weighted aspect.

3 Extracting, Ranking and Rating Service Aspects

3.1 Extracting Candidate Service Aspects

In this work, we use adjectives as candidate opinion words since they can be used in product mining for candidate opinion words [7]. We also adopt noun units as candidate service aspects.

Definition (Noun Unit). *A noun unit is a unit which contains consecutive nouns in a sentences. For instance, both "breakfast" and "breakfast room" are noun units.*

The strategy of candidate service aspect extraction method is straight. We only consider noun units around adjectives as candidate service aspects, so we get the second-order context window around each candidate opinion word in reviews. For example, the $[-3, +3]$ is context window around the candidate opinion word in the same sentence. For example, in the sentence *The staff is nice and helpful*, opinion word *nice*'s [-3, +3] context window contains *staff*. In this work, we extract words with POS tags of *JJ*, *JJR*, *JJS* as candidate opinion words, and words with POS tags of *NN*, *NNS*, *NNP*, *NNPS* as candidate service aspects.

3.2 Scoring Candidate Opinion Words and Weighting Candidate Service Aspect

In our data set, the ratings of reviews range from 1 to 5. These ratings are relevant to consumers' sentiment orientation to service providers. Consumers form their service opinions according to observation of different service aspects with different preferences. Our service aspect weighting methods are based on the hypothesis that a review with higher rating may have more positive opinions than the one with lower rating. Therefore we split our data set into two parts, one part containing reviews that have higher ratings, and the other part with reviews that have lower ratings. In our experiments, we divide global data set into two parts. However, ratings of reviews in Tripadvisor are unbalanced, and the average rating is more than 3. Hence, in this work, one part has rating range $[1, 3]$, and the other part has rating range $(3, 5]$.

Mutual Information-based Candidate Service Aspect Weighting Method. In this method, we firstly divide the review set into negative class C_N and positive class C_P. We present score of opinion word o_j as

$$score(o_j) = MI(o_j) = \sum_{x \in \{0,1\}} \sum_{c \in \{C_N, C_P\}} p(x,c) log \frac{p(x,c)}{p(c)p(x)}, \tag{1}$$

where $p(c)$ is proportion of reviews belonging to the class c, $p(x,c)$ is the proportion that the jth opinion feature takes value x in class c (x=0 means it doesn't appear in class c; x=1 denotes it appears in class c), and $p(x)$ is the proportion that the jth opinion feature take value x in the whole review set. We weight candidate service aspect a_i according to

$$w_{a_i} = \sum_{1 \leq j \leq k} score(o_j), \tag{2}$$

where k denotes the number of candidate opinion words that are associated with candidate service aspect a_i. Here a candidate opinion word associated with a_i means a_i is in the context window around o_j.

Probability-based Candidate Service Aspect Weighting Method. Suppose the global review data set is C, and every part of the global data set is C_i. If we have two parts (positive part C_P and negative part C_N), then $|E(C)| = \sum_{1 \leq i \leq 2} |E(C_i)|$, where $E(C_i)$ denotes review data set labeled with class C_i. Let O be candidate opinion word set, and $o_j \in O$, $n = |O|$. Let $p(C_i|o_j)$ be probability of candidate opinion word o_j appearing in class C_i. $p(C_i)$ is probability of review class C_i, and $p(o_j)$ is probability of candidate opinion word o_j. We can obtain

$$p(C_i|o_j) = \frac{p(o_j|C_i)p(C_i)}{p(o_j)} \propto p(o_j|C_i)p(C_i), \tag{3}$$

where we suppose $p(o_j)$ has uniform distribution and it is constant. We use the Laplacian correction to avoid computing probability values of zero. If we have two classes and $O = \{o_1, o_2, ..., o_n\}$, then we get

$$p(o_j|C_i) = \frac{c(o_j, C_i) + 1}{\sum_{1 \leq j \leq n} \sum_{1 \leq i \leq 2} c(o_j, C_i) + |O|}, \tag{4}$$

where $c(o_j, C_i)$ denotes the frequency of o_j in C_i. We compute $p(C_i)$ in terms of the number of reviews C_i contains. We obtain $p(C_i)$ by

$$p(C_i) = \frac{|E(C_i)|}{\sum_{1 \leq i \leq n} |E(C_i)|}. \tag{5}$$

After we divide the global data set into positive part and negative part, we use log ration Equation (6) to score candidate opinion word o_j. Intuitively if a candidate opinion word is not only mentioned in positive reviews, but also in negative reviews, and if the probability of appearing in positive reviews approximates to the probability of appearing in negative reviews, the candidate opinion word may have low probability to be opinion words. We get the score of opinion word o_j by

$$score(o_j) = |log(\frac{p(o_j|C_P)}{p(o_j|C_N)})| = |log(p(o_j|C_P)) - log(p(o_j|C_N))|, \tag{6}$$

where $p(C_P)$ is probability that positive class C_P contains candidate opinion word o_j, and C_N is probability that negative class C_N mentions o_j. For a candidate opinion word a_i, we obtain its weight using

$$w_{a_i} = \sum_{1 \leq j \leq k} score(o_j), \tag{7}$$

where k denotes the number of candidate opinion words that are associated with candidate service aspect a_i. Here a candidate opinion word associated with a_i means a_i is in the context window around o_j.

IG-based Candidate Service Aspect Weighting Method. Our third service aspect weighting method is based on IG (Information Gain) theory. Let D denote the global data set. If we divide reviews into two subsets, we can obtain the entropy of D using

$$entropy(D) = -\sum_{1 \le i \le 2} p_i \log_2 p_i, \quad (8)$$

where p_i is the probability that a review in D belongs to class C_i and is estimated by $|C_{i,D}|/|D|$. In this work, we divide reviews into two subsets according to their ratings. If the rating of a review in the range $[1, 3]$, the review is in the negative subset, otherwise if the rating in the range $(3, 5]$, it is in the positive subset. If a candidate opinion word o_j has m possible attribute values, o_j can be used to split data set D into m subsets, that is $D_1, D_2, ..., D_m$. Using opinion word o_j, we only split the global data into two parts, one part contains candidate opinion word o_j, the other part doesn't contain the opinion word o_j. Hence here m equals to 2. Then we obtain its $entropy_{o_j}(D)$ by

$$entropy_{o_j}(D) = \sum_{1 \le i \le 2} \frac{|D_i|}{|D|} \times entropy(D_i), \quad (9)$$

where D_i presents one of part of the data set D. Hence the information gain is

$$score(o_j) = IG(D, o_j) = entropy(D) - entropy_{o_j}(D). \quad (10)$$

We weight the candidate service aspect according to

$$w_{a_i} = \sum_{1 \le j \le k} score(o_j), \quad (11)$$

where k denotes the number of candidate opinion words that are associated with candidate service aspect a_i. Here a candidate opinion word associated with a_i means a_i is in the context window around o_j.

3.3 Rating Service Aspects

Customers give ratings to service providers to measure their service quality. Different aspects have different weights in the process of measuring service quality. For a candidate service aspect a, we can get its opinion word set O according to different graph generation modes. The average rating of an opinion word is calculated based on averaging overall ratings of reviews that contain the opinion word. The average rating of an opinion word o_i is calculated using

$$r_{o_i} = \frac{\sum_{1 \le j \le |R|} r_j}{|R|}, \quad (12)$$

where r_j is the rating of a review $review_j$ which contains opinion word o_i, and $review_j \in R$, R is the set of all the reviews that contain o_i. For each candidate aspect a, there may have several opinion words which are related to a_i.

Because a review may mention one or several aspects, and service aspects may have different ratings, we define the aspect rating as

$$r_{a_i} = \frac{\sum_{1 \le j \le k} o_j}{k}, \tag{13}$$

where k is the number of opinion words that are in the context window around a_i in a review, and o_j is computed by Equation (12).

If we know service aspect ratings in a review, we also can predicate review's overall rating using

$$r_p = \frac{\sum_{1 \le j \le n} r_{a_j}}{n}, \tag{14}$$

where n is the number of service aspect mentioned in review r_i.

3.4 Generating Tags for Reviews

An automatically generated tag comprises three parts: candidate service aspect, average rating for the candidate service aspect, and candidate service opinion list associated with the candidate service aspect. These tags are not uncontrolled vocabularies, and they are different from traditional tags for web pages. We propose Algorithm 1 to tag reviews. If two candidate service aspects have the same stems, the candidate service aspect with lower rank number will be a candidate tag. (for example *room* with rank number x and *rooms* with rank number y, if $x < y$, *room* is a candidate tag but *rooms* is not.). We rank candidate service aspects according to their weights in decreasing order.

4 Experiments

4.1 Experiment Setup

We crawled 6559 hotel related web pages from which we extracted 25,625 reviews from *Tripadvisor*. A wrapper has been implemented for extracting service reviews that are related to hotel service. These reviews contain 235,281 sentences that have been gotten by segmenting the reviews using *OpenNLP*[3] which is built based on maximum entropy models. The average sentence number of these reviews is about 9. We also got Part-of-Speech tags and phrases for all the sentences. The average ratings of the reviews is 3.54, and we also find the rating distribution is unbalanced.

We use mutual information-based method, probability-based method, and information gain-based method to score candidate opinion words and weight candidate service aspects respectively. The top 20 candidate opinion words that has the highest scores are listed in Table 1. Candidate opinion words are ranked in decreasing order in terms of their scores computed by Probability-based method using Equation (6) or IG-based method based on Equation (10). We can see some

[3] http://www.opennlp.org

Algorithm 1. Tagging Reviews

Input:

Review set R;

Output:

Tag vector T_v ;

1: Divide review set R into two subsets (classes): positive subset C_p and C_N according to their overall ratings. {In this work, reviews with overall ratings in $[1,3]$ are in negative class, and reviews in $(3,5]$ belong to positive class.}
2: Extract candidate service aspects and candidate opinion words from each $r \in R$;
3: Score candidate opinion words and weight service aspects using one of three proposed methods in Section 3.2.
4: Compute average overall ratings for every candidate opinion word;
5: Rank candidate service aspects in terms of their weights in decreasing order;
6: **for all** $r \in R$ **do**
7: Get distinct candidate service aspects from r;
8: For every ranked candidate service aspect a_i in r, find candidate opinion word set o_i in context substantival window $[-w,w]$ of a_i;
9: Generate tag t_r for r, if we can extract r candidate service aspects, then $t_r = \{< a_i, o_i, r_i > | 0 \leq i \leq m\}$, here r_i is average opinion word ratings for a_i;
10: Push t_r into T_v;
11: **end for**
12: **return** T_v;

candidate opinion words have spelling errors due to the nature of user-generated content. Candidate opinion words in Table 1 are not classified into positive and negative classes, and they are listed indiscriminately together.

Probability-based method can also rank and classify positive and negative candidate opinion words. To obtain the top k positive candidate opinion words, we use $score(o_j) = log(\frac{p(o_j|C_P)}{p(o_j|C_N)})$ to score positive candidate opinion words. This means a candidate opinion word o_j has high score, it has high probability to be a positive opinion word. In the same way, $score(o_j) = log(\frac{p(o_j|C_N)}{p(o_j|C_P)})$ is used to show how likely a candidate opinion words to be a negative opinion word. Table 2 contains the top 20 positive candidate opinion words and the top 20 negative candidate opinion words.

In our experiments, we take FP-tree [4] approach as the baseline, that is we use association mining to find frequent aspect. [7] also employed Apriori [1] to get frequent product features. However, Apriori algorithm needs more computing time than FP-tree on our data set, therefore, FP-tree algorithm is a better choice then Apriori algorithm. In our experiments, FP-tree has support value of 0.005. In Fig. 1(a) and Fig. 1(b) give MAP (Mean Average Precision) and P@N distributions respectively. Since we want to use these weighted candidate service aspects and candidate opinion words to tag reviews, intuitively we do not need all the extracted candidate service aspects. In a product or service domain, there may be hundreds of aspects. We evaluate the precision of the top 100 extracted candidate aspects in terms of the meanings of the candidate aspects by humans. $P@N$ is often used to evaluate results of information retrieval.

Table 1. The top 20 candidate opinion words with the highest scores. The opinion words are listed decreasingly according to their scores.

Method	Candidate opinion words
Probability-based	gracious, mid-range, airless, organic, lighter, copious, efficent, knowledgable, filthy, finest, brill, disgraceful, aweful, torn, impeccable, splendid, effecient, curteous, equiped, moldy
Information Gain	dirty, great, helpful, worst, excellent, comfortable, rude, friendly, quiet, poor, perfect, bad, filthy, spacious, terrible, wonderful, horrible, clean, modern, lovely
Mutual Information	dirty, great, helpful, excellent, comfortable, worst, friendly, quiet, rude, perfect, poor, bad, spacious, wonderful, filthy, terrible, modern, lovely, clean, horrible

Table 2. The top 20 candidate opinion words which are classified as positive or negative opinion words using the probability-based method

Positive or negative	Candidate opinion words
top 20 candidate positive opinion words	gracious, mid-range, organic, lighter, copious, efficent, knowledgable, finest, brill, impeccable, splendid, effecient, curteous, equiped, invaluable, georgian, homemade, ceramic, innovative, sublime
top 20 candidate negative opinion words	airless, filthy, disgraceful, aweful, torn, moldy, redeeming, shoddy, offensive, disgusted, dishonest, scaly, unclean, grimy, unwelcoming, horrid, width, abusive, unrenovated, unkindgood

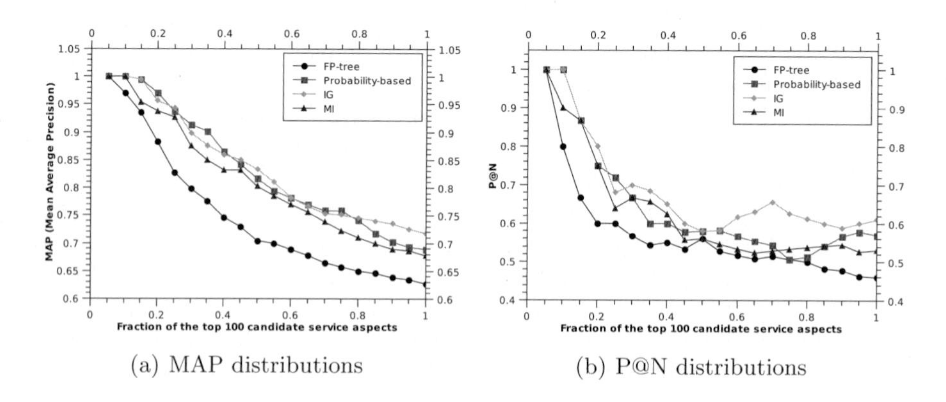

(a) MAP distributions

(b) P@N distributions

Fig. 1. Precision distributions of the top 100 ranked CSAs (CASs are extracted with context window [3,3])

Table 3. The top 20 candidate service aspects which have the highest weights (extracted with context window [3,3])

Method	Candidate service aspects
FP-tree (sp=0.005)	hotel, room, breakfast, staff, rooms, location, night, bathroom, place, time, day, area, bed, station, street, shower, nights, price, stay, floor
Probability-based	hotel, room, staff, location, rooms, breakfast, bathroom, place, bed, area, walk, service, stay, value, restaurants, beds, street, time, night, price
Information Gain	staff, hotel, location, room, rooms, breakfast, bathroom, place, bed, value, service, beds, stay, area, restaurants, time, street, food, price, shower
Mutual Information	room, hotel, staff, rooms, location, breakfast, bathroom, bed, place, time, night, area, walk, hotels, service, price, stay, day, beds, bit

$$P@N = \frac{r_n}{N}, \tag{15}$$

where r_n is the number of relevant document and N is the total number of retrieved document. MAP is often used in evaluation of information retrieval,

$$\text{MAP} = \frac{\sum_{r=1}^{N}(P(r) \times \text{rel}(r))}{r_n}, \tag{16}$$

where r is the rank, N the number retrieved documents, r_n is number of relevant documents, $rel()$ is a binary function on the relevance of a given rank, and $P(r)$ is precision at rank r. Fig. 2(a) and Fig. 2(b) show the MAP and P@N distributions of top 100 candidate service aspects respectively in the same context window size [3,3]. In order to draw distributions clearly, we do not give results of MI and FP-tree because they have lower performance compared to results of Probability-based method and IG-based method that are quite approximated among them in MAP distributions. However, IG-base method outperforms other three methods if we use P@N scheme to evaluate the results. MI-based method has no advantage than IG-based method and probability-based method, but it has better performance compared to FP-tree methods. Table 3 gives top 20 candidate service aspects with the highest weights. The results of FP-tree algorithm have some frequent terms, such as "*night*", "*time*", etc., are not service aspects.

4.2 Experiments with Sentence Coverage

We built index on the sentences that were extracted from the data set using *Lemur*[4]. All the top 100 ranked candidate aspects were submitted as queries to

[4] http://www.lemurproject.org

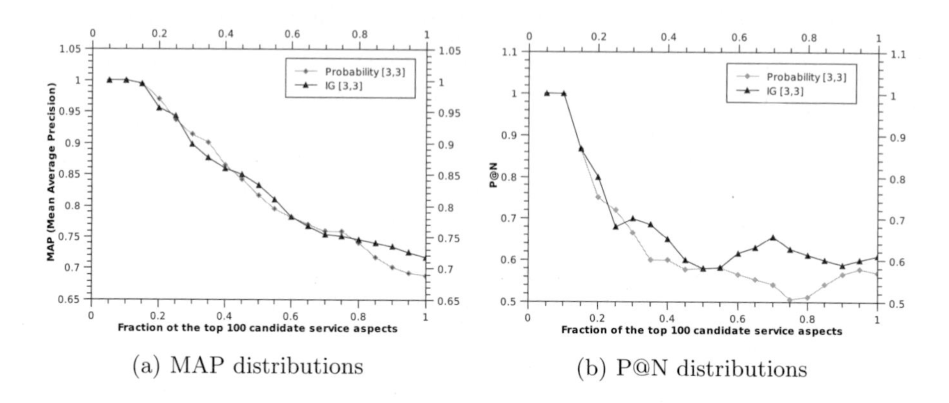

(a) MAP distributions (b) P@N distributions

Fig. 2. Precision distributions of the top 100 weighted CSAs with context window [3,3]. For drawing clarity, we don't draw precision distributions of MI-based results, because MI-based results have lower performance than the other two methods.

Lemur. As Fig. 3 shows the sentence coverage of query results of all the top 100 ranked candidate aspects on the data set. Sentence coverage can be gotten by

$$\text{sentence coverage} = \frac{\text{number of retrieved sentences}}{\text{number of total sentences}}. \tag{17}$$

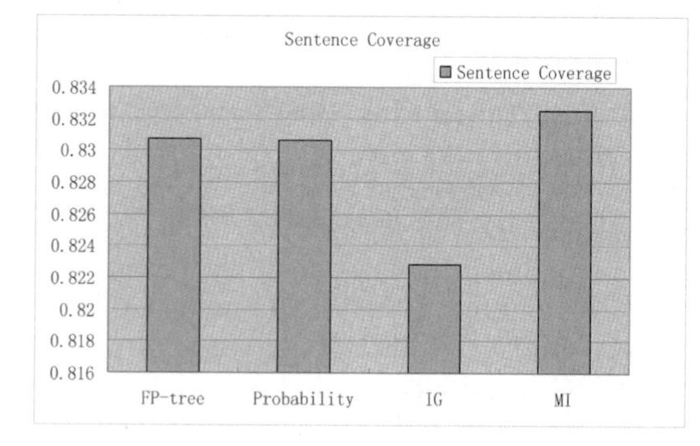

Fig. 3. Sentence coverage of the top 100 candidate service aspects. IG: Information gain; MI: Mutual Information.

We can see FP-tree has the highest sentence coverage, while IG-based method has lowest sentence coverage. The reason lies in that FP-tree algorithm may find frequent items that could not be service aspects, for example word "*night*" and "*time*" are not service aspects but they appear in many reviews.

4.3 Sample Results of Service Aspect Tagging

Fig. 4 is the Web interface of tagging results of randomly sampled reviews from the global data set. In this case, six service aspects are listed on the top of the review, for example "*staff*", "*hotel*", etc. These tags are automatically generated by IG-based method. For each service aspect, we have also computed its average rating score. The rating of service aspect a_i is calculated using Equation (13). When a candidate service aspect is associated with a candidate opinion word, it means the candidate service aspect is in the second-order context windows around the candidate opinion word in this review. In fact, our tags comprise three parts: candidate service aspect, aspect rating, and opinion word list.

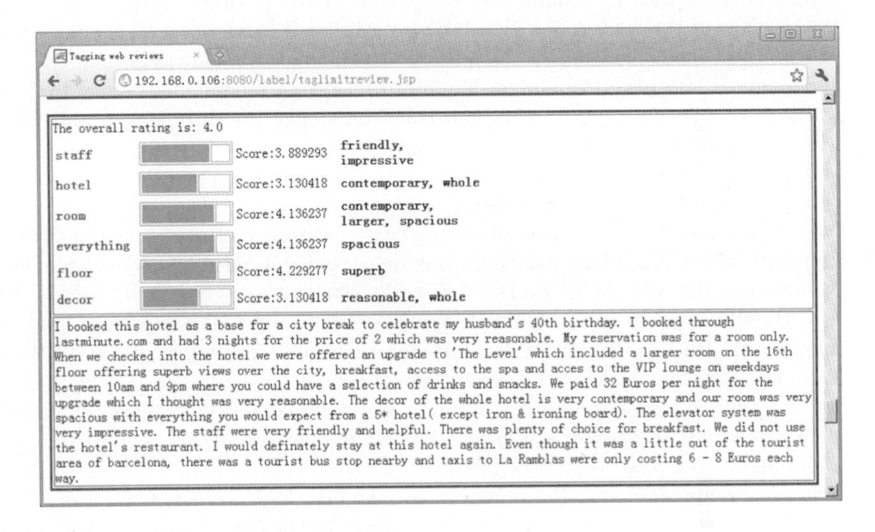

Fig. 4. Web interface for automatically generating service aspect tags. There is one tagged review in the figure.

5 Conclusion and Future Work

In this work, we address the research problem of tagging online service reviews. We tag service reviews using candidate service aspects, aspect ratings, and candidate opinion word lists. Information gain-based method outperforms mutual information-based and probability-based method in finding the top k most important service aspects. Experiments on service aspect weighting show that our tagging algorithm can find major service aspects. Subjective evaluation demonstrates that our methods can generate useful tags for Web reviews.

Acknowledgments. The work was supported by the National Natural Science Foundation of China under the grant No. 60773163 and No. 60911140102, and the Hegaoji project under the grant No. 2009ZX01039-001-001.

References

1. Agrawal, R., Imieliński, T., Swami, A.: Mining association rules between sets of items in large databases. In: Proceedings of the 1993 ACM SIGMOD International Conference on Management of Data, pp. 207–216. ACM Press, New York (1993)
2. Chirita, P.A., Costache, S., Nejdl, W., Handschuh, S.: P-tag: large scale automatic generation of personalized annotation tags for the web. In: Proceedings of the 16th International Conference on World Wide Web, pp. 845–854. ACM Press, New York (2007)
3. Gamon, M., Aue, A., Corston-Oliver, S., Ringger, E.K.: Pulse: Mining customer opinions from free text. In: Famili, A.F., Kok, J.N., Peña, J.M., Siebes, A., Feelders, A. (eds.) IDA 2005. LNCS, vol. 3646, pp. 121–132. Springer, Heidelberg (2005)
4. Han, J., Pei, J., Yin, Y.: Mining frequent patterns without candidate generation. In: Proceedings of the 2000 ACM SIGMOD International Conference on Management of Data, pp. 1–12. ACM Press, New York (2000)
5. Heymann, P., Paepcke, A., Garcia-Molina, H.: Tagging human knowledge. In: Proceedings of the Third ACM International Conference on Web Search and Data Mining, pp. 51–60. ACM Press, New York (2010)
6. Hofmann, T.: Unsupervised learning by probabilistic latent semantic analysis. Machine learning 42(1-2), 177–196 (2001)
7. Hu, M., Liu, B.: Mining and summarizing customer reviews. In: Proceedings of the Tenth ACM SIGKDD International Conference on Knowledge Discovery and Data Mining, pp. 168–177. ACM Press, New York (2004)
8. Liu, B., Hu, M., Cheng, J.: Opinion observer: analyzing and comparing opinions on the web. In: Proceedings of the 14th International Conference on World Wide Web, pp. 342–351. ACM Press, New York (2005)
9. Lu, Y., Zhai, C., Sundaresan, N.: Rated aspect summarization of short comments. In: Proceedings of the 18th International Conference on World Wide Web, pp. 131–140. ACM Press, New York (2009)
10. Song, Y., Zhuang, Z., Li, H., Zhao, Q., Li, J., Lee, W.C., Giles, C.L.: Real-time automatic tag recommendation. In: Proceedings of the 31st Annual International ACM SIGIR Conference on Research and Development in Information Retrieval, pp. 515–522. ACM Press, New York (2008)
11. Wan, X., Xiao, J.: Single document keyphrase extraction using neighborhood knowledge. In: Proceedings of the 23rd National Conference on Artificial Intelligence, pp. 855–860. AAAI Press, Menlo Park (2008)

Developing Treatment Plan Support in Outpatient Health Care Delivery with Decision Trees Technique

Shahriyah Nyak Saad Ali[1], Ahmad Mahir Razali[1], Azuraliza Abu Bakar[2], and Nur Riza Suradi[1]

[1] School of Mathematical Sciences, Faculty of Science and Technology, Universiti Kebangsaan Malaysia, Malaysia
[2] Centre for Artificial Intelligence Technology, Faculty of Information Science and Technology, Universiti Kebangsaan Malaysia, Malaysia
shahriyahali@yahoo.com, mahir@ukm.my, aab@ftsm.ukm.my, nrms@pkrisc.cc.ukm.my

Abstract. This paper presents treatment plan support (TPS) development with the aim to support treatment decision making for physicians during outpatient-care giving to patients. Evidence-based clinical data from system database was used. The TPS predictive modeling was generated using decision trees technique, which incorporated predictor variables: patient's age, gender, racial, marital status, occupation, visit complaint, clinical diagnosis and final diagnosed diseases; while dependent variable: treatment by drug, laboratory, imaging and/or procedure. Six common diseases which are coded as J02.9, J03.9, J06.9, J30.4, M62.6 and N39.0 in the International Classification of Diseases 10th Revision (ICD-10) by World Health Organization were selected as prototypes for this study. The good performance scores from experimental results indicate that this study can be used as guidance in developing support specifically on treatment plan in outpatient health care delivery.

Keywords: User acceptance, Continuous quality improvement, Treatment equity.

1 Introduction

Delivering a high quality health care has been given great attention by health care organizations worldwide. Decision support is seen as one of the solutions to health care delivery improvement that incorporates knowledge from experts or evidenced-based clinical data to support decision-making processes. Past studies have proved that decision supports were capable to support patient treatment [1], improved treatment performance and also improved practitioner performance [2].

Treatment plan refers to management on any interventions consists of treatment and/or examination which will be initiated for each problem based on patient's history, physical examination, provisional diagnosis and differential diagnosis [3]. The TPS proposed in this study acted as a supportive tool that generates treatment suggestions for individual patients at point of care. An optimized treatment decision for patients is expected because both health care practitioners' knowledge and TPS are utilized, rather than either human or support could make on their own.

L. Cao, J. Zhong, and Y. Feng (Eds.): ADMA 2010, Part II, LNCS 6441, pp. 475–482, 2010.

This paper is organized into five sections. Past studies and reports on health care quality that related to treatments and supports are briefly highlighted in the next section. Methodology of our proposed approach is presented in Section 3. The experiments performed and the results obtained are discussed in Section 4. Finally, conclusion of this paper is covered in the final section.

2 Related Research

Decision supports have been used in a wide range of medical applications [1,2,4,5,6], however very few researches to produce outpatient treatment supports were reported. Outpatient is the primary department for patients to receive treatment and involves higher number of transactions per day as compared to inpatient. Despite many decision supports in literature reviews and markets, the use of decision trees techniques has not yet been reported. Many studies have applied this technique and have proved that it gave the best result among other techniques in data mining [7], especially in overall prediction accuracy [8].

Previous studies found that factors that influence treatment varies from patients' demography, morbidity and condition [3,9,10]. However, there are proof that non-clinical factors significant to antibiotic prescription to patients, such as patients' health insurance policy, patients' health support from organization [9], physicians' geographical location place of practice and physicians' gender [10]. Health care delivery equity is emphasized to achieve better health prospects, besides of improving efficiency, quality and accessibility of health care delivery [11]. This issue is very important especially in Malaysia where practice is being made by not only doctors, but also by assistant medical officers.

In line with achieving quality improvement in health care delivery, TPS development is proposed. TPS development framework is based mainly on the same fundamental used for Percuro Clinical Information System [4] which successfully being implemented in more than 20 health centres in Malaysia. The reason to choose a success system is that it is able to meet user requirement as well as user acceptance for it to be applied in their work routine [12].

3 Methodology of the Treatment Plan Support Development

Fig. 1 illustrates our TPS development framework. Patients' demographic information and visits type whether visits related to new cases or follow ups are considered during patient registration. Patient's age, gender and race [9] have been used to differentiate treatment in diseases; for an example acute upper respiratory tract infection. Besides, type of visits must also be considered whether patient's visit is due to a newly reported case or a return visit presumably of treatment failure [9].

Consultation and treatment which are the main important processes in health care delivery utilizes medical record documentation [3]. The S.O.A.P (an acronym for subjective data, objective data, assessment and plan) note format acted as parameters to generate treatment plan for a particular disease specifically. The final disease diagnosed by physician was coded with current International Classification of Diseases

10th Revision (ICD-10) of version 2007 for worldwide universal coding. Six highest reported diagnosed diseases in Percuro database were selected as prototypes in this study: acute pharyngitis unspecified, acute tonsillitis unspecified, acute upper respiratory infection unspecified, allergic rhinitis unspecified, muscle strain and urinary tract infection site not specified, which are coded as J02.9, J03.9, J06.9, J30.4, M62.6 and N39.0 respectively.

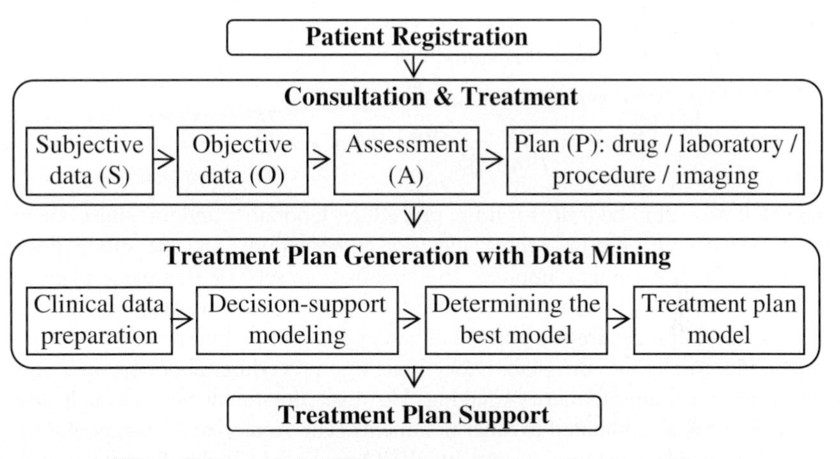

Fig. 1. Framework for TPS development in an outpatient setting

The TPS is developed in two main stages as elaborated below. Firstly, generating TPS predictive modeling and secondly, developing treatment plan support for practical application.

3.1 Generating TPS Predictive Modeling

Throughout generating TPS predictive modeling, following steps are followed: data collection, data preparation, and treatment plan modeling.

Data Collection. Evident-based outpatient clinical data from Percuro database was compiled. A total of 43,409 patients are involved with 278,271 visits for 36 months from January 2006 until December 2008. However, only data for year 2008 was used for modeling, because Percuro was under upgraded development and maintenance and, there were users' data entry inconsistency occurred in year 2006 and 2007.

Data Preparation. The preparation of complex clinical data for treatment plan is presented in this section. Data preparation was done to obtain a good quality data. The task in data preparation involved data cleansing, data modification, data transformation, data representation and data standardization to World Health Organization (WHO) universal coding.

Predictor variables such as patient's gender, racial, marital status, occupation, final diagnosed disease ICD-10, besides physician's identification needed a minor data cleansing. Predictor variable patient's age required preparation, where it is obtained by differencing patient's birth date and consultation date. Predictor variables for patient's visit complaint and clinical diagnosis required data representation because the information was in an open text form.

Data modification and transformation for dependent variable treatment plan are needed, to fulfill flat-file requirement in data mining [13]. Treatment plan given to patient consists of one or combinations of drug, procedure, laboratory and imaging [3]. Representation data was made in order to capture the complete treatment for a single patient. Combination theorem for number of combinations of n distinct objects taken r at a time has been used [14]. The value number of r has been set from 1 to n, instead of from 0 to n, to suite this study. Formula used is shown in (1).

$$\binom{n}{r} = {}^{n}C_{r} = \frac{n!}{r!(n-r)!} \text{ ,for } r = 1,2, \ldots, n \tag{1}$$

Using (1), n was set to be four (for drug, procedure, laboratory and imaging). Therefore, r was replaced with one, two, three and four combinations, and the summation gave total 15 possibilities which implied 15 combination sets of treatment plans: drug, procedure, laboratory, imaging, drug-procedure, drug-laboratory, drug-imaging, procedure-laboratory, procedure-imaging, laboratory-imaging, drug-procedure-laboratory, drug-procedure-imaging, drug-laboratory-imaging, procedure-laboratory-imaging and finally drug-procedure-laboratory-imaging. However, treatment plan through drug accumulated 82.5% as compared to other treatments. The Anatomical Therapeutic Chemical (ATC) drug classification system by WHO is used to make classification within drug. We split drug treatment into: i) drug with antibiotic prescription that consist of *J* classification in ATC code, ii) drug without antibiotics prescription, while other treatment plan combinations are combined together as *others*. External physician validation on the prepared data was finally obtained.

Treatment Plan Modeling. The TPS predictive modeling was generated using Weka 3.6.1 software, with J48 decision trees technique or known as C5 algorithm [13]. Stratified ten-fold cross-validation procedure is carried out in the mining process.

3.2 Development of Treatment Plan Support

Towards TPS development, continuous quality improvement Plan-Do-Check-Act (PDCA) cycle has been adapted. The modified cycle is an iterative four-phase problem-solving process comprises of; i) TPS planning, ii) TPS development, iii) TPS evaluation and iv) deployment. Moving back and forth between different phases might be needed according to current requirements. The cycle continues with subsequent processes focuses on continually improving process performance through both incremental and innovative changes/improvements.

TPS consists of three parts which are the knowledge base, the reasoning engine, and a mechanism to communicate with physicians. Results from the above predictive modeling are used as a knowledge base in TPS. A reasoning engine which contains formulas subsequently combines the knowledge base with actual patient data. Finally through a mechanism in TPS, TPS communicates with physicians by providing suggested list of potential treatments in a graphical user interface.

4 Results

The final data set used for TPS predictive modeling consisted of 766 patient's visits, which involved treatments from seven physicians with 423, 263, 3, 2, 48, 20 and 7 treatments frequencies respectively. Malay race accumulated 740 patients while other races 26 patients. As many as 735 patients are single, whereas 31 patients are married. Majority of patients are still studying, followed by working and other categories with 706, 46 and 14 number of patients respectively. Patient's age, complaint, diagnosis and ICD-10 disease are categorized to 9, 26, 35, and 6 distinct values correspondingly. Dependent variable treatment plan are labeled as drug inclusive *J*, drug exclusive *J* and others with frequencies of 310, 322 and 134.

Four tests are conducted with three conditions of data set. The first test included all 10 variables; the second test excluded race, marital status and occupation; the third test excluded race, marital status, occupation and physician; and the fourth test excluded race, marital status, occupation, physician and age group. The four-tests are done to investigate whether imbalance classes in variables (race, marital status and occupation) and physician might cause differentiation and accuracy in treatment. Data set A contained 766 data of the total number of clinical data that has went through data preparation, data set B contained 764 data after deletion of data from Physician-4 and finally data set C contained 592 data that excluded treatment from Physician-4 and Physician-2 for J06.9 disease. The decision to consider data deletion for Physician-2 and Physician-4 is because there is data entry error occurred in Physician's-4 record, while Physician's-2 treatments to J06.9 patients are varies as compared to other physicians (results obtained from first and second tests for data sets A and B). It is noted that for Physician-2 treatment differences occurred even with similar patient condition. Having experience in less than a year for Physician-2 may contribute to this variation.

Table 1. Generated results

Test	Data set	Correctly classified instances (%)	Incorrectly classified instances (%)	Number of leaves	Size of the tree	Time to build model (seconds)
1	A	417 (54.44)	349 (45.56)	71	81	0.09
	B	426 (55.76)	338 (44.24)	62	72	0.03
	C	362 (61.15)	230 (38.85)	121	130	0.02
2	A	412 (53.79)	354 (46.21)	70	80	0.02
	B	417 (54.58)	347 (45.42)	61	71	0.03
	C	366 (61.82)	226 (38.18)	39	42	0.02
3	A	431 (56.27)	335 (43.73)	58	66	0.02
	B	427 (55.89)	337 (44.11)	51	59	0.03
	C	366 (61.82)	226 (38.18)	39	42	0.02
4	A	429 (56.01)	337 (43.99)	43	50	0.02
	B	420 (54.97)	344 (45.03)	35	41	0.03
	C	367 (61.99)	225 (38.01)	41	46	0.00

Summary for results generated from the twelve experimental tests are shown in Table 1. The effort taken to perform data reduction on several data has resulted improvement in accuracy. In order to reduce treatment differences among physicians [10] and increase treatment equity in data pattern, the final test which is the best achieved accuracy with 61.99% correctly classified treatment is chosen to be used for TPS. Treatment for five diseases other than J06.9 is straight forward. Treatment for J02.9 and J03.9 are through drug that inclusive of *J* classification. While treatment for J30.4 and M62.6 are through drug that exclusive of *J* classification. Treatment for N39.0 is through *others* combination. Whereas treatment for J06.9 relates to patient's complaint and gender, and varies between drug that inclusive of *J* classification or drug that exclusive of *J* classification.

Table 2. TPS performance score

ICD-10	Cases		Items		
	n	%	*n*	Correct top 10	%
J02.9	13	2.2	59	46	77.97
J03.9	57	9.6	272	233	85.66
J06.9	429	72.5	1951	1582	81.09
J30.4	4	0.7	12	11	91.67
M62.6	57	9.6	187	149	79.68
N39.0	32	5.4	128	114	89.06
Total	592	100	2609	2135	

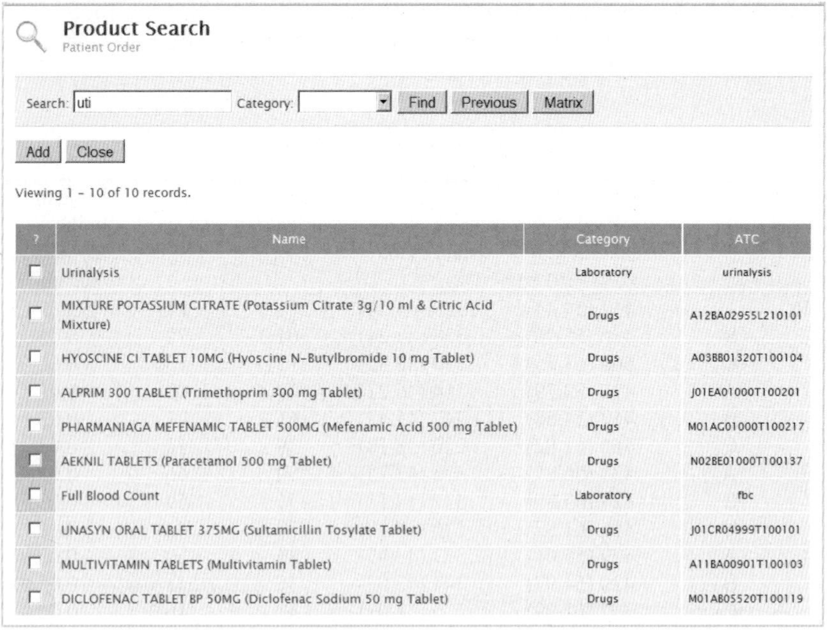

Product Search
Patient Order

Search: uti Category: Find Previous Matrix

Add Close

Viewing 1 - 10 of 10 records.

?	Name	Category	ATC
	Urinalysis	Laboratory	urinalysis
	MIXTURE POTASSIUM CITRATE (Potassium Citrate 3g/10 ml & Citric Acid Mixture)	Drugs	A12BA02955L210101
	HYOSCINE CI TABLET 10MG (Hyoscine N-Butylbromide 10 mg Tablet)	Drugs	A03BB01320T100104
	ALPRIM 300 TABLET (Trimethoprim 300 mg Tablet)	Drugs	J01EA01000T100201
	PHARMANIAGA MEFENAMIC TABLET 500MG (Mefenamic Acid 500 mg Tablet)	Drugs	M01AG01000T100217
	AEKNIL TABLETS (Paracetamol 500 mg Tablet)	Drugs	N02BE01000T100137
	Full Blood Count	Laboratory	fbc
	UNASYN ORAL TABLET 375MG (Sultamicillin Tosylate Tablet)	Drugs	J01CR04999T100101
	MULTIVITAMIN TABLETS (Multivitamin Tablet)	Drugs	A11BA00901T100103
	DICLOFENAC TABLET BP 50MG (Diclofenac Sodium 50 mg Tablet)	Drugs	M01AB05520T100119

Fig. 2. Items listed by TPS for N39.0 disease (urinary tract infection)

The study is further extended to items respective to generated treatment plan. Many systems used ranked list approach to display items [1,4,5], so that physicians can have options to choose appropriate items by themselves. The same approach is employed for the six diseases, and subsequently performance score [15] is made based on the presence of correct prescribed items in the top ten item list (refer Table 2). The result shows that the performance scores percentage for correct top ten items with respective disease cases are high reaching to 91.67%. The good evaluation for the six prototype diseases used in this study reveals that TPS is acceptable to be used for practice. Fig. 2 shows TPS application for physicians that lists the first top ten highest frequencies of items to treat N39.0 patients.

5 Conclusion and Future Research

This paper presents the development of TPS in outpatient health care delivery using decision trees approach. The TPS can provide assistance that aim to optimize physicians' treatment decision making at point of care. This is corresponding to our preliminary study on decision support impact on physicians that showed that 87% (13/15) participated respondents agreed their job will become easier and their service to patients will be better [12]. However, TPS does not intend to replace physician's intuition and interpretive skills, or even encourages any practice by non-qualified or non-authorized person. TPS utilized local evidence-based clinical data and used standard classification for both diseases and drug from World Health Organization, and thus is appropriate to be applied in health centres in Malaysia.

This research can be used as a guidance to develop treatment plan support for other diseases in an outpatient setting. Compilation of treatment plan for all diseases in a single treatment decision support is necessary in order to obtain complete coverage in outpatient treatment. Besides, it is recommended for practicing physicians to develop clinical pathway, where further discussion being made to finalize the appropriate/best treatment based on common given treatment to obtain treatments that are acceptable for all physicians in respective health centres. With this, the accuracy of treatment plan can be higher, and thus treatments for all patients can be monitored and controlled.

Continuous monitoring need to be done to ensure the knowledgebase used in TPS does not get out of date, and that what functioned well in the development process still functions properly in practical usage. Finally, an active support application in practical usage should be encouraged to further improve the quality in health care delivery.

Acknowledgments. This work was supported by the Universiti Kebangsaan Malaysia under the grant UKM-OUP-FST-2010 and UKM-ST-06-FRGS0099-2010.

References

1. Miller, R.A., Waitman, L.R., Chen, S., Rosenbloom, S.T.: Decision Support during Inpatient Care Provider Order Entry: the Vanderbilt Experience. In: Berner, E.S. (ed.) Clinical Decision Support Systems: Theory and Practice, 2nd edn., pp. 215–248. Springer, New York (2007)

2. Garg, A.X., Adhikari, N.K.J., McDonald, H., Rosas-Arellano, M.P., Devereaux, P.J., Beyene, J., Sam, J., Haynes, R.B.: Effects of Computerized Clinical Decision Support Systems on Practitioner Performance and Patient Outcomes: a Systematic Review. J. of the American Medical Association 293, 1223–1238 (2005)
3. Talley, N., O'Connor, S.: Clinical Examination, 2nd edn. MacLennan and Petty Pty Limited, New South Wales (1992)
4. RareSpecies Corporation Sdn. Bhd.: Percuro Clinical Information System, http://www.rarespecies.com.my
5. Open Clinical: Clinical Decision Support Systems, http://www.openclinical.org/dss.html
6. The Informatics Review: Links to Sites Featuring Clinical Decision Support Systems, http://www.informatics-review.com/index.html
7. Delen, D., Walker, G., Kadam, A.: Predicting Breast Cancer Survivability: a Comparison of Three Data Mining Methods. Artificial Intelligence in Medicine 34, 113–127 (2005)
8. Kusiak, A., Dixon, B., Shah, S.: Predicting Survival Time for Kidney Dialysis Patients: a Data Mining Approach. Computers in Biology and Medicine 35, 311–327 (2005)
9. Steinman, M.A., Landefeld, C.S., Gonzales, R.: Predictors of Broad-Spectrum Antibiotic Prescribing for Acute Respiratory Tract Infections in Adult Primary Care. J. of the American Medical Association 289, 719–725 (2003)
10. Mazzaglia, G., Caputi, A.P., Rossi, A., Bettoncelli, G., Stefanini, G., Ventriglia, G., Nardi, R., Brignoli, O., Cricelli, C.: Exploring Patient- and Doctor-Related Variables Associated with Antibiotic Prescribing for Respiratory Infections in Primary Care. European J. of Clinical Pharmacology 59, 651–657 (2003)
11. Ministry of Health Malaysia: Health Indicators: Indicators for Monitoring and Evaluation of Strategy for Health for All. Information and Documentation System Unit, Planning & Development Division, Petaling Jaya (2005)
12. Ali, S., Razali, A.M., Bakar, A.A., Suradi, N.R.: Critical Success Factors for Clinical Decision Support Systems Quality: Medical Practitioners' Perspective. In: International Conference on Quality, Productivity and Performance Measurement, Putrajaya, Malaysia, November 16-18th (2009)
13. Witten, I.H., Frank, E.: Data Mining: Practical Machine Learning Tools and Techniques with JAVA Implementations. Morgan Kaufmann, San Francisco (2000)
14. Miller, I., Miller, M.: John E. Freund's Mathematical Statistics with Applications, 7th edn. Pearson Prentice Hall, New Jersey (2004)
15. Berner, E.S., Jackson, J.R., Algina, J.: Relationships among Performance Scores of Four Diagnostic Decision Support Systems. J. of the American Medical Informatics Association 3, 208–215 (1996)

Factor Analysis of E-business in Skill-Based Strategic Collaboration

Daijiang Chen and Juanjuan Chen

College of Computer & Information Science, Chongqing Normal University, Shapingba 401331, China

Abstract. Based on a comprehensive analysis of skill-based business collaboration in China and abroad, this paper discusses indicators for E-business in skill-based strategic collaboration with field investigation, exploratory factor analysis method and structural equation modeling (SEM). We design the questionnaire which fits China's actual conditions; 1000 pieces were given out randomly and 105 valid pieces were return. Cross validation proves that there are three dimensions in E-business, which are network technology, network management and network marketing. This structure equation is fitted and shows good reliability and validity of the questionnaire.

Keywords: E-business, factor analysis, structure equation model.

1 Introduction

Strategic collaboration is a process in which organization works or combines with another organization to maximize their strategically benefits, moreover when organization contributes its specific know-how for cooperation it is defined as skill-based strategic collaboration[1]. Ever since Chinese economic reform, Chinese enterprises achieve significant development in introduction of foreign capital and technology, establishing new joint ventures with foreign firms. However there is still great distance between domain and abroad, especially in the areas of management, research and development, innovation etc.

Parkde studied the influence which E-commerce has on the resource-based strategic collaboration [2], and Liu et al. researched the theoretical relationship among information technology, strategic collaboration and strategic performance [3-6]. Buyukozkan analyzed the selection of partnership in E-commerce strategic collaboration of supply alliance, which developed only method for partner selection [7]. Peter K. C. et al. conducted empirical studies on the game play of electronic business partnership and their performance [8]-[10]. Although there are many researches about electronic business and its impact on strategic collaboration, there are few literatures about empirical study of skill-based collaboration. Therefore this paper stresses on discovering electronic business factors affected by skill-based collaboration.

In this paper, two-step method is proposed. Exploratory Factor Analysis is used in the first step to explore the strategic structure of the cooperative enterprise e-commerce targets; and Structural Equation Modeling (SEM) is utilized in the second stage to validate our electronic business indicators in skill-based collaboration.

L. Cao, J. Zhong, and Y. Feng (Eds.): ADMA 2010, Part II, LNCS 6441, pp. 483–490, 2010.

2 Exploratory Factor Analysis

This study aims to find out the electronic business indicators and structure in skill-based collaboration, so empirical research and statistical methods, i.e. Exploratory Factor Analysis and Confirmatory Factor Analysis in Structural Equation Modeling (SEM). The first step is the use of exploratory factor analysis to find the basic structure.

2.1 Data Collection

We obtain all data from the 1000 questionnaires mailed out to various enterprises. 339 pieces were mailed back and 105 pieces are valid, most of the questionnaires are filled in by middle or upper level managers from different industries, including manufacturing, IT, finance, real estate and commerce. Those enterprises spread over many central cities like Shanghai, Shenzhen, Chongqing etc.

Considering the fact that most electronic business indicators are referred from relevant literatures, this paper adopts Exploratory Factor Analysis to validate the index structure, and statistical software AMOS is applied.

2.2 Indicator System Exploratory

The results of Exploratory Factor Analysis of indicator system for skill-based collaboration are shown in Table 1, Figure 1-3. According to Figure 1, the KMO value equals 0.744 and the P value in Bartlett's test is less than 0.001, which indicate that E-business indicators, as can be found in Table 1, are suitable for Exploratory Factor Analysis. The results of the factor analysis are listed in Table 2 and Table 3, and there are three principal components whose eigenvalues are over 1 are extracted, total accumulated variance explained is 67.924% and factor loading values are favorable. We named the three factors as Network Technology, Network Management and Online Marketing. Obviously, the loading value of indicator is no less than 0.6 and hence we can conclude that these E-business indicators show superb convergent validity and discriminate validity under the condition that there is no cross loading.

According to Table 4, the CITC values of factors of Network Technology, Network Management and Online Marketing are all over 0.5, and α values are 0.8040, 0.8598, and 0.8096.

KMO and Bartlett's Test

Kaiser-Meyer-Olkin Measure of Sampling Adequacy.		.744
Bartlett's Test of Sphericity	Approx. Chi-Square	562.447
	df	66
	Sig.	.000

Fig. 1. Test results of KMO and Bartlett

Table 1. E-business indicator system

Dimensions	Indicators	
Network Technology	NT1	Internal management network is well constructed by information technology
	NT2	Internet-enabled supply chain system is established
	NT3	Information system is built to improve collaboration among partners
	NT4	Private network platform is built for real time communication among partners
Network Management	NM1	Human resource is assigned to manage the operation and organizational management of network system and information
	NM2	Partners assign staff to manage the operation and organizational management of network system and information
	NM3	Learning platforms, such as forum or virtual community, are built up for better promoting mutual collaboration and skill sharing among partners
	NM4	Firm is flexible and well adapted to different situations
OnlineMarketing	OM1	A large amount of customer information is attained by online B2C marketing
	OM2	Suppliers and retailers base is developing by online B2B marketing
	OM3	Firm releases product information and advertisements based on web technology
	OM4	Customer loyalty and trust is improved in the E-marketplace

Table 2. Total variance explained

Factor	Initial Eigenvalues			Extraction Sums of Squared Loadings			Rotation Sums of Squared Loadings		
	Total	% of variance	Cumu-lative %	Total	% of variance	Cumu-lative %	Total	% of variance	Cumu-lative %
1	4.213	35.107	35.107	4.213	35.107	35.107	2.918	24.314	24.314
2	2.342	19.519	54.626	2.342	19.519	54.626	2.623	21.856	46.170
3	1.596	13.298	67.924	1.596	13.298	67.924	2.610	21.754	67.924
4	0.695	5.790	73.713						
5	0.637	5.306	79.020						
				data below omitted					

Table 3. Factor loadings of E-business measures

E-business Indicators	Network technology	Network Management	Online Marketing
NT1	0.794		
NT2	0.798		
NT3	0.764	0.106	0.189
NT4	0.737		0.334
NM1		0.825	
NM2		0.905	
NM3		0.792	0.235
NM4		0.815	
OM1	0.137	0.360	0.670
OM2	0.107	0.105	0.771
OM3	0.147		0.883
OM4	0.226	0.228	0.754

Note: Blank spaces indicate factor loading less than 0.1 and be omitted.

Table 4. Reliability of E-business measures

Dimensions	Indicator	CITC	α value by deleting this indicator	α
Network Technology	NT1	0.6033	0.7622	0.8040
	NT2	0.6154	0.7575	
	NT3	0.6261	0.7510	
	NT4	0.6357	0.7487	
Network Management	NM1	0.7095	0.8226	0.8598
	NM2	0.8036	0.7786	
	NM3	0.6753	0.8334	
	NM4	0.6482	0.8448	
Online Marketing	OM1	0.5943	0.7763	0.8096
	OM2	0.5247	0.8080	
	OM3	0.7413	0.7015	
	OM4	0.6730	0.7397	

3 Verification Factor Analysis

The second step is the use of structural equation modeling to observe relationships among observed variables and hidden variables; moreover SEM reflects not only the relationships but also the impact degree among variables. According to the results of exploratory factor analysis, a further confirmatory factor analysis is conducted and the data is taken from previous survey. The hypothesis structure is shown in Figure 2 and the goodness fit test result is shown in Table 5.

There are two kinds of test to evaluate goodness fit [11], the first one is test of absolute fit, and the absolute measures include CMIN, DF, CMIN/DF, and P. Generally, the smaller of CMIN and CMIN/DF value (no more than 3), the better fit. Hypothesis won't be rejected only when P value is more than 0.05. The second kind is test of relative fit, and the descriptive measures are GFI, TLI, CFI, RMSEA etc. As the values of GFI, TLI and CFI more approach 1, the better fit. Moreover if those values are larger than 0.9, then observed data might well support the hypothesis; if larger than 0.95, it would be best. Normally RMR value more approaches 0, or is less than 0.100, then better fit [13].

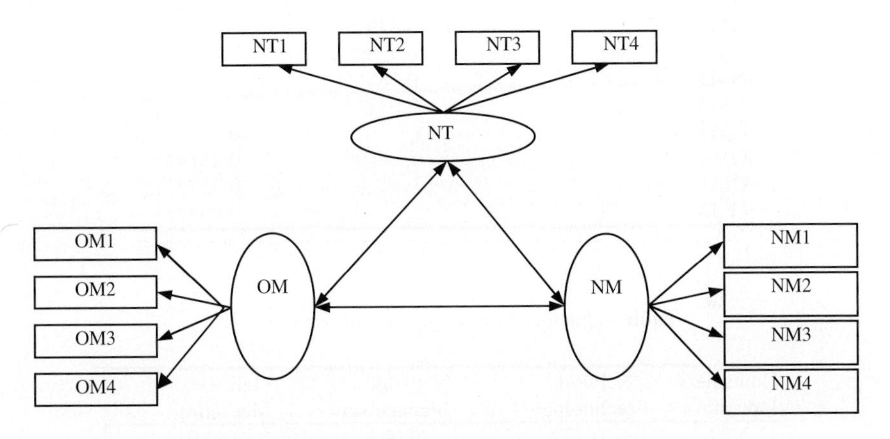

Fig. 2. Relationships among three E-business dimensions

Table 5. The Results of Fit Test for E-business Model

	χ^2	χ^2/df	P	GFI	CFI	IFI	TLI	RMSEA
Model Verification	67.266	1.401	0.262	0.907	0.963	0.965	0.950	0.063

From results in Table 5 we can find that it is an ideal e-business structure, all test measures are fitting for requirements. The value of $\chi2$/df is 1.401, which is close to 1 and less than 5; the value of P measure is more than 0.05, hence null hypothesis is rejected; the value of RMSEA is 0.063 and less than 0.1; the value of GFI is 0.907, CFI is 0.963, IFI is 0.965, TLI is 0.950.

We verify the validity of e-commerce index system from an empirical perspective. Convergent validity can be judged by the load factor, discriminated validity of the weight by the factor score to assess. Model analysis of the factor loadings and factor score weights the results are shown in Table 6 and Table 7.

From Table 6, the 0.01 significance level, the various indicators on their corresponding factor loading factor shows the links of various indicators and factors larger. These factors include the indicators for more information and all describe the various indicators can reflect the corresponding factor. This shows that the index system of e-commerce good convergent validity and convergent validity to electronic commerce has been proved.

Table 6. The factor loading for the model of EC

E-business Indicators	Network technology	Network Management	Online Marketing
NT1	0.564**		
NT2	0.580**		
NT3	0.773**		
NT4	0.826**		
NM1		0.839**	
NM2		0.938**	
NM3		0.668**	
NM4		0.637**	
OM1			0.675**
OM2			0.481**
OM3			0.757**
OM4			0.863**

** $p<0.01$

Table 7. Factor score weights for the model of EC

E-business Indicators	Network technology	Network Management	Online Marketing
NT1	0.122	0.003	0.014
NT2	0.123	0.003	0.014
NT3	0.212	0.005	0.024
NT4	0.256	0.006	0.029
NM1	0.014	0.495	0.014
NM2	0.004	0.224	0.006
NM3	0.002	0.126	0.003
NM4	0.002	0.125	0.003
OM1	0.012	0.004	0.130
OM2	0.011	0.003	0.119
OM3	0.122	0.003	0.014
OM4	0.123	0.003	0.014

Comparative analysis from the Table 7, each of the following factors characterize the various indicators of their corresponding weight factors of the higher ratings are higher than those in other dimensions of the index's weighting factor score. Therefore, the discriminated validity of index system is high and index system has good validity.

Through the empirical relationship between e-commerce index system validations, more fully described in the past exploratory analysis of the effectiveness and reliability of the index system.

4 Discussion and Conclusion

4.1 E-business Structure in Strategic Collaboration

According to the exploratory factor analysis and its result, three dimensions of E-business are well suited for this survey. Firstly, from the aspect of reliability test, the value of α coefficient of each extracted factor is above 0.75, which indicates high reliability. Secondly, from the aspect of factor loading, there are over 80% of loading value exceeds 0.7, and 20% is above 0.6. Thirdly, from the aspect of model goodness fit test, seven measures show positive results, except for TLI is slightly lower, just passes the 0.95 level. As a result, the whole process of three kinds of comprehensive analysis helps to find out the ideal structure.

4.2 Factor Explanation and Reliability

Network platform is composed by intranet and extranet. Access to intranet requires authorization, in which Client/Server architecture is adopted with TCP/IP, FTP, HTML, Telnet and browser. Extranet connects firm with its suppliers and partners, mostly in public network, private network or VPN. The value of Cronbach a is remarkable 0.8040.

Network management contains operational management and organizational management. Operational management refers to supply chain management, procurement, manufacturing, marketing and logistics. Organization management involves with organizational flexibility, human resource information system, integrated coordination and control. Since the 21st century is characterized by informationization, Chinese market economy and organizations still need more development. Therefore network management is a notable dimension to describe E-business index system. The value of Cronbach a is 0.8598.

Particularly, online marketing is separated from network management dimension and is discussed as an independent latent variable. Customers, target markets, distribution channels and tactics are discussed in online marketing. As for skill-based collaboration, the advanced online marketing skills are worth learning by all partners. The value of Cronbach is 0.8096.

5 Some Thoughts

The reality is that most Chinese medium-sized or small-sized enterprises are poor considering their skills, management, marketing, and e-business. So they can hardly meet requirement for strategic alliance and miss a lot of great opportunities, hence they are in inferior position facing global competition. Collaboration will help these firms to improve their competitive advantages under the background of global economical integration. So the first thing is to develop e-business and collaboration with partners, so that the complementary advantages will help firms to reduce cost and improve technology, increase their core competences and upgrade management and operation, and finally achieve better market performance.

Acknowledgments. All data used in this paper is from the project of National Natural Science Foundation (70272066) and this research is founded by Youth Foundation of Chongqing Normal University (08XLQ10).

References

1. Long, Y., Yang, X.T.: Analysis of Foreign Cooperation Mechanism for Chinese Enterprises in Unequal Strategic Alliance. Chonqing Press (2001)
2. Park, N.K.M., Song, J.M.: A Resource-Based View of Strategic Alliances and Firm Value in the Electronic Marketplace. Journal of Management 30(1), 7–27 (2004)
3. Liu, H.J.: Enterprise Organization Capital, Strategic Foresight and Performance: An Empirical Study Based on Chinese Enterprises. Management World 5, 83–93 (2007)
4. Liu, H.C.: Competitive Pressure, Strategic Change and Performance: a Structural Study of Transitional Economy. Journal of Management 5(2), 282–287 (2008)
5. Ireland, R.D., Hitt, M.A., Vaidyanath, D.: Alliance Management as a Source of Competitive Advantage. Journal of Management 28(3), 413–446 (2002)
6. Stuart, T.E.: Inter Organizational Alliances and the Performance of Firms: a Study of Growth and Innovation Rates in a High-Technology Industry. Strategic Management Journal 21, 791–811 (2000)
7. Gulcin, B., Orhan, F., Erdal, N.: Selection of the Strategic Alliance Partner in Logistics Value Chain. Production Economics 113, 148–158 (2008)
8. Gunasekaran, A., Lai, K.H., Edwin, C.: Responsive Supply Chain: a Competitive Strategy in a Networked Economy. International of Management Science 36, 549–564 (2008)
9. Bernstein, F., Song, J.S., Zheng, X.N.: ‘Bricks-and-mortar’ vs. ‘clicks-and-mortar’: an Equilibrium Analysis. European Journal of Operational Research 187, 671–690 (2008)
10. Lee, P.K.C., Yeung, A.C.L., Cheung, E.: Supplier Alliances and Environmental Uncertainty: an Empirical Study. Production Economics 120, 190–204 (2009)
11. Bollen, K.A.: Structural Equations with Latent Variables. John Wiley & Sons, New York (1999)
12. Guo, Z.G.: Application of SPSS in Social Statistic Analysis. Renmin University of China Press (1999)
13. Hou, J.T., Wen, Z.L., Cheng, Z.J.: Application of Structural Equation Modeling. Education and Science Press (2004)

Increasing the Meaningful Use of Electronic Medical Records: A Localized Health Level 7 Clinical Document Architecture System

Jun Liang[1], Mei Fang Xu[1,*], Lan Juan Li[2], Sheng Li Yang[2], Bao Luo Li[3], De Ren Cheng[4], Ou Jin[5], Li Zhong Zhang[6], Long Wei Yang[7], and Jun Xiang Sun[8]

[1] Second Affiliated Hospital of Zhejiang University College of Medicine, Hangzhou 310000, China
[2] Chinese Academy of Engineering, Beijin 100000, China
[3] China Hospital Information Management Association, Beijin 100000, China
[4] Zhejiang University, Hangzhou 310000, China
[5] Hangzhou State Software Industry Base Co., Ltd., Hangzhou 310000, China
[6] Hangzhou Normal University, Hangzhou 310000, China
[7] The First Affiliated Hospital of Zhejiang University College of Medicine, Hangzhou 310000, China
[8] Sir Run Run Shaw Hospital of Zhejiang University College of Medicine, Hangzhou 310000, China

xumf@z2hospital.com, panpanspan4@live.cn

Abstract. The health information systems of most medical institutions in China are isolated. Communications across these systems are generally realized through point-to-point interfaces at the database level, which tend to lack interoperability, extensibility, and security. In the resent study, we developed localized, document-oriented and data-focused clinical document architecture (CDA) templates based on health level 7 (HL7) CDA. Then, by combining these templates with the Service-oriented architecture for HL7 middleware, we accomplished interoperability across multiple heterogeneous systems. Modules of our system have been put into trial use in six medical institutions, including the Second University Hospital (main campus and regional campuses) of the School of Medicine, Zhejiang University and other collaborating hospitals.

Keywords: HL7 CDA, interoperability, localization experience, SOA, structured data.

1 Introduction

The development of hospital information systems (HIS) has been recognized as an imporatnt requirement in the reform of the health care system in China [1]. Although

[*] Corresponding author.

L. Cao, J. Zhong, and Y. Feng (Eds.): ADMA 2010, Part II, LNCS 6441, pp. 491–499, 2010.

84% of the major hospitals in the country are now equipped with locally-developed online HIS, severe limitations remain, including low levels of system integration, high heterogeneity, and low application levels [2]. Trials with internationally-developed technically mature systems, however, have been unsuccessful due to differences in culture, work flow, and user habits [2].

In 2009, the Ministry of Health of China set the goal for HIS in Chinese hospitals as "sharing, not only processing and collection, of information using computers". However, the chief information officers (CIOs) of many hospitals have not yet fully understood the essential role of standardization in the sharing of data across hospitals. Instead, some continue to maintain a closed-shop environment as a convenient shortcut around the establishment of an effective HIS [2]. Past experience in the United States has revealed that this route invariably results in hospital-specific solutions and highly complicated interfacing scenarios. As such, this approach can be considered non-viable overall [3]. At the same time, the experience from Marshfield Clinic Center in United States has shown that structured data form is going to be useful for future exploitation of the documentation [4]. The source of structured clinic data where those come from which will be two different ways. The first type of data is from clinic information data source which has already been structuring. Oppositely, the second type of data is from different data source which has not been structured. Those data have not been structured data would have need reclassification and post process in order to unified data from to be used. This part of the research work is though Localized Health Level 7 CDA template and minimized cost to structuring all clinic data. It is going to be shown as an undeniable truth that unified structuring clinic data is going to be a foundation for the secondary use which includes data mining for digital clinic record and the Interoperability across medical institution.

The China Hospital Information Management Association (CHIMA) [5] has defined three steps for achieving successful sharing of health information, among which the standardization of interoperability is the fundamental step [6]. In this paper, we demonstrate the development of effective exchange and, thus, maximum utilization of health information stored in accord with HL7 CDA R2 [7]. In addition to standardization, the software/service architecture is also a major factor in determining the interoperability of across-institution medical software [8]. Service-oriented architecture (SOA) is an architecture style consisting of consistent but loose coupled components for the building of technology-driven, service-oriented software applications [9]. Web Services provide important methods of realizing SOA, and define mechanisms for providing interoperability between different software in a distributed heterogeneous environment.

Section 2 analyzes the HL7 reference information model (RIM) and its vocabulary system. Section 3 describes the methodology of customizing CDA templates, and the development of an XML schema following the HL7 CDA standard. The schema can be used to create or verify CDA documents consisting of essential data elements and vocabularies. Section 4 discusses the transmission of documents consisting of electronic medical records (EMR) based on Web Services. Section 5 discusses implementation of the system and some experience. The last Section summarizes this work and recommends directions for future studies.

2 RIM and Vocabulary

HL7 RIM is a general, objective-oriented static model of health information that is not limited to the requirements of a specific subject area [10]. RIM includes six important base classes: Act, Participation, Entity, Role, Relationships, and Rolelink. Their interactions are illustrated in Fig. 1.

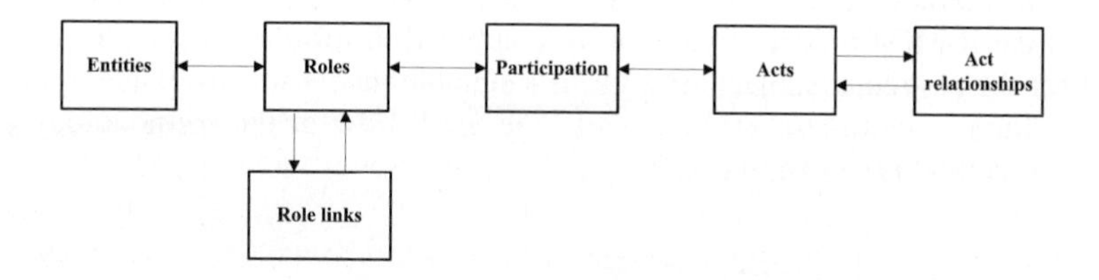

Fig. 1. A schematic showing the structure of RIM

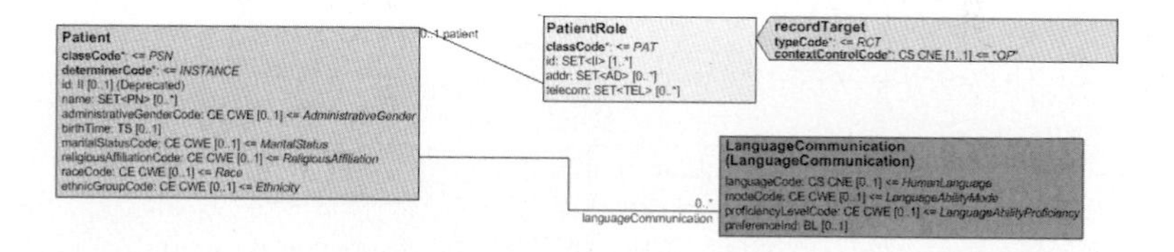

Fig. 2. R-MIM for HL7 CDA Diagnostic Imaging Report; R-MIM is a subset of RIM, and includes class clones for the creation of CDA documents, and fully extended sets of attributes and relations

The data types in RIM and HL7 provides a powerful mechanism that allows CDA to incorporate concepts in other standard coding systems, such as the Systemized Nomenclature of Medicine Clinical Terms (SNOMED CT), Logical Observation Indenters Names and Codes (LOINC), and the International Classification of Diseases (ICD 9-CM). For applications in the health care system in China, the relevant codes must be translated, matched, and extended to meet the requirements of this system. Fig. 2 shows a part of a refined message information model (R-MIM), which describes the HL7 CDA Diagnostic Imaging Report.

3 Locally Customized CDA Templates and Information Architecture

In the development of a locally customized CDA template, each field in the currently used target system was confirmed to follow the standards published by the Ministry of Health of China by checking against four standards set by the Ministry ("Rules for Health Information Dataset Classifying and Coding" [11], Metadata Specification of Health Information Dataset [12], the Rules for Data Element Standardization of Health Information [13], and the Guidelines for Data Schema Description of Health Information [14]), which were published as a framework for regulating the format of metadata and datasets in HIS. It is recommended that these guidelines be followed at all levels of health information subsystems. The metadata were thus updated and reviewed by a panel of experts on hospital computerization.

General CDA templates that meet the requirement of the hospitals were generated following two of the methods recommended in HL7 [6]:

Constraint method. The R-MIM model for the business process of the target subject area and the R-MIM model of CDA were combined to form a general CDA R-MIM model for the particular subject area. Or, if a suitable model was available in the HL7 template library, it was directly adopted. The work flow of the methodology of the constraint method is shown in Fig. 3.

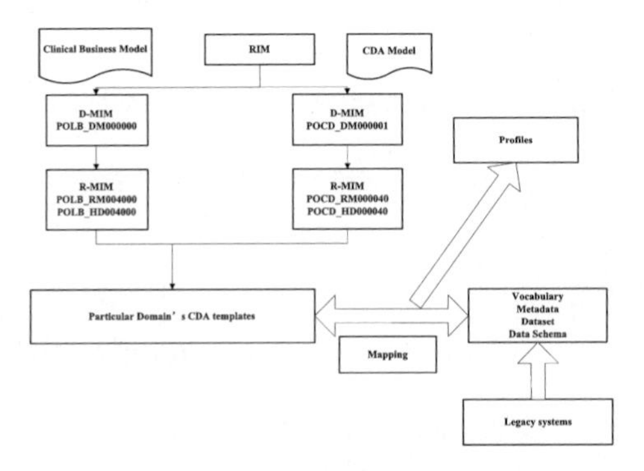

Fig. 3. The methodology of the Constraint method

Transformation method. Clinical report documents observing international standards (e.g. RIS & PACS image reports following the DICOM SR Basic Diagnostic Imaging Reports standards) were transformed into CDA Diagnostic Imaging Reports, compatible with the CDA standards. The work-flow of this method is shown in Fig. 4.

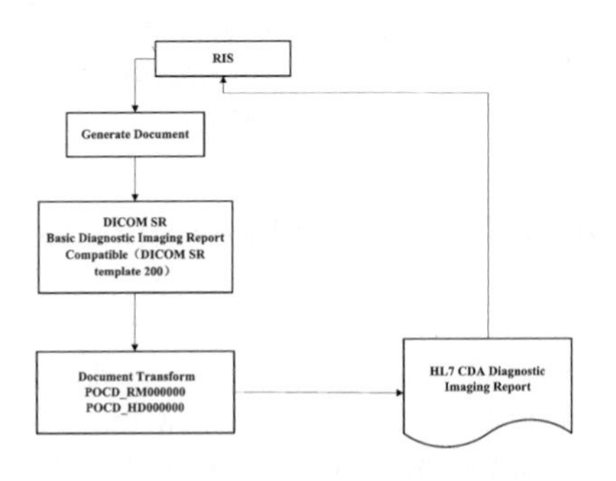

Fig. 4. The methodology of the Transformation method

Subsequently, the relation between general CDA templates for various subject areas and the corresponding metadata in the original systems were mapped. This was a critical and difficult step, and we experimented with the two methods described below.

1. Each element in the template was matched to a metadatum. If a metadatum could not be matched, the template was revised by adding a corresponding CDA section. If a surplus element was found in the template, the match was determined by its frequency of occurrence. It was removed if it was considered to be "optional". And it was unchanged if it was considered to be "required". Then, a CDA schema for the subject area was generated using the HL7 V3 generator [15] based on the R-MIM of that area. Fig. 5 shows an example of element-to-metadatum matching. The resulting CDA Schema and XML documents are shown in Fig. 6.

Patient
classCode*: <= PSN
determinerCode*: <= INSTANCE
id: II [0..1] (Deprecated)
name: SET<PN> [0..*]
administrativeGenderCode: CE CWE [0..1] <= AdministrativeGender
birthTime: TS [0..1]
maritalStatusCode: CE CWE [0..1] <= MaritalStatus
religiousAffiliationCode: CE CWE [0..1] <= ReligiousAffiliation
raceCode: CE CWE [0..1] <= Race
ethnicGroupCode: CE CWE [0..1] <= Ethnicity

CDA element	MetaDatas in Data group and data element dictionary of electronic medical record
id	HR01.01.002.01;HR01.01.002.02;HR02.01.001.01;HR02.01.001.02
name	HR02.01.002
administrativeGenderCode	HR02.02.001
birthTime	HR30.00.001
maritalStatusCode	HR04.00.001.01;HR04.00.001.02;HR04.00.001.03;HR04.00.002
religiousAffiliationCode	
raceCode	HR02.05.001
ethnicGroupCode	

Fig. 5. An example of the mapping between CDA elements and the corresponding metadata

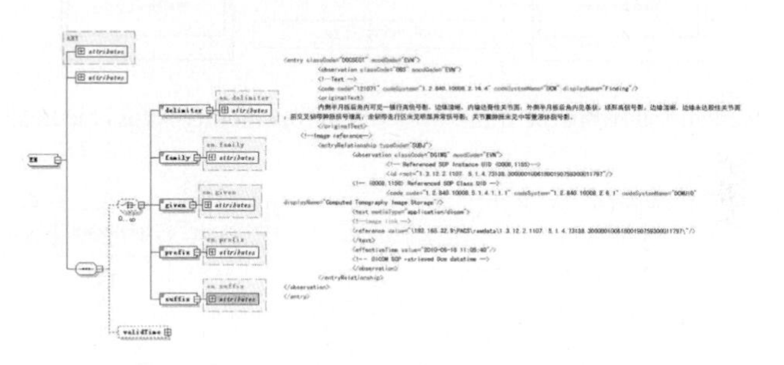

Fig. 6. An example of CDA schema and XML document

2. The metadata were matched to the legacy system data structures. A scheme was extracted from the Legacy System and then matched to the CDA schema.

This work primarily used first method as the basis for realizing the form structure of the clinical documents. Since HL7 CDA Level 2 had already have the identity of completed structuring [7], the compatible-HL7 CDA XML files which follow the methodology of HL7 would meet the requirement of "meaningful use" and Secondary use" for clinical EMR.

4 Transmission of EMR Documents Based on Web Service

Exchange of information across heterogeneous health information systems was realized in a standard style based on SOA. The CDA documents were packaged using

XDS to allow each heterogeneous system to continue functioning autonomously and independently. As a mature technique, Web Service provides a complete framework of services and has been commonly adopted in the development of SOA for distributed software [9]. In the SOA developed in this work, each heterogeneous system communicated with other systems via Web Services interfaces; the CDA documents were packaged following the XDS standard and exchanged following the HL7 specification. Based on these techniques, a framework of across-institution interoperability was developed corresponding to the current status of HISs in China see Fig. 7. The framework of the integrated SOA is shown in Fig. 8.

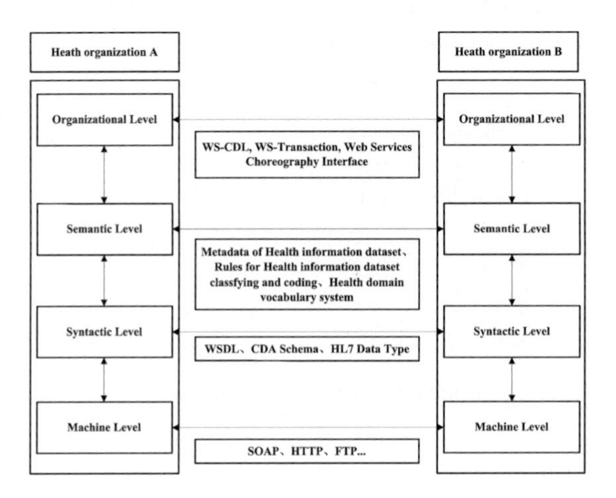

Fig. 7. A schematic showing the framework of interoperability across medical institutions

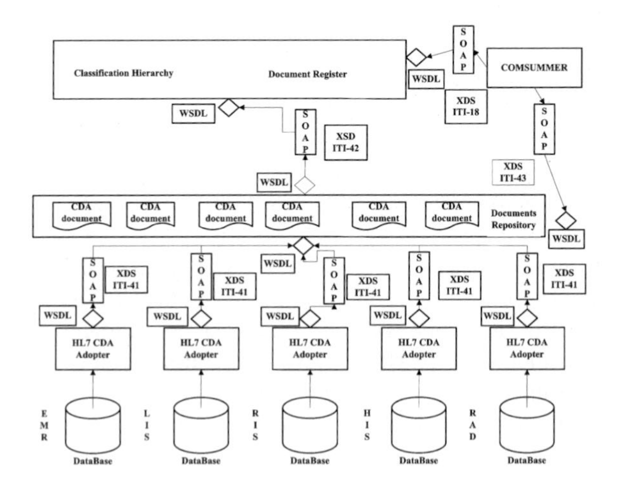

Fig. 8. A schematic showing the unified SOA framework; the design of the HL7 CDA Web Services adapter is similar to a simple Web Service interface, but the configuration file of HL7 Web Services and the HL7 methodology (RIM→D-MIM→R-MIM→HMD→MT) requires a top-down approach, in which the WDSL protocol is defined first and then the proxy and bottom codes are generated

The system was trialed in several hospitals, and the business processes and operation conditions in each heterogeneous system, including a subset of relevant data from other systems, were evaluated. The "engineer" mode was used, and a HL7 CDA adopter layer was added to the system without changing its internal structure or workflow. After a medical report was verified by the heterogeneous system, the HL7 CDA adopter then retrieved the relevant data from the business database and created a digitally signed CDA document by combining with an XSLT according to the specific requirement of the hospital. The CDA document was then packaged, submitted to a document database and registered.

5 Analyses of Experimental Applications

In this work, we demonstrated the development of an EHR that is able to cover the life span of a patient and maintain and exchange medical documents in a patient-centered fashion. Our experiences in this work suggest that other types of CDA can also be successfully locally customized. In the future, we hope to extend the work to all clinical observation documents in 18 heterogeneous health information subsystems. Fig. 9 shows an example document.

In terms of practical applications, the system is able to search the XDS-based registry for the ID code of the patient to obtain an index number. Based on this index number, the clinical document can then be retrieved in the document database. This document was created on XML. It follows HL7 CDA R2 standards, and is directly machine-processable [16] without the need for semantic transformation through additional interfaces. The complexity of system integration is shown below, with the original P2P mode for comparison:

$$\text{P2P Model: } T(n) = n\,(n-1) \qquad \text{SOA Model: } T(n) = n - 1\,. \tag{1}$$

where n is the number of the heterogeneous systems.

Fig. 9. An example showing the generation of CDA and applied XSLT based on a document from a hospital in China; the CDA Schema was mapped to relevant metadata and combined with XSLT to generate a document that meets the target clinical requirements (CDA version: CDA R2)

6 Conclusion and Future Directions

In this work, we realized the mapping between CDA Schemes of some clinical documents and the corresponding metadata using a flow process-based, document oriented, top-down approach and following the methodology "RIM→ domain message information model (D-MIM) → R-MIM → Hierarchical message descriptions (HMD) → message type (MT)" of the HL7 standard. In future, we plan to extend our system to all clinical observation documents in 18 heterogeneous health information subsystems, improve the security and privacy of the CDA documents, and develop a unified patient identifier cross-referencing profile (PIX) control center based on the WS/PIDS standard. Our ultimate goal is to achieve a system of effective transmission of clinical documents across medical institutions based on structured CDA templates and to realize "Secondary Use" of clinical medical data.

Acknowledgments. This research was supported by the key projects of the National Science and Technology program during the eleventh five-year plan period (Contract No.2008BAH27B01; Contract No.2008BAH27B03).

References

1. Chen, Z.: Overall Strengthening of Health Care Quality Management. Ministry of Health of the People's Republic Of China (2008)
2. Li, B.L.: China Health White Paper. In: China Health Information Technology Conference & International Forum (2008)
3. Kraft, G.: Healthcare IT Standardization: Will the New Stimulus Package Help. IT Professional 11, 4–5 (2009)
4. Kate, K.: Marshfield Clinic Data Warehouse/Analytics. In: China Health Information Technology Conference & International Forum, pp. 16–20 (2009)
5. China Hospital Information Management Association, http://www.chima.org.cn
6. Li, B.L.: EHR Meaningful Use. China Hospital Information Management Forum, Hang Zhou (2010)
7. Dolin, R.H., Alschuler, L., Boyer, S., Beebe, C., Behlen, F.M., Biron, P.V.: HL7 Clinical Document Architecture, Release 2. Journal of the American Medical Informatics Association 13, 9–20 (2005)
8. Lewis, G.A., Morris, E., Simanta, S., Wrage, L.: Why Standards Are Not Enough to Guarantee End-to-End Interoperability. In: 7th IEEE International Conference on Composition-Based Software Systems, pp. 164–173. IEEE Press, New York (2008)
9. Newcomer, E., Lomow, G.: Understanding SOA with Web Services. Pearon Education, Upper Saddle River (2005)
10. Beeler, G.W.: HL7 Version 3 - An object-oriented methodology for collaborative standards development. International Journal of Medical Informatics 48, 151–161 (1998)
11. Ministry of Health of the People's Republic of China: WS/T 306-2009. Rules for Health Information Dataset Classifying and Coding (2009)
12. Ministry of Health of the People's Republic of China: WS/T 305-2009. Metadata Specification of Health Information Dataset (2009)

13. Ministry of Health of the People's Republic of China: WS/T 303-2009. Rules for Data Element Standardization of Health Information (2009)
14. Ministry of Health of the People's Republic of China: WS/T 304-2009. Guidelines for Data Schema Description of Health Information (2009)
15. Bonney, W.: HL7 Tools: The Comprehensive Guide. HL7 Tooling Committee (2005)
16. Dogac, A., Laleci, G.B., Aden, T., Eichelberg, M.: Enhancing IHE XDS for Federated Clinical Affinity Domain Support. IEEE Transactions on Information Technology in Biomedicine. 213–221 (2007)

Corpus-Based Analysis of the Co-occurrence of Chinese Antonym Pairs*

Xingfu Wang[1,2], Zhongfu Wu[2],Yan Li[3], Qian Huang[3], and Jinglu Hui[3,4]

[1] Research Center of Language, Cognition and Language Application, Chongqing University
[2] College of Computer Sciences, Chongqing University
[3] College of Foreign Languages, Chongqing University
[4] School of Foreign Languages, Nanyang Normal University
wxfcqu@yahoo.com.cn

Abstract. Chinese antonym pairs pattern differently than antonyms in English. The opposites in a Chinese antonym pair often co-occur in a sentence. In a phenomenon unique to Chinese, most antonym pairs provide a good basis for the formation of four-character phrases. The opposites in a pair can usually be interpolated by an arbitrary number of characters to form new larger collocations, in such a way that the same number of characters precedes or follows each element of the antonym pair to keep the new collocation symmetric. All compounds and phrases with Chinese antonym pairs have the expanded form of a+X+b+!X, where !X denotes the opposite of X, a and b are the interpolated elements having the same character length. In our work, we characterized patterning of the interpolated elements and analyzed typical interpolations of one character in canonical Chinese antonym pairs, and identified the patterns involved in the separation and linkage of the antonym pairs.

Keywords: antonym pairs, co-occurrence, CCL corpus.

1 Introduction

Antonymy has been a popular research topic for centuries in many languages due to its remarkable ability to activate two contrasting members in a pair. In English literature Lyons[13] claimed that there exists a 'general human tendency to categorize experience in terms of dichotomous contrast'. Justeson & Katz[10] defined antonymy as a special lexical association between word pairs. Jones[9] also said "antonymy holds a place in society which other sense relations do not occupy." In a new research Paradis and Willners [14] concluded that "Antonymy is a ubiquitous phenomenon and comes in different guises in linguistic communication." The omnipresent antonym pairs are a universal phenomenon in world languages.

Chinese antonymy is also an important research topic in recent years. Yan & Zhang [18] found that there were 98 Chinese papers involving Chinese antonymy in

* Project (No. CDJSK100115) supported by the Fundamental Research Funds for the Central Universities.

L. Cao, J. Zhong, and Y. Feng (Eds.): ADMA 2010, Part II, LNCS 6441, pp. 500–507, 2010.

the past ten years. We searched "antonymy" and "corpus|corpora" as keywords in CNKI, an e-resource containing almost all Chinese journals, but no matches were retrieved. Therefore there is a need for further analysis of Chinese antonymy in this way. As Liu [12] observed, Chinese antonym needs to be studied with a large-scale corpus rather than out of experimentation. This is the main motivation for our research into Chinese antonymy in a large-scale corpus in Chinese.

Chinese antonym pairs pattern differently than antonyms in English. With Chinese parataxis they can form compounds without any other connected words, such as 左右 (left and right), 问答 (question and answer), 真假 (true or false), 阴阳 (ying and yang), etc. Even though in English there are still some very close co-occurrence cases, they do not function as typical Chinese antonym pairs do, a similar word in English is bitter-sweet. The following examples are extracted from Mark Davies's Corpus of Contemporary American English (COCA—http://www.americancorpus.org/):

> I could make a pretty *good bad* guy if I wanted to.
> I'd rather be a great bad poet than a *bad good* poet.

In the above sentences we find the two opposites 'good' and 'bad' modify each other directly without any connecting words. In these cases, though, the two words do not function as an antonymous unit; instead, the second element has closer relation with the head noun. The opposites in a Chinese antonym pair often co-occur in a sentence, but not in the usual way in English as found by Justeson & Katz [11], Fellbaum [5] and Jones [7].

Chinese antonym pairs are unique for their co-occurrence in sentences which are somewhat different from those in English. In ancient Chinese there are some literatures about the description of opposites in daily life and mind. In his famous Taode Jing(Daoism) Laozi put the opposites as follows.

老子《道德经》	Thomas Cleary's English Version of Daoism[1]
有无相生, 难易相成， 长短相形， 高下相倾， 音声相和, 前后相随。	Being and nonbeing produce each other, difficulty and ease complement each other, long and short shape each other, high and low contrast with each other, voice and echoes conform to each other, before and after go along with each other.

We know that antonym pairs in Chinese usually co-occur in context because they are the closest two opposite concepts in mind. But what is outstanding is that usually they are put one after another directly, and they are regular in characters numbers and they make a symmetric structure. As Bosley[1] observed these sentences here: "In addition to listing two terms, there is also an indication of a relationship."

Our research essentially involved two steps: 1) identifying the canonicity of Chinese antonym pairs by the frequency of stand-alone antonym pairs and their expanded forms with other characters within the CCL corpus; 2) retrieving paratactic four-word constructions within which antonym pairs co-occur. The aim of our research is not to

[1] http://www.davemckay.co.uk/philosophy/taoism/taoism.php?name=taoteching.cleary

elicit the most possible opposite of a word, not to count the frequency of co-occurred antonym pairs (A-Pair) to identify the phrase types, we focus on B-pair in the sentences by checking whether A-pair is canonical or not first.

2 Typicality of Antonymous Compounds and Their Verification in Corpus

Chinese scholar Tan [15] listed a "total" of 371 antonymous compounds in Chinese; this list was compiled partly from dictionaries and partly intuitively (but apparently untested against data). Being conducted before access to corpora was possible, this is understandable. Deese's work in choosing fundamental antonym pairs in English was based entirely on the more systematic method of using word association tests (Jones [7]). Our research uses the modern Chinese section of the CCL (Corpus of the Center of Chinese Linguistics, Peking University) corpus to verify the accuracy of Tan's list of antonym pairs in Chinese. And we focus on spatial antonym pairs to extract the possible patterns in expanding Chinese antonym compounds. We thus aim to verify the list (whether it consists of canonical paired terms within a corpus) and to analyze the formation of a special construction involving antonym pairs.

Jones [7] said that "Corpus data have allowed for the co-occurrence frequency of antonym pairs to be quantified and the various discourse functions of antonymy to be identified". We can easily access large corpora to study what we could not find in the past. The canonicity of Chinese antonym pairs and their expanded properties are what we want to explore in this paper, based on the CCL. Tan's "total" 371 pairs of Chinese antonym compounds was also cited by some Chinese scholars (Wei [17], Wang [16], Zeng [19]) in arguing for related phenomena.

In our work we have counted Tan's list and found that in fact there are only 370 pairs instead of 371. Most importantly, 8 repeated pairs are listed without any changes in the pair's form, they are 褒贬，东西，多少，教学，来往，冷热，利害，兄弟, but the intonation of the second character of the repeated forms in the six pairs is silent to demonstrate that the pair has changed a little bit or been combined into a fixed phrase with the second word losing its contrasted feature, they have become noun compounds. Two pairs (教学&冷热) are listed repeatedly without any phonetic annotation and the readers do not know why there is some difference between the same pairs. And the most strange thing is that the pair 冷热 is repeated after several other pairs are inserted between them and the pair利害 and its repeated pair is separated by another antonym pair for no reason. In the aspect of compounds, they could be regarded as two pairs, but in the aspect of antonym pairs they are not. So we need to disregard these 8 repeated pairs, thus reducing the total number to 362 pairs.

In our research we focus on the modern Chinese section which contains more than 264 million Chinese characters, with only 9552 types. We want to judge whether Tan's antonym pairs are canonical or not in Chinese. Jones et al.[9] use 14 frames to retrieve the canonical antonyms of some seed words in English. Our research is to search the B pairs as Jones[7] proposes and try to find their common features. That is, we assume that two members of an antonym pair are opposite in meaning and that they co-occur in the sentences, and we hope to find out that the relation between the interpolated elements in the B pair. Generally in English the distance between the B pair elements are farther than the elements of B pairs in Chinese, i.e. there is only one character between the two opposites in our research. And this is the common feature because of Chinese preference to four-character idioms or phrases.

In Tan's list 17 antonym pairs are listed in their reverse order so another 17 pairs are made. This is a Chinese tradition of making new phrases by reversing existing ones. But the difference between the original pairs and their reverse ones are great, some are not very common in usage. Three pairs (怯勇, 瘦胖, 寡众) do not have any trace in this 264 million words CCL corpus, and two (勇怯, 夭寿) have cases less than 10. Because some of them are ruled out in this procedure, only another two pairs do not satisfy the criterion of 10 cases. Jones[7] proposes 8 antonym sequence rules in his analysis: morphology, positivity, magnitude, chronology, gender, phonology, idiomaticity, frequency and markedness. What matters in deciding which of the two pairs is better than the other should be the one with the higher frequency in Chinese. In analyzing the relation of the two characters in the pair with more frequency we find that 'positivity', 'magnitude' and 'chronology' are the most influential factors in the sequence of the two opposites in Chinese antonym pairs. This shows that Chinese are free to use these antonyms in different orders even though the two pairs have great difference in usage.

After these analyses we arrive at 317 stand-alone pairs or 302 pairs interpolated with one character whose frequency exceeds the threshold of 10 instances. Thus we conclude that the list of 317 pairs could act as a standard for Chinese stand-alone antonym compounds because of their cohesiveness in putting two members together. Alternatively, the list of 302 pairs interpolated with one character can be taken as the canonical list in Chinese antonym pairs for their expanded productivity. The expanded structures of antonym compounds are useful in proving that the two members of an antonym pair are used in their opposite meanings. We do not claim that the rest are not correct as antonym compounds or antonym pairs, they are just not canonical in representation. We have also added some twenty antonym pairs and have tested their presence in CCL, which will be studied in another paper.

3 Patterns of the Interpolated Characters

3.1 Constructions to Be Analyzed

In our research we focus on the four-character phrases containing the antonym pairs. Different number of characters can be inserted to expand the antonym pairs, but we only focus on the phrases interpolated by one character between the elements of an antonym pair, and meantime we chop the string containing another character before the first opposite and one more character after the second element of the antonym pair. That is, we focus on the string of a+X+b+!X+c in which a, b and c denotes one Chinese character and !X is the opposite of X; the structure of a+X+b+!X or X+b+!X+c are what we want to probe into, and we name ab as B1 Pair and bc as B2 Pair. B1 Pair is characteristic to Chinese and Jones[7] has no suitable cases to analyze in English. When B1 Pair is meaningful, it excludes the possibility of B2 Pair because B2 Pair is non-constructional noise for Chinese preference in making phrases into four-character idioms, and vice versa.

Chinese idioms are mostly made up of four Chinese characters and four-word construction is the special feature of Chinese idioms, and entries in Chinese idioms dictionary are arrayed regularly. Many idioms involve stories. As mentioned above, the stand-alone antonym pairs are good evidence to Chinese strong cohesiveness in combining the opposites together into short structures. Four-character idioms are usually the compressed forms of some larger phrases. Chinese are always omissible in less important section in a phrase in depicting or describing ideas. Only the important parts are chosen to form the traditional four-character idioms or phrases in order to elicit a special rhetorical effect, such as symmetry and rhythm. It is like the way to shuffle a pack of poker; other elements could be interpolated into the antonym pairs. But these elements are not chosen randomly, our purpose is to find the relation between the interpolated elements and the relation between the interpolated section and the antonym pair.

3.2 Patterns Available

In the analysis made by Jones et al. [9] seven search frames such as 'X and Y alike', 'X as well as Y', 'both X and Y', 'either X or Y', etc. were used to extract the canonical antonym pairs with seed words in English. In Chinese the situation is somewhat different because the four-character phrases are mostly connected by content words; we have to rely on abstract patterns to summarize the constructions of Chinese antonym pairs.

We adapt Chen's[3] classification and Han's[6] analysis of Chinese antonym pairs to conclude six patterns for antonym pairs to be interpolated: 1) Interleaved with a 2nd antonym pair: X+Y+!X+!Y; 2) Two juxtaposed antonym pairs: X+!X+Y+!Y; 3) Interleaved with synonymous words: X+Y1+!X+Y2; 4) Followed by a fixed phrase: X+!X+m+n 5) Interleaved with grammatical particles: 半/忽/亦+X+半/忽/亦+!X; 6) Repetition of antonym pairs: X+X+!X+!X. These short strings can effectively filter the false antonyms pairs occurring in different syntactic environments as Fellbaum [5] found. In this way we can sort out the false co-occurred antonym pairs in a sentence because of the long span between the members of the antonym pairs. We think that a long span between the members of an antonym pair is an uncertain factor to claim that the sentence has parallel syntactic slots to fill in B-pair information, both in English and in Chinese.

In our research we choose the spatial terms as the objects to analyze. We follow Zhu et al.'s[20] 18 categories in dividing Chinese words because their classification is more exhaustive and executable. The 18 parts of speech in contemporary Chinese include: nouns, pronouns, verbs, adjectives, adverbs, quantitative terms, spatial terms, locational terms, temporal terms, distinct terms, stative terms, numerals, quantifiers, conjunctions, auxiliaries, modal particles, onomatopoeia and interjections. Because of Chinese parataxis it is not possible to follow patterns first and analyze the frames second. We will reverse Jones's method in our investigation of the phenomenon in Chinese. This is reasonable because spatial antonym pairs in Chinese form a category which contains the most productive pairs which are representative in searching the features of Chinese antonym pairs. In Jones et al.'s[9] analysis of antonym pairs they focused on the structural words such as 'both…and', 'neither…nor', 'from…to', etc to conclude the frames with antonym pairs. In Chinese it is somewhat difficult to follow this procedure because empty words in Chinese do not matter too much in four-character idioms except for pattern 5). The interpolated elements are mostly content words and they are important parts in making new phrases. So we cannot implement the method in analyzing English antonym pairs, but instead choose another method to reach our goal in identifying the patterns of Chinese antonym pairs.

3.3 Distribution of Spatial Antonym Pairs

There are 8 spatial antonym pairs in Chinese: 东西 (east-west), 南北 (south-north), 上下 (up-down), 左右 (left-right), 前后 (before-after), 远近 (far-near), 里外 (inside-outside), and内外 (home-abroad). Their distribution deserves our analysis first. All 8 spatial antonym compounds are among the first 80 compounds in occurrence in CCL, and their interpolated pairs are among the top 73 pairs, so spatial antonym pairs are the most frequently used terms in writing and spoken materials, and they are productive in making new expanded phrases as a category.

Among these 8 pairs, we extract the character before the first opposite, the mid character between the two opposites, the post-character after the second character, the pre-mid string, and the mid-post string. That is, in the frame of a+X+b+!X+c, we got the data of a, b, c, ab, and bc to be analyzed.

In this analysis we can see the productivity of the spatial antonym pairs with one character between them. Two pairs 上下 and里外 have the "apparently" high productivity when interrupted by one character. Our first impression is that上下should be the king or queen in the interpolated group because it ranks 2nd , and 里外ranks 14th. But our further research found much noise in the interpolated items of these two pairs. Firstly, in Chinese two-character phrases there are many phrases which contain 下 to denote the result of an action, that is, it seems to function as the suffix or auxiliary to a verb; and in other cases it acts as a verb to be modified by an adverb. And 上 can be used frequently in the prepositional phrase '在…上'(on, about, etc.), therefore these meanings could result in many strings containing 上 and 下 though they are not antonymous in meaning. In the CCL corpus antonym pairs interrupted with one character average 28.5% compared to the stand-alone antonym pairs. And the ratio for spatial antonym pairs is 25%, close to the average and therefore the analysis of them is typical.

Can the problem be solved if the CCL corpus was segmented? We tried segmenting the data with 里外 but found that new problems still occur. For example, 对里对外 and 宫里宫外 are two typical cases to be segmented wrongly because they do contain 里外 as an antonym pair. These two phrases are used by the authors in their opposite meanings but the first phrase was segmented as对/里/对外, the second phrase宫里宫外 was divided into宫里/宫/外, and another Chinese parser put it into宫/里/宫外. In these two four-character phrases the interpolated elements are repeated by itself, and our analysis claim that this case is absolutely the use of the opposite meanings of the antonym pairs. And our analysis also proves such a presumption. Therefore the wrong segmentation of the typical examples just tells us that the segmented data are not much useful and helpful in analyzing the antonymity.

From this we conclude that there are false expanded antonym pairs, or noise. Our hypothesis is that if we interpolate an antonym pair with another two characters, being antonyms or synonyms to each other, we can safely say that the antonym pair (A-pair) inserted by other characters (B-pair) is used in their opposite meanings. This is also a good way to reduce the noise in choosing the true antonym pairs in their expanded strings.

4 Conclusion

We have analyzed Chinese antonym pairs in this paper, and also highlighted some problems in successfully identifying interpolation of words into the antonym pairs. There is some noise in choosing meaningful strings to analyze, so further work will reduce the noise in the expanded strings of antonym pairs to get better symmetric antonym pairs strings. This also might be an effective way to disambiguate the false antonym pairs. We have shown that Jones et al.'s[9] frames are too limited to extract highly frequent canonical antonym pairs because they rely on seed words first. In Chinese it is challenging to retrieve the classical and canonical antonym pairs in this way because many of them are not connected by the obvious frames that Jones proposed for English, due to the phenomenon of Chinese parataxis. The canonicity of Chinese antonym pairs is not straightforward if we follow their method.

We revised Tan's list of antonym pairs mainly by expanding them with one character within the CCL corpus data. We then verified the abstract patterns for spatial antonym pairs because of their representativeness among all antonym pairs. The interpolated elements of the expanded antonym pairs were analyzed to retrieve the common properties of this category. The conclusion we draw in this research is as follows. In addition to a few frames composed of grammatical words mentioned above for Chinese spatial antonym pairs, the content words interpolated into the pair have the following common features: A) The two meaningful interpolated words analyzed in our research are often either synonyms or antonyms, which is found by matching ab or bc to the entries in the Contemporary Chinese Dictionary. B) When the two interpolated words are the same word they definitely indicate the usage of the antonym pair in their opposite meanings. C) Pattern 5) is not a typical construction for Chinese spatial antonym pairs because they are interleaved by many different content words, and content words almost always govern the filled information because of the properties of Chinese four-character idioms.

References

1. Bosley, R.: Sources of Skepticism and Dogmatism in Ancient Philosophy East and West. Journal of Chinese Philosophy 29, 397–413 (2002)
2. Center of Chinese Linguistics, Peking University, http://ccl.pku.edu.cn:8080/ccl_corpus/
3. Chen, X.Y.: The Coexistence of Anti-morphemes in Modern Chinese Word Construction. Journal of Yancheng Technology College 1, 56–60 (2004)
4. Contemporary Chinese Dictionary. Foreign Language Teaching and Research Press, Beijing (2002)

5. Fellbaum, C.: Co-Occurrence and Antonymy. International Journal of Lexicography 8, 281–303 (1995)
6. Han, H.X.: Comparison of Chinese and English Antonym Co-occurrence. Journal of Hangzhou Teachers College 1, 134–140 (1993)
7. Jones, S.: Antonymy: a corpus-based perspective. Routledge, Amsterdam (2002)
8. Jones, S.: A lexico-syntactic analysis of antonym co-occurrence in spoken English. Text & Talk 26, 191–216 (2006)
9. Jones, S., Paradis, C., Murphy, M.L., Willners, C.: Googling for 'opposites'—a web-based study of antonym canonicity. Corpora 2, 129–154 (2007)
10. Justeson, J.S., Katz, S.M.: Co-occurrences of antonymous adjectives and their contexts. Computational Linguistics 17, 1–19 (1991)
11. Justeson, J.S., Katz, S.M.: Redefining Antonymy. Literary and Linguistic Computing 7, 176–184 (1992)
12. Liu, G.H.: Rethinking of Research on Antonymy for Thirty Years and its Asymmetric Features. Foreign Languages Research 109, 1–7 (2008)
13. Lyons, J.: Semantics, vol. 2. Cambridge University Press, Cambridge (1977)
14. Paradis, C., Willners, C.: Antonyms in Dictionary Entries: Methodological Aspects. Studia Linguistica 61, 261–277 (2007)
15. Tan, D.R.: On Antonymous Compounds. Linguistic Research 30, 27–33 (1989)
16. Wang, Y.: The Event-Domain Cognitive Model. Modern Foreign Languages 28, 17–26 (2005)
17. Wei, D.C.: Study of Juxtaposed Opposite Elements in Yan's Family Admonition. Journal of Northeast Normal University 171, 75–79 (1998)
18. Yan, Q.M., Zhang, J.Y.: Reviewing and Looking Forward to the Research on Ancient Chinese Antonyms. Journal of Zhejiang University 36, 53–59 (2006)
19. Zeng, D.: Cognitive Study of Chinese Opposite Compounds. Unpublished Ph.D Dissertation. Zhejiang University (2007)
20. Zhu, X.F., Yu, S.W., Wang, H.: The Development and Application of Dictionary of Modern Chinese Grammar. Communications of COLIPS 5, 81–86 (1995)

Application of Decision-Tree Based on Prediction Model for Project Management

Xin-ying Tu[1] and Tao Fu[2]

[1] Institute of Spacecraft System Engineering, China Academy of Space Technology, 100094 Beijing, China
[2] Beijing Shenzhou Aerospace Software Techology Co., Ltd, 100094 Bejing, China
futao@bjsasc.com

Abstract. In recent years, with rapid development of China's space program and aerospace technology, massive data of project management have been collected. However, provided massive data, there is plenty room to improve our project management standards, and the inheritable and applicable knowledge on data. This paper reports on methodology based on data mining techniques to tackle such issues, in order to promote project management standards and technical reformation.

Keywords: Data mining, Decision tree, Project management.

1 Background

China's development in aerospace technology still lags behind some western countries. The main reason is that our research and development in this field took off in a later stage, and we are currently in a phase to make further explorations. Satellites mark one of the most comprehensive and advanced contemporary technology, and it has a longer development cycle with higher complexity in project management. We have made astonishing achievement in our aerospace program development in the past few years. Our experience and lessons learned from project management are invaluable, and it is necessary to summarize and analyze them so that these experiences may help to boost the development of satellite technology to reach another milestone.

The Institute of Spacecraft System Engineering, China Academy of Space Technology(ISSE of CAST) has taken charge of numerous technical reformation projects. "The Fixed Assets Management System" is initiated as a highly specialized project. It is intended to understand the states of project management thoroughly and systematically, coordinate the constructive resource of all projects, take full life cycle project management, improve productivity in the technical reformation process, ensure the completion of tasks with high satisfaction and the steady progress of our technical strength.

Along with the growing popularity of information, the project technology reform administration board has built extensive databases for project management, gathered massive data from business operation. Yet, these are mostly "irrelevant" data, and the majority of information processing is still limited to update, query and statistical

L. Cao, J. Zhong, and Y. Feng (Eds.): ADMA 2010, Part II, LNCS 6441, pp. 508–513, 2010.

analysis. Thus, in-depth analysis of these data is inadequate. The tendency of having correlation lurks under large scale of business data gathered, and is waited to be mined and extracted. For example, in the Fixed Assets Management System, there is strong correlation between project equipments and device component list. Some devices are re-usable, but technical experts and superiors from higher management may not know the details of all equipment due to the large scale of equipment involved. Nevertheless, applicant to a certain project usually have little notion about other projects. Therefore, this sometimes leads to a duplicated purchase which is a waste of money. We can solve this kind of issues by applying data mining techniques.

2 An Overview of the Methodology

2.1 Data Mining

Data mining is the process to extract unknown but possibly useful information hidden in large datasets. It employs various analyzing tools to find patterns and inter-relations from massive data. These patterns and inter-relations may be used to make predictions. Data mining commonly involves tasks such as classification/prediction pattern discovery, generalization, clustering, regression, association rule learning, sequence pattern learning, dependency relation/ pattern learning, and exception/ tendency learning.

The Fixed Assets Management System involves large scale of project information, for example, single piece device, device component list, device utility, project funding, project flow, project monitoring, and resource scheduling. Every step in the operation needs to be approved by superior technical staff, in order to ensure quality of the entire development process; however, these technical experts are sometimes in charge of multiple tasks, and they usually work on site. Thus, together with other uncertain causes, this sometimes leads to disjointed project flow, or even project suspension. Since some devices are shared resources which are used in a variety of models, it is inevitable to come across resource scarcity and conflict, but these issues can be solved by effective resource scheduling. Under some complex circumstances, human being appears to show little capability to coordinate resources. With massive data, it is hash to accept the fact that we can't extract useful knowledge from the data. Data mining techniques uncover unknown information and knowledge from large datasets which is possibly useful to us, and are commonly employed to promote project management standards.

Data mining tools can be used to predict tendency and action in the future, therefore, help human beings with their decision making. Widely used methods include neural networks, generic algorithms, decision- tree method, etc. The decision- tree method model is based on the concept of entropy used in information theory.

2.2 Decision Tree

Decision tree method extracts and justifies data characteristics, by classifying the extracted data with respect to their main characteristics; thus, it builds failure diagnosis rules that serve the purpose for failure detection and diagnosis. A decision tree looks similar to flow chart. Every interior node represents an attribute test. Every branch represents an output from test, whereas each leaf node represents a class or distribution of classes. The knowledge representation of decision tree is to embed expert knowledge among interior and leaf nodes, and obtain from leaf node the conclusion part of the knowledge by looking at the attribute values and the conditions of expression, saved in the interior nodes. The goal for constructing decision trees is to find relationship between attributes and classes, which is taken to further predict the class of the unknown record. The system capable of making predictions is called "decision tree classifier".

3 System Design Concept

3.1 Construction of the Decision Tree

The key to perform fault diagnosis based on decision tree method is building the decision tree. Usually, it takes two steps to build a decision tree: growing a decision tree and pruning it.

Growing a decision tree works as a top-down process, and looking for the best extended attribute is the key to grow a decision tree. It will firstly determine the form of extended attributes, and locate the all extended attributes associated to this form, and then make assessment to these attributes, selecting the one with strongest judgment to be the best extended attribute. Algorithm C4.5 selects the one with highest information gain to be the best extended attribute. It can process discrete values of attributes, as well as contiguous ones, so it is well-suited to perform data mining for fault checking in on-orbit satellites.

Pruning a decision tree. As a decision tree is being built, many branches might represent noises or singleton points from training data. The pruning process attempts to identify and eliminate this kind of branches, so that we can solve the problem of data overfit. By replacing a sub-tree with a leaf node, we obtain better accuracy in the classification of position data. We can use cross-verification, or select a portion of the data to make assessment of the decision tree.

In The Fixed Assets Management System, among all factors that may have impact to the project, we need to pay attention to the factors that may cause severe consequences such as suspension. Therefore, we can preset impact-factor value for each factor by assigning weight values, and it is also possible remove factors with very little impact by adjusting the weight values in a repetitive way. In practice, integrating additional relevant information is required to make synthesized judgment.

3.2 Algorithm Selection

A decision tree is in general built from top to bottom. There are multiple methods addressing how to split a tree, while the goal is the same – to split the target class in an optimized way. The concept is that we treat it as leaf node and node content is the classifier, if all examples from the set of training data belong to the same class. Otherwise, we should choose an attribute base on some strategy, and split the example set into several smaller subsets according to the values associated to this attribute, so that all examples in each subset have the same value in this attribute. In the next step, the subsets are processed recursively. This is a typical divide-and-conquer method. There is always a path connecting the root and a leaf node, and this path is also called a "rule".

ID3 is a decision tree classification algorithm. It selects example's class based on values in the attribute set. The core idea is to select attribute attached to the tree node at each level, and the selection criteria is information gain, so that we can obtain maximal class information about the example tested when a non-leaf node is visited and tested. The entropy value reaches its minimum after we use such attribute to split the example set into subsets, expecting the average path connecting this non-leaf node and its successor nodes to be shortest. This algorithm can be described as follows:

The expectations of any sample can be described as follows:

$$I(s_1, s_2, \cdots\cdots, s_m) = -\sum p_i \log_2(p_i) \quad (i = 1...m). \tag{1}$$

where S is the dataset, m is the number of classification of S, $p_i \approx \frac{|s_i|}{|s|}$

c_i is the classification label, p_i is the probability of a random sample, and c_i is the number of samples of class c_i.

The entropy of subsets classified by A is :

$$E(A) = \sum (s_{1j} + \cdots\cdots + s_{mj}) / s \times I(s_{1j} + \cdots\cdots + s_{mj}). \tag{2}$$

A is the attribute, which has V distinct values.

Information gain: $Gain(A) = I(s_1, s_2, \ldots\ldots, s_m) - E(A)$

Therefore, the information gain rooted at A is $Gain(A) = I(p, n) - E(A)$.

Select attribute A such that $Gain(A)$ is maximal (that is $E(A)$ is minimal) to be the root node. Applying the above procedure recursively to the V subset of E_i associated with distinct values of A, we can generate child nodes of A, $B_1, B_2, \cdots, B_v$.

3.3 Building the Model

(1) Choosing and building a model

Which variables are closely related to project risk? We can find correlation between project risk rate and each input variable by using the functionality for correlation comparison feature provided in data mining tools. Through this comparative selection process, we can eliminate those variables that are less correlated to the project risk rate, and reduce the number of modeling variable and model needed. This will not only shorten the time-span for building the model, reduce the model complexity, but also in some extent help up building a more accurate model. Oracle's data mining tool offers various modeling method such as decision tree, Bayes Discrimination, neighbor learning, neural network, regression, correlation, clustering, etc. It is feasible to firstly build several models based on multiple modeling methods, and then evaluate the pros and cons of these models to select the best suited model for project management predictive analysis. Before designing a model, the model parameters can be reset by the users, so as to build a model well-suited to a certain circumstance.

(2) Model building and Adjustment
Model building and adjustment is the core process in data mining. The strategy for model building and adjustment changes along with the specific problems presented, data distribution and attribute variation. Also, many approximation algorithms may be applied to simplify the optimization process in modeling. All these methods mentioned may have impacts on the model's prediction result. So we seek technical experts' help in making model adjustment strategy during the entire process of model building and adjustment, to keep us away from information loss due to inappropriate optimization.

In project management, the influence factors of the project progress and predictable factors mainly includes two aspects. One is researchers aspect and the other is resource aspect. The researchers aspect can be changed frequently by any reason, which can't under control. But there will be a cyclical variation, such as someone attend meetings or movement in a certain time every month, etc. In this case, we can build the model like this, gather the information of everyone's effective time spent, the role played in the project, whether the key position or not. Using these information as the meta data to establish the model.

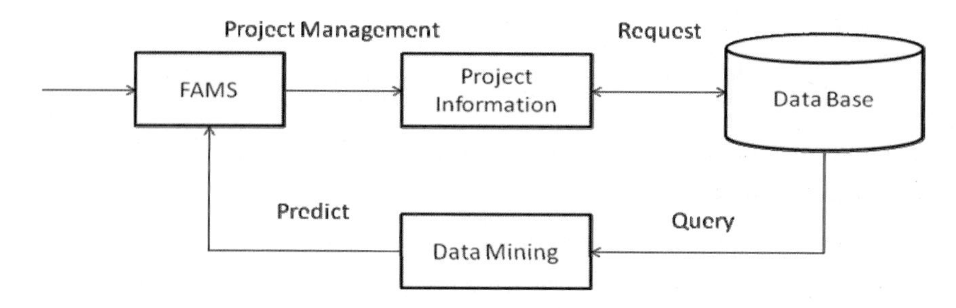

Fig. 1. Sytem Flow Chart

(3) Verification of model
Verifying a model is a critical phase in data mining process. We need to verify the correctness of the predictive pattern procedure, as well as the correctness of other input and output procedure which uses these models. The verification method is that we first input some historical data, and compare the historical data mining result with the result obtained from the current model. If there is a significant difference, we should consider re-designing or modifying this model.

3.4 The Interpretation and Application of Model

After we have built the optimized model, a technical interpretation to the obtained model given by technical staff is always needed. Through these interpretations, we can discover some common rules that were unknown to us before. After revealing these rules, we can gain knowledge about a company's business activity. On the other hand, if we are able to derive mathematical models based on professional knowledge, this approved the rationality of such mathematical model in business. That means we have more confidence to apply such model in our business activities.

The model's bias may become amplified after the model has been in use for awhile, or after significant changes has occurred in the environment. We can then consider of building a new, better suited model.

4 Conclusion

Data mining techniques are among the cutting edge research and development of the database and information systems. It draws great interest from both academia and industry, and quickly becomes a research hotspot. In the Fixed Assets Management System, the decision-tree-based prediction model for project management, which employs data mining techniques, help us to make predictions on the implementation of a new project's life cycle, by utilizing massive data collected in project management, ensuring a smooth project development, the development process of satellites, scientific project management, and promote the total level of technical reform program proposed by the ISSE.

References

1. He, Q., Hu, J.: A Surver of Data Mining. Journal of Southwest University for Nationalities 14, 36–39 (2003)
2. Han, Q.H., Hu, X.: Application of Decision Tree and SVM Methods in Fault Diagnosis of Liquid Rocket Engine. Computer Science 33, 84–86 (2006)
3. Wang, Y.: The Fault Diagnosis of Locomotive Converter Based on Data Mining. Southwest Jiaotong University 21, 120–130 (2002)
4. Tang, X., Cai, Q.: Application of Data Mining in Telecommunication. Computer Engineering 30, 36–37 (2004)
5. Yang, J., Zhang, N.: Research and Application of Decision Tree Algorithm. Computer Technology and Developmen 20, 114–116 (2010)

Management Policies Analysis for Multi-core Shared Caches

Jianjun Du[1], Yixing Zhang[2], Zhongfu Wu[1], and Xinwen Wang[1]

[1] College of Computer Science and Engineering, Chongqing University, Chongqing, 400030, China
[2] School of Mathematical Sciences, Fudan University, ShanHai 200433, China
{dujianjun,lh,wzf,zjstud}@cqu.edu.cn

Abstract. To improve performance and fairness of the LLC shared among the multiple cores, the recent *Promotion/Insertion Pseudo-Partitioning* (PIPP) that combines dynamic insertion and promotion into the cache management policy. Compared with PPIP, in this work we propose a new *Homologous Promotion Insertion Policy* (HPIP) which can determine the insertion position when a *new* core situation occurs and balance the cache resource allocation simultaneously. HPIP depends on the existing cache structure and require negligible change overhead. In addition, we analyze *Dynamic Insertion Policy* (DIP) and maintain that the sampling sets selection for *Set Dueling Monitors* (SDM) should be according to a processor's cores number rather than the running applications. Finally, our experiments with multi-programmed workloads for 2-core, 4-core CMPs based on M5 simulator show that the performance of HPIP approximate to PPIP and its adaptive capability is enhanced.

Keywords: Cache Management, Multi-core, Insertion, Promotion, Set Dueling.

1 Introduction

The multi-core processor architectures have two typical cache organizations for the on-chip Last-Level Caches (LLCs): the *shared* and the *private*. For better flexibility and dynamic allocation, many recent processors have opted for a *shared* LLC. Past researches [1, 2] show that the de-facto standard management policy LRU used for shared L2 often incurs a few pathological behaviors such as thrashing, unfairness. The PPIP and DIP are the novel designs for the cache management. Compared against PPIP, we propose a new HPIP which can implicitly partition the cache resource similar to PPIP and solve a special *new* core situation not considered in PPIP. Since directly extending DIP would cause the hardware logic more change, we design a new structure for *Set Dueling Monitors* (SDM) sample selection.

The main contributions of this work are:

- We first discuss the *new* core situation in this domain, and HPIP not only can deal with it but also improve the LLC performance and fairness.
- The number of cores in a multi-core processor is first suggested to be better choice in SDM design for DIP implementation.
- A new simulator M5 is employed to do experiment for L2 cache research.

L. Cao, J. Zhong, and Y. Feng (Eds.): ADMA 2010, Part II, LNCS 6441, pp. 514–521, 2010.

2 Motivation

Many related works show that more than half of the blocks retained in the cache are never reused before getting eviction and name them *dead blocks* [3] or *zero reuse lines* [4]. To reduce the time of dead blocks occupying cache without contributing to hits, *LRU Insertion Policy* (LIP) places all incoming blocks in the LRU position to substitute traditional LRU policy which inserts all incoming blocks in the MRU position. PIPP suggest compute the insert position rather than insert blocks into LRU position directly [5]. PPIP needs extra monitoring mechanism and its algorithm can't determine the insert position when a *new* core situation occurs.

The commonly used LRU policy which treats all misses and demands uniformly is unable to know whether or not an application benefits from the cache when multiple applications compete for a shared cache. The cache miss rate is a function of cache size for different applications classified into four types: cache-friendly, Cache-fitting, Cache-thrashing and Streaming. So *Dynamic Insertion Policy* (DIP) mechanism dynamically estimates the number of misses incurred by the two competing insertion policies and selects the policy alternatively [4]. To differentiate the behaviors between competing applications, *Thread-Aware Dynamic Insertion Policy* (TADIP) extends DIP to choose sampling sets by means of the amount of running applications. We propose a new structure design for SDM sample selection, so as to alleviate the logic complexity for DIP implementation in practice.

3 Homologous Promotion and Insertion Policy

In this section, we review cache insertion policies and the concept of promotion, and then detail how our HPIP manage a shared cache compared against PPIP.

3.1 Cache Insertion and Promotion Polices

The conventional replacement policies include the victim decision and the insertion.

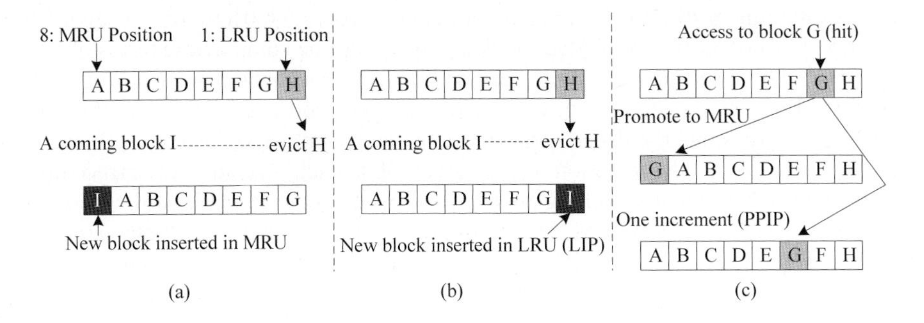

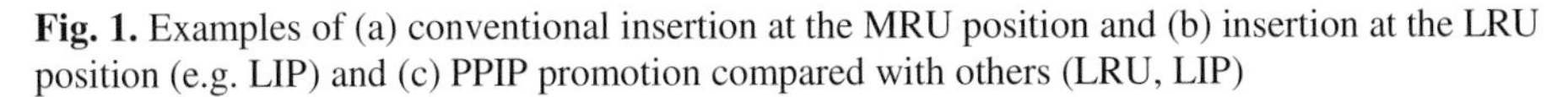

Fig. 1. Examples of (a) conventional insertion at the MRU position and (b) insertion at the LRU position (e.g. LIP) and (c) PPIP promotion compared with others (LRU, LIP)

Figure 1(a) illustrates an example of a cache set comprising eight blocks, logically organized left-to-right from Most Recently Used (MRU) position 8 to Least Recently Used (LRU) position 1. MRU with the highest priority would be kept and LRU with the lowest priority would be evicted. An incoming block will be inserted in position 8 that causes block H to be chosen as the victim in LRU policy. For the reason of dead block, paper [4] first introduces the concept of separating a cache replacement policy into victim selection and insertion independently. As shown in Figure 3(b), LRU Insertion Policy (LIP) which direct places all incoming block in LRU position. The blocks are promoted only if they are reused. LIP reduces so-called dead blocks through minimizing the time of them remain in cache. Bimodal Insertion Policy (BIP) updates LIP and infrequently (throttled by a parameter ε) inserts some incoming blocks into the MRU position. The promotions of theirs all simply move the hit block to the MRU position. Except the victim selection, PPIP insertion decision and promotion are both different with them. Figure 3(c) illustrates PPIPs one-increment-promotion and a traditional promote-to-MRU policy. The priority of a block on hit will be promoted only one-increment position in PPIP that relatively hold some useful blocks in cache longer. HPIP promotion mechanism is similar to PPIP (assuming p_{prom}=1) and its insertion algorithm is distinct.

3.2 Modified Insertion Algorithm Compared with PPIP

PPIP insertion decision requires support of *utility-based cache partitioning* (UCP) policy [1] monitoring mechanism. For a processor with n cores, a given target partitioning $\Pi=\{\pi_1,\pi_2, \ldots,\pi_n\}$such that $\Sigma\pi_i=\omega$ where ω is the total set associativity of the cache. The insertion policy places $core_i$ new incoming block at priority position π_i. For example, a quad-core processor with a 16-way cache $\Pi=\{4,5,6,1\}$ leads to the blocks never being inserted higher than priority 6. But the hypothesis of PPIP is that all cores run simultaneously. In other word, there is a question how to calculate the insertion position when a new core sends the first request to cache. A new core means it has no any block retained in the cache. For instance, before a core recovers from suspending, the other cores already occupy all capacity of the cache. Even at the operating system getting initial start, all cores don't run synchronously. A new core gets partitioning π_i=0 and the total partitioning is $\Pi=\{5,5,6,0\}$ ($core_4$ is new). The insertion position of first requested block by $core_4$ can't be computed in terms of PPIP algorithm. The insertion position is seemed good to choose the lowest priority LRU position. But once a streaming application runs on $core_4$, it will incur thrashing seriously. PPIP similarly discuss that the streaming application's blocks should be inserted into a specific position π_{stream}. Even if a cache-friendly or cache-fitting application rather than a streaming application runs on $core_4$, it's still hard to *steal* the cache occupied by other cores. HPIP simply employs a static parameter $\psi=[\omega/n]$, where ω is the total set associativity and n is the number of cores in a processor. HPIP insertion algorithm is that when a *new* core appears, the first incoming block insertion position is ψ, otherwise it is π_i similar to PPIP. This simple change not only gets resolution for the new core situation but also provides a chance to the new core fairly compete for the cache resource with other cores. The example is same as above that $\Pi=\{5,5,6,0\}$, where ω=16 and n=4. HPIP inserts the first incoming block of core4 into the priority position ψ=4. After that, since a new running application usually has higher cache miss rate, $core_4$ obtains an opportunity to faster occupy the priority position 1 to 4 of the cache.

3.3 Example

Show two examples of an 8-way cache shared between two cores with Π={8,0}, where $core_2$ is new. $Core_1$'s blocks are represented by numerals in squares, and $core_2$ by letters in black circles. The operations in figure 2(a) are determined by PPIP. $Core_2$ sends a request for block A, but π_2=0 means the cache set is entirely occupied by $core_1$. PPIP doesn't consider this situation, so assume to insert block A into LRU position 1 according to LIP [4]. Next access for $core_2$ B leads to miss and insert it in position π_2=1. Then $core_2$ access B block and hit. LRU policy and LIP will promote block B to the MRU position 8. PPIP only promotes one-increment p_{prom}=1. Working for same request sequence, the operations managed by HPIP in figure 2(b) are distinct. The first step inserts block A into position ψ=4 because $core_2$ is new. The 4-step inserts core1's block 7 into position π_1=6 that hits in figure 2(a).

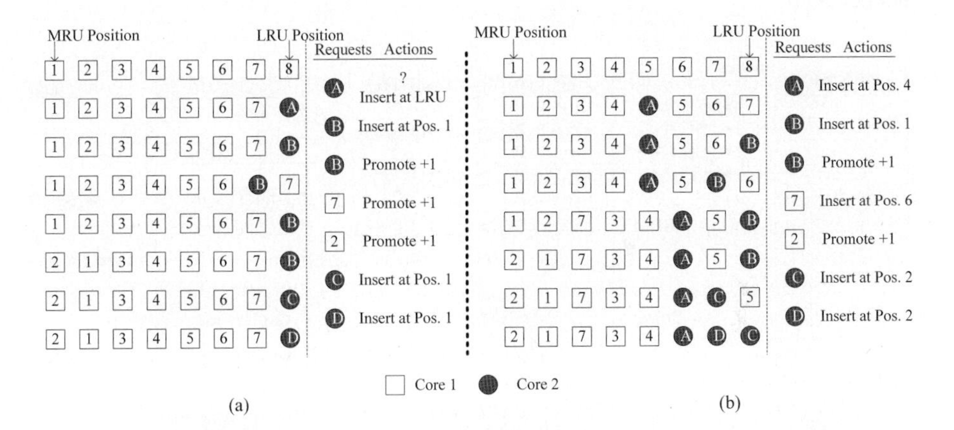

Fig. 2. Examples of (a) operations of PPIP for a variety of cache misses and hits and (b) operations of HPIP for the same request sequence. Always choose the LRU position as eviction.

The comparison in same request sequence illustrates some similarities between PPIP and HPIP. Note that the request sequence shown as in figure 2 is not a special instance. Assume block B become dead after one reused, promote it directly to MRU position maximize the dead block resident time. HPIP and PPIP only promote it by a single position that allows it to be evicted more quickly. HPIP inserts the first request of a *new* core into a static position ψ=4 but PPIP can't. With initial condition that Π={8, 0}, at most of the time the cache allocation matches Π={7, 1} in figure 3(a) but Π={6, 2} in figure 3(b). Providing a relative longer interval to a *new* core to *rob* cache resource, HPIP can keep the cache allocation trade-off in a certain degree.

4 Adaptive Management Policy Election

Recent researches show that dynamically choosing between two insertion policies can improve the cache performance. DIP estimate the misses incurred by the two competing insertion policies and select one incurs the fewest misses [4]. SDM design resorts to Dynamic Set Sampling (DSS) concept [6] that the cache behavior approximates a high probability by sampling few sets. A saturating counter called Policy Selector (PSEL) tracks the misses. The Most Significant Bit (MSB) of PSEL indicates which one of two policies incurs fewer misses and then makes decision. TADIP extends DIP to sense the application diversity [7]. For N applications the TADIP-Isolated use N+1 SDMs that first SDM using policy P_0 (denoted by 0) for all applications is baseline, the others are bimodal sets use policy P_1 (denoted by 1) for one application and policy P_0 for others. In figure 3(a) <0,0,0,0> is baseline and <1,0,0,0>,…,<0,0,0,1> are bimodal for 4 applications. TADIP-Feedback requires 2N SDMs, a pair of SDMs owned by everyone application. In figure 3(b) the first pair SDMs is denoted by binary strings <0,P_1,P_2,P_3> and <1,P_1,P_2,P_3>, where always use P_0 and P_1 for the first application and P_x is the MSB of $PSEL_x$ for others applications independently.

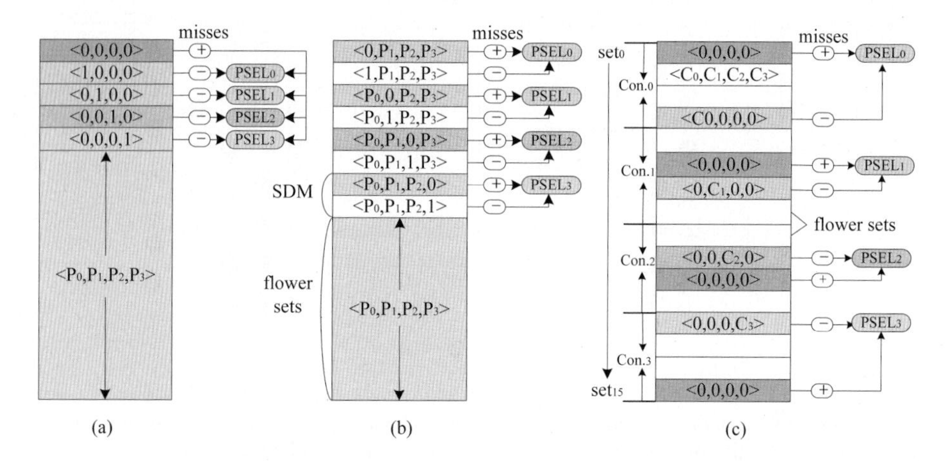

Fig. 3. Examples of (a) TADIP-Isolated and (b) TADIP-Feedback design for 4 applications and (c) Adaptive Policy Election (APE) design for 16 sets cache shared by 4-core

TADIP is only fit for one core in CMP, and the amount of applications varying dynamically makes it hard to be implemented. Since one core situation approximately feedbacks the application characteristic running on it and a processor's core number is fixed, our Adaptive Policy Election (APE) mechanism choose 2N SDMs for one N-core processor. In figure 3(c) 4 pairs of SDMs, one is baseline denoted by<0,0,0,0> and others are bimodal denoted by <C_1,0,0,0>,…, <0,0,0,C_4>, where C_i means using P_1 for $core_i$ and P_0 for others cores. The dedicated sets are selected by means of complement-select approach [4], so sets 0, 5, 10, 15 are permanent used for P_0 and sets 3, 6, 9, 12 for P_1, the others remaining sets are followers.

5 Experimental Methodology

We use a new simulator M5 to execute the combination workloads on the multi-core processors [8]. Table 1 lists the simulation configuration, and the baseline is 2-core/4-core CMPs with two levels cache hierarchy managed by conventional LRU policy. Without loss of generality, we arrange that each application runs on one core.

Table 1. The Experimental Environment and Configuration

Experiments Environment		M5 Simulator Configuration	
Operating System	Ubuntu 9.04 (Linux kernel 2.6.04)	Simulating Processors	Alpha SimpleScalar Out-of-order,1GHz 2/4-Core, Snooping MESI
Processor	Intel® Core(TM)2 Duo P8400 2.26GHz	L1 Cache	Private 2-way 256KB/512KB
Main Memory	DDR 3 2GB two banks	L2 Cache	Shared 8-way 512KB/1MB/2MB

We use combination workloads including the 10 SPEC2006 benchmarks. The workloads are classified into two groups listed in table 2.

Table 2. The Combination Workloads

Dual-Core Workloads		Quad-Core Workloads	
Mix2-01	specrand+mcf	Mix4-01	specrand+mcf+hmmer+star
Mix2-01	specrand+mcf	Mix4-01	specrand+mcf+hmmer+star
Mix2-02	hmmer+astar	Mix4-02	gcc+soplex+ specrand+mcf
Mix2-03	gcc.+soplex	Mix4-03	hmmer+astar+gcc+specrand
Mix2-04	gcc+astar	Mix4-04	milc+hmmer+ mcf+gcc
Mix 2-05	milc+hmmer	Mix4-05	bzip2+specrand+ milc+hmmer
Mix 2-06	lbm+ milc	Mix4-06	bzip2+specrand+ lbquantum+lbm
Mix 2-07	lbquantum+lbm	Mix4-07	specrand+astar+ mcf+gcc
Mix 2-08	specrand+astar	Mix4-08	lbquantum+lbm+ mcf+hmmer
Mix 2-09	hmmer+specrand	Mix4-09	hmmer+specrand+ mcf+gcc
Mix 2-10	mcf+hmmer	Mix4-10	soplex+bzip2+ specrand+astar
Mix 2-11	mcf+gcc	Mix4-11	mcf+gcc+ milc+astar
Mix 2-12	soplex+bzip2	Mix4-12	mcf+gcc+ soplex+bzip2
Mix 2-13	soplex+mcf	Mix4-13	lbm+ milc+ lbquantum+lbm
Mix 2-14	bzip2+specrand	Mix4-14	bzip2+specrand+ gcc+astar

Figure 4 and figure 5 show the results measured by total IPC throughput relative to LRU. The results of HPIP are very close to PPIP since its promotion and insertion mechanism are all similar to HPIP except when the *new* core situation incurs. Both of them provide higher raw throughput than LIP frequently. When the combination applications Mix2-06 and Mix2-07 run on the dual-core, the LIP total throughput is higher than HPIP and PPIP. The reason is that the combination benchmarks in Mix2-06 and Mix2-07 are typical streaming. In general, the raw throughput of PPIP and HPIP outperform LIP (dual-core: 17.2% and quad-core: 15.6%).

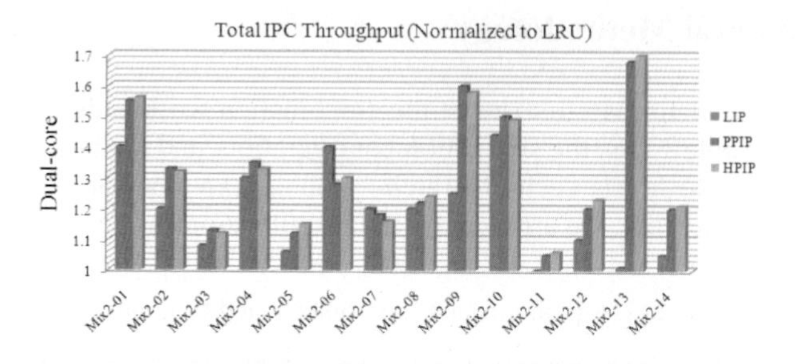

Fig. 4. Performance results for the IPC throughput (Dual-Core in 1MB L2 Cache)

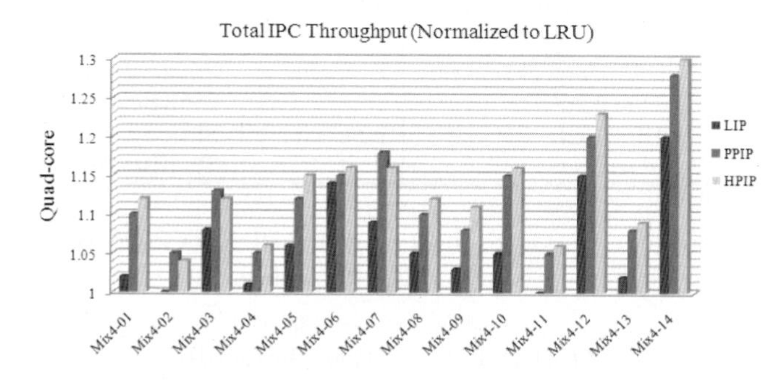

Fig. 5. Performance results for the IPC throughput (Quad-Core in 2MB L2 Cache)

Figure 6 illustrates HPIP MPKI of four combination workloads. Similar to others policies it drop remarkably when working cache size is bigger than a certain value, so HPIP keep the allocation trade-off depending on total cache size in a certain degree.

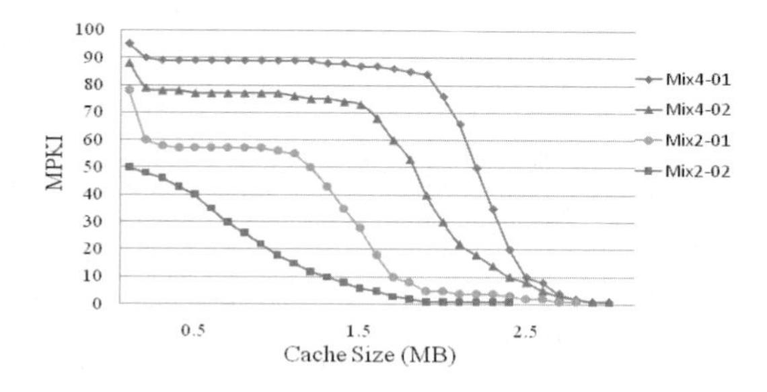

Fig. 6. MPKI vs. Cache Size in MB

6 Conclusion

Being different to many related works such as DSR and SNUG aim to reduce shared cache access latency [10, 11] and CCP try to make private cache to be shared [12], HPIP focus on cache management policy design. Compared with others policies, it approves high performance and benefits for cache trade-off. The *new* core situation in shared cache is first to be discussed by us. Furthermore, APE design makes DIP mechanism closer to implementation. Deeper exploring interaction between insertion and promotion and more experiments would be our future work.

Acknowledgments

We appreciate the constructive comments from the reviewers. This work is supported by the Key Projects in the National Science & Technology Pillar Program (Grant No. 2008BAH37B04) and the Third Stage Building of "211 Project" (Grant No. S-10218).

References

1. Qureshi, M.K., Patt, Y.N.: Utility-Based Cache Partitioning: A Low Overhead High-Performance Runtime Mechanism to Partition Shared Caches. In: IEEE 39th International Symposium on Microarchitecture, pp. 423–432. IEEE Press, Orlando (2006)
2. Suh, G.E., Rudolph, L., Devadas, S.: Dynamic Partitioning of Shared Cache Memory. Journal of Supercomputing 28(1), 7–26 (2004)
3. Haiming, L., Michael, F., Jaehyuk, H., Doug, B.: Cache Bursts: A New Approach for Eliminating Dead Blocks and Increasing Cache Efficiency. In: IEEE 41th International Symposium on Microarchitecture, pp. 222–233. IEEE Press, Washington (2008)
4. Qureshi, M.K., Jaleel, A., Patt, Y.N., Steely, J., Emer, J.: Adaptive Insertion Policies for High-Performance Caching. In: 34th IEEE/ACM International Symposium on Computer Architecture, pp. 381–391. ACM, San Diego (2007)
5. Yuejian, X., Gabriel, H.L.: PIPP: Promotion/Insertion Pseudo-Partitioning of Multi-Core Shared Caches. In: 36th IEEE/ACM International Symposium on Computer Architecture, pp. 174–183. ACM, Austin (2009)
6. Qureshi, M.K., Lynch, D.N., Mutlu, O., Patt, Y.N.: A Case for MLP-Aware Cache Replacement. In: 33th IEEE/ACM International Symposium on Computer Architecture, pp. 167–178. ACM, New York (2006)
7. Aamer, J., William, H., Qureshi, M.K., Sebot, J., Simon, S.J., Emer, J.: Adaptive Insertion Policies for Managing Shared Caches. In: 7th International Conference on Parallel Architecture and Compilation Techniques, Toronto, pp. 208–219 (2008)
8. The M5 Simulator System Information, `http://www.m5sim.org/wiki/index.php/Main_Page`
9. Qureshi, M.K.: Adaptive Spill-Receive for Robust High-Performance Caching in CMPs. In: 15th International Symposium on High-Performance Computer Architecture, pp. 45–54 (2009)
10. Dongyuan, Z., Hong, J., Seth, S.C.: Exploiting Set-Level Non-Uniformity of Capacity Demand to Enhance CMP Cooperative Caching. In: IEEE International Parallel & Distributed Processing, pp. 1–10 (2010)
11. Jichuan, C., Gurindar, S.S.: Cooperative Cache Partitioning for Chip Multiprocessors. In: 21th Annual International Conference on Supercomputing, pp. 242–252 (2007)

Multi-core Architecture Cache Performance Analysis and Optimization Based on Distributed Method

Kefei Cheng[1], Kewen Pan[1], Jun Feng[1], and Yong Bai[2]

[1] College of Computer Science, Chongqing University of Posts and Telecommunications, Chongqing, China, 400065

[2] Chongqing GuoHong Technology Development Co., LTD, Chongqing, China, 400065

chengkf@cqupt.edu.cn, byong@cqghong.com

Abstract. With the rapid development of computing performance on multi-core era, the capacity of shared cache has been increasing. System architects need make maximum usage of shared resources to improve system performance. This paper mainly rebuilt free lists based on page coloring for achieving their privatization by a distributed method, which could really achieve page-level parallelism at the operating system level and decrease cache thrashing among applications. Experimental results show that if the paper uses matrix computing as working load, L2 Cache Misses Rate is reduced by about 12%, IPC increased by 10%.

Keywords: Shared Cache, Page Coloring, Page-level Parallelism, Cache Thrashing.

1 Introduction

According to Moore's Law, processor clock speed increases twice every year, while memory speed only upgrades double every six years. Such difference is main bottleneck to realize HPC (High Performance Computation). Caching technology could increase the speed of accessing data and reduce latency on account of time and spatial locality principle. In modern CMP (Chip Multi-core Processor) system, Cache generally apply hierarchy framework that each processor core has independent L1 Cache, shared L2 Cache. On-chip shared Cache has a better advantage of providing a larger storage space. All processor cores can directly access data of shared cache instead of main memory. So cache resource can be fully exploited. But shared cache on CMP architecture also brings about other problems, such as data contention, cache data consistency, error sharing. These problems have a side-effect on cache performance. So the effective data sharing and optimized resource allocation on shared cache play an important role to enhance cache performance.

2 Related Work

Traditionally, the research on shared cache had focused on hardware methods to improve performance. It is seldom on the point of software view, in particular operating system level. Due to chip-hardware architecture complexity, it could further add to the complexity for using hardware method to improve performance. In addition, another disadvantage is that the cost of cache hardware tuning is much larger. By contrast, soft-

L. Cao, J. Zhong, and Y. Feng (Eds.): ADMA 2010, Part II, LNCS 6441, pp. 522–528, 2010.

ware method could be more flexible. These are also similar study on the field in abroad. Tam and Amizi [2] in University of Toronto proposed that they managed shared L2 cache on so-level in page coloring, which could decrease cache contention among applications. But it has a shortage that each processor core has global multiple free lists. Memory management need utilize the mechanism of synchronous communication and locking to maintain page data consistency, avoiding free list contention among cores. However, these mechanisms have a greater cost, particularly frequent page allocation and release in multi-core environment.

Some of UNIX currently uses page coloring to improve cache utilization, typically such as Solaris and AIX [3]. The mainstream Linux is class UNIX. It applies the "Buddy" algorithm to manage page allocation and release [4]. But the buddy algorithm could not make full use of free pages as a result of powers of 2 in size on page allocation and release. The coarse-grained page management will result in a serious waste of pages and the lower utilization of shared L2 Cache in multi-core architecture. This paper utilizes distribution to rebuild free lists for realizing their privatization instead of globalization in Linux, which decreases page contention and communication cost.

3 Memory Management Optimization

3.1 Page Coloring Principle

Page coloring is a classical page allocation method. It is the basis of achieving fine-grained static L2 Cache division. In modern operating system, operating system accesses L2 Cache and physical memory by physical address. In limited cache associative, there should be overlapped on bit field between L2 Cache set number and physical page number [5]. Taking the following Fig.1 for example, physical page is the size of 4KB. And there is at least 12 bits to represent page offset. The remaining bits are physical page number. L2 Cache has the size of 512KB, 16-way associative, cache line 64B. So the lower 3 bits of physical page number is overlapped with the higher 3 bits of cache set number. The overlapped parts are called "page color". Figure 1 shows the mapping relationship between physical address and cache line.

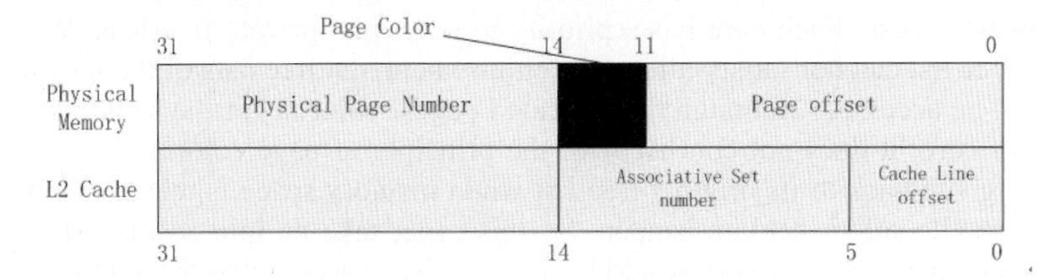

Fig. 1. Mapping Relationship between Physical Address and Cache Line

Generally OS decides how virtual page maps to physical page. The mapping of physical page into cache is fixed by hardware itself, which is actually an important requirement for the page coloring technique. OS can use its control of virtual to physical page mapping to indirectly control physical page to cache line mapping. Different color pages could be mapped into different cache sets. The mapping between Page and Cache Line is shown in Figure 2 [2]:

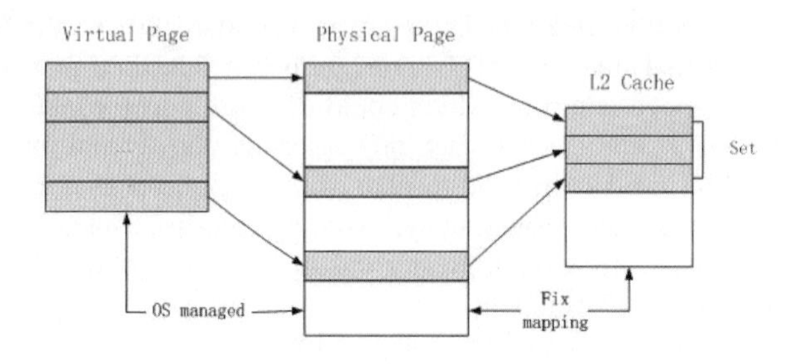

Fig. 2. Mapping Relations between Page and L2 Cache Line

Physical memory pages will be divided into N free lists of different colors by page coloring when operating system begins into initialization. Each core has corresponded with N color lists. Besides, operating system sets up color vector for core, which will be identified with bitmap. If i bit of bitmap is 1, which the core could use the free pages with color i [5]. The allocated pages should be located into different partitions of L2 Cache due to different color vector.

3.2 A Distributed Method for Rebuilding Free Lists

When core requests free page, firstly system call must fulfill interaction with memory management. Then system accesses free list which has the same color with core's color vector. Due to highly degree parallelism in multi-core architecture, the operation of page allocation and release should frequently happen. Synchronous communication and locking mechanism will lead to a larger overhead. So the global idea of multiple free lists turns into performance bottleneck. In order to enhance page-level parallelism and decrease communication cost among cores, this paper propose a distributed approach which is used to rebuild free lists to make free lists private. The specific ideas are shown as follows:

Privatization of Free Lists: The whole free lists are divided into N parts. N is the number of the processor core. Since each core has small private free lists, the cost of synchronous communication will be reduced and cache thrashing will be eliminated.

Priority Access: Each core has a priority to access its private free lists. When the private free list can not satisfy allocation requirement, the free lists of the neighboring core will be accessed. Although the allocated pages' color is not the same with color vector of core, it does not conflict with the principle of page coloring. These pages must be given back to its original free list when memory space is release. Generally, such page allocations seldom happen. So this paper take no into account it and just makes sure that processor core would have a priority to access its private free list.

According to the above description, the basic principle is that the private free list has a higher priority to page operator. This paper respectively defines page operation as follows:

1) Page Allocation
   Page allocation complies with the maximizing principle of private free lists. Firstly memory management searches the free list whose color is consistent with core's color vector in its private free lists. And then pages in the free list would

be allocated out. If its private free lists can not satisfy page allocation requirement, memory management will allocate free pages for requirement from adjacent core instead of local one. The flow diagram of page allocation is shown in Figure 3:

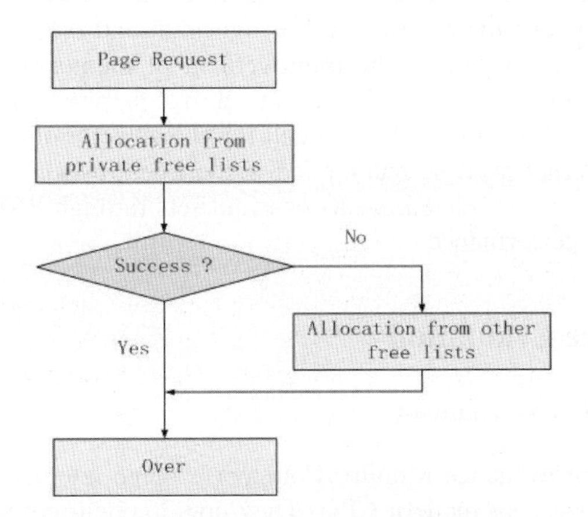

Fig. 3. Page Allocation Operation

2) Page Release

Memory management checks out whether the released pages belong to its private free list. If so, the pages will be given back to its private free lists of local core. Otherwise, they are back to the free lists of the adjacent core. The flow diagram of page release is shown in Figure 4:

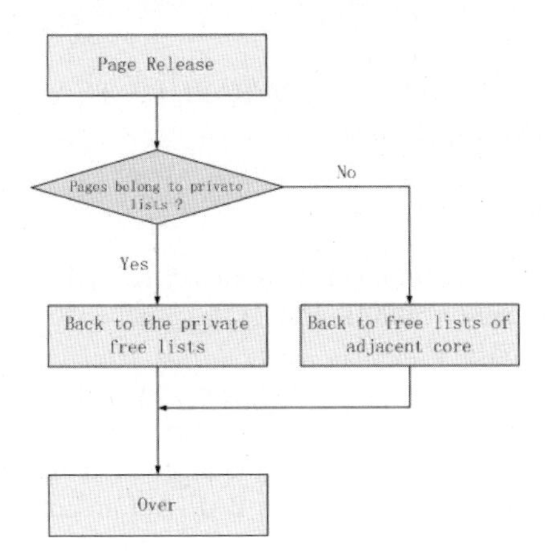

Fig. 4. Page Release Operation

Page allocation and release are executed on their private free list. OS can achieve page-level parallelism and have an optimization on L2 Cache in multi-core architecture in distributed method.

Memory management apply "buddy" algorithm to allocate and release successive pages in Linux. But the mechanism of page coloring only allocates or releases a single page. This paper retains the original buddy algorithm and reserves parts of memory for allocating successive pages. The improvement of memory management mainly rebuilds free lists to make them private by the above distributed method. In addition, memory management should add system call for color vector of processor core. After modifying code, kernel need be recompiled. Experiment uses matrix computation as working load and collects related events or event sets through PAPI-C mechanism to evaluate L2 Cache performance.

4 Performance Evaluation

4.1 Data Collection Mechanism

HPM (Hardware Performance Monitor Counter) is some internal registers which record CPU performance in modern CPU. They could concurrently read a number of hardware performance criterions, such as Cache Misses, Instruction Cycles [6] .This paper exploits PAPI-C technology to collect data for evaluating cache performance in multi-core architecture. PAPI (Performance Application Interface) which is developed by ICL laboratory of Tennessee University is application interface based on Perfctr. PAPI-C can collect the related event or event sets from HPM through the interface in user level. The experiment needs to collect some events, which are shown as follows:

PAPI_L2_TCM: L2 Cache Misses;
PAPI_L2_TCA: L2 Cache Accesses;
PAPI_TOT_INS: Total Instructions;
PAPI_TOT_CYC: Total Cycles;
L2 Cache Misses Rate=PAPI_L2_TCM/PAPI_L2_TCA;
IPC=PAPI_TOT_INS/PAPI_TOT_CYC;

4.2 Data Experiment

Software and Hardware Environment: Fedora 10, Kernel Version 2.6.32, AMD Athlon(tm) Dual Core Processor 5000B, 1.8GH. L2 Cache is shared, 512KB, 16-way set associative, cache line 64B.

Testing Method: Experiment applies matrix multiplication as working load. The reason why experiment chooses it is that it runs shorter and has a higher reliability on cache data. Due to using the PAPI-C mechanism to collect the related events, experiment need insert the code of matrix multiplication into PAPI before collecting events. Experiment executes different working loads of matrix computation under the condition of different memory management algorithms. The working loads are matrixes of 5*5, 10*10, 15*15. The element of matrix is 2. Each experiment respectively collects the following events: PAPI_L2_TCM、PAPI_L2_TCA、PAPI_TOT_INS、PAPI_TOT _CYC. Matrix multiplication code is shown as follows:

```
Void matrix (int a [I] [M], int b [] [N])
      {
        int i, j, k;
         for (i=0; i<I; i++)
              for (j=0; j<M; j++) {
                   for (k=0; k<N; k++)
                     c [i][j]=a [i][k]*b [k][j];
                   }
              }
```

In order to analyze the effect of various memory management changes we need to derive a metric for comparison. The obvious and simplest metric for evaluating cache performance is average IPC and L2 Cache Misses Rate. The experiment data are gained through many collections and then calculate their average value as experiment result. Experimental results are shown as follows:

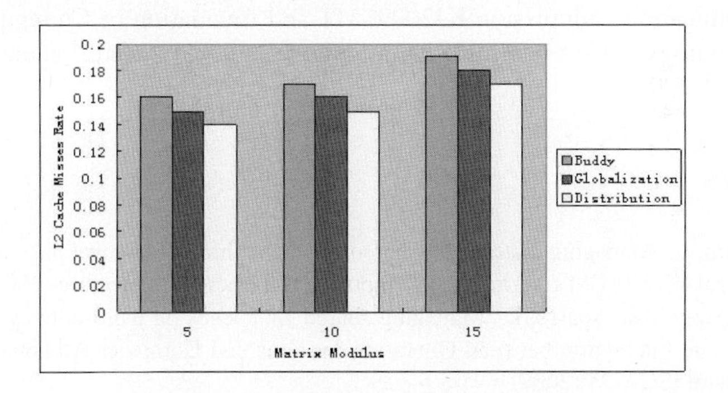

Fig. 5. Comparison of L2 Cache Misses Rate among Different Algorithms

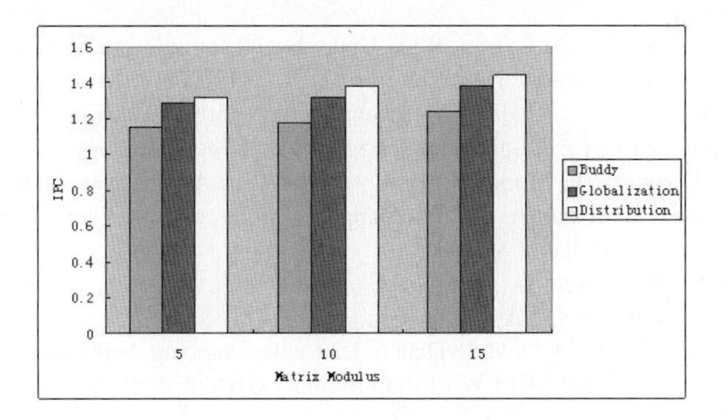

Fig. 6. Comparison of IPC among Different Algorithms

According to the analysis for raw data, there is reduced by about 12% on L2 Cache Misses Rate, increased by 10% on IPC.

5 Conclusion

This paper mainly rebuilds the structure of free lists to achieve their privatization based on the principle of page coloring by a distributed method, which could achieve page-level parallelism and reduce cache thrashing among applications. With the advent of multi-core systems, the optimization based on software method on shared cache has been greatly paid attention in recent years. It ultimately leads to unfair utilization on CPU if memory management has an unfair use for shared cache, which has a side- effect on system performance. In future work we will study on how to achieve a fair utilization on shared cache. Moreover, memory management based on page coloring has also a good market prospect for the development of smart phone system in the embedded field.

Acknowledgments. This paper is sponsored by science foundation of Chongqing municipal education commission KJ20090511 and foundation of Chongqing Science and Technology Commission-Smart phone R&D and industrialization, CSTC2010AB2003.

References

1. Cho, S., Jin, L.: Managing distributed, shared L2 cache through os-level page allocation. In: 37th Annual IEEE/ACM International Symposium on Micro Architecture (2006)
2. Tam, D., Azimi, R., Soares, L.: Managing shared L2 caches on multi-core systems in software. In: The Interaction between Operating Systems and Computer Architecture, in Conjunction with ISCA-34 (2007)
3. Jin, L., Lee, H.: A Flexible Data to L2 Cache Mapping Approach for Future Multi-core Processors. In: Proc. ACM Workshop Memory Systems Performance and Correctness (October 2006)
4. Defoe, D.C., Cholleti, S.R., Cytron, R.K.: Upper bound for defragmenting buddy heaps. In: Assoc. Computing Machinery, New York, pp. 222–229 (2005)
5. XinXin, J., Haogang, C., Xiaolin, W., Zhenglin, W.: Virtual Machine Design and Implementation of the cache partition. Computer Science and Technology 4(1), 36–45 (2010)
6. London, K., Dongarra, J., Moore, S.: End-user Tools for Applications Performance Analysis, Using Hardware Counter. In: International Conference on Parallel and Distributed Computing Systems, Dallas, TX (2001)
7. Qureshi, M.K., Thompson, D., Steely, S.C., Emer, J.: A daptive insertion policy for high performance caching. In: ISCA-34 (2007)
8. Jin, L., Lee, H., Cho, S.: A Flexible Data to L2 Cache Mapping Approach for Future Multi-core Processors. In: Proc. ACM Workshop Memory Systems Performance and Correctness (October 2006)

The Research on the User Experience of E-Commercial Website Based on User Subdivision

Wei Liu[1,2], Lijuan Lv[1], Daoli Huang[2], and Yan Zhang[1]

[1] School of Automation,
Beijing University of Posts and Telecommunications, Beijing 100876, China
[2] Key Lab of Information Network Security of Ministry of Public Security
twhlw@163.com, lijuan_lv_0120@163.com, huangdaoli@stars.org.cn,
yydelp.2010@yahoo.com.cn

Abstract. Aiming at the problem that it is difficult to accurately define the quality of the user experience, based on the usability testing we analyzed the elements of user experience during the use process of E-commercial website and divided the users into two types (planned users and impulsive users). Furthermore, we constructed the user evaluation system from the three aspects of behavior, cognition and emotion, quantified the quantitative indicators, and established a user experience comprehensive evaluation model. Finally, we verified the validity of the model through some cases.

Keywords: E-commerce, user experience model, usability testing, AHP, fuzzy comprehensive evaluation.

1 Introduction

User experience is the users' feelings when they visit the website. By constructing user experience evaluation system based on the user behavior, cognition and emotion, this study will use fuzzy comprehensive evaluation method to analyze the user experience of C2C E-commercial websites, overcome the ambiguity and uncertainty of user experience, and thus obtain a quantitative comprehensive evaluation.

2 The Establishment of User Experience Quality Index System

2.1 The Description of Index System

Firstly, after the data collection of the shopping websites' user experience, from the traditional evaluation view, we usually construct the user experience evaluation system from the three aspects (user behavior indicators, user emotion indicators and user cognition indicators).

2.2 The Determination of the Indicator Factors' Weigh

The determination of the indicators factors' weight a_i is one of the most critical parts of fuzzy evaluation. Whether fuzzy subset A is appropriate or not directly affects the comprehensive evaluation results.

L. Cao, J. Zhong, and Y. Feng (Eds.): ADMA 2010, Part II, LNCS 6441, pp. 529–536, 2010.

2.2.1 The Structure of Judgment Matrix

Supposing that there is n evaluation factors' weigh needed to be determined, the matrix can be expressed as:

$$A = \begin{bmatrix} a_{11} & a_{12} & \cdots & a_{1n} \\ a_{21} & a_{22} & \cdots & a_{2n} \\ \cdots & \cdots & & \cdots \\ a_{n1} & a_{n2} & \cdots & a_{nn} \end{bmatrix} \quad (1)$$

In this equation, a_{ij} shows the importance degree of A_i to A_j, which is following the principles: (1) $a_{ij} = W_i/W_j > 1$ (a_i is more important than a_j); (2) $a_{ij} = 1$ (i=j); (3) $a_{ji} = 1/a_{ij}$ (aij ≠0). Saaty and some other people's experiment proved that it was more appropriate to use 1-9 ratio scale method [6].

2.2.2 The Calculation of Index Weighs

The square root method is used to calculate the weight value of each index. Matrix calculation process: (1) to obtain the product of all elements in each row $P_i = \prod_{j=1}^{n} a_{ij}$ (I = 1, 2, ..., n); (2) to calculate the n-th root of Pi $\overline{W}_i = \sqrt[n]{P_i}$; (3) to obtain the weights $W_i = \frac{\overline{W}_i}{\sum_{j=1}^{n} \overline{W}_j}$ by the vector $(\overline{W}_1, \overline{W}_2, \cdots, \overline{W}_n)^T$ normalization; (4) W_i is the weight coefficient of each index. Consistency test is used as the matrix consistency deviation index CI =. $\frac{\lambda_{max} - n}{n-1}$. In the formula: λmax is the largest characteristic root to determine the matrix. $\lambda_{max} = \sum_{i=1}^{n} \frac{(A\overline{w})_i}{n w_i}$, and $(A\overline{w})_i$ represents the i-th element of the vector. Random consistency ratio CR, CR = CI / RI, in the formula: RI (Random Index) is the average random consistency index. $RI = (\tilde{\lambda}_{max} - n)/(n-1)$, among this equation, $\tilde{\lambda}_{max}$ is the average of m n-order positive and negative matrixes' largest characteristic root. If CR<0.1, the matrix has satisfactory consistency. If else, the matrix should be adjusted.

After the comparison of these four evaluation index by six experts, we use Delphi method to create the final matrix till the consistency test results can meet the standard.

Table 1. The judgment matrix and weight of user behaviour data evaluation index

	Validity	Effectiveness	Weight
Validity	1	2	0.667
Effectiveness	1/2	1	0.333
$\lambda_{max} = 2$, $CI = 0$ $CR = 0 < 0.1$			

3 The Fuzzy Comprehensive Evaluation of User Experience Quality

3.1 The Establishment of Evaluation Factor Set U

Firstly, we should establish the corresponding evaluation factor set, U= $(u_1, u_2, \ldots, u_n)$, and n is the number of influencing factors.

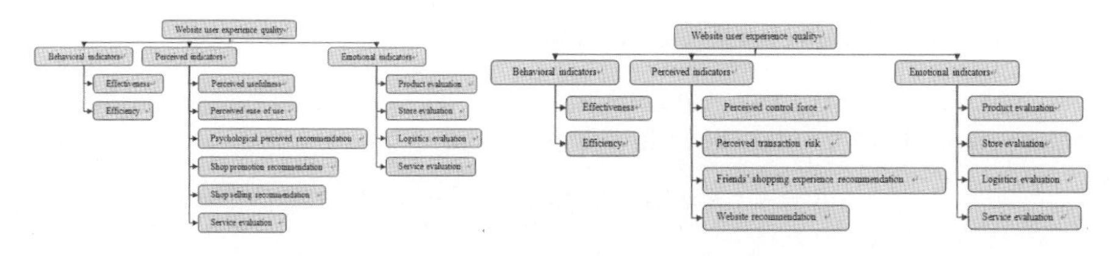

Fig. 1. User experience evaluation index (planned users and impulsive users) structure scheme

3.2 The Establishment of Comment Set V and the Membership Function of Factors on V

After building user experience quality index system, we should establish comment rate set $V=\{v_1, v_2, \ldots, v_n\}$ and the membership function of evaluation factors on V. First of all, the evaluation criteria of index is divided into five levels (very poor, poor, ordinary, good, excellent); then the threshold of indicators is determined by experts (v_1, v_2, v_3, v_4, v_5, v_6), finally, according to the membership, we need to compare the metric value with threshold to obtain the quality level of all data. Assuming that Q is the fuzzy set on V, the membership functions constructed are:

$$Q_1(v)=\begin{cases}1 & v_1 \le v \le v_2\\ \dfrac{c_2-v}{c_2-v_2} & v_2<v<c_2\\ 0 & \text{else}\end{cases}\qquad Q_i(v)=\begin{cases}1 & v_i \le v \le v_{i+1}\\ \dfrac{v-c_{i-1}}{v_i-c_{i-1}} & c_{i-1}<v<v_i\\ \dfrac{v-c_{i+1}}{v_{i+1}-c_{i+1}} & v_{i+1}<v<c_{i+1}\\ 0 & \text{else}\end{cases}\qquad Q_5(v)=\begin{cases}1 & v_5 \le v \le v_6\\ \dfrac{v-c_4}{v_5-c_4} & c_4<v<v_5\\ 0 & \text{else}\end{cases}$$

In the formula, c_1, c_2, c_3, c_4 and c_5 are the center values of interval (v_1, v_2), (v_2, v_3), (v_3, v_4), (v_4, v_5), and (v_5, v_6). After determining the threshold, the five-level value of the specific indicator v can be obtained by the membership function. This article uses 0-1 to express the task completion extent. Perceived indicator score is collected by 7-point questionnaire. Emotional indicator data is gained by the scores of shopping websites' evaluation system. Task completion time refers to the time from the beginning to the end of the task. Due to the different task time supported by different systems, in order to determine the threshold, we converted the task time by equation 2.

$$X'=1-\frac{X-T}{T}=2-\frac{X}{T} \tag{2}$$

Where: x is the original value; T is the shortest task completion time done by the product experts and usability engineers before usability testing (the average completion time of the subjects); x' is the conversion value.

Following the opinion of Delphi method's experts, the threshold is determined. Through average prediction we can arrive at the results: to task completion correct rate and time ($0 \le x' \le 1$), the threshold values are: 0, 0.4, 0.6, 0.8, 0.92, and 1; to perceived index questionnaire score, the threshold values are: 1, 2, 4, 6.5, 8.5, 9; to emotional evaluation score after shopping, the threshold are: 1, 1.5, 2.5, 4, 4.5,5.

The establishment of membership matrix r_{ij} is the possibility that several subjects make a Vj evaluation to the Ui aspect of a particular object. The membership vector is Ri= (ri1, ri2, …, rim), i=1,…,n, and $\sum_{j=1}^{m} r_{ij}=1$. The membership matrix is R= (R1, R2, … ,Rn) T=(rij).

3.3 The Fuzzy Comprehensive Evaluation

If weight vector W and the evaluation matrix R is known, we can obtain the total evaluation results B through the following equation.

$$B = W * R = \begin{pmatrix} b_1 & b_2 & b_3 & \cdots & b_m \end{pmatrix} \tag{3}$$

In the equation, the composite operator * has a variety of algorithms. Because the weighted average model does not only retain the evaluation information of single factor, but also take into account the effect of all factors, it is used as the evaluation model[8], that is,

$$b_j = \sum_{i=1}^{n} (a_i \cdot r_{ij}) = \min\left\{1, \sum_{i=1}^{n} a_i r_{ij}\right\} \quad (j=1,\ldots,m) \tag{4}$$

When we evaluate multi-layer index, the same layer's comment set B_{ij} (i is the number of the layer, and j indicates that every layer has j sub-goals) can constitute a new fuzzy matrix R_i, and B is the total comprehensive evaluation results.

4 The Application and Results Analysis

This paper takes a C2C E-commercial website's user experience as an example to illustrate the method's application in practice.

4.1 Test Evaluation

Test methods: usability testing is a number of ways to improve the ease of use in the design process.

Test purpose: (a)To access the operation indicators of task completion during the specific operation scenarios (task completion time, task completion correct rate) and user preference data by interviews after finishing the tasks (perceived preference, emotional preference). (b)To verify the realization and effectiveness of the shopping website's user experience fuzzy comprehensive evaluation model by the processing statistics of user behavior and preference data. (c)To observe and conclude the differences of two sub-users during operation.

Testing task: By the task analysis conducted at the early design stage, we take one typical task as test task: to buy a 10-yuan mobile phone recharge card at the C2C E-commercial website and charge for a specified number. Task flow: (a)to login to the E-commercial website; (b)to find out the phone recharge zone; (c)to select on seller; (d)to use his account to pay for the card; (e)to conform the payment and make the evaluation of the transaction.

4.2 Data Collection

User behavior data can be obtained in usability testing, in which the correct rate are gained by the observers' scores on the user task completion situation and the task completion time are given by the time between the beginning and the end of the tasks.

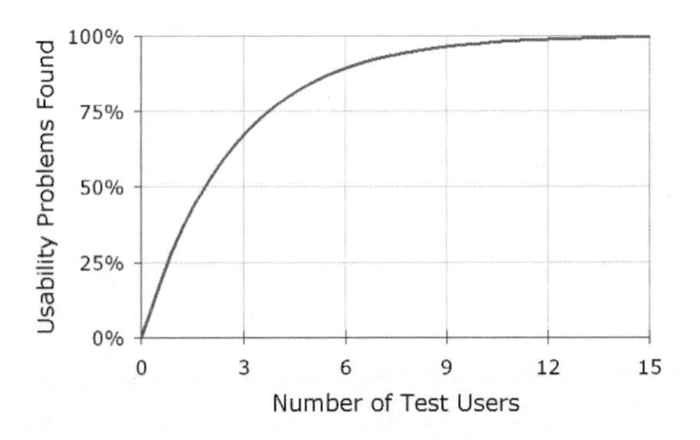

Fig. 2. The relation graph of test user number and usability problems

User perceived data is mainly based on the "user perceived characteristics investigation of C2C E-commercial websites". The observation on the user behaviors and the interview after finishing the tasks can help users to get these data a lot.

User emotion data mainly depends on the evaluation of users after they finish online shopping.

4.3 Results Analysis

4.3.1 The Evaluation of Planned Users on C2C Shopping Website User Experience

According to the index system in figure 1, the data is sorted. Take the data validity analysis of planned users' behavior data as an example. The correct rate is the task average and the task completion time is the total time to finish the task. Furthermore, we need to calculate the corresponding level of every user's every index value according to the membership function and to obtain the membership extent r_{ij}. Next, we establish the fuzzy matrix to do the evaluation. Combined with the data processing method illustrated in table 3, we use the minimum time of task completion of every subject as the original index to implement the normalization following equation (2). Finally we take the task completion time threshold into membership function to obtain the membership r_{pj} ($p = 1, 2, \ldots, 12; j = 1, 2, 3, 4, 5$), and then obtain p_j according to equation (5):

$$p_j = \sum_{m=1}^{12} r_{mj} \quad (j=1,\ 2,\ 3,\ 4,\ 5) \tag{5}$$

After normalization, the membership extent r_j whose correct rate is v_j can be obtained:

$$r_j = \frac{p_j}{\sum_{j=1}^{5} p_j} \tag{6}$$

Similarly, we can access to membership degree r_{ij}. The first level fuzzy evaluation matrix can be determined by figure 1, and then we can carry on the fuzzy evaluation

Table 2. The processing and analysis of validity data

user	Original time	normalization	Very poor	Poor	General	Good	Excellent
U1	9.8	1.292	0	0	0	0	1
U2	11.6	1.153	0	0	0	0	1
U3	12.7	1.073	0	0	0	0	1
U4	14.5	0.942	0	0	0	0.45	1
U5	16.4	0.803	0	0	0.95	1	0
U6	17.5	0.723	0	0	1	0.23	0
Total p_j			0	0	1.950	1.680	4
Normalization p_j			0	0	0.256	0.220	0.524

according to equation (3) and (4): B_1 is (0 0 0.178 0.402 0.421); B_2 is (0 0.083 0.407 0.369 0.142); B_3 is (0 0.009 0.319 0.256 0.405). Take the comprehensive evaluation results of first-level index as the second-level evaluation fuzzy matrix, and thus we obtain product comprehensive user experience quality through re-calculation.

$$R = \begin{bmatrix} 0 & 0 & 0.178 & 0.402 & 0.421 \\ 0 & 0.083 & 0.407 & 0.369 & 0.142 \\ 0 & 0.009 & 0.319 & 0.256 & 0.405 \end{bmatrix} \tag{7}$$

B= (0 0.036 0.309 0.350 0.302)

The comment set can be made assessment as following: excellent (95), good (82), general (67), poor (50), very poor (31). After the conversion of B, we can obtain more intuitive centesimal evaluation results, and we can access to the evaluation results of the E-commercial websites with the application of equation.

$$S = \frac{\sum_{j=1}^{m} b_j^{\,2} s_j}{\sum_{j=1}^{m} b_j^{\,2}} \tag{8}$$

In this equation, b_j is the component of the corresponding evaluation results and s_j matches along with the grade value B_j.

Among the type, s_1=31, s_2=50, s_3=67, s_4=82, s_5=95. That is: S=81.1, so we can see that the ultimate evaluation results on the C2C E-commercial websites' user experience quality of planned users are good.

4.3.2 The Evaluation of Impulsive Users on C2C Shopping Website User Experience

Similarly, the data is sorted according to the index system of figure 1. The method of processing the impulsive users' data is same as that of planned users.

$$R' = \begin{bmatrix} 0 & 0 & 0.159 & 0.386 & 0.455 \\ 0.104 & 0.315 & 0.363 & 0.178 & 0.041 \\ 0 & 0.049 & 0.445 & 0.340 & 0.166 \end{bmatrix} \quad (9)$$

B'= (0.043 0.143 0.318 0.288 0.209)

Finally, the evaluation results on E-commercial websites of impulsive users are obtained with the help of equation (9), which are: S'= 75.22.

4.3.3 The Two Types of Users' Comprehensive Evaluation Index Comparison Analysis

User behavior data and user perceived data and user emotion data collected by experiment can be taken into the shopping website fuzzy comprehensive evaluation system, and then by calculation we can see the evaluation on the shopping website of planned users and impulsive users. We can draw the conclusion that the evaluation on the C2C shopping website of planned users is higher than that of impulsive users, so this shopping website can make further improvement to achieve a better shopping experience for impulsive users.

5 Conclusions

The evaluation of user experience on C2C E-commercial websites does not only need to consider the website usability, but also demand to take into account the websites' friendliness and emotional experience. The user experience model established in this article is based on the two typical types of users. Combined with the decision-making cognition, shopping behavior and the emotional evaluation after shopping during the three stages of shopping and considered the different online shopping user experience of different users, the model is more complete and reliable. The perceived index is derived by the empirical regression equation of cognition-recommendation factor model, which makes the index more representative and credible. To behavior indicators and emotional evaluation indicators that have personal attributes, the data is easy to obtain, which makes the model easier to implement. Due to the unclear characteristics of the user experience quality, the application of fuzzy evaluation seems more effective, because it can not only make the comprehensive quantitative evaluation on user experience, but also make up the possible bias on the evaluation brought by the usability and user experience fuzzy attribute.

The application of practical cases shows the realization and expansion of shopping website user experience fuzzy comprehensive evaluation. By the collection of objective behavior data and subjective bias data, with the AHP and fuzzy comprehensive evaluation method, different angles of views can be transformed into quantitative models, which can help site builders measure the different types of users' behavior bias from user types and indicator content to establish an effective user model for building different sites.

References

1. Jinling, C., Guoping, X.: B2C E-commercial website usability evaluation. Information Learned Journal, 237–242 (2005)
2. Ronggang, Z.: Research on fuzzy comprehensive evaluation of IT product's user experience quality. J. Computer Engineering and Applications 43(31), 102–105 (2007)
3. ISO 9241- 11 Ergonomic requirements for office work with visual display terminals (VDTs) part 11: guidance on usability. International Organization for Standardization (1998)
4. Davis, F.D., Bagozzi, R.P., Warshaw, P.R.: User acceptance of computer technology: A comparison of two theoretical models. J. Management Science 35(8), 982 (1989)
5. Shubai, X.: Analsis Hierarchy Process Principle, pp. 160–165. Tianjin University Press, Tianjin (1988)
6. Li, H., Wang, Q., et al.: Engineering Fuzzy Mathematics Method and Application. Tianjin Science and Technology Press (1993)
7. Wei, L., Xiugan, Y., Zhongqi, L.: Pilots' Situation Awareness Fuzzy Comprehensive Evaluation [J]. Acta Psychologica Sinica 36(2), 168–173 (2004)
8. Qun, L.: Uncertainty Mathematics Method Studying and Its Application in Social Science. China Social Sciences Press, Beijing (2005)

An Ontology-Based Framework Model for Trustworthy Software Evolution

Ji Li [1], Chunmei Liu [1], and Zhiguo Li[2]

[1] Department of Computer, Chongqing University, Chongqing 400044, China
[2] Shanghai Baosight Software Corporation, Chongqing 400039, China
Leedge@cqu.edu.cn

Abstract. In this paper, a framework model for trustworthy software evolution is constructed. It could not only solve semantic problems but also guide the dynamic evolution of service composition. First of all, it adopts a method of ontology space to solve the interactive problem among users, system and environment. Then, a set of pre-defined rules are used to evaluate the credibility of software behavior and the necessity for self-adjustment. According to these results, we make adjustment, reconfiguration and revision to the software in software life cycle, from rule guidance in micro-level to man-machine cooperation in macro-level. Finally, the instances and test results prove the proposed framework model to be effective and feasible.

Keywords: trustworthy software, dynamic evolution, ontology.

1 Introduction

In recent years, the research on trustworthy and trustworthy software evolution models has become a hot topic in software revolution. In the past, the trustworthy software mainly concerns about the reliability of hardware or software, system availability and real-time. However, with the changement of computing environment, some new factors that influence trustworthy software have emerged. For example, with the opening of advances of network environment and the socialization of software engineering, the concepts of "trustworthiness" and "context" have been gradually applied in the field of software trustworthy evolution.While as for software evolution, these context have different ways of expression, which means that they could not be directly utilized in software. Therefore, a descriptive framework is needed to provide a unified way of expression. On the other hand, software evolution is an inference process in which a series of adaptive rules act on the system. Thus, the key point is the description and organization of the reasoning rules.

Aiming at solving these problems, this paper introduces ontology. Ontology in computer research, which is quite suitable to express the concepts in environment factors and their relations, is a formal specified illustration of sharing conceptual model. By means of ontology, various functions can be achieved, such as knowledge sharing, logic inference, knowledge reuse and so on. Based on these, in this paper we propose a framework for trustworthy software evolution based on ontology. This framework, using Web Ontology Language (OWL) as description language, describes

L. Cao, J. Zhong, and Y. Feng (Eds.): ADMA 2010, Part II, LNCS 6441, pp. 537–544, 2010.

the strategy of trustworthy software evolution through the inference rules defined in OWL and make sure these concepts and relations can be manipulated in a unified frame, achieving the consistency and controllability of evolution.

2 Ontology Space Model Based on "Environment – Behavior"

Factors of environment have increasing influence on software behavior. In order to guarantee that software behavior in open environment accords with expectations, it should be on the basis of real-time monitoring information whose contents are rich and various. In the view of their source, they can be divided into three types, ranging from requirement related factors, architecture related factors and the context of applications related factors. The purpose of constructing these factors is to reach mutual understanding of the information, which makes the operation among different subjects (human, machine, system) available. Ontology is a formal description of sharing concepts, and when properly selecting ontology descriptive language, efficient automatic reasoning could be achieved by using relatively mature logic instruments. We construct an ontology space model which contains requirement ontology, architecture ontology, and the trustworthy of context ontology. They can be employed to respectively describe the environment of requirement target, software architecture and software system.

2.1 Requirement Ontology

Requirement ontology refers to extracting knowledge relevant to trustworthy software evolution from requirement analysis results, and it provides premise for software evolution. In this paper we adopt the thought of goal-oriented requirement engineering [5] and organize these entities in two levels.Meta level is defined as a set of abstract concepts that has no concerns about domain, and it contains goal, task, action, resource and so on. Among them, the goal is the general requirements or state to be achieved by software and this is also the basis of software evolution. Goal can be further divided into functional and non-functional goal, of which the latter is mainly used to describe system performance, system maintenance and other features like reliability, security and maintainability.Domain level includes the public concepts in certain application field and it realizes the reification of concepts through inheriting from Meta level. For example, in the application of "Education Resources", to achieve the goal of sharing digital resources, while media material can be subdivided into material of text, image, audio, video and so on.

2.2 Architecture Ontology

Referring to SOA [6] and combining with OWL-S [7], we construct ontology of architecture, which describes not only their static configuration among components, but also their dynamic behaviors. In the model of SOA, the core concept is service, while the implementation body of service is component which includes the underlying components and cooperative relationships among them. Software evolution, based on service composition, primarily means to promote software reliability through service addition, deletion, replacement, upgrading or changing the structure of services, which are driven by monitoring information. And through describing components and their cooperative relations, it also provides the structural basis. So we divided these entities into three levels: top, middle and instance level. The hierarchical relationship of concepts in architecture ontology is shown in figure 1.

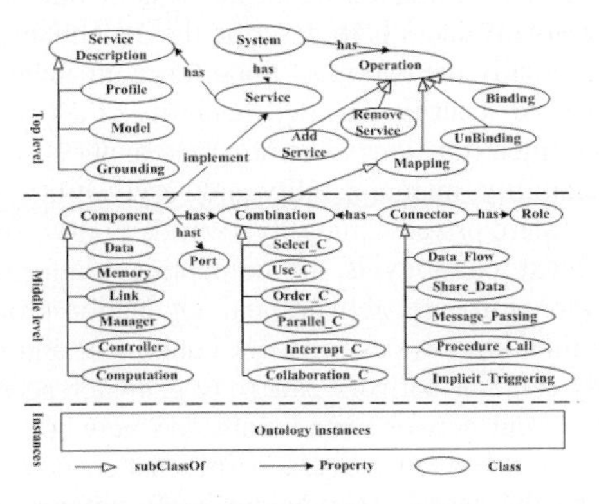

Fig.1. Ontology of architecture entities

Top level mainly includes the concept of service and corresponding operation. And Service covers all the service instances, while the specific function or performance features of each service are characterized by describing ontology. Describing ontology has three sub-ontologies: profile, model and grounding, which are used to respectively describe the semantic relation of what to do, how to do and how to access the service. Further information can be found in reference [6].

The middle level is to increase the rate of reusing, and extract components, connectors and their modes of combination which are commonly used. In reference [8] these components and connectors are listed such as data structure, memory, computation, manager, controller, etc. The commonly used connectors are procedure call, message passing, shared data, etc. the service composition contains Order_C, Select_C, Use_C, Parallel_C, and Collaboration_C. Among them, Order Combination : only implement S_1 then do S_2; Select Combination: it can use either S_1or S_2 rather than using both of them.These components and connectors can be combined in different ways to improve the speed of system development.

Instance level consists a series of instances that compose a knowledge base for specific application.

2.3 Trustworthy of Context Ontology

The evolution of trustworthy software aims to enhance its credibility, but the concept of trustworthy software has the duality of both the subjective and the objective [10]. The credibility of software is not simply equivalent to its quality, but also depends on the degree of identification by users. Therefore, besides concerning about the theories and technology of quality, trustworthy software should also pay attention to its context. In this paper, the context is all the environmental information that impacts the credibility of software through operation monitoring. It may be a parameter of hardware in the deployment phase, or user's feeling in the process of using.

Trustworthy of context ontology mainly contains user, environment, CompEntity, and their relations. As for user, it not only includes description on attributes but also evaluation on software credibility from the perspective of using, which covers satisfaction and trustworthiness. Satisfaction, more inclined to reflect users' feelings, refers to the users' subjective evaluation about the system functionality, usability and performance; while trustworthiness is to describe the dependability that can be measured and perceived. Actually the two most importance attributes of trustworthy software are dependability and security. The dependability of a system is that the system can complete the determined functions in certain time under specific conditions, and it covers availability, reliability, maintainability and survivability; while the security of a system refers to the system privacy, integrity, confidentiality, creditability, etc.

CompEntity in context ontology is the computing platform of service system, which includes hardware, software, network and application protocols. Hardware can be subdivided into computing devices, network equipment and other physical facilities; network primarily contains network pattern (e.g. switch access, wireless network access), network delay, and network bandwidth and network security (the security level). The network delay refers to the average time that costs from sending request to receiving response, and the network bandwidth is the amount of data that can pass through a network interface per unit time. The ontology is shown in figure 2.

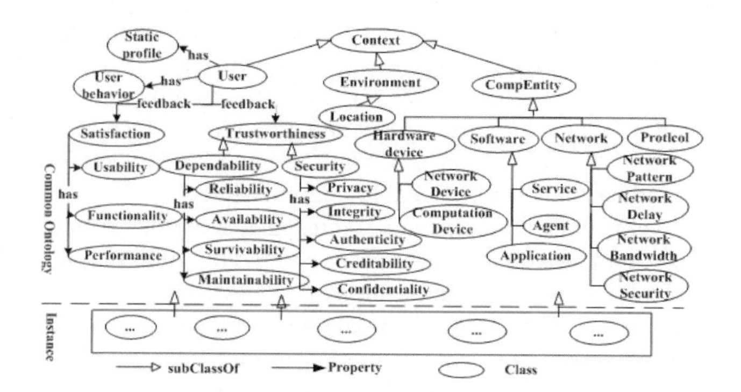

Fig. 2. The trustworthy of context ontology

3 BP-Based Weights Learning Algorithm

As for context-aware, lots of researches have been carried out, but the contextual parameters in Current approaches have the same weights for all users. We propose an approach to learn the weights of contextual parameters for every user based on back-propagation (BP) neural network (NN).

According to Kolmogorov theorem, three-layer BP with sigmoid function can represent any linear or non-linear relationship between the input and output. So in our research, a three-layer network and sigmoid is used. It consists of an input layer (IL), a hidden layer (HL), and an output layer (OL). It takes two steps to get the data of input layer.

Step1. Given a training dataset D, we extract subset $\{D_{c1},D_{c2}...D_{cn}\}$ from D according to different contexts. Every subset D_{ck} includes the data belongs to *ck*.

Step2. For every items *i* and *j* of D_{ck}, calculate average deviation $d_{i,j,ck}$ of them by(1).$r_{u,i}$ is a unknown rating, while $U_{ck(i)}$ includes users who have rated i in D_{ck}. All the $d_{i,j,ck}$ form the context DM_{ck} for *ck*.

$$d_{i,j,ck} = \sum_{u \in U_{ck}(i) \cap U_{ck}(j)} \frac{(r_{u,i} - r_{u,j})}{|U_{ck}(i) \cap U_{ck}(j)|} . \quad (1)$$

3.1 Weight Adjustment

During training, the adjustment of the weights uses error back-propagation. The weight connected with two nodes is adjusted according to the amount proportional to the strength of the signal and the error. The error at the OL is reallocated backwards through the HL until the IL is reached. This process is repeated until the error for all data is sufficiently small. For the node in the OL, we compute E_{rri} using formula 2.

$$Err_i = O_i (1 - O_i) (T_i - O_i) , \quad (2)$$

Here O_i is the computed output at node *i*. T_i is the real output instance value (in TT), and $(T_i\text{-}O_i)$ is the actual output error.For nodes in the HL, we reallocate the errors Err_i using formula 3.

$$Err_i = O_i(1 - O_i)\textstyle\sum_j Err_j \, w_{i,j} , \quad (3)$$

Here *j* means that there are *j* outputs of node I; Err_j denotes the error of node *j*, and $w_{i,j}$ is the weight of the link from node *i* to *j*. After getting all errors, we adjust the weights $w_{i,j}$ using formula 4. Here η is the learning rate parameter. It is 0.8 at the beginning. And it differs with the deviation of the computed output and the real value.

$$w_{i,j} = w_{i,j} + \eta * Err_j * O_I . \quad (4)$$

3.2 Weights Learning Algorithm

Input: D_{ck}, DM_{ck}
Output: The weights of contextual parameters for every user
Procedure:

Step1. Prepare instances for inputs and outputs for BP-based weight learning All predictions $p_{u,\ i,\ ck}$ form input instance datasets and all true ratings $r_{u,\ i}$ form output instance dataset.

Step2. Initialization, activation, weight computation

1) Initialize the input and output instances. Use random values between 0 and 1as the weights of contextual parameters.
2) While not (terminal condition)
   { a) For a set of input instances
   { Compute the weighted sum of the inputs as x

Compute the outputs of hidden layer using sigmoid function
Regarding the outputs of hidden layer as the inputs of output layer and compute the weighted
Compute the outputs of output layer using sigmoid function
}
b) For the output instances
{ Compute the error of the node in output layer by formula 2
Compute the error of the node in hidden layer by formula 3
Update weights of the links by formula 4
}
}
At last, we get the weights wck,u of the contextual parameters for every user.

4 The Evolution Mechanism Based on Ontology

When system runs in an open environment, the application scene will be: users using system, system serving users[3]. Also in the process of using, people or system will take advantage of external environment to improve the quality of service. The abstract model for complex distributed system contains the three parts, as shown in figure 3.

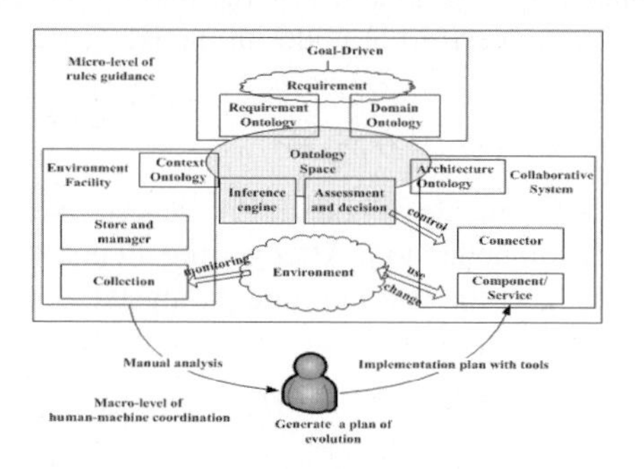

Fig. 3. The framework for trustworthy software evolution based on ontology

This framework is composed of rule guidance and man-machine cooperation, respectively belonging to micro-level and macro-level. In the micro-level, it establishes real-time environment abstraction (i.e. ontology space model), depending on monitoring information. Then it evaluates the credibility of software behavior and the necessity for self-adjustment through pre-defined rules. According to the results of evaluation, the software implements dynamic adjustment, reconstruction and other operations autonomously. In macro-level, man, as a factor, is taken into consideration in software control loop from the perspective of "society – technology". Human plays an active role in extracting dynamic information, defining monitoring rules and generating plans of dynamic evolution, so as to ensure that software behaviors are consistent with people's expectations. Thus, the reliability of services operation can be guaranteed and the in this case software failure and aging can be efficiently prevented.

Furthermore, with the interaction among system, users and environment, people gradually realize some new environmental factors and their new effects on system behavior. And in the framework, all main components support on-line expansion, such as deploying new information detectors, upgrading the architecture to support non-predetermined. And the loose coupling in ontology space is suitable for adding or deleting ontology and upgrading the rule engine. This open framework could deal with the dynamic changes of complex network environment.

5 Instance

We have developed a public platform for digital education which would supply material downloading, on-line lecture, on-line answer and many other teaching services. According to the C/S style, the platform is constructed which will initialize each component after searching and assembling the corresponding service components. The process includes loading ontology space, setting pre-defined rules, starting monitoring mechanism and so on. As for this platform, the response time and throughput are the fundamental performance requirements that we must consider. Suppose that the response time of server should be less than 2000 milliseconds and the maximum of service request is 500.Based on above description, we have to define some inference rules,and the instance is experimented under a pressure test by this method: generating a servlet document and deploying it with website together.The result is shown in fig4.

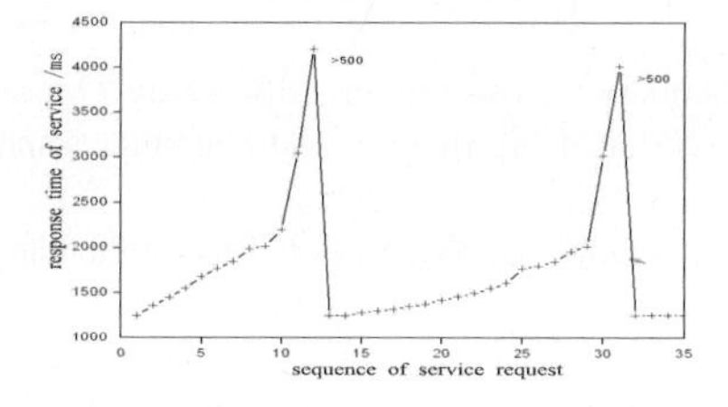

Fig. 4. The respond time of service processing center under a pressure test

As we can see from figure 5, for initial simulation, the total number of service requests is less than 500 and the response time is about 2 seconds, which meet the requirement of system performance. When processing the 11th service request, the response time is extended, but it is not long enough to trigger an evolution and then application is continued. But when dealing with the 12th request, the total number of requests is over 500 and response time exceeds the threshold, which trigger the operation of evolution. According to the pre-defined rules, a new server component is added and starts to work normally when processing the 13th request. That is, the response time has returned to a normal level and the adjustment of system has been completed. As the text goes, the total number of requests continues to grow and the software system needs further adjustment to meet the demand. From this instance, we can draw the conclusion that this framework for software evolution is effective.

6 Summary and Future Research

People have carried out a lot of evolution researches to promote software credibility from different perspectives and levels. However, for researches on applying network environment to trustworthy software evolution is at beginning. In this paper, constructing relevant factors aims to provide semantic foundation for software evolution.Further research mainly includes two aspects. First, we plan to analyze the current techniques and tools for software evolution. Based on these ontologies, we wish to do some improvements. Second, we plan to enhance the adaptive mechanism of software evolution and its explicit implementing scheme. Through the organic combination of ontology space model and inference mechanism, the rationality and feasibility of trustworthy software evolution are expected to be achieved in further research.

Acknowledgments. The work reported in this paper has been supported by Major Research Plan of the National Natural Science Foundation of China（No.90818028）and the Third Stage Building of "211 Project" (Grant No. S-10218).

References

1. Wang, H.M., Yi, G.: Trust-based Software evolution of the network time. J. Communications of the CCF 6, 28–33 (2010)
2. Lv, J., Ma, X.X., Tao, X.P.: Research on environment driven model and Underlying Technique oriented internetware. J. Science in China(Series E): information science 38, 864–872 (2008)
3. Anita, S.: A survey of collaborative tools in software development. Institute for software research, Donald Bren school of information and computer science, University of California, Irvine (2005)
4. OWL Web Ontology Language overview W3C recommendation (February 10, 2004), http://www.w3.org/TR/owl-features/
5. Van, L.: Goal-oriented requirements engineering: a guided tour. In: Proceedings of the 5th IEEE International Symposium on Requirements Engineering, Toronto, pp. 249–262. IEEE computer society, Los Alamitos (2001)
6. W3C OWL Web Ontology Language for Services, http://www.w3.org/submission/2004/07/
7. Ren, H.M., Qian, L.Q.: Research on Component Composition and Its Formal Reasoning. J. Journal of Software 14, 1074–1077 (2003)
8. Shaw, M., Garlan, D.: Software architecture: perspectives on an emerging discipline. J. New Jersey Prentice Hall 19, 97–127 (2003)
9. Trustie research Team, Trustie series technical criterion(V2.0), http://www.trustie.net, 2009.9

Multi-level Log-Based Relevance Feedback Scheme for Image Retrieval

Huanchen Zhang, Weifeng Sun, Shichao Dong, Long Chen, and Chuang Lin*

School of Software of Dalian University of Technology,
116620 Dalian Liaoning, China
to_hanson@sina.com, wfsun@dlut.edu.cn,
dongshichao1988@gmail.com, hermitBaby@gmail.com,
linchuang_78@sina.com

Abstract. Relevance feedback has been shown as a powerful tool to improve the retrieval performance of content-based image retrieval (CBIR). However, the feedback iteration process is tedious and time-consuming. History log consists of valuable information about previous users' perception of the content of image and such information can be used to accelerate the feedback iteration process and enhance the retrieval performance. In this paper, a novel algorithm to collect and compute the log-based relevance of the images is proposed. We utilize the multi-level structure of log-based relevance and fully mine previous users' perception of content of images in log. Experimental results show that our algorithm is effective and outperforms previous schemes.

Keywords: multi-level, log-based relevance, content-based image retrieval.

1 Introduction

Content-based Image Retrieval (CBIR) encounters difficulties such as the semantic gap between high-level concepts and low-level features like color, texture and shape [1]. Because of the complexity of images, describing an image only with those low-level features is always semantic ambiguous. To bridge the semantic gap, relevance feedback was introduced to CBIR and has attracted a lot of research interests. Users' feedback consists of valuable information about users' perception on the content of the image. Such information is high-level and can be used to help bridge the semantic gap. However, such interaction process is tedious and time-consuming, so if too many times of iteration of feedback are asked, users may be impatient to interact with the CBIR system [2]. The feedback log of previous users consists of valuable information about the users' common perception of images. Such information can help understand the high-level concepts of images and reduce the times of feedback. In this paper, we develop a novel algorithm to compute log-based relevance of the images. We name our algorithm multi-level log-based relevance (MLLR). In MLLR, an implicit link is built between two images which are positive feedback in one search session. Those implicit links between images build a multi-level structure. The upper level images'

* Corresponding author.

L. Cao, J. Zhong, and Y. Feng (Eds.): ADMA 2010, Part II, LNCS 6441, pp. 545–552, 2010.

log-based relevance to the user's desired target can be transmitted to the lower level images level by level through the implicit links between the upper level image and lower level image. Experimental results show that our algorithm fully mines the log-based relevance between images and is effective.

The rest of this paper is organized as follows. We review the related work in Section 2. Section 3 proposes an effective algorithm fully mining the multi-level log-based relevance of images. Section 4 presents detailed experiment and performance comparison. Section 5 discusses the conclusion and future work.

2. Related Work

To reduce the times of iteration and accelerate the relevance feedback, some research focused on the active learning techniques [3]. However, in an active learning process, users are asked to label additional images selected by the system which are considered as the most informative ones and this additional feedback often causes users' impatience. To learn the users' logs effectively for CBIR, Hoi did a pioneering job and introduced a log-based relevance feedback scheme [4].

Compared to the previous schemes, our algorithm solves the problems in [4] when computing the log-based relevance. We fully mine the deeper multi-level relevance between images. The valuable information about the previous users' perception of the content of the images is fully and correctly mined.

3 Multi-level Log-Based Relevance

In [4], user log information is used in two aspects. First, log-based relevance of images to user's desired target is computed. Second, log-based relevance of images to user's desired target is used to look for more training samples for SVM whose performance can be degraded with insufficient training samples. Images having strong positive or negative log-based relevance are chosen to be additive positive or negative training samples. Then these training samples are engaged to train a soft label support vector machine (SLSVM) classifier which incorporates the label confidence degree of data in the regular SVM. Finally, combination of the relevance score based on the low-level feature computed using SLSVM and log-based relevance of images to user's desired target are used to rank the images in the dataset. In [4], when computing the log-based relevance, both the positive and negative feedback are collected and utilized. If two images are fed back as positive samples in one feedback session, these two images are viewed as relevant '1' while if one image is fed back as positive and the other as negative, these two images are viewed as irrelevant '-1'. The log-based relevance degree between two images is computed by adding their corresponding relevance in every feedback session. Based on the previous user log, the log-based relevance of image i to user's desired target in the database is the difference between image i's log-based relevance to user's positive feedback and log-based relevance to user's negative feedback. To put the thought of [4] in a simple and direct way, a figure Fig. 1 is made.

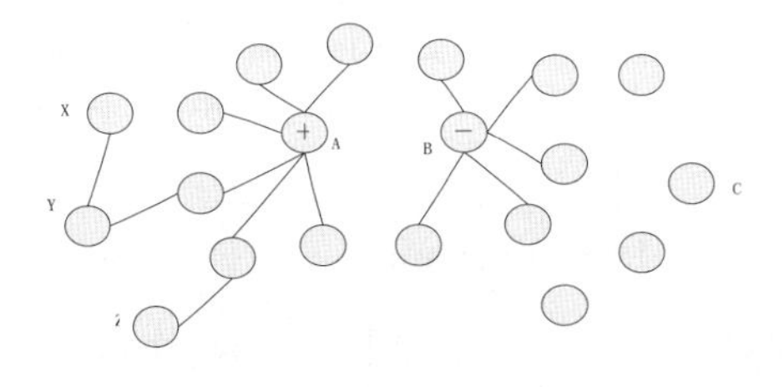

Fig. 1. Log-based relevance between images

The CBIR system returns the initial result based on the low-level feature. We suppose the user only gives two feedback samples, one is the image A with "+" representing the positive sample and the other is the image B with "-" representing the negative sample. In Fig. 1, images with positive log-based relevance are linked to each other. There are two problems in this log-based relevance computation method. First, there may be some images which are also relevant to A but linked to A indirectly e.g. X, Y, Z. This phenomenon can be caused by following reasons. First, users only label the first few images in the initial result, so the images ranked a little lower initially are ignored by the previous users. Second, the log is not mature in the beginning so some relevant relationships are not reflected in the log. In [4], the log-based relevance of X, Y, Z to user's desired target is set as zero unfairly because they have no direct log-based relevance with A. We call this phenomenon one-level relevance unfairness. Another problem in [4] is caused by the diversity of negative samples. All positive examples are alike; each negative example is negative in its own way [5], i.e. images which are dissimilar to user's negative feedback may be of another kind of negative sample rather than positive samples. However, images which are dissimilar to negative feedback are viewed as positive in the log-based relevance computation in [4]. As a result, images irrelevant or weakly relevant to positive feedback but strongly irrelevant to negative feedback are given positive log-based relevance unfairly.

We propose a novel algorithm named MLLR which solves the problems mentioned above. We assume that when a user feeds back two positive samples in one feedback session, an implicit link is built between these two images. The implicit links between images form a multi-level structure of log-based relevance. Supposing image A is one of the positive samples fed back by the user, if D is relevant to C, C is relevant to B and B is relevant to A, then D, C is relevant to A indirectly. We call B is in an upper level than C and C is in an upper level than D. The upper level images' log-based relevance to the user's desired target can be transmitted to the lower level images level by level through the implicit links between the upper level image and lower level image. Instead of only using the direct relevance, the indirect multi-level relevance between images is also used in our algorithm. So the log-based relevance between images is fully mined. Besides the negative effects caused by the negative feedback mentioned above, manually collected negative samples could be biased because of human's unintentional prejudice and could be detrimental [6]. Only asking the user to click the most relevant images can also enhance the user experience. So we only use positive feedback in our algorithm.

In this paper, we only discuss how to fully mine the log-based relevance of images and solve the problems in [4]. To develop a similar soft label one-class svm [7] only using positive training sample is beyond this paper's scope and it may be the future work. We first discuss how to organize user log information. Then a method to evaluate the log-based relevance degree between two images is proposed. After that, an algorithm to compute the relevance of images to user's desired target is given.

3.1 Organization of User Log Information

After a user first launches a query in a CBIR system, the CBIR system retrieves the top X images with the low-level features vector nearest to that of the user's query. If the user cannot obtain the desired targets from the initial results, the user may choose to begin a relevance learning procedure. Each relevance feedback round is viewed as a unit of user feedback session. The feedback information is stored into the log database after one feedback session is over. To manage the log information well, an implicit link vector (ILV), $ILV = \{i, j, c_{ij}\}$, is constructed to represent the implicit link information. In the vector *ILV* there are three elements. The first two elements i and j denote the image i and image j which are the endpoints of the implicit link. The third element c_{ij} denotes the number of session in which image i and j are fed back as two positive samples, i.e. the implicit links between i and j.

3.2 Relevance between Two Images

To evaluate the relevance degree between two images, a value c that represents 'the same' needs to be found, i.e. if c_{ij} exceeds c, then this image i and j are viewed as the same; if c_{ij} is less than c, then c_{ij} / c represents the relevance degree between image i and image j. If we set c with a constant value, there may exits a problem that we name as image topic bias, i.e. if image i and image j are about a hot topic, the chance of building implicit links between them will be much higher than that between two images about cold topic. As a result, evaluating the similarity degree between any two images based on the same c is not reasonable. The problem of image topic bias is solved by setting c for each feedback session dynamically. Let $I = \{i_1, i_2, ...i_n\}$ denote the positive images fed back by the user. We assume those images which the user feeds back as positive samples equally represent the users' desired target. Based on this assumption, the images in *I* are the same or extremely similar. So we take c computed as formula (1) which is the average number of implicit links between every pair of images in *I* as the value that represents 'the same'.

$$c = (\sum_{x=1}^{n-1} \sum_{y=x+1}^{n} c_{i_x i_y}) / C_n^2 \qquad (1)$$

There may be another problem which we name as minority user noise. This means that c_{ij} is quite small. This phenomenon can be due to users' misoperation or special perception of images. Such noise can affect the process of the relevance computation process. To solve this problem, we give a threshold value t=α*c for c_{ij} .α is a coefficient ranging from 0 to 1. For each c_{ij}, if the value of c_{ij} is less than the threshold value t, then the similarity degree between these two images will be set as 0. To conclude, the relevance degree between two images r_{ij} can be computed as formula (2).

$$r_{ij} = \begin{cases} 1 & c_{ij} \geq c \\ c_{ij} / c & t < c_{ij} < c \\ 0 & c_{ij} \leq t \end{cases} \tag{2}$$

3.3 Computation of Relevance to Desired Target

Each image will be given a level label which represents the level in the structure after the image's log-based relevance to the desired target is computed. The first reason why we give each image a level label is that each image can only in one level, i.e. in each iteration, only the log-based relevance of images without a level label is computed. The second reason is that the log-based relevance of each image to the desired target of the user is only influenced by that of the upper level images. Fig. 2 shows the multi-level structure of the log-based image relevance to the user's desired target.

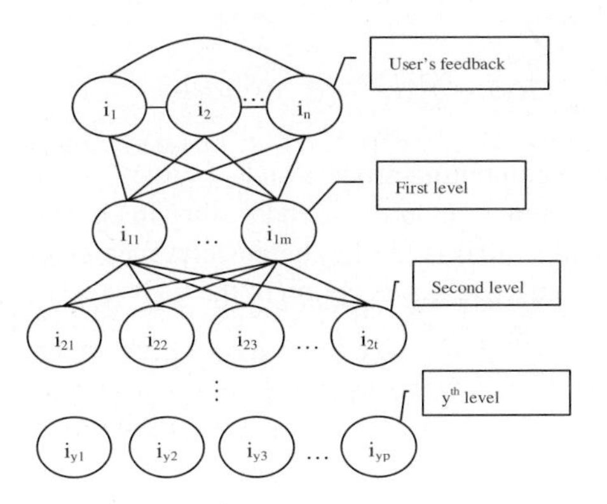

Fig. 2. Multi-level structure of log-based relevance

Let R(i) denote the log-based relevance of image i to the desired target of the user.

Set $R(i_1) = R(i_2)... = R(i_n) = 1$ as the initial value, which means the log-based relevance of the positive feedback images to the user's desired target is 1. And give image $i_1, i_2, ..., i_n$ a level label 0 which represents the 0 level in the multi-level structure of log-based relevance.

In the first iteration of the computation of log-based relevance to the desired target, for each image i without a level label in the database, the log-based relevance between i and the desired target of the user is computed as formula (3).

$$R(i) = (\sum_{x=1}^{n} (r_{ii_x} * R(i_x))) / n \tag{3}$$

$r_{ii_x} * R(i_x)$ is the log-based relevance transmitted from i_x to i through the implicit links between i_x and z. The final log-based relevance of image i to user's desired target R(i) is the average log-based relevance transmitted from all the images in *I* to i.

Those images i that R(i)>0 will be labeled 1 which denotes the first level or the first iteration. Let $I_1 = \{i_{11}, i_{12}, ..., i_{1m}\}$ denote the images which are labeled 1.

In the second iteration, for an image i in the database without a level label, the log-based relevance between i and the desired target of the user is computed as formula (4).

$$R(i) = (\sum_{x=1}^{m} (r_{ii_{1x}} * R(i_{1x}))) / m \tag{4}$$

And those images i that R(i)>0 will be labeled 2 which denotes the second level or the second iteration.

The iteration goes on in this way until:

Let $I_y = \{i_{y1}, i_{y2}, ..., i_{yp}\}$ denote the images which have the level label y. If for all the unlabeled images i,

$$R(i) = (\sum_{x=1}^{p} (r_{ii_{yx}} * R(i_{yx}))) / p = 0 \tag{5}$$

the log-based relevance computing process ends.

When the log-based computing process ends, every image i in the database has a R(i) value ranging from 0 to 1. R(i) is the log-based relevance of image i to user's desired target according to previous users' perception of the high-level concepts of images.

4 Evaluation

4.1 Log Data Simulation

To make a comparison with method in [4], both positive and negative feedback need to be collected, but only positive feedback is used in our method. In our experiment, log data is simulated as follows. We use 20 categories of images and in each category, there are 100 images. First, we randomly select a query from the image dataset. Then we randomly select n images of the same category of the query as positive feedback and 20-n images of a different category from the query as negative feedback (n is a random value ranging from 5 to 15). To simulate the real world log data, we inject about 15% noise into the feedback sessions. We insert 200 such feedback into the log data, i.e. the log simulating process mentioned above are repeated 200 times.

4.2 Experimental Results and Analysis

In order to evaluate our method's effectiveness, we compare the performance of MLLR with the log-based relevance feedback algorithm used in [4] (LRF). Let q denote the query the user selects from the database.

In our experiment, compared schemes are evaluated on 50 queries randomly selected from the data set. Three performance measure metrics are employed in our experiment to evaluate the quality of log-based relevance.

Let Num1(q) denote the number of images which are of the same category as q and are given positive log-based relevance. The first measure metric is average ratio of Num1(q) over the number of images which are of the same category of q, i.e. 100, over the 50 queries. From Table 1, we can see that MLLR links more images of the same category as q together than LRF. This is because MLLR fully mines the direct and indirect relevance while LRF only mines the direct relevance.

Let Num2(q) denote the number of images which are of a different category from q and are given log-based positive relevance. The second measure metric is the average Num2(q) over the 50 queries. From Table 1, we can see that MLLR gives less irrelevant images positive log-based relevance than LRF. This is because LRF utilizes negative feedback and image strongly irrelevant negative feedback are viewed as positive unfairly while MLLR only uses positive feedback and voids negative effects caused by negative feedback.

Table 1. Average Num1/100 and Average Num2

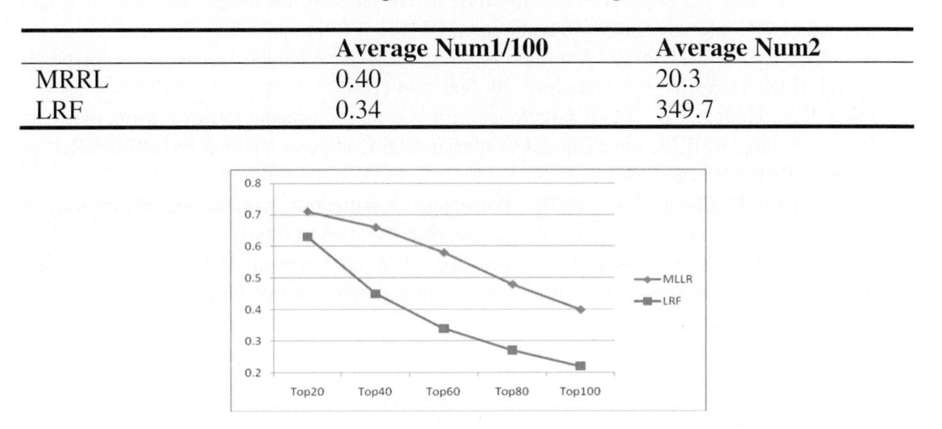

	Average Num1/100	Average Num2
MRRL	0.40	20.3
LRF	0.34	349.7

Fig. 3. Average Precision

The third measure metric is based on the Average Precision, which is defined as the average ratio of the number of relevant images of the returned images over the number of total returned images over the 50 queries. The returned images are ranked based on log-based relevance of images. Fig. 3 illustrates the visual comparison of the experimental result. From this figure, we can observe that MLLR shows promising improvement on the retrieval performance compared with the Hoi's method in [4].

5 Conclusion and Future Work

An algorithm MLLR which fully and correctly mines the valuable information in previous user log has been proposed. Analysis and experimental results have confirmed that MLLR not only mines the direct log-based relevance but also indirect log-based relevance. It also solves the negative effects caused by negative feedback in previous study. However, we must address the limitation of our scheme that the noise in the log data can be amplified when transmitted from upper level to lower lever. This problem can be solved by carefully choose the thresh hold value t in formula (2). Although our empirical approach for choosing t has resulted in satisfactory performance, we plan to investigate other approaches in principle for tuning this parameter effectively.

Acknowledgements. Nature Science Foundation of China under grant No.: 60673046, 90715037. NSFC-JST under grant No.:51021140004.

References

1. Arnold, S., Marcel, W., Simone, S., Amarnath, G., Ramesh, J.: Content-Based Image Retrieval at the End of the Early Years. IEEE Transactions on Pattern Analysis and Machine Intelligence 22, 1349–1380 (2000)
2. Hoi, S.C., Lyu, M.R.: A novel log-based relevance feedback technique in content-based image retrieval. In: Proc. ACM Int. Conf. Multimedia (MM 2004), New York (2004)
3. Tong, S., Chang, E.: Support vector machine active learning for image retrieval. In: ACM International Conference on Multimedia, pp. 107–118 (2001)
4. Hoi, S.C., Lyu, M.R., Jin, R.: A unified log-based relevance feedback scheme for image retrieval. IEEE Trans. Knowl. Data Eng. 18, 509–524 (2006)
5. Zhou, X.S., Huang, T.S.: Small sample learning during multimedia retrieval using biasmap. In: Proceedings of IEEE International Conference on Computer Visiona and Pattern Recognition (CVPR), Hawaii (2001)
6. Yu, H., Hen, J., Chang, K.C.: PEBL: Web page classification without negative examples. IEEE. Trans. on Knowledge and Data Engineering 16, 70–81 (2004)
7. Chen, Y., Zhou, X.S., Huang, T.S.: One-class SVM for Learning in Image Retrieval. In: IEEE Int'l Conf. on Image Proc. (ICIP 2001), Thessaloniki, Greece (2001)

A Distributed Node Clustering Mechanism in P2P Networks

Mo Hai and Shuhang Guo

School of Information, Central University of Finance and Economics,
100081 Beijing, China
haimozhi@gmail.com, guoshuhang@hotmail.com

Abstract. A P2P network is an important computing model because of its scalability, adaptability, self-organization, etc. How to organize the nodes in P2P networks effectively is an important research issue. The node clustering aims to provide an effective method to organize the nodes in P2P networks. This paper proposes a distributed node clustering mechanism based on nodes' queries in P2P networks. In this mechanism, we propose three algorithms: maintaining of node clusters, merging of node clusters and splitting of node clusters. Theoretical analysis shows the time and communication complexity of this clustering mechanism is low. Simulation results show that the clustering accuracy of this clustering mechanism is high.

Keywords: P2P network, node clustering, maintaining of node clusters, merging of node clusters, splitting of node clusters.

1 Introduction

P2P networks are designed for the sharing of computer resources (content, storage, CPU cycles) by direct exchange, rather than requiring the intermediation or support of a centralized server or authority. Peer-to-peer architectures are characterized by their ability to adapt to failures and accommodate transient populations of nodes while maintaining acceptable connectivity and performance [1]. Except for these advantages, P2P networks face some challenges. How to organize peer nodes in P2P networks effectively is a challenge. Most existing P2P networks organize peer nodes in two ways: unstructured and structured. Unstructured P2P networks will generate huge network traffic when resolving a query. Structured P2P networks incur a large number of communication messages and routing hops when processing a complex query, like multiple-keyword query.

Clustering is a network management technique, for it creates a hierarchical structure on top of a flat network. Clustering of nodes in P2P networks is an effective method of organizing nodes, which can help improve the system performance. [2][3][4] propose a method to construct a super node network by node clustering to optimize the structure of P2P networks. In a super node network, each node cluster is made up of a super node and a set of ordinary nodes, and the super node is in charge of the management of the node cluster. The formation of node clusters help decrease the number of communication messages and routing hops in P2P networks. However,

L. Cao, J. Zhong, and Y. Feng (Eds.): ADMA 2010, Part II, LNCS 6441, pp. 553–560, 2010.

how to cluster nodes in P2P networks in an effective way is a challenging problem. In previous study, the node clustering methods are mainly based on node connectivity, link delay, and traffic flow patterns or require global knowledge. In this paper we propose a new node clustering mechanism in P2P networks. Nodes with similar queries are clustered together. When a new node joins, a node leaves, or any node's query behavior changes, node clusters will reorganize. Each cluster is made up of one or several super nodes and several ordinary nodes. The size of each node cluster is controlled by the maximum size threshold s_{max} and the minimum size threshold s_{min}. If the size of a cluster is larger than s_{max}, this cluster will be splitted into two clusters; if the size of a cluster is smaller than s_{min}, this cluster will be merged with another cluster, and after merging, the size of the new cluster lies between s_{min} and s_{max}.

2 Related Work

Distributed node clustering has been used in the field of wireless ad-hoc networks and P2P networks. There are many researches in ad-hoc networks, in which node clustering helps decrease the power consumption of node, such as[5][6][7][8][9]. In previous study, researchers propose many methods to cluster nodes, such as Max-Min D-Cluster[10], MCL[11], CDC(Connectivity-based Distributed Node Clustering) [12], a node clustering algorithm based on link delay[13], a node clustering algorithm based on the traffic flow patterns[14], etc. Max-Min D-Cluster discovers node clusters based on *d*-hop dominating set. It has proved that the minimum d-hop dominating set problem is NP-complete, and Max-Min D-Cluster algorithm is an approximate solution of the optimal solution for the minimum d-hop dominating set problem. In MCL algorithm, the global information about the entire P2P network, such as the number of nodes, the number of node connections etc, is stored in a central location. CDC clusters nodes with high connectivity together by random routing of weighted messages. Experiments show CDC effectively handles the node dynamics, but improper choice of initiators may incur bad clustering results. Node clusters in CDC help decrease the number of communication messages and the routing hops. [13] proposes an approach to node clustering based on link delay of node communications in P2P networks. In this approach, the link delay of communications between peer nodes and super nodes can be restricted to a time limit, which will improve the overall performance of P2P networks. [14] proposes a node clustering algorithm for global positioning system (GPS)-based mobile ad hoc networks that takes into consideration the direction of the overall traffic flow in the network. The proposed cluster leader logic algorithm is motivated by the GPS quorum hybrid routing algorithm, where cluster heads are positioned on the terrain upon a conceptual cellular grid. The proposed distributed clustering algorithm chooses the cluster heads based on the traffic flow patterns, i.e. the nodes best suited to forward and route network traffic are selected. Different from the above node clustering algorithms, our proposed node clustering algorithm is based on similarities of nodes' queries. Furthermore, in our algorithm, the size of each cluster is restricted by the maximum size threshold s_{max} and the minimum size threshold s_{min}.

3 Problem Description

In this paper we define a P2P network as a connected and weighted graph, denoted by G = (E, V, W). Here V denotes the set of nodes in a P2P network, and any vertex $v \in V$ represents an actual node in a P2P network. If the corresponding nodes of vertex u and v have logical connection in P2P overlay network, then there is an edge $(u, v) \in E$ between them. The edge weight w(u,v) is the similarity of vertex u and v's corresponding nodes' queries.

In a P2P network represented by the graph G, a node cluster is a set of nodes corresponding to a subset of V. Each node cluster is made up of a super node and several ordinary nodes. In order to control the clustering process, a similarity threshold S is defined. The problem of node clustering based on similarity of nodes' queries in a P2P network is to find a covering of vertices with the minimum value of k, which can be denoted by the following:

$$\{V_1, V_2, V_3, \ldots, V_k\}(V_i \subseteq V, 1 \le i \le k, V_1 \cup V_2 \cup V_3 \cup \ldots \cup V_k = V, V_i \ne \phi) . \quad (1)$$

In V_i, the edge weight of vertex s, whose corresponding node in P2P network is a super node, and any other vertices of V_i is not smaller than S. That is, a super node and any ordinary node in the same cluster have similar queries.

4 Node Clustering Algorithm

In order to increase the fault tolerance, there can be configured several super nodes in a node cluster, whose quantity is decided by applications. In each cluster, there is an active super node that works and serves this cluster. The other super nodes in this cluster are called backup super nodes. Each active super node knows addresses of all active super nodes in the P2P network and addresses of all ordinary nodes in the same cluster, and each ordinary node stores addresses of all super nodes in the same cluster.

4.1 Maintaining, Merging and Splitting of Node Clusters

When a new node q arrives, firstly it is inserted into the P2P overlay network using P2P's join protocol. Then it will discover a cluster it should join. The process of discovering is: q firstly asks its neighbors whether they are super nodes by an application multicast protocol. If there are super nodes in q's neighbors, q will know all super nodes' addresses from its neighbor which is a super node. But if all of q's neighbors are not super nodes, q will get all super nodes' addresses from a neighbor's super node. Then q will compute the similarity of its queries and any super node's queries one by one until it finds such super node*s*, which satisfy: the similarity of whose queries and q's queries is not smaller than S. If there are several super nodes which satisfy this condition, then a super node t is selected, the similarity of whose queries and q's queries is the largest. q will join the cluster charged by t. If there is only a super node which satisfies this condition, q will join the cluster charged by this super node. But if no super node which satisfies this condition, q will creates a new cluster with itself as the super node. Then the address of the new super node will be notified to all super nodes in the P2P network using an application multicast protocol. When one super node fails, one backup super node in the same cluster can take its place and serve this cluster.

Each node *u* will periodically compute the similarity between its queries and each super node's queries by the method of multicasting. Once *u* finds a super node *v*, whose queries is more similar with *u*'s queries than *u*'s current super node, *u* will firstly notify its super node its departure, and then become an ordinary peer of the cluster charged by *v*. If no such super node is found and the similarity between *u*'s queries and *u*'s super node's queries is smaller than *S*, *u* will firstly notify its super node its departure, then create a new cluster with itself as the super node, and the address of the new super node *u* will be notified to all super nodes in the P2P network using an application multicast protocol.

Similarity threshold *S* will control the number of clusters. One extreme case is that *S* is so small that few clusters are formed, and the super nodes become bottlenecks; the other extreme case is that *S* is so large that a large number of clusters are generated, and the communication traffic for maintaining these clusters will be significant. Proper value of *S* will be decided by applications.

In order to avoid one situation that each node becomes a cluster with itself as the super node, the minimum size threshold s_{min} is set. When there is a cluster *c* whose size is smaller than s_{min}, *c* will be merged with another cluster *l*, and after merging, the size of the new cluster lies between s_{min} and s_{max}. The new cluster's super node is *c* or *l*'s original super node, and the other nodes of *c* and *l*'s cluster become ordinary nodes of this new cluster.

In order to avoid the other situation that only one cluster is formed, the maximum size threshold s_{max} is set. When there is a cluster *u* whose size is larger than s_{max}, this cluster will be splitted into two clusters. *u*'s super node is as the super node of one new cluster; and one of *u*'s backup super nodes is as the super node of the other new cluster. This new super node's address will be notified to all super nodes using an application multicast protocol. Any other node of *u* becomes an ordinary node of one new cluster.

4.2 Performance Analysis

When any one of the following conditions occurs, the communication traffic is generated.

1. When a new super node is generated, the address of this new super node will be notified to all super nodes;
2. When a super node leaves, all super nodes will be notified to delete its address.

So in order to decrease the communication traffic of maintaining node clusters, we can select the node who keeps alive longer to be a super node of a cluster.

Next we will analyze the communication complexity and time complexity of maintaining clusters. The communication complexity is measured by the number of generated messages. Supposed there are *m* clusters with *n* nodes, and no messages are lost.

Theorem 1. The time complexity of maintaining clusters is $O(\log(m))$.

Proof: When a new node q joins, it will firstly acquire the addresses of all super nodes. The process is: q firstly asks its neighbors whether they are super nodes by using an application multicast protocol. If there are super nodes in q's neighbors, q will get all super nodes' addresses from its neighbor which is a super node. But if all of q's neighbors are not super nodes, q will get the addresses of all super nodes from a neighbor's super node. These steps will take $O(\log(m))$ time. Then q will compute the similarity of its queries and any super node's queries one by one until it finds such super nodes, which satisfy: the similarity of whose queries and q's queries is not less than S. If no such super node is found, q will create a new cluster with itself as the super node, and q's address will be notified to all super nodes using an application multicast protocol. This process will take $O(\log(m))$ time. So the time complexity of maintaining clusters when a new node joins is $O(\log(m))$. Each node u will periodically compute the similarity between its queries and each super node's queries by the method of multicasting. Once u finds a super node v, whose queries is more similar with u's queries than u's current super node, u will firstly notify its super node its departure, and then become an ordinary peer of the cluster charged by v. The process will take $O(\log(m))$ time. If no such super node is found and the similarity between u's queries and u's super node's queries is less than S, u will firstly notify its super node its departure, and create a new cluster with itself as the super node, then u's address will be notified to all super nodes using an application multicast protocol, which will take $O(\log(m))$ time. So the time complexity of maintaining clusters when any node's query behavior changes is $O(\log(m))$.

Theorem 2. The communication complexity of maintaining clusters is $O(m)$.

Proof: When a new node joins, it will discover a cluster it should join. The discovering process described above will generate $O(m)$ messages. If no such cluster is found, it will create a cluster with itself as the super node, and then this new super node's address will be notified to all super nodes using an application multicast protocol, which will generate $O(m)$ messages. So the communication complexity of maintaining clusters when a new node joins is $O(m)$. Each node u will periodically compute the similarity between its queries and each super node's queries by the method of multicasting. Once u finds a super node v, whose queries is more similar with u's queries than u's current super node, u will firstly notify its super node its departure, and then become an ordinary peer of the cluster charged by v. This process will generate $O(m)$ messages. If no such super node is found and the similarity between u's queries and u's super node's queries is less than S, u will firstly notify its super node its departure, then create a new cluster with itself as the super node. The address of u will then be notified to all super nodes using an application multicast protocol, which will generate $O(m)$ messages. So the communication complexity of maintaining clusters when any node's query behavior changes is $O(m)$.

From the theorem 1 and 2, we can see that our proposed node clustering algorithm has low time complexity and communication complexity.

5 Simulation Results

We implement a simulator on top of FreePastry[15]. In FreePastry, the base of node ID is 16, the size of leaf set is 24, and the length of node ID is 160. In our simulations, CPU is AMD Opteron ™ 2.2GHz, the RAM is 3GB, and the operating system is Linux2.4.21. The adopted application multicast protocol is Scribe[16]. In order to measure the clustering quality, we use the metric called clustering accuracy, which is defined as the ratio of the number of nodes which are put in proper clusters and the total number of nodes. In the first group simulations, no node joins or fails when the clustering algorithm is running. All nodes start concurrently, and each node generates 5,000 queries. Every query is made up of one or several keywords, which are selected randomly from a given keyword set. We measure the clustering accuracy of our proposed node clustering algorithm when all queries have been processed. Figure 1 draws the relation between the clustering accuracy and the total number of nodes. From this figure, we can see that with the increase of the number of nodes, the clustering accuracy decreases gradually. However, the clustering accuracy is not less than 90%. So the clustering algorithm has a high clustering accuracy.

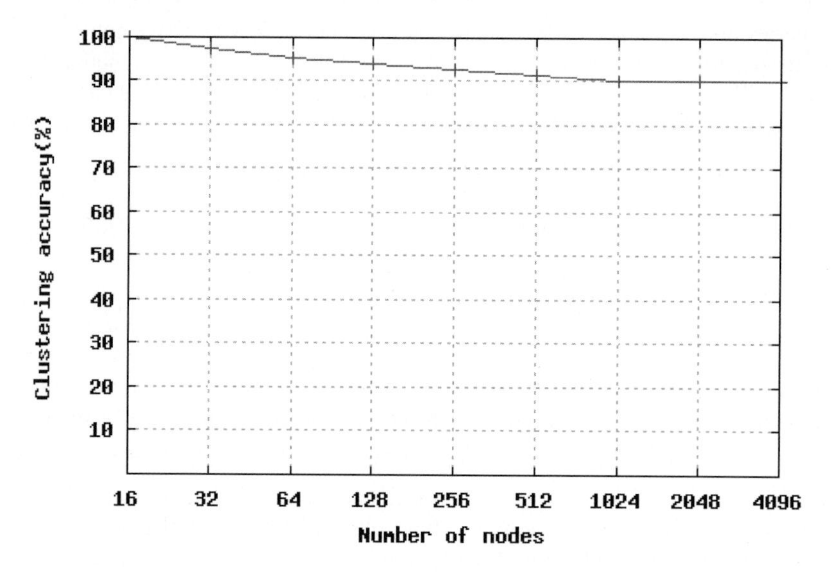

Fig. 1. Clustering accuracy as a function of the number of nodes

In the second group simulation, supposed new nodes join when the clustering algorithm is running. For this simulation, we consider pre-clustered nodes, to which new nodes join. The total number of nodes, including pre-clustered nodes and joining nodes, is 1024. Each joining node only knows its neighbors and joins a proper cluster through the mechanism described above. We measure the clustering accuracy when the percentage of joining nodes varies. Figure 2 shows the relation between the clustering accuracy and the percentage of joining nodes. We can observe that the influence of the percentage of joining nodes on the clustering accuracy is little. So our proposed node clustering algorithm has a high clustering accuracy even in the face of joining nodes.

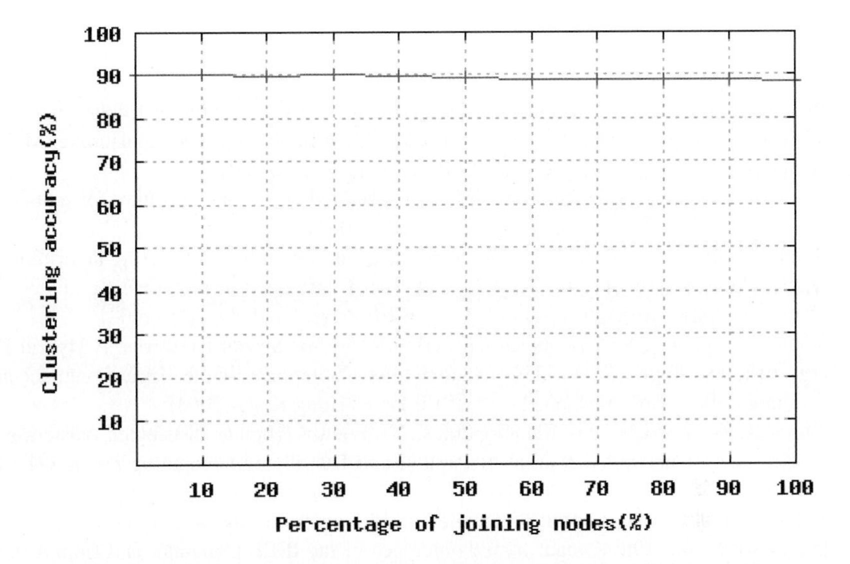

Fig. 2. Clustering accuracy as a function of the percentage of joining nodes

6 Conclusions

In this paper we propose a novel node clustering algorithm in P2P networks. This algorithm is fully distributed, and it needs no global information. The main idea of this clustering algorithm is: nodes with similar queries will be clustered together. When a new node joins, a node fails or any node's query behavior has changed, node clusters will be reorganized. Theoretical analysis shows the time complexity and communication complexity of this algorithm is low. Simulation results show that the clustering accuracy of this algorithm is high.

Acknowledgments. This paper is supported by 211 Project for Central University of Finance and Economics(the 3rd phrase), Project of Ministry of Education Humanities and Social Sciences Foundation of China(Project Number: 09YJC630240) and Project of National Social Science Foundation of China(Project Number: 09CTQ015).

References

1. Androutsellis, T.S., Spinellis, D.: A Survey of Peer-to-Peer Content Distribution Technologies. ACM Computing Surveys 36(4), 335–337 (2004)
2. Yang, B., Garcia, M.H.: Designing a super-peer network. In: 19th International Conference on Data Engineering, pp. 49–74. IEEE Press, Bangalore (2003)
3. McDonald, A., Znati, T.: A mobility based framework for adaptive clustering in wireless ad-hoc networks. IEEE Journal On Selected Area of Communications 17(8), 1466–1487 (1999)

4. Das, B., Bharghavan, V.: Routing in ad-hoc networks using minimum connected dominating sets. In: IEEE International Conference on Communication, pp. 376–380. IEEE Press, Montreal (1997)
5. Bettstetter, C., Krausser, R.: Scenario-based stability analysis of the distributed mobility-adaptive clustering (DMAC) algorithm. In: 2nd ACM International Symposium on Mobile ad hoc Networking, pp. 232–241. ACM Press, Long Beach (2001)
6. Chatterjee, M., Das, S., Turgut, D.: WCA: A weighted clustering algorithm for mobile ad hoc networks. Cluster Computing 5(2), 193–204 (2002)
7. Bandyopadhyay, S., Coyle, J.E.: An energy efficient hierarchical clustering algorithm for wireless sensor networks. In: 22nd Annual Joint Conference of the IEEE Computer and Communications Societies, pp. 1713–1723. IEEE Press, San Francisco (2003)
8. Younis, O., Fahmy, S.: Distributed Clustering in Ad-hoc Sensor Networks: A Hybrid Energy-Efficient Approach. In: 23nd Annual Joint Conference of the IEEE Computer and Communications Societies, pp. 23–35. IEEE Press, Hong Kong (2004)
9. Dimokas, N., Katsaros, D., Manolopoulos, Y.: Energy-efficient distributed clustering in wireless sensor networks. Journal of Parallel and Distributed Computing 70(4), 371–383 (2010)
10. Amis, A., Prakash, R., Vuong, T., Huynh, D.: Max-min d-cluster formation in wireless ad hoc networks. In: 19th Annual Joint Conference of the IEEE Computer and Communications Societies, pp. 32–41. IEEE Press, Tel Aviv (2000)
11. Van, D.S.: A cluster algorithm for graphs. Technical Report, National Research Institute for Mathematics and Computer Science in the Netherlands (2000)
12. Ramaswamy, L., Iyengar, A., Liu, L., Douglis, F.: Connectivity based node clustering in decentralized peer-to-peer networks. In: 3rd International Conference on Peer-to-Peer Computing, pp. 66–74. IEEE Press, Linkoping (2003)
13. Zheng, W., Zhang, S., Ouyang, Y., Makedon, F., Ford, J.: Node clustering based on link delay in p2p networks. In: ACM Symposium on Applied Computing, pp. 744–749. ACM Press, Santa Fe (2005)
14. Lacks, D., Chatterjee, M., Kocak, T.: Design and Evaluation of a Distributed Clustering Algorithm for Mobile ad hoc Networks. The Computer Journal 52(6), 656–670 (2009)
15. FreePastry, `http://freepastry.org`
16. Castro, M., Druschel, P., Kermarrec, A.M., Rowstron, A.: Scribe: A large-scale and decentralized application-level multicast infrastructure. IEEE Journal on Selected Areas in Communications 20(8), 1489–1499 (2002)

Exploratory Factor Analysis Approach for Understanding Consumer Behavior toward Using Chongqing City Card

Juanjuan Chen[1] and Chengliang Wang[2]

[1] College of Computer & Information Science, Chongqing Normal University, Shapingba 401331, China
[2] College of Computer Science & Engineering, Chongqing University, Shapingba 400044, China
{cameroncjj,wangcl55}@gmail.com

Abstract. This paper examines the attitude and consumer behavior toward using Chongqing City Card by surveying 202 respondents with a self-administered questionnaire. With exploratory factor analysis, there are seven major factors found out, which are defined as general dimension, marketing dimension, use cost dimension, technology dimension, utility dimension, convenience dimension and E-commerce dimension. Then some suggestions for better operation of Chongqing City Card are provided according to these findings. Limitation and future researches are given also.

Keywords: City Card, Exploratory Factor Analysis, Empirical Study, Customer Behavior.

1 Introduction

Golden Card Project, which was officially initiated in 1993, has popularized Integrated Circuit Card (ICC) in China. ICC, also known as a type of smart card or chip card, is any pocket-sized card with embedded integrated circuits. Based on this technology, City Card was developed for applications in various industries like public transportation, subway, taxi, health care and charges of public services such as water, gas and electricity. The multi-application City Card facilitates citizens in micro-payment areas and improves service level, showing great vitality in city informalizaiton process. After a slow start, the smart-card industry has grown steadily over the last 15 years as new applications have been developed using this technology and subsequently rolled out to new countries. Figure 1 shows the way in which the geographic balance of the market has changed; whereas the technology originated in Europe, the main feature of the first few years of the 21st century has been the growth of Asian, and particularly Chinese-speaking markets [1].

Globally, big cities like Singapore, Hongkong and Beijing are devoting efforts to build up the City Card network and integrate resources to digitalize their cities. Started in 2005, Chongqing City Card project is developing very fast and the number of card holder succeeds 3 millions. However, number of inactive card is also growing quickly, accounting for more than 20% of all issued cards.

L. Cao, J. Zhong, and Y. Feng (Eds.): ADMA 2010, Part II, LNCS 6441, pp. 561–567, 2010.

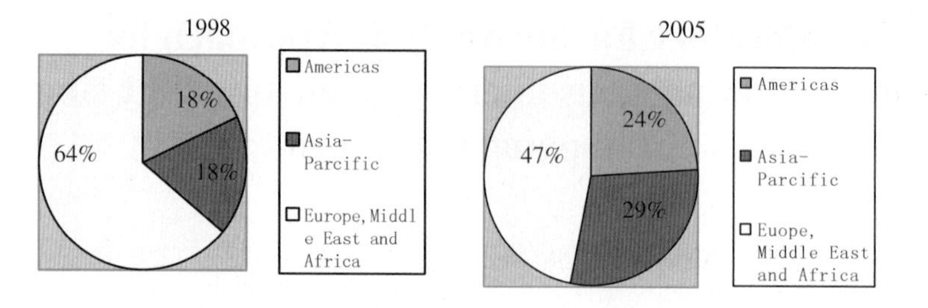

Fig. 1. Smart Card Market by Region 1998 and 2005 (Source: EuroSmart)

It is very necessary to understand customer behaviors and perceptions toward using Chongqing City Card. This paper empirically researched hundreds of Chongqing citizens and tries to analyze by statistical methods according to the survey results. In order to accurately understand customer behavior, especially in Chongqing area, the research relied on an inductive approach. The survey examines the attitude and perceptions of consuming behavior of City Card. The intention is to gain more understanding of how consumers select the micro-payment cards.

Factor analysis was selected to create measurement scales. In order to develop these scales, Explaratory factor analysis (EFA) with varimax rotation was employed. EFA is used to uncover the underlying structure of a relatively large set of variables, it is a widely utilized and broadly applied statistical technique in the social sciences. The priori assumption is that any indicator may be associated with any factor. This is the most common form of factor analysis. There is no prior theory and one uses factor loadings to intuit the factor structure of the data.

In published studies, EFA was used for a variety of applications, including developing an instrument for the evaluation of school principals [2], determining what types of services should be offered to customers [3], etc. Basically, the factor model is motivated by the following argument: suppose variables can be grouped by their correlations, then it is conceivable that each group of variables represents a single underlying construct or factor, that it is responsible for the observed correlations [4]. Factor analysis can be considered as attempts to approximate the covariance matrix $\sum$. The factor model is given in formula (1).

$$\begin{aligned} X_1 - \mu_1 &= l_{11}F_1 + l_{12}F_2 + \cdots + l_{1m}F_m + \varepsilon_1 \\ X_2 - \mu_2 &= l_{21}F_1 + l_{22}F_2 + \cdots + l_{2m}F_m + \varepsilon_2 \\ &\vdots \\ X_p - \mu_p &= l_{p1}F_1 + l_{p2}F_2 + \cdots + l_{pm}F_m + \varepsilon_p \end{aligned} \quad (1)$$

The observable random vector X, with p components, has mean μ and covariance matrix $\sum$. The factor model postulates that X is linearly dependent upon a few unobservable random variables F_1, F_2......, F_m, called common factors, and p additional sources of variation $\varepsilon_1, \varepsilon_2, \cdots, \varepsilon_p$, called errors. The coefficient l_{ij} is called the loading of the ith variable on the jth factor.

2 Questionnaire Design

According to present researches and literature overview, many scholars studied digital cards from different perspectives. Many involved literatures remain optimistic to smart card technology after trial experiments and empirical study in United States and Hong Kong [5-6]. In the light of the presumed advantages by using city card, consumer and merchant acceptance were identified as key to attaining wide scale use and eventual success [7]. According to Truman et al. [8], consumers' card use involved three distinct procedures—obtainment, transaction and replenishment, Zhang studies impact factors on consumers toward using online payment platform, convenience factor, charge acceptability factor, and cost saving factor were discovered [9]. However, many researches focus only on the technical platform construction and network design; there are very few empirical studies about the customer behavior toward using City Card. So this paper tries to analyze by statistical methods and present some useful suggestions for the city card operators.

There are many measures reflecting customer behavior and perception toward using smart card and other forms of micro-payment cards. We design the measuring construct and try to analyze the internal relationships among different indicators; the construct is listed below.

X_1: Before using City Card, I need to deeply aware of its brand name

X_2: The payment platform of City Card should maintain excellent operation.

X_3: The website presence of City Card and its image will affect my purchase decision.

X_4: As for operator, the types of enterprise, i.e. private-owned or state-owned, will affect my purchase decision.

X_5: Software and hardware for City Card payment system should be advanced.

X_6: Private platform for instant communication is necessary.

X_7: Internal management network should be well supplied by advanced information technology.

X_8: City Card system should be equipped with advanced security technology.

X_9: The larger scale of marketing campaign conducted by operator, the better chance to purchase City Card.

X_{10}: It is very necessary to advertise City Card in mass media like TV or newspaper.

X_{11}: Online advertisements should be placed.

X_{12}: Comprehensive customer services, such as discount and bonus reward, are highly valued.

X_{13}: The number of recharge points affects my purchase decision.

X_{14}: Various recharging types, such as online bank recharging, should be provided.

X_{15}: The number of consuming points available for City Card affects my purchase decision.

X_{16}: The ability of money stored in City Card be converted to other forms of e-cash is highly valued.

X_{17}: It is high concerned whether money stored in City Card could be consumed online.

X_{18}: It is important whether City Card could be used in any public transportation vehicles, such as light way and subway.

X_{19}: Whether it is convenient in public services payment areas, such as medical registration and water toll payment, affect my purchase decision.

X_{20}: The opening cost to activate City Card concerns me.

X_{21}: The less registration information is required, the better chance to purchase City Card.

X_{22}: Whether the use of City Card could reduce the time cost affect my purchase decision.

All question items are measured by the most commonly used 5-point Likert scale, in which points can be labeled, 1 means "disagree strongly", 2 means "disagree somewhat", 3 means "neutral" , 4 means "agree somewhat" and 5 means "agree strongly".

In order to accurately examine customer behavior, especially those potential customers, the questionnaire survey was mainly conducted in Shapingba District of Chongqing, which has higher population density and better educational background. The first stage involved interviews with practitioners working for City Card centre and some consumers to gain more detailed understanding. Before surveying, the wording of questions was validated by expert's opinions and customers.

We sent out 202 pieces of the questionnaire and 182, which accounts for 90.1%, were replied. Despite 19 invalid pieces, the effective questionnaires make up for 89.56% of all, and male respondents account for 46% and female 54%.

3 Data Analysis

Reliability test reflects the consistency of a set of measurements of a measuring instrument, i.e. whether the measurement data acquired in survey reflects the real situation of subject investigated. Cronbach Alpha is a statistic measure that is used for reliability test, the formula is given in formula (2).

$$\alpha = \frac{p}{p-1}\left(1 - \frac{\sum_{i=1}^{p} \sigma_{Y_i}^2}{\sigma_X^2}\right) \tag{2}$$

Here p is the number of variables, σ_X^2 is the variance of the observed total test scores, and $\sigma_{Y_i}^2$ is the variance of component i for the current sample. Normal, Alpha can take on any value less than or equal 1, but a reliability of 0.7 or higher is required. In our study about City Card, the reliability statics Cronbach's Alpha is 0.789, which means our study is quite reliable.

Validity refers to whether a study is able to scientifically answer the questions it is intended to answer. Factor analysis is applied here to verify the data from survey, and the greater the communality value the more related of factors to measurement item, which means factors are valid to measure those questions.

Normally, communality value which is greater than 0.4 represents factors can well explain indicators. In our study most communalities are greater than 0.4, hence these indicators are statistically significant in the factor analysis of Chongqing City Card. To assess the construct validity, here the Kaiser-Meyer-Olkin measure of sampling adequacy and Bartlett's test of sphericity are used, the KMO measure is defined in formula (3), in which r_{ij} is the correlation coefficient of variable i and variable j, and s_{ij} is the partial correlation coefficient.

$$KMO = \frac{\sum\sum_{i=j} r_{ij}^2}{\sum\sum_{i=j} r_{ij}^2 + \sum\sum_{i=j} s_{ij}^2} \tag{3}$$

KMO value above 0.7 indicates test items are suitable for factor analysis. According to figure 2, the KMO value equals 0.761 and the chi-square value in Bartlett's test is 1067, the degree of freedom is 7, p value is less than 0.000, which indicate that the measuring indicators are suitable for exploratory factor analysis.

Kaiser-Meyer-Olkin Measure of Sampling Adequacy.		0.761
Bartlett's Test of Sphericity	Approx. Chi-Square	1067.14
	df	231
	Sig.	0.000

Fig. 2. Test results of KMO and Bartlett

The objective was to obtain fewer dimensions that reflected the relationships among inter-related variables. The eigenvalue greater than one rule was applied in identifying the number of factors. The variables that had large loadings on the same factors were grouped. Factor loadings value of 0.50 and above is considered good and significant. Initially there were 22 scaled variables that were measured. After factor analysis, 6 of these variables with loadings of less than 0.50 were deleted, and 7 factors were created. The cumulative percent of variance explained was over 50% (table 1).

The first factor has significant loading values to indicator X_1(.737), X_7(.791), X_8(.825), X_{12}(.618), X_{14}(.544). According to Table 1, indicator X_1 measures brand influence toward using City Card; indicator X_7 and X_8 measure information technology toward using City Card; indicator X_{12} measures comprehensive customer service and indicator X_{14} measures convenience of recharging toward using City Card. As this factor mostly represents the customer perception of operator's overall strength, so is named as General Dimension (F_1).

The second factor has significant loadings on X_9(.671), X_{10}(.537) and X_{11}(.573), which mainly describe the effect of marketing strategy and various tactics conducted by operator will have on customer's purchase decision, so is named as Marketing Dimension (F_2).

The third factor has very significant loading on X_{20}(.922), which involves with opening cost to activate a card, so is named as Use Cost Dimension (F_3).

The fourth factor has significant loading on X_5(.930), which involves software and hardware technology and security technology, so is named as Technology Dimension (F_4).

Table 1. Rotated Factor Matrix (a)

Indicators	F_1	F_2	F_3	F_4	F_5	F_6	F_7	Communalities
X_8	0.826							0.795
X_7	0.791							0.762
X_1	0.737							0.706
X_{12}	0.618							0.567
X_{14}	0.544							0.539
X_9		0.617						0.608
X_{11}		0.573						0.535
X_{10}		0.537						0.489
X_{20}			0.922					0.781
X_5				0.930				0.999
X_{18}					0.776			0.340
X_{19}					0.550			0.499
X_{13}						0.679		0.525
X_{15}						0.642		0.530
X_{13}							0.710	0.996
X_{15}							0.508	0.439
percent of cumulative variance	16.98	23.97	29.72	35.1	40.3	45.48	50.55	

The fifth factor has significant loading on X_{18}(.776) and X_{19}(.550), which relate with various applications and functions in micro-payment areas, so is named as Utility Dimension (F_5).

The sixth factor has significant loading on X_{13}(.679) and X_{15}(.642), which relate with number of consuming points and recharging points, so is named as Convenience Dimension (F_6).

The seventh factor has significant loading on X_{16}(.710) and X_{17}(.508), the first indicator measures how money stored in City Card could be spent online will effect customer's purchase decision; the second indicator measures City Card in related with other forms of e-cash. So this could be named as E-commerce Dimension (F_7).

4 Discussion and Conclusion

This study found out that there are mainly seven types of factors may have great impact on customer's decision of using City Card, i.e. general dimension, marketing dimension, use cost dimension, technology dimension, utility dimension, convenience dimension and E-commerce dimension. When the operator considers its overall developing strategy, these factors and variables should be carefully examined to ensure the final success of City Card project.

We can find out that customers in Chongqing are very sensitive to the opening cost to activate a card, so operator should reduce any unnecessary opening cost at the brand awareness stage, moreover issue totally free cards to some potential customers. Respondents also strongly expressed their preference on well branded City Card, so it is essential to build positive brand image by large scale advertising on Internet, radio, TV and local newspapers. Information technology and system security are highly concerned by most customers, and so the operator should build up internal management platform and external communication platform well supported by advanced information technology, improving safety of customers' benefits. Nevertheless, as various unique function IC cards are popular everywhere in our daily life, City Card network should be extensive enough and equipped with all kinds of basic functions to replace those IC cards. Customers in our study showed great interests on multiple utilities and conveniences of City Card, for online and offline usage.

In conclusion, this study has contributions for the operator of City Card to better understand customer behavior toward using City Card. Several significant factors are found out with the exploratory factor analysis, and they need special address by the operator. However, this study scale is limited in Chongqing, therefore a broader perspective should be employed for future researches.

Acknowledgments. The work described in this paper is supported by the National Natural Science Foundation of China (61004112) and the Soft Science Project of Chongqing (2010YK0218).

References

1. Hendry, M.: Multi-application Smart Cards: Technology and Applications. Cambridge University Press, Cambridge (2007)
2. Lovett, S., Zeiss, A.M., Heinemann, G.D.: Assessment and Development: Now and in the Future. Issues in the Practice of Psychology, 385–400 (2002)
3. Moris, S.B.: Sample Size Required for Adverse Impact Analysis. Applied HRM Research 6(1-2), 13–32 (2001)
4. Johnson, R.A., Wichern, D.W.: Applied Multivariate Statistical Analysis. Pearson Education, London (2007)
5. Hove, L.V.: The New York City Smart Card Trial in Perspective: A Research Note. International Journal of Electronic Commerce Winter 5(2), 119–131 (2001)
6. Westland, J.C., Kwok, M., Shu, J., Kwok, T., Ho, H.: Customer and Merchant Acceptance of Electronic Cash: Evidence From Mondex in Hong Kong. International Journal of Electronic Commerce 2(4), 5–26 (1998)
7. Clemons, E.K., Croson, D.C., Weber, B.W.: Reengineering Money: The Mondex Stored Value Card and Beyond. International Journal of Electronic Commerce 1(2), 5–31 (1997)
8. Truman, G.E., Sandoe, K., Rifkin, T.: An Empirical Study of Smart Card Technology. Information and Management 40, 591–606 (2003)
9. Zhang, Y., Du, B., Li, W., Zhang, X.M.: Empirical Study of Influential Factors Toward Using Online Payment Platform. Science Technology and Management Research 11, 194–196 (2006)

Author Index

Printing: Mercedes-Druck, Berlin
Binding: Stein+Lehmann, Berlin